Springer Proceedings in Physics 38

Springer Proceedings in Physics

Managing Editor: H. K. V. Lotsch

Volumes 1 – 29 are listed on the back inside cover

Electroluminescence

Proceedings of the Fourth International Workshop
Tottori, Japan, October 11-14, 1988

Editors: S. Shionoya and H. Kobayashi

With 360 Figures

Springer-Verlag Berlin Heidelberg New York
London Paris Tokyo Hong Kong

Professor Shigeo Shionoya, Ph. D.
Department of Electronics, Tokyo Engineering University
1404 Katakura, Hachioji, Tokyo 192, Japan

Professor Hiroshi Kobayashi, Ph.D.
Department of Electronics, Tottori University
Koyama, Tottori 680, Japan

ISBN-13: 978-3-642-93432-2 e-ISBN-13: 978-3-642-93430-8
DOI: 10.1007/978-3-642-93430-8

Softcover reprint of the hardcover 1st edition 1989

2154/3150-543210 – Printed on acid-free paper

Preface

The Fourth International Workshop on Electroluminescence (EL-88) was held at the Hotel Holiday, Tottori, Japan, October 11–14, 1988. This workshop was sponsored by the 125 Research Committee on Mutual Conversion between Light and Electricity, Japan Society for the Promotion of Science, in cooperation with SID (Society for Information Display) Japan Chapter, Tottori Prefecture, the Tottori Industrial Technology Association, and the Foundation for Advancement of International Science (FAIS).

The workshop EL-88 was a continuation of the series of international workshops held successively at Liège (Belgium) in 1980, Bad Stuer (DDR) in 1983, and at Warm Springs (Oregon, USA) in 1986. It brought together scientists and engineers from universities and industry who shared a common interest in discussing electroluminescence and related topics. The number of participants reached 253; 49 from abroad (10 countries) and 204 from Japan. This is almost four times as many as in the previous workshop in 1986, reflecting the recent rapid development and progress of electroluminescence research.

The phenomenon of electroluminescence (EL) was discovered by the late Professor Destriau in Paris in 1936. In the 1950s, a great deal of related research was carried out in a considerable number of universities and industrial laboratories. The aim in those days was mainly to obtain a flat type of light source for illumination, using ZnS powder phosphors. The progress of the research and development, however, was not as fast, or promising, as expected. It was found to be difficult to fabricate EL panels with a high luminance and efficiency and a long life. The research on EL declined temporarily, but in 1974 a new era began. A novel attractive information display was developed by the Sharp Corporation, Japan, using ZnS thin-film EL panels with a high luminance level and a long life. Since then, the research on EL display panels has been renewed and is now active again.

In the workshop EL-88, the topics included under EL research have expanded remarkably. The scientific program consisted of 20 invited papers and 60 contributed papers, of which 23 were presented in oral sessions and 37 in poster sessions. A panel discussion on "Electroluminescence – Present and Future" was also held. Discussions were very active and animated in all sessions, so that the aim of the workshop to provide a stimulating and informal atmosphere for discussion was fully achieved.

These proceedings of the workshop collect together invited and contributed papers presented at the workshop. The contributions are arranged under the following headings: Basic Physics, Luminescence Characteristics and Materials, Color Electroluminescence, Processing Technology, Thin Film Electroluminescent Panels, Powder Electroluminescent Panels, and Light Emitting Diodes. We hope that the proceedings vividly reflect the present status of this field.

The following points characterize work reported at EL-88. Our understanding of EL mechanisms seems to be making steady progress. A new quantum-mechanical approach to the excitation mechanism was proposed and discussed. Efforts to obtain color EL panels are being continued. The progress makes us optimistic for the future. The luminance of green EL phosphors has already reached practical levels. Red and blue EL phosphors have been widely investigated using rare-earth-doped CaS and SrS, together with a ZnS host. A prototype of matrix-addressed full-color EL panels was demonstrated. A number of new and useful processing technologies have been used in an attempt to obtain high-performance EL panels; for example, atomic layer epitaxy (ALE), multi-source deposition (MSD) and metal organic chemical vapor deposition (MOCVD). A new EL panel with a memory effect was presented, consisting of stacked photoconductor and EL layers. This panel structure will simplify driving electronic circuitry, and also reduce power consumption. A novel EL device using organic materials that can emit full visible color was reported. These organic materials might challenge traditional inorganic EL materials for the manufacture of color EL panels in the future.

At the beginning of the workshop, opening remarks were made by one of us (S.S.). These included a brief summary of the history of the research and development of EL and pointed out important topics which were expected to be discussed in the workshop. The second opening talk was given by Prof. Hiroshi Sasakura of Tottori University, who explained the history of EL research in Japan, and particularly in Tottori University. In the late 1950s he was the first person in Japan to successfully make powder EL devices. During the workshop, some special events were organized. A "Tottori Evening" was held on October 11. A welcome speech was given by one of us (H.K.). We were honored by the presence of the Governor of Tottori Prefecture, Mr. Yuji Nishio. He gave a welcome speech, expressing his pride that all the leading scientists in this field were in Tottori, and hoping that they would also enjoy the natural beauty of Tottori during their stay. Participants and their families then watched traditional Tottori folk dances, which are rich in local character. An excursion to the San-in Coast National Park was held on October 13.

Numerous people contributed to the success of the workshop. First, we would like to thank the members of the International Advisory Committee and the Japanese Organizing Committee for their helpful advice and great effort in organizing the workshop. Special thanks are due to Prof. Shosaku Tanaka, Dr. Hideki Yoshiyama, students of the Kobayashi laboratory of Tottori University, Mrs. Akiko Kobayashi and Mrs. Yumiko Tanaka for performing the laborious task of making arrangements for the meeting. In addition, we would like to express our deep appreciation of financial support from a number of Japanese industrial companies and firms.

The next (fifth) International Workshop on EL will be held in June 1990 in Helsinki, Finland. We hope that scientists and engineers working in this field will gather again in Helsinki and that our EL community will develop further.

Tokyo, Tottori
December 1988

Shigeo Shionoya
Hiroshi Kobayashi

International Advisory Committee

J. Benoit	Université de Paris, France
F.J. Bryant	University of Hull, UK
W.E. Howard	IBM Corp., USA
C.N. King	Planar Systems Inc., USA
J.M. Langer	Institute of Physics, Poland
G.O. Müller	Zentralinstitut für Elektronenphysik, DDR
J.P. Noblanc	CNET, France
S.J.T. Owen	Oregon State University, USA
J.I. Pankove	University of Colorado, USA
E. Schlam	Sigmatron Nova Inc., USA
T. Suntola	Lohja Corp., Finland
D. Theis	Siemens AG , F.R. Germany
A. Vecht	Thames Polytechnic, UK
N.A. Vlasenko	Institute of Semiconductors, Kiev, USSR
Xu Xurong	College of Science and Technology, Tianjin, P.R. China

Japanese Organizing Committee

Chair

Shigeo Shionoya	Tokyo Engineering University

Secretary

Hiroshi Kobayashi	Tottori University

Members

Koichi Dazai	Fujitsu Ltd.
Masakazu Fukai	Mie University
Yoshihiro Hamakawa	Osaka University
Sumiaki Ibuki	Setsunan University
Toshio Inoguchi	Sharp Corp.
Akinobu Kasami	Toshiba Corp.
Hideaki Kawakami	Hitachi, Ltd.
Hiroshi Kukimoto	Tokyo Institute of Technology
Eiichiro Nakazawa	NHK
Kohji Nihei	Oki Electric Industry Co., Ltd.
Mitsuhito Sakaguchi	NEC Corp.
Hiroshi Sasakura	Tottori University
Bunjiro Tsujiyama	Nippon Telegraph and Telephone Corp.

Associate Members

Haruki Kozawaguchi	Nippon Telegraph and Telephone Corp.
Tomizo Matsuoka	Matsushita Electric Industrial Co., Ltd.
Shoshin Miura	Fujitsu Ltd.
Hideomi Ohnishi	Ehime University
Yoshimasa A. Ono	Hitachi, Ltd.
Shosaku Tanaka	Tottori University
Masaru Yoshida	Sharp Corp.

Contents

Part III Luminescence Characteristics and Materials

Part IV Color Electroluminescence

Part VI Thin Film Electroluminescent Panels

Part VII Powder Electroluminescent Panels

Part VIII Light Emitting Diodes

Part I

Opening Invited Paper

Retrospect and Prospect on Research and Development of Electroluminescent Panels

T. Inoguchi

Energy Conversion Laboratories, Sharp Corporation,
Shinjo-cho, Kitakatsuragi-gun, Nara, 639-21, Japan

1. Development of the First Generation EL

Electroluminescence (EL) discovered in 1936 [1] was recognized as a new physical phenomenon. However, EL remained unexplored as far as the device was concerned, due to the immature development of both peripheral materials and technologies.

In the early 1950's, EL attracted much interest for its practical use following the development of transparent electrode material with both natures of transparency in the visible light region and high electrical conductivity.

Worldwide research and development were actively done from the latter half of the 1950's to the first half of the 1960's. In this period in Japan, the 125th Research Committee on EL was organized under the Japan Society for the Promotion of Science and then the research and the development were promoted covering a wide range from fundamental study on materials and their nature to the development of a practical device.

The EL cell fabricated in this period was called powder-type EL or dispersion type EL, because ZnS powder was bound with the appropriate polymer and was formed into a thin layer. The first target of its practical application was the realization of a new light source for wall illumination, and then as the next target, various optoelectronic devices were considered such as a light intensifier and a flat display panel. However, as is well known, the brightness of the EL cell was insufficient and too unstable for the practical use.

The so-called half life, which is defined as the time when the brightness decreases to a half of initial brightness, remained in the range of a few hundred hours to a thousand hours. In order to improve these points, the research on materials, panel configuration and fabrication process were energetically continued. However, the EL cell could not be essentially improved and could not secure the position as a general application of flat panel illumination, except in special applications such as an all-night lamp and an illumination lamp of measurement instrument in ship pilothouse.

For the application of the EL cell in a flat diplay panel, powder type EL was used only for displaying a certain rigid pattern because of its poor contrast ratio due to the light scattering by powder of the active layer and the above mentioned problems. So, powder type EL was not realized as practical information panels which can display changing information. Thus, the First Generation of EL development gradually faded away toward the end of 60's.

Springer Proceedings in Physics, Vol. 38 Electroluminescence
Editors: S. Shionoya and H. Kobayashi

2. Development of the Second Generation EL

A sign of emergence from the stagnation appeared in the following two papers presented in 1968. One paper presented by A.VECHT et al. [2] showed that a DC drive of powder type EL panel is available by surface treatment of phosphor powder. The other paper presented by D.KAHNG [3] announced that thin film EL with high brightness is attained using molecules of rare-earth floride as the luminous centers.

With the spread of information transaction equipment headed by computers, the information display device attracted growing interest as a man-machine interface, and the realization of flat display device in solid state was required in this period. Considering their cost and size, polycrystal EL devices had to be looked at again instead of mono-crystalline EL devices.

At the SID International Symposium in 1974, the author's paper [4] "Stable High Brightness Thin Film Electroluminescent Panels" announced that the problems such as low brightness and short life, which remained unsolved in the First Generation EL, could be solved at a stroke, and C. SUZUKI et al. [5] also announced that this EL panel is capable of displaying moving TV pictures with half tone. In the fall of 1974 at the 6th conference of Solid State Devices, M.TAKEDA et al. [6] presented a paper "Inherent Memory Effects in ZnS:Mn Thin Film EL Device". Looking back at EL development, it is possible to say that worldwide research and development on thin film EL, so-called "The Second Generation EL" was triggered by the aforementioned three papers by the Sharp Corporation. Analysing our success, we can cite the following three points from the technical view point. First, the development of process technology was challenged for the formation of II-VI compound semiconductor thin films with an homogeneous constituent element toward growth direction. The vacuum deposition method in which a ZnS pellet was evaporated by electron beam bomberdment was established as a practically useful method. Second, the symmetrical device structure of the EL active layer sandwiched by insulating layers with a high breakdown voltage, so-called double insulating layer structure as shown in Figure 1, was adopted in order to apply stably a sufficiently high voltage to the EL active layer and also in order to simplify the analysis of driving mechanisms and characteristics. Third, we tried to eliminate the mobile ions such as the Cu^+ ion which can move easily under high electric field.

In the initial development stage of the above mentioned EL cell, we were annoyed by the instability phenomenon of decreasing brightness with the aging time under constant applied voltage.

In the spring of 1971, we untiringly pursued the variation of luminance - voltage (L-V) characteristics under constant driving voltage, and we fortunately

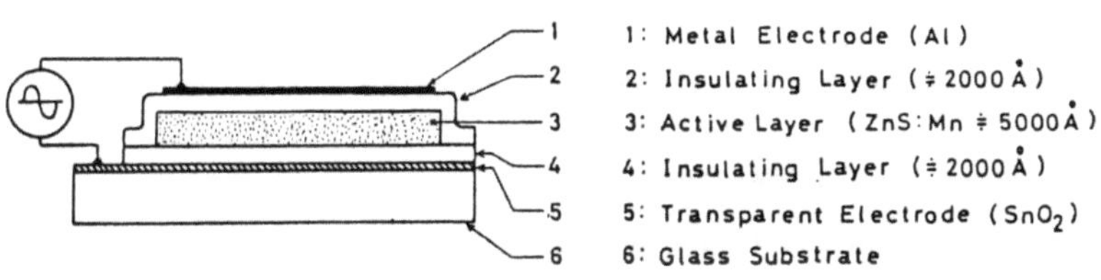

Fig.1 Schematic structure of the thin film EL device

found out that with increasing aging time, the L-V characteristic simply shifted to the higher voltage side retaining its shape and height, and decreasing gradually its shifting rate , then finally stabilized after about 80 hours as shown in Fig.2.

We concluded from the above observation that this phenomenon is not the degradation, but the relaxation process of stress induced during the fabrication process of EL cell.

This conclusion became a strong motivating factor in our development of "The Second Generation EL". After the relaxation process was completed, we confirmed stable L-V characteristics without any degradation, and as a result of continuous driving, we could present the result in the SID symposium that its life was more than twenty thousand hours as shown in Fig.3.

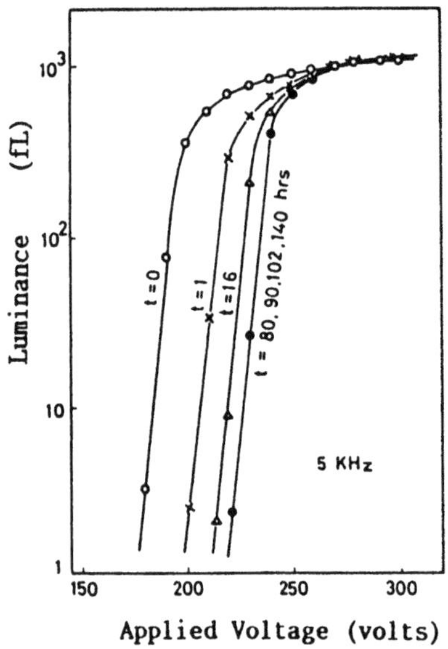

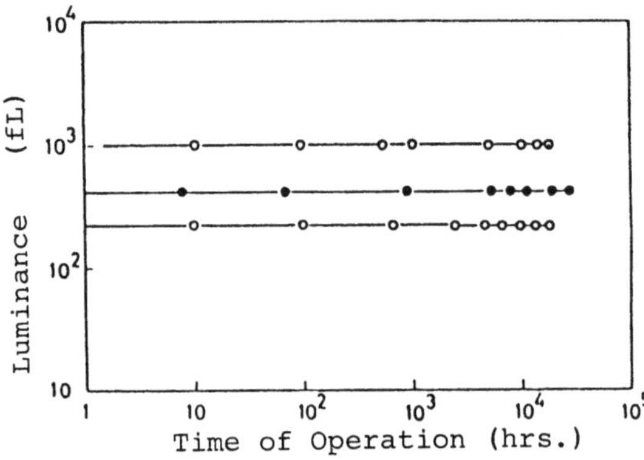

△ Fig.3 Luminance vs. time of operation at a constant voltage operation. Three samples under different operating conditions are shown

◁ Fig.2 Luminance vs. applied voltage characteristics of the thin film EL device

In the subsequent development of the second generation EL, a lot of effort was put into the establishment of fabrication technology and the planning of EL products.

On the other hand, fundamental research was continued into new structure, new material and new process propositions and towards a total understanding of driving and degradation mechanisms. For example, the attempt to reduce the driving voltage using high permitivity materials such as $PbTiO_3$ and $BaTiO_3$, the proposal of ALE (Atomic Layer Epitaxy) as a new fabrication process and its practical use [7], and the proposal and trial of CaS and SrS as new host materials instead of ZnS [8]. We can find new progress in the above mentioned research. As these matters are familiar to all the members attending the Workshop, only two typical papers [9] and [10] are presented as references.

3. EL panels as Information Display Devices

From the period of the first generation EL development, there was a great demand to realize "Wall Hanging TV" as a large flat display panel without a vacuum tube, and its situation to date remains unchanged.

As shown in the paper [5] in the aforementioned SID symposium, which triggered the Second Generation EL, the thin film EL (TFEL) is expected to satisfy the above mentioned requirements. Large information display devices with an X-Y matrix configuration as shown in Fig.4 can be reevaluated as follows.

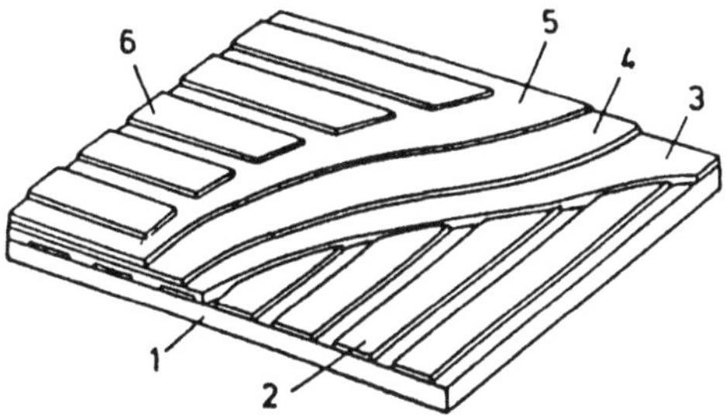

Fig.4 Schematic construction of the thin film EL matrix type information display panel. 1 Glass substrate, 2 Transparent electrodes, 3 Insulating layer, 4 Active layer, 5 Insulating layer, 6 Rear electrodes

First, concerning the contrast ratio of the display, the brightness of a selected pixel is required to be suitably high and the scattering light in non-selected pixel adjacent to the selected pixel must remain at a low level. In the TFEL consisting of all transparent thin films except back metal electrode, high contrast ratio can be easily obtained because of an essentially low light scattering level.

Second, from the view point of the cross talk inherent to the X-Y matrix configuration , a distinct threshold and steep increase of brightness for the input is required. TFEL with a double insulating structure has no problem of such crosstalk in the practical use because of a distinct threshold as shown in Fig.2.

Third, with respect to the response time, rising up time is required within a selected period of refresh display cycle, and falling down time is within one frame time. Luminescence of TFEL can respond sufficiently to the signal of a TV picture mode because the rising up time is a few μsec, and falling down time is a few msec in a general TFEL. Therefore, TFEL is suitable for the information display device with a large capacity.

In order to realize the information display panel, the first encountered problem was the driving voltage. The driving voltage of thus-far developed devices to ensure high brightness and high stability was in the vicinity of 200V because the driving mechanism of TFEL is high field emission. So, TFEL could not be driven by the generally used semiconductor IC. At that time, the above mentioned problem was thought to be a fatal weak point in the practical application of TFEL, as compared with the light emitting diode (LED) which acts under the same driving voltage as that of the semiconductor IC.

Keeping the brightness and stability, the reduction of the driving voltage is very difficult. Fortunately, Sharp Central Research Laboratories developed high voltage MOS transistors with high break down voltage of 1000V and also the MOS IC [11,12]. Thereafter, TFEL made strong progress in its practical use.

There were many problems to be solved in order to industrialize the TFEL while keeping appropriate production yield. For example, adhesion problems between constituent thin film layers with different materials, and the development of a packaging method to protect mechanically and chemically thin film structure from enviromental conditions etc. Taking into consideration these matters, the TFEL panel with the configuration as shown in Fig.5 was realized for practical use.

Nowadays, at least four companies in the world have stepped forward to mass produce TFEL for information display. Six inch diagonal EL panel units with dot capacity of 320 x 240 were first commercialized, and resently, the EL display module has been enlarged to 10 inches in diagonal with 720 x 400 dots capacity, and has been commercialized including driving circuits. A recently developed EL panel module is shown in Fig.6 as an example.

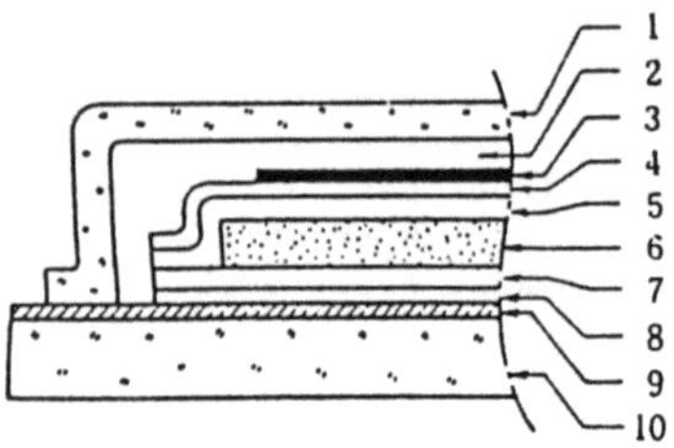

1: Back sheeling glass
2: Insulating oil
3: Back electrode
4: Al_2O_3
5: SiO_2-Si_3N_4
6: EL active layer
7: Si_3N_4-SiO_2
8: SiO_2
9: Transparent electrode
10: Glass substrate

Fig.5 Practically used configuration of thin film EL panel

Fig.6 9 inch size EL panel unit with dot capacity of 640 x 400

4. Future Development and Expectations for the Third Generation EL

As concerns future TFEL development, the most urgent and important subjects are full coloring in order to confirm a solid position in the flat display panel and to get a wider market including general consumer goods. The brightness of red and green color EL have reached to practical level. But, the brightness and purity of blue color have to make a further step.

In order to produce a full color EL panel, we have to first make fundamental research on EL materials of red, green and blue colors in well balanced brightness, and then efforts will be made to develop the panel structure and process technologies. It would be appropriate to get three primary colors using the same host material. For the moment, different host materials will be used in the present developing stage.

After the acomplishment of the above mentioned development, the Second Generation EL will leap ahead to its final goal. The drastic reduction of driving voltage in the Second Generation EL could not be attained without sacrifice of brightness and stability as long as the mechanism is high electric field excitation. Therefore, research of EL goes ahead to "The Third Generation EL".

What is the Third Generation EL ? I expect it to act by a different mechanism than that of high field excitation. I think it is gradually maturing and I hope the birth of the Third Generation EL will be announced in a future EL work shop.

References

1. G. Destriau: J. Chem. Phys. 33, 620 (1936)
2. A. Vecht et al.: Brit. J. Appl. Phys. (J.Phys.D), 1, 134 (1968)
3. D. Kahng: Appl. Phys. Letters, 13, 210 (1968)
4. T. Inoguchi et al.: '74 SID International Symposium Digest, 84 (1974)
5. C. Suzuki et al.: '74 SID International Symposium Digest, 86 (1974)
6. M. Takeda et al.: Proc. 6th'Conf. of Solid State Devices (Tokyo), 103 (1974)
7. T. Suntora et al.: '80 SID International Symposium Digest, 108 (1980)
8. S. Tanaka et al.: '85 SID International Symposium Digest, 218 (1985)
9. T. Inoguchi and S. Mito: Chap 6, Phosphor Film, "Electroluminescence", Topics in Applied Physics vol 17, ed. by J.I.Pankove, 197 (Springer-Verlag, Heidelberg 1977)
10. R. Mach and G. O. Muller: Phys. Stat. Sol. (a) 69, 11 (1982)
11. K. Awane et al.: '78 IE³ Intern'l Solid State Circuit Conf., 224 (1978)
12. K. Fujii et al.: '81 IE³ Intern'l Solid State Circuit Conf., 46 (1981)

Part II

Basic Physics

Developments in the Theory of Electroluminescence Mechanisms

J.W. Allen

Wolfson Institute of Luminescence, Department of Physics and Astronomy, University of St. Andrews, Fife KY16 9SS, Scotland

1. Introduction

It has been a tradition that high-field electroluminescence devices have been developed on an empirical basis, while the development of devices in silicon or III-V compounds rests on a detailed understanding of the relevant physics. Quantitative theories of the mechanisms of high-field electroluminescence are in a primitive state by the standards of semiconductor physics. Partly this is because very hot carriers and deep levels are involved and partly because there are few experimental data, obtained under well-defined conditions, with which to compare theory. Here a few particular areas in which advances are being made are surveyed. They are the hot electron distribution, impact and ionization cross-sections, Auger effects, the position of rare-earth energy levels relative to the energy gap, and the rôle of hot holes.

2. Energy Distributions

Our interest in energy distributions and cross-sections lies in the fact that the rate of excitation or ionization R, and hence the light intensity, depends on a convolution of them:

$$R = n N \sum_{\nu} \int_{BZ} \sigma(\nu, \underset{\sim}{k}) \; v(\nu, \underset{\sim}{k}) \; f(\nu, \underset{\sim}{k}, V, x) \; d^3\underset{\sim}{k} \; . \tag{1}$$

Here n is the carrier density, N the concentration of centres which can be ionized or excited (eg. the density of centres in the ground state), ν a band index, $\underset{\sim}{k}$ the carrier wave-vector, σ the cross-section, v the velocity, and f is the distribution function which in general is a function of potential and position. Often a simple approximation to (1) can be used in terms of energy:

$$R \cong n N \int_0^{\infty} \sigma(E) \; v(E) \; g(E, \varepsilon) \; dE \tag{2}$$

where the distribution function g depends on the local electric field ε. Figure 1 shows schematically how the product $g\sigma$ in the integrand behaves at a given field strength.

The early view, following the ideas of Fröhlich, that electrons at high fields break away from optical phonons and thus achieve quasi-ballistic motion is too simplistic to have any useful validity in the energy range with which we are concerned. Other scattering mechanisms intervene. In contrast, some authors attempt to interpret their experimental data in terms of a Maxwell-Boltzmann or Druyvesteyn distribution, usually within a single valley. This implies strong scattering by some mechanism. It is then difficult to explain the observed efficiency of practical devices. For example, the quantum efficiency of good ZnS:Mn panels suggests that the mean cross-section for impact excitation is ~ 10^{16} cm^2. Theoretical estimates of the impact cross-section suggest that $\sigma(E)$ is unlikely to attain values much greater than this. Hence if the mean value (after convolution with g(E)) and the actual value have the same order of magnitude, the distribution function g(E) must be such that a large fraction of the electrons must have energy greater than 2eV, rather than being in the tail of a Maxwell-Boltzmann or Druyvesteyn distribution.

Springer Proceedings in Physics, Vol. 38 **Electroluminescence**
Editors: S. Shionoya and H. Kobayashi

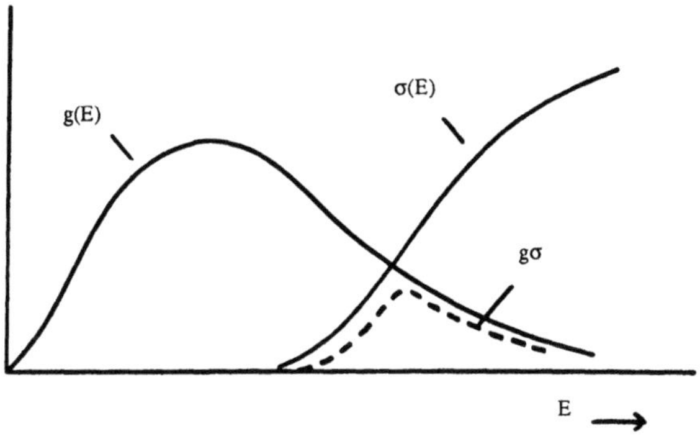

Fig. 1. Schematic diagram of the variation with energy of the cross-section σ, the distribution function g and the product gσ

Another possibility is that electrons are transferred to higher minima, as in the Gunn effect, and thus have a peaked distribution with width much less than the mean energy. Above the field necessary for transfer, many electrons then have energies of 2 eV or more requisite to produce the observed optical effects. We have observed [1] that at a field of about 4×10^5 V cm^{-1} in ZnS there is a rapid increase in three effects, namely interband emission by hot electrons, impact excitation of manganese and impact ionization of deep centres. We interpret this as the field at which transfer of electrons to the higher minima occurs. (Of course Fröhlich breakaway in the lower valleys may be an essential precursor of this transfer.) In this respect it is interesting that the calculated band structure of SrS [2] also has extensive higher conduction band minima at a suitable energy.

It is sometimes, but not always, observed that in ZnS doped with a rare earth the rate of increase of light output with voltage for transitions from different levels increases with the energy of the level, as described notably by KRUPKA [3]. When this occurs over a wide range of energies, a model of impact excitation by transferred electrons with a comparatively narrow energy range is not applicable. We return to this later.

3. Cross-sections

Calculations of impact and ionization cross-sections in free atoms or ions are now sophisticated and accurate whereas in solids they are still crude. Among the particular problems is the form of the wavefunction of the luminescent centre, especially when covalent bonding with the host crystal is considered. Also there is dielectric screening which properly should be taken as dependent on the carrier energy and wave-vector, or approximately as a function of distance. In addition, the carrier velocity is not a simple function of its energy. A simple Born-Bethe treatment [4] of the direct Coulomb term in impact excitation gives

$$\sigma = \left[\frac{1}{\epsilon^2 n \left(\frac{\varepsilon_{eff}}{\varepsilon_o} \right)} \right]^2 \times \left[\frac{2\pi\, m_\ell^x\, e^2\, \hbar\, c^3\, S_{\ell u}^2}{v_u\, E_{ge}^2\, k_u} \,\ell n \left| \frac{k_u + k_\ell}{k_u - k_\ell} \right| \right] \times \left[\frac{1}{\tau_{ED}} \right] . \qquad (3)$$

Here σ has been explicitly written as a product of three terms. The first describes screening. The second is a function of the properties of the carrier when it is incident in its upper state u and scattered in its lower state ℓ and of the energy E_{ge} between the ground and excited state of the centre. The third term is the electric dipole radiative transition rate of the centre. For centres with radiative lifetimes in the range 10 µs - 1 ms the cross-section is estimated to be 10^{-18} - 10^{-20} cm^2, too small to be useful. It is true that the Born approximation is inaccurate near threshold, but the inaccuracy introduced by its use is not greater than that produced by a simplified treatment of the dielectric screening. There may be centres with allowed radiative transitions rather than the forbidden ones of $3d^n$ and $4f^n$ impurities, in which case the cross-section could be usefully large. Exchange interaction is likely to be the cause of the large cross-section for electrons in ZnS:Mn. Inelastic scattering, whether direct or exchange, is enhanced by a low incident carrier velocity, and the exchange effect is enhanced near threshold. In ZnS:Mn the exchange cross-section is then doubly enhanced, because *in a solid it is possible to have electrons in upper minima with high energy but low velocity*. (The calculation for ZnS:Mn by BERNARD et al [5] cannot be taken too seriously, for it assumes rectilinear motion of the electron despite its interaction with the centre and also assumes that a $^6A_1$ - $^4T_1$ transition is fast while a $^4T_1$ - $^6A_1$ transition is slow. As a result the calculated cross-section does not depend on energy and is not zero below the threshold required by energy conservation.) One concludes that the number of combinations of host and centre giving efficient electroluminescence by impact excitation will be small because of the need for a host in which the hot carriers have a suitable energy distribution and a centre with an unusually large cross-section near threshold.

Impact ionization of deep centres has received less detailed attention. BALTENKOV and BELORUSETS [6] have made a model calculation for a centre with a square well potential with a Coulomb tail but its applicability to real systems has not been tested. Alternatively one could start with the LOTZ empirical formula [7] for the impact ionization cross-section of atoms and modify it for centres in crystals, although this has not yet been attempted. The formula, for energies not too far from threshold, is

$$\sigma \cong a\, n_a \frac{\ln\left(\frac{E}{E_I}\right)}{E\, E_I} \tag{4}$$

where $a \sim 3 \times 10^{-14}$ $cm^2\ eV^2$, n_a is the number of electrons in the atomic shell, E is the incident electron energy and E_I is the ionization energy. Close to threshold this becomes

$$\sigma \cong a\, n_a \frac{\left(\frac{E}{E_I}\right) - 1}{E_I} \, . \tag{5}$$

In a solid the ionization energy is 2-3 eV, much less than in a free atom. Hence, even allowing for screening, cross-sections of 10^{-16} cm^2 or greater are possible. Also, symmetry selection rules are less important in ionization so cross-sections should not vary from centre to centre by several orders of magnitude as they do in impact excitation. We have unpublished experimental data that show that deep-level impact ionization rates in ZnS can indeed be large. It therefore seems likely that in the search for new and better materials impact ionization will be at least as important as impact excitation.

4. Auger Effects

Auger effects are the inverse of impact effects, so knowledge of one gives information about the other. For excitation within a centre, the impact cross-section and Auger coefficients are proportional [4]. For ZnS:Mn both rates are usually large. Unfortunately this in principle puts a limit on high-intensity operation. If we make the approximation that the Auger coefficient for hot electrons is the same as that for thermal electrons, the efficiency of light production from ZnS:Mn is expected to be reduced by 50% at a current density ~ 1A cm^{-2}. As with a laser, a three-level system could solve the problem. Figure 2 illustrates an example. A centre is ionized, with large cross-section and large Auger coefficient. If the

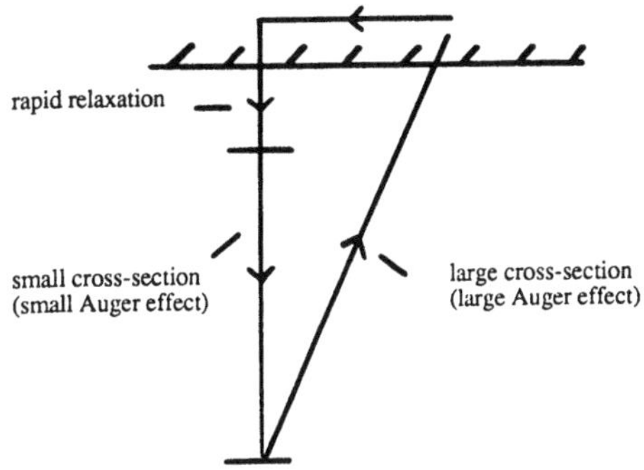

Fig. 2. A possible three-level mechanism for excitation to avoid Auger quenching at high excitation levels

electrons rapidly relax to a comparatively shallow level, Auger recombination is avoided. The electrons can then return radiatively to the ground state by a transition for which the impact cross-section and hence Auger coefficient are small. It is interesting that the group at Tottori University have suggested a mechanism of this type for SrS and CaS rare-earth doped electroluminescent films [8].

5. Rare-earth Ionization Levels

If impact ionization is to be important the relevant deep level should lie in the forbidden gap. One therefore would like to be able to predict the positions of the levels. ALLEN [9], and in a more refined manner FAZZIO et al [10], have given semi-empirical procedures for predicting the ionization energy-levels of $3d^n$ transition metal impurities. The extension to rare earths can be made without great difficulty. The requisite matrix elements for the multiplet splittings are available in eg. NEILSON and KOSTER [11] or McCLURE and KISS [12]. Detailed application to specific materials of interest in electroluminescence has not yet been made because of the scarcity of experimental data. Figure 3 shows a first rough try for the divalent/trivalent ionization level of rare earths in ZnS. Developments of the method would be valuable in predicting the usefulness of new materials. A complication is that the rare earths are often associated with other centres such as charge compensators, but this is not expected to change the ionization energy greatly.

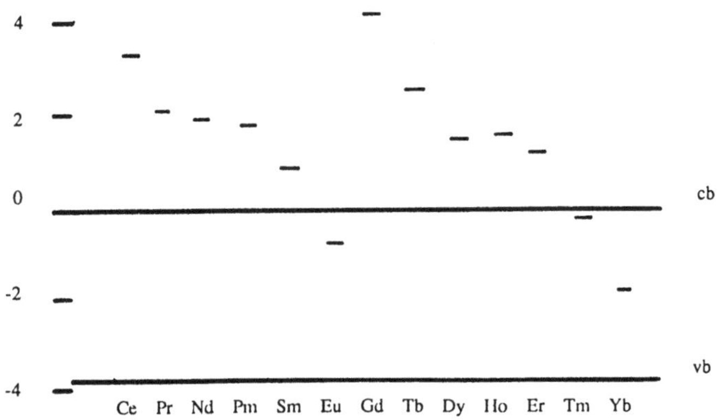

Fig. 3. A rough estimate of the position of the divalent/trivalent ionization level for rare earths in ZnS

6. The Rôle of Hot Holes

It is customary to ignore the effects of holes on impact rates because holes with thermal energies are rapidly trapped at deep centres. This is not necessarily true for hot holes. The presence of free holes in high-field regions in ZnS is shown by experimental observations on band-to-band impact ionization, deep level impact ionization and photocapacitance. Holes can be produced by band-to-band ionization or by a two-step process involving a deep centre whereby one hot carrier removes an electron from the

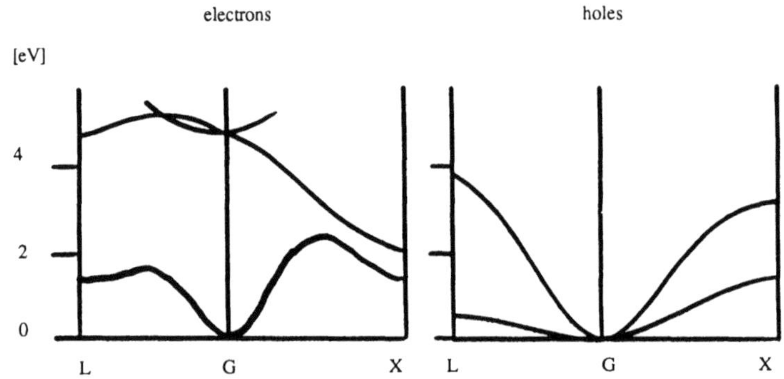

Fig. 4 The conduction and valence bands of cubic ZnS drawn so that electron energies in the conduction bands and hole energies in the valence bands both increase upward from the band edge.

centre to the conduction band by an impact process, then a second hot carrier removes an electron from the valence band onto the centre, again by impact. The structure of the conduction and valence bands, and hence the kinetics of electrons and holes, can be quite different. This is illustrated in fig. 4 for cubic ZnS, where the bands are drawn for comparison by having both the electron energy in the conduction bands and the hole energy in the valence bands increase upwards. It follows that in the same material at the same field strength the electrons and holes can have quite different energy distributions. For example, the electrons could have a fairly narrow distribution in the upper maxima while the holes have approximately a single-valley Maxwell-Boltzmann or Druyvesteyn distribution. This could be a cause for the apparent contradictions concerning carrier distributions in the literature, since the relative proportions of electrons and holes may be different in different experimental arrangements. Therefore when extracting information about energy distributions or cross-sections from experiment it is important to ascertain the relative importance of hot holes and hot electrons.

One experimental arrangement in which the existence of hole impact effects may remove a difficulty of interpretation is that in which an n-type crystal is ion-implanted with a luminescent centre, such as a rare earth, and a Schottky contact is made on the implanted side. Light emission is seen when the Schottky contact is negative. If the centres are impact-excited by electrons, it is difficult to see how the electrons acquire sufficient kinetic energy in the small depth of the implanted region. However, carrier multiplication by the two-step deep level impact mechanism can occur in the region beyond the implanted layer. (The deep levels are either already present in the initial materials or are lattice defects produced by the ion bombardment beyond the implantation depth.) Holes then move in the high field towards the cathode, ie. towards the implanted region where they can produce impact excitation.

7. Conclusions

It can be seen that we are still a long way from having precise physical theories of the various mechanisms operative in high-field electroluminescence. However, progress is being made. The ideas presented here are still largely tentative but at least we can see ways in which they can be refined. Experience in other areas of solid state physics suggests that theory progresses more rapidly when it can go hand-in-hand with experiment, but for this the experiments must be done in well-defined conditions. Understanding the physics of the mechanisms has its own inherent interest but also it can be of value in defining the limitations of existing materials and in pointing the way to better materials.

8. References

1. N.E. Rigby, T.D. Thompson and J.W. Allen: J. Phys. C: Solid State Phys. 21, 3295 (1988)
2. A. Hasegawa and A. Yanase: J. Phys. C: Solid State Phys. 13, 1995 (1980)
3. D.C. Krupka: J. Appl. Phys. 43, 476 (1972)
4. J.W. Allen: J. Phys. C: Solid State Phys. 19, 6287 (1986)
5. J.E. Bernard, M.F. Martens and F. Williams: J. Luminescence 24/25, 893 (1981)
6. A.S. Baltenkov and E.D. Belorusets: Sov. Phys. Semicond. 12, 1283 (1978)

7. W. Lotz: Zeit. für Physik 216, 241 (1968)
8. S. Tanaka, S. Ohshio, J. Nishiura, H. Kawakami, H. Yoshiyama and H.Kobayashi: Appl. Phys. Lett. 52, 2102 (1988)
9. J.W. Allen: In Proc. 7th Int. Conf. on Physics of Semiconductors (Dunod, Paris 1964) p. 781
10. A. Fazzio, M.J. Caldas and A. Zunger: Phys. Rev. B 30, 3430 (1984)
11. C.W. Nielson and G.F. Koster: Spectroscopic Coefficients for p^n, d^n and f^n Configurations (M.I.T. Press, Cambridge, Mass. 1964)
12. D.S. McClure and Z. Kiss: J. Chem. Phys. 39, 3251 (1963)

Auger Effect in Semiconductors: Why Does It Matter for Electroluminescence?

J.M. Langer

Institute of Physics, Polish Academy of Sciences,
02-668 Warsaw, Al.Lotnikow 32/46, Poland

The Coulomb interaction between a free carrier and the electrons at a localized center may lead to either impact excitation, or to a reverse effect - the Auger effect. This review summarizes a current status of the experiment and theory of the Auger effect and its relationship to the quantum efficiency of high-field electroluminescence. Mechanisms (electric-dipole versus exchange) governing the Auger-type energy transfer processes are discussed. I also comment on limits imposed by the effect on a possible laser action in systems utilizing intra-center radiative transitions in order to achieve laser action.

1. INTRODUCTION

Electroluminescence (EL) in semiconductors is generated either by direct recombination of the injected minority carriers (in, e.g., a forward biased p-n junction) or by the impact excitation process. The latter is the dominant EL mechanism in all devices employing carrier acceleration in high electric field (usually well above 10^5V/cm) [1,2]. Its proper description as well as the experimental investigation, are very challenging tasks as the total rate is a convolution of the two relatively poorly known functions of energy (E): the carrier energy distribution in a high electric field and the impact excitation cross-section $\sigma(E)$ [2,3]. Theoretical computation of both quantities rarely produces a satisfactory result. To get a reasonable estimation of at least σ well above the threshold, one may use, however, the well-known reciprocity between the impact excitation and the Auger effect (AE) (see Fig.1) [2,4-7]. Since the interaction leading to both effects is the same, the two processes may be described within the same formalism. According to the detailed balance principle,the knowledge of either of the two processes is adequate for a full description of the second one [5,6].

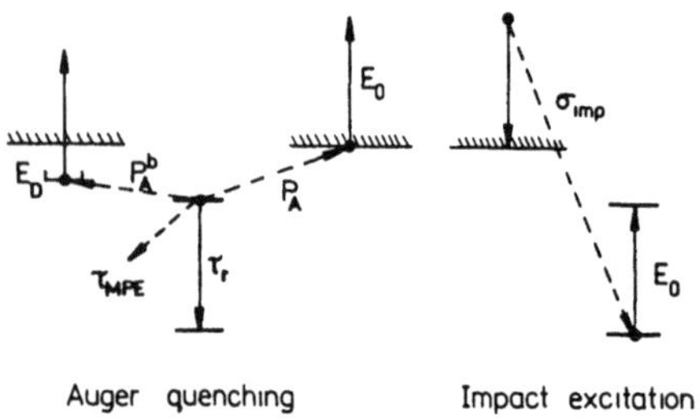

Fig. 1. The reciprocity of impact and Auger processes for localized defects.

In high-field EL, emission occurs mostly at localized centers, such as Mn or the rare earths (RE) in ZnS. For rough estimations it is reasonable to assume a step function for $\sigma(E)$, i.e., a constant value σ_0 above the

Springer Proceedings in Physics, Vol. 38 **Electroluminescence**
Editors: S. Shionoya and H. Kobayashi

threshold. For these centers the Auger quenching probability P_A is proportional to the carrier concentration n:

$$P_A = C_A n \quad (1)$$

From a detailed principle argument, one gets that [2,4,7]

$$\sigma_o = C_A/V_{hot} \quad (2)$$

where the velocity V_{hot} corresponds to the impact excitation threshold and is between 10^7 and 10^8 cm/s. This simple relationship indicates the usefulness of studying the Auger effect, as providing at least a hint as to the best choice of the host and impurity system for high field EL. Studies of the Auger effect may also provide information about the detailed mechanism of the interaction, as well as provide a new tool in profiling an impact excitation process in an EL device [2,4,8]. All these problems are briefly reviewed here.

2. AUGER EFFECT IN LOCALIZED DEFECTS

The mechanism of the nonradiative recombination of the excited states of impurities in insulating crystals and semiconductors has long been a topic of fundamental interest in solid state physics [9]. It has been found that multiphonon processes usually dominate nonradiative recombination for localized defects. If the energy to be dissipated is large compared to the highest energy of phonons in the host or the temperature is sufficiently low, the probability of the multiphonon nonradiative recombination becomes small and other types of recombination become more competitive.

In semiconductors, the presence of carriers (free or weakly bound on shallow donors or acceptors) opens a new recombination channel - the Auger Effect (AE). It belongs to a general class of energy transfer processes and the energy acceptor in this case is either the free carrier or the carrier bound to some defect center. The AE involving either free or weakly bound carriers was proved to be responsible for the luminescence quenching for various types of bound excitons or donor - acceptor pairs [9] but it was commonly neglected as an efficient nonradiative deexcitation mechanism of localized centers in semiconductors. In 1981 Gordon and Allen [10] in ZnS:Mn and our group [4,11] in CdF_2:Mn have found that the intrashell Mn^{2+} luminescence is very efficiently quenched in conducting crystals. The efficiency of this mechanism has been confirmed in the intrashell luminescence quenching of several rare earth dopants [11,12] as well as of transition metals [13,14] in semiconductors.

In contrast to the free-carrier AE, the AE due to energy transfer to weakly bound carriers has been found to be much weaker. It can be observed only at low temperatures, at which the carriers freeze out on shallow donors. Until now the only unambiguous case of this type of the AE has been found in the luminescence of Mn^{2+} in conducting CdF_2 [15].

Independent evidence of the intrashell Mn^{2+} luminescence quenching by the AE is given by the observation of the reverse-bias dependence of the Mn^{2+} cathodoluminescence generated under a thin gold gate evaporated on a semiconducting CdF_2:Mn crystal [2,15] (no cathodoluminescence is observed in this crystal if excited directly , i. e. not through the metal gate). The increase of the reverse bias causes the increase of the intensity of the cathodoluminescence (Fig. 2). This increase is proportional to the volume of the electron-depleted region beneath the metal gate. This region is equivalent to the insulating crystal, i. e. the region where there is no AE due to either

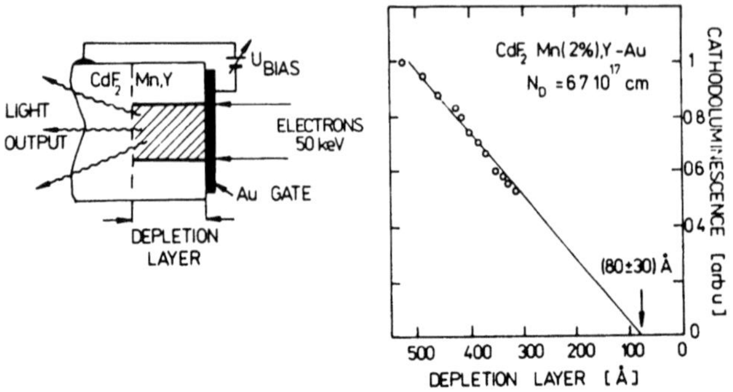

Fig.2
Cathodoluminescence of CdF_2:Mn,Y/Au Schottky diode versus the width of the depletion layer

free or weakly bound electrons. This is the reason for a linearity of a total Mn^{2+} emission versus the depletion layer width.

2.1 Theory of the Auger Quenching of the Localized Centre Luminescence

The Auger coefficient C_A for the luminescence quenching by the free carriers equals [16,17]:

$$C_A = \frac{1}{2\pi h}\frac{g_o}{g_e}\int d\mathbf{k}\left|M_C(\mathbf{k}) - M_{ex}(\mathbf{k})\right|^2 \delta\{E(\mathbf{k})-E_o\} \tag{3}$$

where E_o is the quenched energy, g_o and g_e are the degeneracies of the localized center ground and excited states, and M_C and M_{ex} are the Coulomb and exchange matrix elements of the electron-electron interaction. The integration proceeds over all the band states conserving energy.

2.1.1 Electric-Dipole Contribution in the Free Carrier Auger Effect

A poor knowledge of the localized center wavefunction makes direct calculation of the exchange term unreliable. To get a lower boundary of C_A a much simpler calculation of the Coulomb term suffices, however. Due to the localized nature of the emitting center, the standard expansion of the Coulomb interaction in a multipole series is adequate. The dipole-dipole contribution is therefore

$$M_C = \frac{e^2}{\varepsilon_\infty}\left(\frac{\mathcal{E}_{eff}}{\mathcal{E}_o}\right)\left\langle \phi_i \left| \frac{\mathbf{rR}}{r^3} \right| \phi_f \right\rangle \tag{4}$$

Here $\mathbf{r}$ and $\mathbf{R}$ are the free and localized center electron coordinates, respectively and the form of dielectric screening follows the Agranovich arguments [18]. Since the initial and final wavefunction ϕ_i and ϕ_f can be written as symmetrized products of the localized center wavefunctions Ψ_e or Ψ_o and the electron (or hole) Bloch wavefunction in the conduction (valence) band, the matrix element in (4) separates into the product of the optical electric- dipole (ED) matrix element for the localized center $\langle\Psi_o|\mathbf{R}|\Psi_e\rangle$ and a function dependent only on the band electron $\mathbf{k}$-vector. On the other hand the ED matrix element is directly related to the ED component of the measurable radiative lifetime of the intra- center transitions τ_r. Therefore:

$$C_A = (\tau_r n_o)^{-1} \tag{5}$$

where the critical carrier concentration n_o depends only on the quenched transition wavelength $\lambda_o = hc/E_o$ and not on the details of the localized center wavefunction. It is equal to

$$n_o = \frac{2k_o^2}{e^2} n_r^5 \frac{1}{|F_c|^2 \mathcal{N}_c(E_o)} \lambda_o^{-3} \qquad (6)$$

where F_c is the overlap of the Bloch oscillatory parts of the electron in the conduction band at its bottom and at the energy E_o (the exact values of F_c are very close to unity), k_o corresponds to the Brillouin zone region contributing mainly to the density of states $\mathcal{N}_c(E_o)$ at this energy. From this equation the enhancement of the AE by the high density of band states is evident. This equation can be simplified further by employing the parabolic approximation, i. e assuming that ($E(\mathbf{k})=\hbar^2k^2/2m^*$) and assuming a value of unity for F_c. In this approximation [12,16,17]:

$$n_o = 4\pi^{5/2} n_r^5 \sqrt{\frac{m_o}{m^*}\frac{a_o}{\alpha_f}}\, \lambda_o^{-7/2} \qquad (7)$$

where $\alpha_f=1/137$, $a_o= 0.529$ Å, and n_r is a refractive index. For a more complicated band structure n_o can be shown to be inversely proportional to the band density od states at energy $E(\mathbf{k}_o)=E_o$. For a broad band emission a spectral averaging similar to that performed by Dexter [19] must be made.

For most semiconductors $2<n_r<4$, hence for the emission at $\lambda_o=1\mu m$, the values of n_o must be within the limits: $2\times10^{14}<n_o<6\times10^{15}cm^{-3}$. A further lowering of n_o can be expected if the exchange mechanism would be more efficient or the resonant enhancement due to an anomalously high density of states in a band occurs. On the other hand if a quenched transition is only partly ED, the calculated value of n_o must be multiplied by an inverse fraction of the ED contribution to the radiative lifetime.

Table 1 presents a comparison of the experimental values of the critical carrier concentration and those calculated within the parabolic approximation taking into account only a partial ED character of the optical transitions. Agreement for the wide-gap highly ionic CdF_2 is surprisingly good. A notable discrepancy occurs for much more covalent II-VI compounds. One of the reasons already pointed out by ALLEN [7,10] is a very high density of states $\mathcal{N}_c$ in the conduction band of ZnS and ZnSe at the energy corresponding to the Mn emission. Detailed calculations using (6) show that this enhancement does not

Table 1. Experimental and calculated (ED contribution) values of the critical carrier concentration $n_o(cm^{-3})$ in the Auger quenching of the intracenter emission

Center	λ_o(nm)	$n_o^{exp}(cm^{-3})$	$n_o^{calc}(cm^{-3})$
CdF_2:Gd	312	2.6×10^{16} [12]	3.3×10^{16}
CdF_2:Mn	≈520	$(1-2)\times10^{15}$ [11,14]	8×10^{14}
CdF_2:Tb	≈520	$<2\times10^{16}$ [12]	$>2\times10^{15}$
CdF_2:Er	green&red	$\approx 10^{14}$ [15]	-
ZnS :Mn	≈590	$\approx 10^{12}$ [10]	2×10^{14}
ZnSe:Mn	≈600	$\approx 10^{12}$ [13]	2×10^{14}

account for the anomalously small value of n_0. More possible reason is the much more significant contribution of the exchange in this case.

2.1.2. The Exchange Term

In contrast to the Coulomb term, the exchange term can not be presented as the product of the functions dependent on the impurity center and the free carrier coordinates. Due to rather poor knowledge of the impurity wavefunction in solids, only very rough estimates of M_{ex} are possible. Such estimate has been made by MAJEWSKI and LANGER [20] for the single electron d-like impurity in the T_d environment to allow for the p-d mixing. For most optical transitions the Coulomb term is dominant. In RE impurities shielding of the 4f shell should even more quench the exchange term. There is, however, the one major difference between them, which may in turn enhance the exchange term. The difference comes from the spin selection rules for both terms. In the exchange a double spin-flip is allowed, while in the Coulomb term it is not. They may be important for the spin-forbidden transitions, for which the optical transition allowance comes from the spin-orbit interaction. It is also worth pointing out that in the dipole approximation [20] the ratio of the exchange and Coulomb matrix elements scales as k_0^2, i.e, exchange prefers higher energy transitions.

Quite helpful in estimating the exchange terms are the already mentioned results of our study of the Auger quenching of Mn emission in CdF_2 [13]. The very detailed low temperature measurement of the concentration dependence of kinetics and efficiency indicated that the only mechanism that can explain consistently the experimental data is the exchange. MAJEWSKI and LANGER [20] have, however, shown that for the delocalized quencher wavefunction $\Psi(R)$, the exchange matrix element is proportional to the same term for the free-electron case:

$$|M_{ex}^{bound}(R)|^2 = |\Psi(R)|^2 \; |M_{ex}^{free}|^2 \tag{8}$$

Since for the large distance $\Psi(R) \propto \exp(-R/a)$, with a being the Bohr radius of the electron bound to a shallow donor, a well known exponential dependence of the energy transfer rate for the exchange mechanism [19] is obtained. From this result it is clear that the energy dependence of the exchange term for the bound electron case is the same as for the free electron case, that is $P_{ex} \propto E^{-1.5}$. The dipole-dipole term has much stronger energy dependence, namely $E^{-7.5}$ [2,20]. This explain why for the same impurity (e.g. Mn in CdF_2) the bound electron Auger effect may be governed by the exchange interaction, while the free electron case by the simple dipole term. From (8) it is obvious that the probability of the AE by the bound carrier is proportional to the free-carrier AE, i.e. $1/(n_0\tau_r)$, where n_0 can be taken directly from the experiment. If we define R_0 as the average distance between the occupied donors characterized by a Bohr radius a:

$$N_D = 3/(4\pi R_0^3) \tag{9}$$

then the following relation is obtained:

$$\pi a^3 n_0^{exch} = \exp(-2R_0/a). \tag{10}$$

For ZnS and ZnSe:Mn n_0 is about 10^{12} cm^{-3}, which corresponds to N_D in the range of $10^{17} cm^{-3}$, which is quite a reasonable value judging from the electrical data for this material [10] and consistent with the lack of appreciable quenching by the shallow donors in these materials.

3. ELECTROLUMINESCENCE PROFILING

The free electron Auger effect can be used to determine where the junction emits light. In reverse polarized MS junction emitting centers are excited on impact by the energetic carriers accelerated in a high electric field. Since they must gain enough energy before the collision, there must be a "dead" zone within the depletion layer at which no EL is generated.

In a first approximation the depletion layer edge separates two regions of the crystal: the strongly conducting bulk and the insulating depletion layer, that is, the region with and without the Auger quenching. In a bulk of the crystal the lifetime of the emitting centers is very short due to the Auger quenching. Within a depletion layer, on the contrary, it is limited only by the radiative lifetime if other non-radiative processes can be neglected.

The EL kinetics consists of two parts: a very fast part corresponding to the portion of the emitting region crossed by the depletion layer edge and a slow part corresponding to a non-crossed region. Therefore a simultaneous measurement of the depletion layer width from a junction capacitance, corresponding to various U_p voltages, and kinetics of the EL gives direct information on the spatial profile $N^*(x)$ of the excitation process.

EL kinetics of the two representative diodes differing in carrier concentration, shown in Fig.3, reveal all the three characteristics discussed above. For a given diode the accelerating voltage U_o was kept constant just above the EL threshold. In the "thinner" junction ($N_D=2.4\times10^{18}cm^{-3}$) rapid quenching is observed for U_o, but in the "wider" ($N_D=5.2\times10^{17}cm^{-3}$) a four-fold decrease of U_o is necessary to see this effect. This means that the width of the excited region is limited in the "thinner" junction by the depletion layer edge, and hence by the Auger quenching, contrary to the case of the "wider" junction, where carrier scattering is the limiting factor. In the first case the kinetics always contain the non-exponential portion, while in the second case substantial rapid voltage reduction is necessary for observation of the kinetics non-exponentiality in accordance with the experiment.

Intensity of the unquenched emission at $t\approx0$ is proportional to the total number N of the unquenched emitting centers

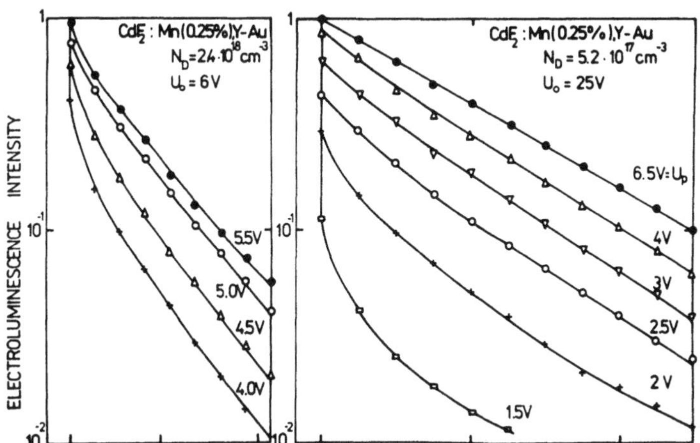

Fig.3 The electroluminescence kinetics after bias reduction of two CdF_2 EL diodes differing by a donor doping level [8].

$$I^{0}_{slow}(W_p-\delta) \propto \int_0^{Wp-\delta} N^*(x)\,dx \qquad (11)$$

where W_p is the depletion edge position at a given voltage U_p and the factor δ, being close to the Debye screening length [2,4] takes into account the fact that the carrier concentration does not drop abruptly at the edge of the depletion layer. Therefore the spatial distribution of the excited centers $N^*(x)$ is given by the derivative

$$N^*(W-\delta) = \frac{dI^{0}_{slow}(W)}{dW} \qquad (12)$$

The excitation profile $N^*(x)$ determined from the experimental kinetics is shown in Fig.4. The results clearly prove that acceleration and excitation are spatially separated processes in high-field electroluminescence. Moreover, the width of the emitting region is a fraction of the total junction width. It is interesting to note the very large distance crossed by high-energy electrons without appreciable scattering (up to 100 nm), much larger than the cold electron free path (< 1nm) in the case of CdF_2 in which mobility is below 10 cm^2/Vs at room temperature. The data taken on the "wider" diode (Fig.3) prove that the high electric field is necessary to support the ultra-hot electron transport. It is to be noted, however, that there is no indication that this high-electric field affects appreciably the excitation process itself and the localized centre recombination.

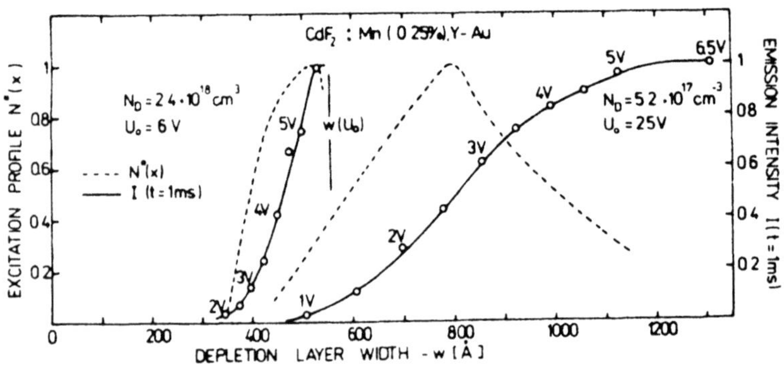

Fig.4 A spatial profile of the excitation (---) in two CdF_2 diodes [8].

4. SUMMARY

The experimental and theoretical results presented in this paper show that the Auger quenching of the localized center luminescence may be the dominant non-radiative mechanism for localized defects even in weakly conducting solids. The Auger effect due to free carriers should generally be stronger than that due to bound carriers. The difference between them should be diminished with the increased delocalization of the weakly bound carriers, i.e. in semi-conductors characterized by a large dielectric constant and small effective mass. In case of CdF_2:Mn the difference is large (about a factor of 400) due to the very small value of the Bohr radius of the shallow donor (a≈7Å). The dominant mechanism for the AE cannot be generally predicted, since it depends on the particular energy structure of the quenched emitter, the energy to be dissipated, as well as the host band structure (possibility of a resonance condition between the quenched emission and the energetic position of the high density of states [7,10]). In any case the rate of the free-carrier AE must be larger than the dipole contribution, which can be estimated from (6) or (7). For most semiconductors ($\lambda\approx1\mu m$, $n_r\leq4$) n_o must be smaller than 10^{16} cm^{-3}. This clearly indicates the importance of the AE in quenching the emission of the localized centers in semiconductors. Quite often it has been hoped that an injection type laser could be constructed by incorporating into the junction region of the laser diode localized impurities (especially rare-earth dopants establishing a very narrow-line emission), which could be the active center for the laser action [21]. A high probability for the AE described here makes this hope quite hard to realize, especially in the d.c. pumping regime [16]. A similar comment applies to high-pumping-level cathodoluminescence devices in which the AE has already been seen and suspected for the saturation effects in emission [22] .

It is also worth pointing out that the reciprocal relationship between the AE and the impact excitation allows estimation of the impact cross section and thus the efficiency of the light emitting devices employing this mechanism solely from the Auger effect (2) [2]. It also explains why ZnS:Mn is such such efficient material for high field EL devices.

I would like acknowledge most sincerely A. Suchocki who has been my principal collaborator in the research of the Auger effect. This study has been financially supported by research programs CPBP 01. 04 and CPBP 01.12.

REFERENCES

1. P. J. Dean: J. Luminescence 23, 17 (1981)
2. J. M. Langer: In *Optoelectronic Materials and Devices*, ed. by M. A. Herman (PWN-Polish Scientific Publishers, Warsaw 1983) p.303
3. J. W. Allen and S. G. Ayling: J. Phys. C 19, L369 (1986)
4. J. M. Langer: J. Luminescence 23 141 (1981)
5. P. T. Landsberg: Proc. Roy. Soc. London A 331, 103 (1972)
6. D. J. Robbins: phys. stat. sol. (b) 97, 9, 387 and 98, 11 (1980)
7. J. W. Allen: J. Phys. C 19, 6287 (1986)
8. J .M. Langer, A. Lemańska-Bajorek and A. Suchocki: Appl. Phys. Lett. 39, 385 (1981)
9. A. M. Stoneham: Rep. Progr. Phys. 44, 1251 (1979); P. T. Landsberg: phys. stat. sol (b) 41, 457 (1970); J. M. Langer: in *Defects in Crystals*, ed. by E. Mizera (World Scientific, Singapore, 1987), p. 50
10. N. T. Gordon and J. W. Allen: Solid State Commun. 37, 1441 (1981)
11. J. M. Langer, A. Suchocki, Le Van Hong, P. Ciepielewski and W.Walukiewicz: Physica 117 & 118, 152 (1983)
12. J. M. Langer and Le Van Hong: J. Phys. C. 17, L923 (1984)
13. S. G. Ayling and J. W. Allen: J. Phys. C. 20, 4251 (1987)
14. P. B. Klein, J. E. Furneaux and R. L. Henry: Phys. Rev. B29, 1947 (1984)
15. A. Suchocki and J. M. Langer: Phys. Rev. B39, (1989), to be published
16. J. M. Langer: in *Rare Earth Spectroscopy*, ed.by B. Jeżowska-Trzebiatowska, J. Legendziewicz and W. Stręk, (World Scientific, Singapore 1985) p.523
17. J. M. Langer: J. Luminescence 40 & 41, 589 (1988)
18. V. M. Agranovich: Uspekchi Fiz. Nauk 112, 143 (1974)
19. D. L. Dexter: J. Chem. Phys. 21, 836 (1953).
20. J. A. Majewski and J. M. Langer: Acta Phys. Polon. A67, 51 (1985)
21. W. T. Tsang and R. A. Logan: Appl. Phys. Lett. 49, 1686 (1986); J. P. van der Ziel, M. G. Oberg and R. A. Logan: Appl. Phys. Lett. 50, 1313 (1987)
22. S. Kuboniwa, H. Kawaji and T. Hoshina: Jap. J. Appl. Phys. 19, 1647 (1980)

The Impact Cross Section of Electroluminescence Centers

Yu Jiaqi, Shen Yongrong, Xu Xumou, Luo Baozhu, and Zhong Guozhu

Changchun Institute of Physics, Academia Sinica, P.R. China

1. Introduction

ZnS thin films doped with Mn^{2+} ions or rare-earth ions are promising material for electroluminescent planar panels, which might be strong candidates for the next generation solid planar display devices to partly replace cathode-ray tube displays. They are attracting an increasing amount of research.

In these devices the dominant excitation mechanism is hot electron impact excitation. Electrons are accelerated in a high electric field (10^6 v.cm^{-1}). When the electrons have gained sufficient energy, they impact manganese centers or rare -earth centers. Light is emitted when the centers return to the ground state. The impact cross sections of luminescent centers are very important parameters in the understanding of the impact excitation process and in the estimation of the efficiency of impact excitation . Some researchers estimated approximately the orders of the magnitude of the cross sections of Mn^{2+} ions and trivalent rare-earth ions. [1--4] Muller and Mach obtained the cross section of Mn^{2+} in ZnS thin films experimentally.[5]

In this paper we report our experimental and theoretical research on the impact cross section in electroluminescence. We have used high resolution spectroscopy under laser selective excitation and electric field excitation to study ZnS thin films. The relative ratios of impact cross sections of different Er^{3+} centers, the relative ratios of that of Er^{3+} and ErF_3 centers to that of Mn^{2+} centers in ZnS thin films were obtained experimentally. We calculated the inelastic scattering cross sections based on the Born approximation by considering the coulomb interaction between accelerated free electrons and the 4f electrons of rare-earth ions.

2. Comparison Between the Impact Cross Sections of Different Centers

2.1 Basic Principles

Under hot electron impact excitation, the number of centers excited to an upper state within unit time is $CN\int n_0 f(E)v(E)\sigma(E)\,dE$, where no, f(E), v(E) are the number, distribution function and velocity of hot electrons respectively, (E) is the impact cross section of luminescent centers, C is a constant related to the sample's structure and field's distribution, N is the number of centers. The dynamical equation is

$$\frac{dn}{dt} = CN\int n_0 f(E)v(E)\sigma(E)\,dE - nP_r. \qquad (1)$$

where n is the number of centers in excited state, Pr is radiative transition probability from the excited state to the ground state.

Under steady excitation, dn/dt = 0, therefore transition intensity

$$I \propto N\int n_0 f(E)v(E)\sigma(E)\,dE. \qquad (2)$$

Springer Proceedings in Physics, Vol. 38 **Electroluminescence**
Editors: S. Shionoya and H. Kobayashi

The impact cross section depends on energy of hot electrons. But only average value corresponding to certain transition can be obtained from experimental data, so we use average value σ instead of $\sigma(E)$. The above formula may be rewritten as

$$I \propto N\sigma \int n_0 f(E) v(E)\, \mathrm{d}E. \tag{3}$$

If the differences of transition energies for different centers are very small, the integrands in formula (3) for different centers can be considered equal. The constant C are equal for different centers in the same sample. Then for center a and b in the same sample we have

$$\frac{\sigma_a}{\sigma_b} = \frac{I_a N_b}{I_b N_a}. \tag{4}$$

In order to find σ_a/σ_b we need to know I_a/I_b and N_b/N_a. The procedures are as follows. Using laser selective excitation we can identify different centers in a electroluminescent sample and obtain emission spectrum for each center. Decomposing the electroluminescence spectra of the sample into the emission spectra of the above identified centers, the relative electroluminescent intensities contributed by different centers are obtained, so does the ratio I_a/I_b. From laser selective excitation we can also find the relative photoluminescence intensities I' of different centers. The ratio of the concentrations of center a and b can be deduced as follows:

$$\frac{N_a}{N_b} = \frac{I'_a \nu_a^3 P_{ra}^{-1} F_b}{I'_b \nu_b^3 P_{rb}^{-1} F_a} \tag{5}$$

Where ν is the transition frequency, Pr the radiative transition probability, F the laser excitation intensity, a and b denote center a and center b respectively.

2.2 The Impact Cross Section of Er^{3+} Center

The following experiments are designed to determine the absolute value of the impact cross section of Er^{3+}. The principle of the experiments is that the relative ratio of the impact cross sections of Er^{3+} and Mn^{2+} is obtained first using ZnS thin film samples co-doped with Er^{3+} and Mn^{2+}. Then the cross section of Er^{3+} is deduced using the value of the cross section of Mn^{2+} obtained by Muller.

In order to avoid significant energy transfer between Er^{3+} and Mn^{2+}, the concentrations of Er^{3+} and Mn^{2+} in the samples are low, within the range of 5×10^{-5} --- 5×10^{-4} mol/mol ZnS.

For Mn^{2+} we take the zero phonon line 17891 cm^{-1} in the emission spectrum to be the lower integral limit in the formula (3), for Er^{3+} we take the strongest emission line 18135 cm^{-1} of the $^4S_{3/2}$ -- $^4I_{15/2}$ transition as the lower integral limit. We take ∞ as the upper integral limit. The energy difference of the two lower limits ΔE = 250 cm^{-1} = 0.03ev, is quite small compared with the average energy of hot electron, 0.15ev [6], so that the integrations for Er^{3+} and Mn^{2+} may be regarded as equal to each other.

Electroluminescence spectra of ZnS:Er^{3+},Mn^{2+} thin films were measured, they have broad emission band from Mn^{2+} and emission lines corresponding to the transition from $^2H_{11/2}$ and $^4S_{3/2}$ to $^4I_{15/2}$ of Er^{3+}. Decomposing the electroluminescence spectra of $ZnS:Er^{3+},Mn^{2+}$ samples, ratios of electroluminescence intensity of Mn^{2+} to that of Er^{3+} have been obtained [7]. The concentrations of Er^{3+} and Mn^{2+} in these samples were determined by the induction coupled plasma atomic emission spectroscopy. Using formula (4), the ratio of the impact cross sections of Er^{3+} to that of Mn^{2+} is calculated to be approximately 0.5. Using the impact cross section of Mn^{2+} in ZnS, 4×10^{-16} cm², measured by Muller, the impact cross section of Er^{3+} is obtained to be about 2×10^{-16}cm². This is the average value for different centers formed by Er^{3+} in ZnS, deduced from emission from $^2H_{11/2}$ and $^4S_{3/2}$ to $^4I_{15/2}$.

The relative ratios of the impact cross section of Er^{3+} and Mn^{2+} in $ZnS:Er^{3+}$, Mn^{2+} have been measured from threshold voltage to saturation voltage. The results are listed in table 1. It can be seen that the ratio does not vary with voltage.

Table 1 Relationship between impact cross section and voltage

Voltage	193	199	204	217	221	225
I_{Er} (r.u.)	1.0	1.0	1.0	1.0	1.0	1.0
I_{Mn} (r.u.)	2.3	2.4	2.4	2.3	2.3	2.3
σ_{Er}/σ_{Mn}	0.46	0.44	0.44	0.46	0.46	0.46

2.3 The Impact Cross Section of ErF_3 Center

We prepared $ZnS:ErF_3$,$ZnS:Er^{3+}$ thin films by means of a rotating substrate holder and appropriate masks to separate ZnS and c (c=ErF_3,Er) beams from separate boats before they reach the substrate, in order to avoid the interaction between ZnS beam and ErF_3 beam and to reduce the decomposition of the latter .

Some lines are common in spectra of $ZnS:ErF_3$ and $ZnS:Er^{3+}$, some lines only appear in the spectrum of $ZnS:ErF_3$. We expect that common lines are related to Er^{3+} centers, the latter lines are related to ErF_3. It means that there are ErF_3 centers and Er^{3+} centers in $ZnS:ErF_3$ samples. Using laser selective excitation,emission spectra of ErF_3 center and Er^{3+} center in sample $ZnS:ErF_3$ were distinguished. By formula (5), we find the ratio of the concentration of ErF_3 to that of Er^{3+} is approximately 0.48. It implies that in our sample one third Er exist in ZnS in form of ErF_3, the rest are in the form of Er^{3+}. The electroluminescence spectrum of ZnS :ErF_3,Er^{3+} thin film can be decomposed into emission spectrum of ErF_3 centers and that of Er^{3+} centers. The ratio of emission intensity of ErF_3 to that of Er^{3+} was obtained to be about 0.4. Using formula (4), we find the ratio of the impact cross section of ErF_3 to that of Er^{3+} to be about 4, thus the impact cross section of ErF_3 is about 8×10^{-16}cm²[8].

2.4 Comparison Between the Impact Cross Sections of Different Er^{3+} Centers

Rare earth ions in ZnS will form different centers. It is interesting to compare impact cross sections of different Er^{3+} centers in $ZnS:Er^{3+}$ samples.

We have identified four different Er^{3+} centers a,b,c and d in $ZnS:Er^{3+}$ samples. Using the emission intensities of centers a,b,c and d under laser selective excitation, the relative concentrations of these centers were deduced using formula (5). Decomposing electroluminescence emission spectra into the spectra of centers a, b,c and d by iteration and best fitting methods, the relative electroluminescent intensities from center a,b,c and d were found. Using formula (4), it is found that

$$\sigma_a : \sigma_b : \sigma_c : \sigma_d = 3.6 : 2.3 : 1 : 2.5 \text{ [7]}.$$

Table 2 Values of Pr^{-1} and σ of four centers of Er^{3+} in ZnS

center	a	b	c	d
Pr^{-1}(us)($\pm$-20%)	5.7	14	22	16
σ (r.u.)($\pm$45%)	3.6	2.3	1	2.5

3. Theory

Considering inelastic direct Coulomb scattering of an electron by an atom or an ion in a crystal, we have

$$\frac{d\sigma}{d\Omega} = (2\pi)^4 \frac{m^2}{g_0 \hbar^4} \frac{k_n}{k_0} \sum_{i,f} |T_{if}|^2 \tag{6}$$

where σ is the impact cross section corresponding to the transition from state i to state f, m the mass of an electron, g_0 the degeneracy of state i, K_0 the wavevector of the incident electron, K_n the wavevector of electron after scattering, T_{if} the T-matrix element of scattering.

In the first Born approximation

$$T_{if} = \langle \varphi_f | V | \varphi_i \rangle = \frac{e^2}{(2\pi)^3} \iint \psi_f^* \left(-\frac{Z}{r_0} + \sum_j^z \frac{1}{r_{0j}} \right) \psi_i \exp[\mathrm{i}(\boldsymbol{k}_0 - \boldsymbol{k}_n) \cdot \boldsymbol{r}_0] \, d\boldsymbol{r}_0 \, d\boldsymbol{r}_j \tag{7}$$

where r_0 is the position vector of the incident electron, $r_{0j} = |r_0 - r_j|$, r_j is the position vector of the jth electron of the target, Z is the charge number of nucleus, z is the number of electrons of the target, ψ_i is the state of the target before scattering and ψ_f is the state of the target after scattering. We use the plane-wave approximation for the incident electron.

Because of the orthogonality of the functions ψ_i and ψ_f the integration of $(-Z/r_0)$ over r_j vanishes.

Let us expand $1/r_{0j}$ in spherical harmonics. Assume that the incident electron does not penetrate into the 4f shell; then $r_0 > r_j$,

$$T_{if} = \frac{2e^2}{(2\pi)^2} \sum_l \sum_j \frac{\mathrm{i}^l \Delta^{l-2}}{(2l-1)!!} \langle \psi_f | C_{l_j}^0 r_j^l | \psi_i \rangle \tag{8}$$

Where C_{lj}^0 is the spherical tensor operator.

$$C_{l_j}^0 = \left(\frac{4\pi}{2l+1} \right)^{1/2} Y_l^0(\theta_j, \varphi_j)$$

$$\sigma = \int \frac{d\sigma}{d\Omega} d\Omega = \int_0^{2\pi} \int_0^{\pi} \frac{(2\pi)^4 m^2}{g_0 \hbar^4} \frac{|\boldsymbol{k}_n|}{|\boldsymbol{k}_0|} \sum_{i,f} |T_{if}|^2 \sin\theta \, d\theta \, d\varphi$$

$$= \int_0^{2\pi} \int_{k_0-k_n}^{k_0+k_n} \frac{(2\pi)^4 m^2}{g_0 \hbar^4} \frac{|\boldsymbol{k}_n|}{|\boldsymbol{k}_0|} \sum_{i,f} |T_{if}|^2 \frac{1}{\boldsymbol{k}_n \boldsymbol{k}_0} \Delta \, d\Delta \, d\varphi$$

$$= \frac{(2\pi)^5 m^2}{g_0 \hbar^4} \frac{1}{k_0^2} \int_{k_0-k_n}^{k_0+k_n} \sum_{i,f} |T_{if}|^2 \Delta \, d\Delta \tag{9}$$

where

$$\boldsymbol{\Delta} = \boldsymbol{k}_0 - \boldsymbol{k}_n$$

$$\Delta = (k_0^2 + k_n^2 - 2k_0 k_n \cos\theta)^{1/2}$$

$$\Delta\, \mathrm{d}\Delta = k_0 k_n \sin\theta\, \mathrm{d}\theta.$$

Neglecting high-order terms in (9),we obtain

$$\sigma = \frac{8\pi m^2 e^4}{g_0 \hbar^4} \frac{1}{k_0^2} \int_{k_0-k_n}^{k_0+k_n} \sum_{i,f} \left| \mathrm{i}\Delta^{-1} \langle \psi_f | \sum_j C_{1,j}^0 r_j | \psi_i \rangle - \tfrac{1}{3} \langle \psi_f | \sum_j C_{2,j}^0 r_j^2 | \psi_i \rangle \right|^2 \Delta\, \mathrm{d}\Delta$$

$$= \frac{8\pi m^2 e^4}{g_0 k_0^2 \hbar^4} \sum_{i,f} \left| \langle \psi_f | \sum_j r_{jz} | \psi_i \rangle \right|^2 \ln\left(\frac{k_0 + k_n}{k_0 - k_n}\right)$$

$$+ \frac{16\pi m^2 e^4}{9 g_0 \hbar^4} \frac{k_n}{k_0} (\langle r^2 \rangle)^2 \sum_{i,f} \left| \langle \psi_f | \sum_j C_{2,j}^0 | \psi_i \rangle \right|^2 \qquad (10)$$

where $\langle r^2 \rangle = \int R_f^* r^2 R_i r^2\, \mathrm{d}r$ (R is the radial part of the wavefunction ψ). We denote the two terms in (10) as σ_1 and σ_2 respectively.

The first term σ_1 in (10) can be connected with the spontaneous transition probability A_{n0} corresponding to the transition from state f to state i. Considering the local-field correction, we obtain

$$A_{n0} = \frac{1}{g_n} \sum_{i,f} \frac{4e^2\omega^3}{3\hbar C^3} \frac{(n^2+2)^2}{9} n |\langle f | r | i \rangle|^2$$

$$= \frac{1}{g_n} \sum_{i,f} \frac{4e^2\omega^3}{\hbar C^3} \frac{n(n^2+2)^2}{9} |\langle f | r_z | i \rangle|^2 \qquad (11)$$

$$\frac{1}{g_n} \sum_{i,f} |\langle f | r_z | i \rangle|^2 = A_{n0} \frac{9\hbar C^3}{4e^2\omega^3 n(n^2+2)^2} \qquad (12)$$

where n is the refractive index of the crystal, substituting (12) into (10) and assuming a parabolic energy band, we find that

$$\sigma_1 = \frac{18\pi m^2 e^2 C^3}{\hbar^3 \omega^3 n(n^2+2)^2} \frac{1}{k_0^2} \ln\left(\frac{k_0 + k_n}{k_0 - k_n}\right) \frac{g_n}{g_0} A_{n0}$$

$$= \frac{18\pi^2 m e^2 h^2 c^3}{n(n^2+2)^2} \frac{1}{E_{n0}^3} \frac{1}{E} \ln\left(\frac{E^{1/2} + (E - E_{n0})^{1/2}}{E^{1/2} - (E - E_{n0})^{1/2}}\right) \frac{g_n}{g_0} A_{n0} \qquad (13)$$

where E_{n0} is the energy difference between the initial state and the final state, and E is the energy of the incident electron above the conduction band.

The second term σ_2 in(10) can be calculated by the irreducible tensor operator method.

In our previous work[7], using laser selective excitation, four types of Er^{3+} center in $ZnS:Er^{3+}$ thin films were identified, their radiative transition probabilities A_{if} for $^2H_{11/2}$ and $^4S_{3/2}$ to the ground state are listed in table 3 and 4.

Using (13), the impact cross sections σ_1 of the four centers corresponding to the impact excitation from $^4I_{15/2}$ to $^2H_{11/2}$ and $^4S_{3/2}$ were calculated. The results are listed in table 3 and table 4. Using radiative transition probabilities from our previous work[9], the impact cross sections σ_1 from the ground state $^5I_8$ to excited state $^5G_5$, $^5G_6$, $^5F_5$ of Ho^{3+} in ZnS were also calculated (table 3).

Using (10), we calculate the impact cross section σ_2 for $^2H_{11/2}$ of Er^{3+}:

$$\sigma_2 = 6.29 \times 10^{-19} [(E - E_{if})/E]^{1/2} \text{ cm}^2$$

it is clear that σ_2 is much less than σ_1.

The impact cross section σ_1 varies with the electron energy E as

$$(1/E) \ln ((E^{1/2} + (E - E_{if})^{1/2})/(E^{1/2} - (E - E_{if})^{1/2}))$$

when $E \leq E_{if}$, $\sigma_1 = 0$, when E exceeds E_{if}, σ_1 increases first with increasing E, reaches a maximum and then decreases slowly (figure 1).

Table 3. Calculated values of σ_1 for the four different centers a,b,c and d corresponding to the $^4I_{15/2}$--$^2H_{11/2}$ transition for Er^{3+} in ZnS (E=4.0ev)

Centre	σ_1 (10^{-16} cm^2)	A_{if}^{-1} (μs)
a	2	5.7
b	0.7	14
c	0.5	22
d	0.6	16

Table 4. Calculated values of σ_1 for four different centers a,b,c and d corresponding to the $^4I_{15/2}$--$^4S_{3/2}$ transition for Er^{3+} in ZnS (E=4.0ev)

Centre	σ_1 (10^{-17} cm^2)	A_{if}^{-1} (μs)
a	2	50
b	0.7	138
c	0.5	226
d	0.6	173

Table 5. The calculated impact cross sections of $^5G_5$, $^5G_6$ and $^5F_5$ for Ho^{3+} in ZnS

State	Energy E (eV)	A_{if}^{-1} (μs)	σ_1 (10^{-17} cm^2)
$^5G_5$	2.91	203	0.2
$^5G_6$	2.67	34	2
$^5F_5$	1.89	569	0.4

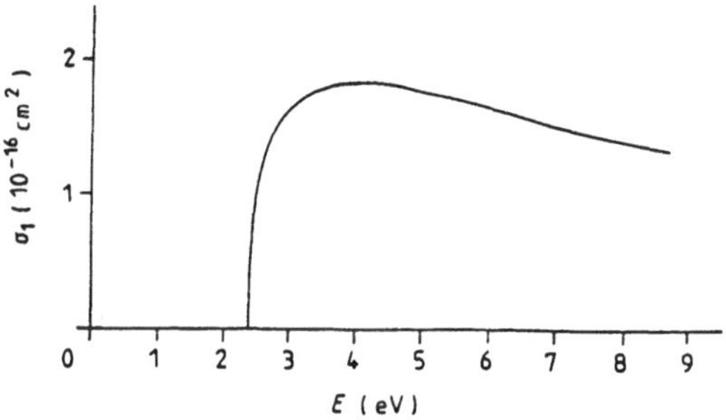

Fig.1. The energy dependence of the impact cross section of the a center in ZnS:Er^{3+}, corresponding to the transition from $^4I_{15/2}$ to $^2H_{11/2}$

4. Discussion

In paragraph 2.2, the average value of the impact cross section determined from the emission from $^4S_{3/2}$ and $^2H_{11/2}$ of Er^{3+} in ZnS was $2x10^{-16}cm^2$. The calculated values of σ_1 corresponding to the excitation from $^4I_{15/2}$ to $^2H_{11/2}$ of Er^{3+} in ZnS are listed in table 3. We see that the calculated values of σ_1 are in reasonable agreement with the measured result, considering the complexity involved in calculating the impact cross sections of rare earth ions in a crystal.

From our theory, the impact cross section σ is mainly determined by σ_1. According to the formula (13), σ_1 is proportional to the radiative transition probability. The larger the radiative transition probability, the larger the impact cross section. From table 2, we see that within the error, the relationship between σ and Pr has the same trend as expected from the theory.

σ_1 is determined by electric dipole transition. Since σ_2 is much less than σ_1, the electric dipole transition makes the main contribution to the impact cross section of Er^{3+} in ZnS. If we do not consider the mixing of the configuration with opposite parity into the $4f^n$ configuration, the electric dipole transition is forbidden. However, it is well known that most radiative transitions of rare earth ions in solids are electric dipole transitions in nature. This implies that mixing between opposite parity configurations is important in determining the radiative transition probabilities. Since the electric dipole transition probabilities of rare earth ions are sensitive to the environmental situation (symmetry, crystal field strength, charge compensator, etc), so is the impact cross section.

Different excited states have different radiative transition probabilities, so their impact cross sections are different. For instance, σ_1 of $^4S_{3/2}$ of Er^{3+} in ZnS is one order less than that of $^2H_{11/2}$. From table 3 and 4, we also see that different luminescent centers formed by Er^{3+} ions have different impact cross sections. $^5G_5$, $^5G_6$ and $^5F_5$ of Ho^{3+} in ZnS have σ_1 less than $^2H_{11/2}$ of Er^{3+} in ZnS (table 5).

Some one suspected that the impact cross section of Mn^{2+} in ZnS was mainly determined by exchange process, rather than direct process, and the impact cross section increased with electron energy faster for exchange process than for direct process. If it is so, the impact cross section for exchange process will increase faster with applied voltage than for direct process. However, from table 1 we see that the ratios of the impact cross section of Er^{3+} to that of Mn^{2+} in ZnS:Er^{3+}, Mn^{2+} do not vary with applied voltage. It implies that the mechanism of impact process for Mn^{2+} may be the same as that for Er^{3+}, which is direct coulomb impact process.

References

1. J.W.Allen: J. Lumin. 23, 127 (1981)
2. J.E. Bernard, M. Martens, F. Williams: J. Lumin. 24/25, 893 (1981)
3. W.E. Hagston: Phys. Status Solidi, (a)81, 687(1984)
4. J.W. Allen: J. Phys. C: Solid State Phys. 19, 6287(1986)
5. R. Mach, G.O. Muller: Phys. Status Solidi,(a)81, 609(1984)
6. G.Z. Zhong, F.J. Bryant: J. Lumin. 24/25, 909 (1981)
7. Xu Xumou, Yu Jiaqi, Zhong Guozhu: J. Lumin.36, 101(1986)
8. Luo Baozhu: Thesis, Changchun Institute of Physics, Academia Sinica, People's Rep. China
9. Shen Yongrong, Zhang Hong: Acta Phys. Sin. 35, 1574(1986)

Impact Excitation Cross Section in Electroluminescence

Shen Mengyan and Xu Xurong

Institute of Material Physics, Tianjin Institute of Technology,
Tianjin, 300191, P.R.China

1. INTRODUCTION

Recombination radiation and luminescence from discrete centers appear to have different characteristics. In general, the lifetime of recombination radiation is much shorter than that of luminescence from discrete centers. The concentration of discrete centers is also limited. Before 1970, the phosphor of electroluminescence (EL) was mainly in the form of powder. The recombination through impurities, e.g. Cu^{+}, was preferred. After 1974, thin film and discrete center were preferred, e.g. Mn^{+}? But the saturation of EL is often observable. And the reported highest energy efficiency is obtained for samples which is designed to guarantee the ultimate return of free electrons to the region of ionized centers. The direct impact excitation is believed to be the machanism of excitation of the luminescent centers in the phosphor. Hence, the impact cross section is a very important parameter for the excitation process. It reflects the excitation probability of the luminescent center, and has been investigated both theoretically and experimentally /1,2,3,4/. Most of the results are rather rough and the variation of impact cross section with hot electron energy is not clear. In this paper, by means of the simple but effective method developed by CHEW /5/ and subsequently used by AKERIB and BOROWITZ /6/ in their study of inelastic scattering problem, we investigated the impact cross section σ and the variation of σ with hot electron energy. And taking ZnS:Mn (discrete center) and ZnS:Cu (recombination center) as examples, we performed the numerical calculation.

2. MODEL AND CALCULATION

As a model system, we consider the luminescent center as the neutral atomic target which is impacted by hot electrons in the upper conduction band. The impact excitation process in the phosphor is much more complex than that in the free space. The structure of the upper conduction band is very complex and cannot be determined quantatively; different approaches can get different results /7,8,9/. For simplicity, we suppose that the hot electron is in an average parabolic monoconduction band, and the effective mass m_{hc}. The Hamiltonian of the system in which the hot electron impact-excites the luminescent center is composed of three terms. The first term is the Hamiltonian of the luminescent center system, the second one is that of hot electron, the last one the interaction V between the hot electron and the luminescent center system.

Use of standard techniques converts the Schrodinger equation into an integral equation. The scattering term of the wavefunction is given by,

Springer Proceedings in Physics, Vol. 38 Electroluminescence
Editors: S. Shionoya and H. Kobayashi

$$\psi_s(r_1,r_2) \underset{r_1\to\infty}{\widetilde{\quad}} -\frac{m_{hc}}{2\pi\hbar^2}\sum_n \phi_n(\vec{r}_2)\cdot\frac{\exp(i\vec{k}_n\cdot\vec{r}_1)}{r_1}\cdot$$

$$\cdot\iint d\vec{r}_1 d\vec{r}_2\, \phi_n^*(\vec{r}_2)\exp(-i\vec{k}_n\cdot\vec{r}_1)\; V\psi(\vec{r}_1,\vec{r}_2) \qquad (1)$$

We shall, in the following, follow the method of CHEW and AKERIB /5, 6/ for solving the analogous scattering problem. Assume $\psi(\vec{r}_1,\vec{r}_2)$ in (1) to be expressed in the following approximate form:

$$\psi(\vec{r}_1,\vec{r}_2) \approx \psi_a(\vec{r}_1,\vec{r}_2) = \int d\vec{k}_2 g_o(\vec{k}_2)\psi_{\vec{k}_1,\vec{k}_2}(\vec{r}_1,\vec{r}_2) \qquad (2)$$

$\psi_{\vec{k}_1,\vec{k}_2}(\vec{r}_1,\vec{r}_2)$ represents the wavefunction of an electron of momentum $\hbar\vec{k}_1$ moving in the field of an electron $\hbar\vec{k}_2$. (2) is equivalent to the interaction of the hot electron with a free electron whose momentum distribution is given by the Fourier transform of the ground state wave function of the luminescent center system.

Substitution of (2) into (1) leads to the V matrices:

$$\langle\vec{k}_n,n|V|\vec{k}_1,0\rangle = \int d\vec{k}_2 g_o(\vec{k}_2)\cdot\iint d\vec{r}_1 d\vec{r}_2 \exp(-i\vec{k}_n\cdot\vec{r}_1)\phi_n^*(\vec{r}_2)V\psi_{\vec{k}_1,\vec{k}_2}(\vec{r}_1,\vec{r}_2) \qquad (3)$$

Then the differential cross section for excitation from ground state to the state n will be

$$d\sigma_n/d\Omega = m_{hc}k_n/m_c k_1\cdot\left|(m_{hc}/2\pi\hbar^2)\langle\vec{k}_n,n|V|\vec{k}_1,0\rangle\right|^2 \qquad (4)$$

where m_c is the effective mass of electron at the conduction band bottom.

Further, we will use the formula (4) to perform numerical evaluation for ZnS:Mn and ZnS:Cu. As is known, the first excited state of Mn^{+2} is a discrete state, and the energy level is isolated very far from the bottom of the conduction band. The shape of the light emission spectra from both blue center B-Cu^+ and green center G-Cu^+ are symmetrically Gaussian-like, with maximum at 4450 Å and 5230 Å respectively /10/. If the light were emitted by the electric transition from conduction band to the ground state of the center Cu^+; the shape of the spectra would be asymmetric. Therefore the first excited state of Cu^+ center is discrete state isolated from the conduction band. We assume that the first excited state energy level of Cu^+ is just below the bottom of the conduction band, and the energy difference between the bottom of the conduction band and the first excited state energy is so small that the electron can easily go from the first excited state into the conduction band at normal temperature or due to electric field. When the hot electron impacts the Cu^+ center, the electron of the center can be excited both to the isolated discrete state and to the ionization state (into the conduction band). For simplicity here we consider only the transition from the ground state to the isolated discrete state impact excitation. The lifetime of Mn^{+2}emission is of the order of msec, while the lifetime of Cu^+emission is of the order of μsec /11/. For hydrogen atoms, the 2S state is a metastable state, with a lifetime of several msec, while the lifetime of 2P is only several μsec /12/. Because the luminescent center system is very complex, here using adiabatic approximation, we consider the luminescent system as a hydrogen-atom-like system. Hence, we can assume transition from the ground state to the first excited state of Mn^{+2} and that of Cu^+ to be similar to the transition 1S→2S and 1S→2P in hydrogen atom respectively. With these in mind, the differential cross sections for Mn^{+2} and Cu^{+} can be easily given as follow:

$$d\sigma_{Mn^{+2}}/d\Omega = a_o^2 \cdot \frac{2^{19} a_o^2 \pi^2}{a_1^2} \cdot \frac{m_{hc} k_1'}{m_c \; k_1} \cdot \frac{\pi/2a_1k_1 \; e^{-\pi/2a_1k_1}}{[9+4a_o^2(k_1-k_1')^2]^6 \; \sinh(\pi/2a_1k_1)} \qquad (5)$$

$$d\sigma_{Cu^+}/d\Omega = a_o^2 \cdot 9 \cdot 2^{17} \cdot \frac{a_o^2 \pi^2}{a_1^2} \; \frac{m_{hc} k_1'}{m_c \; k_1} \cdot \frac{a_o^2(\vec{k}_1-\vec{k}_1')^2}{[9+4a_o^2(k_1-k_1')^2]^6 \; [a_o(k_1-k_1')]^4} \cdot \frac{\pi/2a_1k_1 \; e^{-\pi/2a_1k_1}}{\sinh(\pi/2a_1k_1)} \qquad (6)$$

We choose m_{hc} varying between $0.5m_o$ and $2.0m_o$ (m_o is the mass of an electron in free space). The impact cross sections for Mn^{+2} G-Cu^+ and B-Cu^+ versus the hot electron energy at various effective mass m_{hc} can be given by (5) and (6). It is shown that the larger the effective mass is, the larger the impact cross section is. If the energy of the hot electron runs between 2ev and 4ev, and m_{hc} is about $1.4m_o$, the expected value for the Mn^{+2} impact cross section can fit the experiment result /3/. Therefore, for energy of hot electron from 2ev to 4ev, we choose the effective mass as $1.4m_o$, and calculate the value of the three impact cross sections (see Fig. 1). From the Fig.1, we can see that the variation of impact cross sections for Mn^{+2} and Cu^+ with the hot electron energy is different. For Mn^{+2}, when the energy of the hot electron increases, the cross section increases at first, passes a maximum at 3.5ev, then decreases. However, for G-Cu^+ (B-Cu^+), the cross section monotonously increases with the increase of the energy of the hot electron.

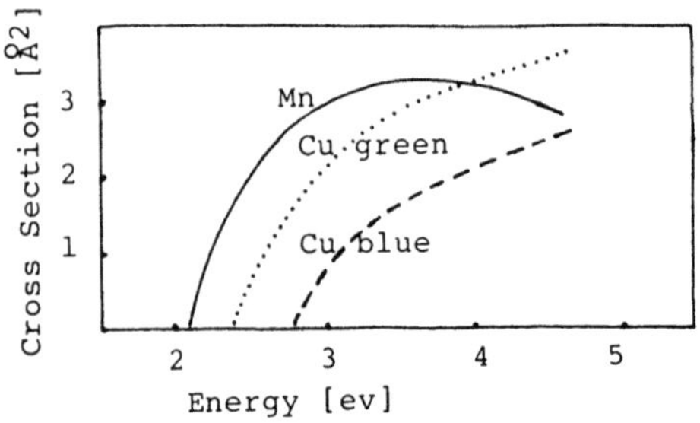

Fig.1. The impact excitation cross sections for Mn^{+2}, G-Cu^+ and B-Cu^+ versus the energy of hot electron energy

3. DISCUSSION

We have said that when the hot electron impact Cu^+ center, the electron in Cu^+ center system can be excited both to the isolated discrete excited state and to the ionization state. For simplicity, only the cross section related to the former process is calculated in the above formula. But that related to the latter cannot be neglected. From the results of scattering of electrons by hydrogen atoms /12/, we can estimate that the cross section related to the latter process, in shape and magnitude, is close to that related to the former one. So the magnitude of impact cross section in the Fig. for Cu^+ may be doubly increased. Thus the impact cross section for Cu^+ is possibly larger than that for Mn^{+2}. In the above calculation, we assume the luminescent center to be a neutral atomic system. Obviously the charge of the luminescent center will influence the impact excitation process. Now we give a qualitative estimate for the influence of the charge of Cu^+-like center on the cross section. Assume the center to be Cu^{+2}, the lattice in the vicinity of Cu^{+2} is deformed, this results in that Cu^{+2} center as a positive center screened by the negative deformed lattice. Then we simply suppose the negative charge distributes on a spherical surface with radius R. Cu^{+2} can absorb an electron and turns into Cu^{+}. When a hot electron is coming and excites the Cu^{+} system, the negative deformed lattice must repulse the hot electron as it is outside the sphere, and the negative deformed lattice does not interact with the hot electron as it is inside the sphere. So if the free path l of the hot electron (l is the average distance between two successive excitation collisions for a hot electron) is larger than R, the negative lattice makes the cross section smaller, and if l is smaller than R the influence of the negative lattice on the section can be neglected. Therefore, the larger the concentration of $Cu^{\pm}$like luminescent center, the smaller the influence of the negative deformed lattice on the impact excitation cross section . When we study TFELD, we must consider the recombination cross section. For Mn^{+2} the recombination cross section is determined mainly by the emission lifetime. For Cu^{+}-like center, the recombination cross section is affected by the electric field and the energy distribution of electron in conduction band, the two factors can be artificially controlled by getting optimum TFELD structure.

4. CONCLUSION

a. A general formula for impact excitation corss section is given. The variation of the cross section with the hot electron energy can be known.

b. Numerical evaluation is performed for ZnS:Mn and Zn:Cu . For Mn^{+2}, when the energy of the hot electron increases, the section passes a maximum at 3.5ev, then decreases, However, for Cu^{+}, the section monotonously increases with the increase of the energy of the hot electron.

The authors wish to thank Pro.Pan Jinsheng for a lively discussion.

1. J.W.Allen: J. Luminescence 23, 127 (1981)
2. J.E.Bernard, et.al.: J. Luminescence 24/25, 893 (1981)
3. R.Mach and G.O.Muller: phys. stat. sol. (a) 81, 609 (1984)
4. W.E.Hagston: phys. stat. sol. (a) 81, 687 (1984)
5. G.F.Chew: Phys. Rev. 80, 196 (1952)
6. R.Akerib and S.Borowitz: Phys. Rev. 122, 1177 (1961)
7. M.L.Cohen and T.K.Bergstresser: Phys. Rev. 141, 789 (1966)
8. P.Eckelt, O.Madelung and J.Treusch: Phys. Rev. Lett. 18, 656 (1967)
9. M.Z.Huang and W.Y.Ching: J. Phys. Chem. Sol. 46, 977 (1985)
10. H.K.Henisch: Electroluminescence. (Pergamon Press Ltd. 1962) p.23
11. Geng Ping and Zhang Xinyi: Chinese J. Luminescence 8, 163 (1987)
12. P.G.Burke and K.Smith: Rev. Modern Phys. 34, 458 (1962)

Electroluminescent Mechanisms of Rare-Earth-Doped ZnS Thin Films

P. Benalloul

Laboratoire d'Optique de la Matière Condensée, Université P. et M. Curie, 4 place Jussieu, F-75252 Paris Cedex 05, France

1. INTRODUCTION

A.C. thin films electroluminescent (E.L.) devices with ZnS host matrix and Inoguchi structure (MISIM) [1] have a simpler and more definite structure than Destriau ones. These devices work under high a.c. electric field. Müller [2] has listed and discussed the various processes which occur when the device is operating.

We can register three kinds of processes :

1. electrical processes : release and acceleration of electrons by the applied electric field ;
2. excitation process of the luminescent center ;
3. optical processes : radiative and non radiative deexcitations of this center.

If the release of electrons only depends on the insulator-host matrix interfaces, all other processes depend on the host matrix and also on the luminescent center.

In the case of ZnS:Mn active layer, Mn^{2+} is an isoelectronic dopant in substitution to Zn^{2+} ; concerning ZnS:TbF_x active layer, the nature of the center is an important question, the knowledge of which has made significant progress these last years [3→7]. According to the preparation conditions of the active layer, one gets two different complex centers :
- TbF_3 as a complex center with 3 F^- ions near Tb^{3+} ion, in weak interaction with ZnS matrix ;
- TbF as a complex center, in stronger interaction with ZnS matrix ;

ZnS:TbF active layers are more efficient than ZnS:TbF_3 in E.L.

This paper is devoted to the analysis of the excitation process of the luminescent center. This question is not yet solved and, in literature, the conclusions are contradictory [7→10].

Two different mechanisms are possible :

1. A one step direct excitation by impact of energetic electrons on the Tb^{3+} center.
2. A two step excitation process involving energy transfer : generation of electron-hole pairs by impact ionization of ZnS, followed by an energy transfer to the Tb^{3+} center.

In order to get a clear understanding of this question, we present results obtained by three kinds of experiments :

1. Dependence of the intensity of the emissions from the $^5D_3$ and $^5D_4$ levels of Tb^{3+} on the applied voltage.
2. A comparison of photoluminescence (P.L.) and E.L.
3. An analysis of the different time response emissions to a short voltage pulse.

Springer Proceedings in Physics, Vol. 38 Electroluminescence
Editors: S. Shionoya and H. Kobayashi

The results concern $ZnS:TbF_3$ active layers made by electron-beam coevaporation. For preparation conditions, see references [3] and [10]. For these layers, the complex center is TbF_3.

2. THE RATIO $I(^5D_3)/I(^5D_4)$ VERSUS THE APPLIED VOLTAGE

Tb^{3+} has been used as a probe able to give information upon the kind of the mechanism of excitation. The energies of the $^5D_3$ and $^5D_4$ levels are respectively 3.25 and 2.54 eV. The increase (or not) of the ratio $R = I(^5D_3 \rightarrow {}^7F_4)/I(^5D_4 \rightarrow {}^7F_5)$ of the emissions from the $^5D_3$ and $^5D_4$ levels with increasing applied voltage V_p could give an answer to this problem. It was assumed that if V_p increases, the active layer electric field $\mathcal{E}$ increases ; therefore the energy function of the electrons f(E) is shifted towards higher energy.

The increase of R shown by Krupka [12] was an attractive demonstration of an impact mechanism. Kobayashi et al. [13] confirmed this conclusion with $ZnS:ErF_3$ active layers. This question seemed to be solved when contradictory results were published by Marello et al. [14] who have found no variation of R for $ZnS:TbF_3$ devices, and by Szczureck et al. [15] who have found a decrease of the ratio for $ZnSe:ErF_3$ devices.

We have repeated this kind of experiments for Tb^{3+} ion ; for Er^{3+} ion, the results are difficult to interpret due to the multiplicity of energy levels and also to cross-relaxation processes between Er^{3+} ions as shown by Li-Jian Meng et al. [16].

We must be coutious : usually, the two insulator layers devices have a very sharp B(V) curve ; therefore the applied voltage and the active layer electric field variations are limited. This sharp B(V) curve is related to insulator-ZnS interface properties, and as for ZnS:Mn devices [17,18,19], our devices present an enhancement of E.L. when irradiated by 365 nm UV, and a shift of the threshold voltage (Fig. 1a). Under particular conditions of preparation of the interfaces, we have succeeded in getting devices with weak slope B(V) curves. These devices do not present any UV enhancement of E.L. (Fig. 1b).

Moreover, we must adjust the dopant concentration to get a $^5D_3$ emission not too weak compared to $^5D_4$ emission, and larger than ZnS emission. With such

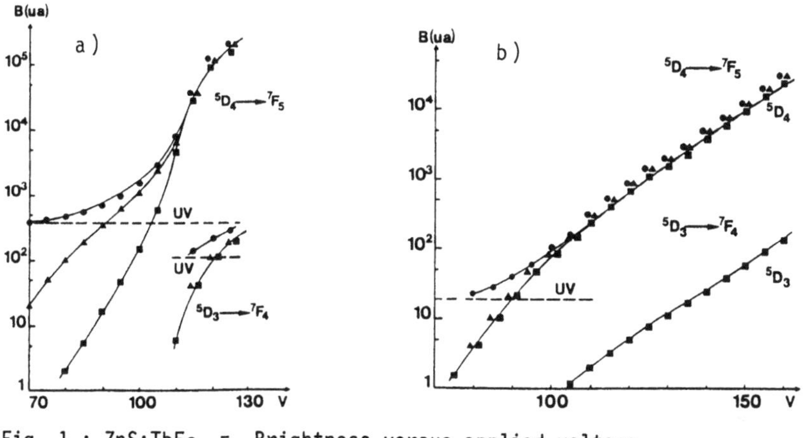

Fig. 1 : $ZnS:TbF_3$ = Brightness versus applied voltage.
a = X 241 b = X 249
o EL + UV 365 nm --- PL UV 365 nm
△ EL (o) - PL (---) □ EL

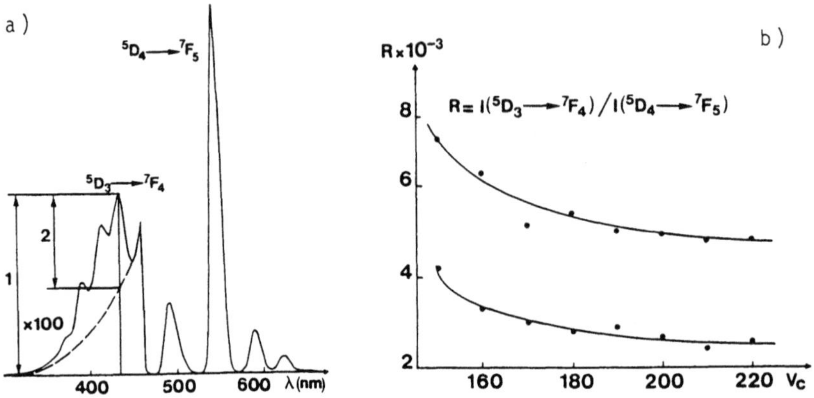

Fig. 2 : Ratio of the emissions from the $^5D_3$ and $^5D_4$ levels versus applied voltage.
X249 ; ZnS:TbF$_3$ c = 0.2 mol %
a) emission spectra : pulse excitation ν = 5 kHz ; L = 30 μs
b) variation of R : R_1 = taking account all the emitted light at 437 nm.
R_2 = evaluated ZnS emission is taken off.

devices, we have studied the variation of R ; it decreases slightly when V_p increases (Fig. 2).

In order to interpret this kind of experiment, it seems necessary to us to come back to the basic hypothesis : we must know the excitation functions of the $^5D_3$ and $^5D_4$ levels (for example, the result can be different if we take step functions or gauss distribution functions [15]) ; the energy distribution function f(E) of the electrons accelerated by the active layer electric field $\mathcal{E}$ is required. But $\mathcal{E}$ depends on insulator-ZnS interfaces and on the quality of insulators, as shown recently by Müller et al. [20].

Our conclusion will be that the results can depend on the way of devices production. Up to now, this kind of experiment cannot solve the problem of the excitation mechanism of the rare-earth center in E.L. ZnS devices.

3. COMPARISON OF P.L. AND E.L. OF ZnS:TbF$_3$ ACTIVE LAYERS

Contrary to E.L., in P.L. we can use an excitation of well defined energy ; therefore we can excite in the absorption band of ZnS or in the $^5D_3$ or $^5D_4$ ones. Time resolved spectroscopy (T.R.S.) in P.L. and E.L. and excitation spectrum of Tb^{3+} luminescence give enlightened results allowing us to draw conclusions about excitation mechanism.

3.1. Time resolved spectroscopy

First, we make clear how the energy is transfered to Tb^{3+} when we excite in the ZnS absorption band. P.L. studies on ZnS:Tb^{3+} crystals [21,22] have shown that there are an acceptor level E_A at 1 eV above the valence band and a donor level E_D at 0.33 below the conduction band. Therefore the energy of the donor acceptor associated center ($\Delta E = E_G - E_A - E_D$) is just higher than the excitation energy of the $^5D_4$ level. This explains why the $^5D_3$ emission is not present, or very weak. Moreover, Shionoya [23] has presented T.R.S. of ZnS:Tb^{3+} crystals. At 77 K, a band centered at 430 nm and with 50 nm half-width appears at the time of pulse excitation, then decreases quickly in some μs, when the

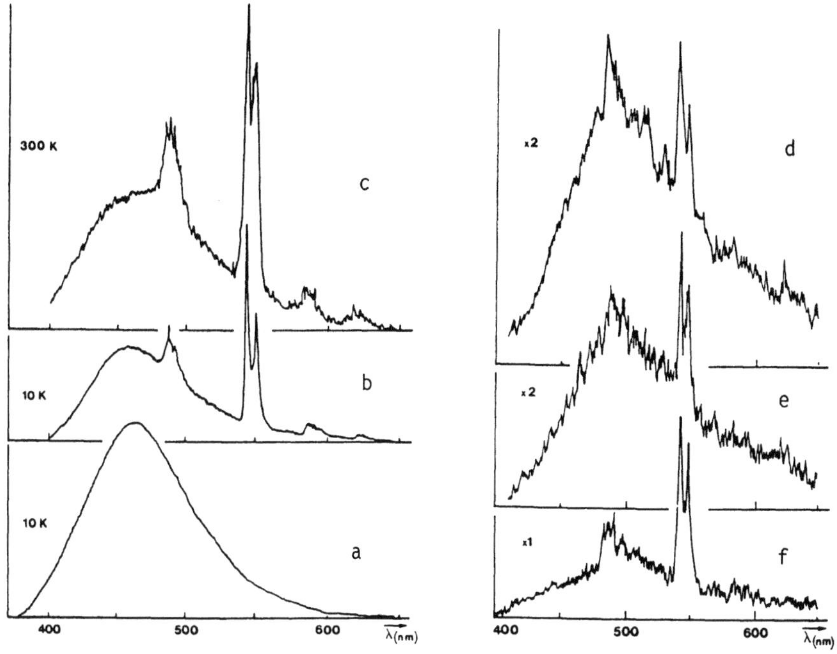

Fig. 3 : T.R.S. of ZnS:TbF_3 layer ; e = 3 mole % gate = 0.6 μs.
a) delay 0.5 μs ; R_C = 50 Ω
b-c) delay 66 μs ; R_C = 50 kΩ
d) delay 0.5 μs, e) delay 1.5 μs: 77 K
f) delay 8 μs; R_C = 1 kΩ

$^5D_4$ emission increases. These data are explained as an excitation of Tb^{3+} by an energy transfer from the 430 nm center.

We present some P.L. T.R.S. of ZnS:TbF_3 active layers. The excitation was a N_2 10 ns pulse laser (λ = 337.1 nm) (Fig.3) and the gate width 0.6 μs. Just after the pulse excitation, the spectrum shows only a band centered at 460 nm which extends beyond 600 nm, hiding the start of the $^5D_4$ emission. This emission is very sizeable and decreases quickly, τ_e being of the order of 45 ns at 77 K. T.R.S. shows clearly that $^5D_4$ emission is excited by an energy transfer from this "blue center". Moreover, it was not possible for us to detect $^5D_3$ emission.

T.R.S. in E.L. (Fig.4) are quite different. The pulse voltage had a pulse width of 3 μs, the shorter pulse we can get up to now. The time response for the measurement system was about 0.1 μs. From the start, we can record weak ZnS emission, and more intense $^5D_3$ and $^5D_4$ emissions. As ZnS and $^5D_3$ emissions decrease quickly, the $^5D_4$ emission increases, the maximum being reached about 3 μs after the two other ones.

These results are deciding in favour of the direct impact excitation mechanism as the dominant process for ZnS:TbF_3 E.L. devices.

3.2. Excitation spectrum

We present also the excitation spectrum, at 300 K, of a ZnS:TbF_3 layer ; we register the P.L. at 545 nm which is the $^5D_4 \rightarrow {}^7F_5$ emission. The excitation is a 450 W Xenon source followed by a monochromator. This spectrum shows two bands at 380 nm and 488 nm due to $^7F_6 \rightarrow {}^5D_3$ and $^7F_6 \rightarrow {}^5D_4$ absorption bands

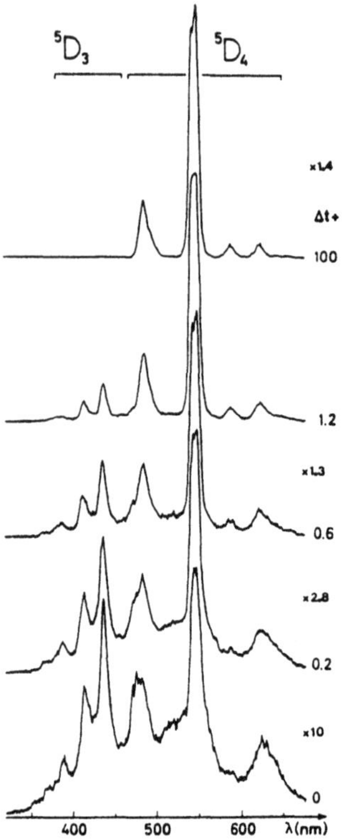

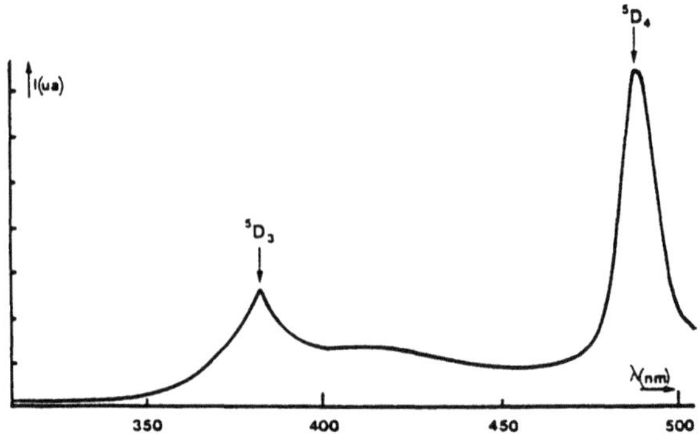

Fig. 5 : Excitation spectrum of $ZnS:TbF_3$ layer
c = 3 mol % T = 300 K
analysis : 545 nm

Fig. 4 : E.L. TRS of $ZnS:TbF_3$ device
c = 0.25 mol % gate = 0.6 μs
for time delays, see Fig. 6

(Fig. 5). The efficiency of Tb^{3+} P.L. is more important by a factor higher than 10 when Tb^{3+} is directly excited than by energy transfer from ZnS. This result is in agreement with [24,25,7], and supports this one step excitation mechanism in E.L. We must quote that Chen et al. [25] had compared energetic efficiency of E.L. and indirect ZnS excited P.L. of $ZnS:TbF_3$ devices ; as $\eta_{E.L.} > \eta_{P.L.} \cdot \eta_{electron\text{-}hole\ pairs}$, they had concluded for this direct excitation process.

This last argument cannot be used for ZnS:TbF devices. Recently it has been shown [7,10] the results are quite different for TbF_x center according to x ; for direct P.L. excitation, the P.L. efficiency is independent of x, while for indirect P.L. excitation via ZnS host, this efficiency is highly dependent on x ; for x = 1, this indirect excitation being very efficient.

3.3. Optimal concentration

At least, we note another result favourable to this direct process : we list the optimal concentration of the dopant for ZnS:TbF layers in P.L. for different excitation processes of Tb^{3+} and in E.L.

	P.L. 325 nm	P.L. 380 nm	E.L.
$ZnS:TbF_3$ [26]	1 mol %		3 mol %
ZnS:TbF [10]	0.2 mol %	2 mol %	2 mol %

In conclusion, if we look for all these data concerning a comparison between P.L. and E.L., we can consider that the direct excitation mechanism by energetic electron is the dominant process for $ZnS:TbF_3$ E.L. devices.

4. ANALYSIS OF THE TIME RESPONSE EMISSIONS TO A VOLTAGE PULSE

In order to get a more precise understanding of the excitation process of $^5D_3$ and $^5D_4$ levels of Tb^{3+}, we have considered the time response emissions to a voltage pulse (Fig.6) :

. ZnS emission at 400 nm
. $^5D_3 \longrightarrow {}^7F_4$ emission at 437 nm
. $^5D_4 \longrightarrow {}^7F_5$ emission at 545 nm.

The pulse has 3 μs width and a rise time of 2.10^{-7} s. According to the 3 nF device capacity and the 30 Ω electrode resistances, the electric time constant of the device is about 10^{-7}s. The emissions are recorded with a digital memory scope interfaced to a microcomputer to accumulate signals ; the digitizing rate is 20 ns. The time response for the measurement system is about 0.1 μs. The $^5D_3$ emission has been corrected in order to take into account the contribution of the ZnS emission at 437 nm ; this emission extends overall the spectrum. This contribution has been evaluated on T.R.S. The TbF_3 dopant concentration in the active layer is 0.25 mol %.

The main results are :

1. ZnS emission : the decay of the ZnS emission is very fast, less than 100 ns [27]. According to the time constant of the device, we can consider that this emission and the dissipative current i(t) can been superimposed. For such a short pulse excitation we used, it was not possible for us to take i(t) with enough accuracy by a compensator method. Just at the end of the pulse, there is a small ZnS emission due to the depolarization current [28]. 2 μs after the ZnS emission maximum, this emission has decreased by a factor 100 and does not show any slow decay component, which is a probe that in these active layers, there is not any unknown impurity from which an energy transfer can occur.

2. Excitation process : the beginning of ZnS, $^5D_3$ and $^5D_4$ emissions don't show any delay, result in agreement with E.L. T.R.S. study. Now we are looking only the increase part of the emission curves. As they are very short with time, especially the ZnS and $^5D_3$ ones, we can consider that the derivatives versus

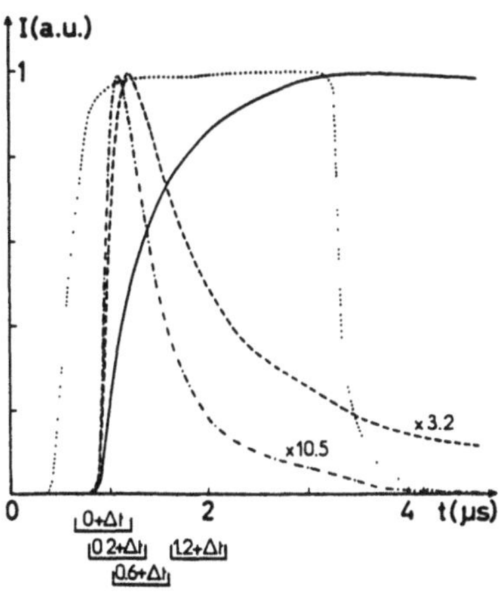

Fig. 6 :
Time response emissions to a voltage pulse.
.... voltage pulse
_._. ZnS emission λ = 400 nm
_ _ _ _ $^5D_3 \rightarrow {}^7F_4$ emission λ = 437 nm
—— $^5D_4 \rightarrow {}^7F_5$ emission λ = 545 nm

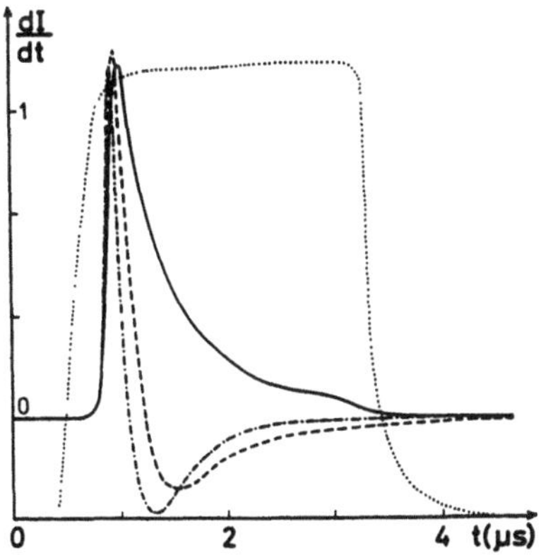

Fig. 7 :
Derivatives with time of the emissions
_._. dI(ZnS) / dt
_ _ _ _ dI($^5D_3$) / dt
—— dI($^5D_4$) / dt

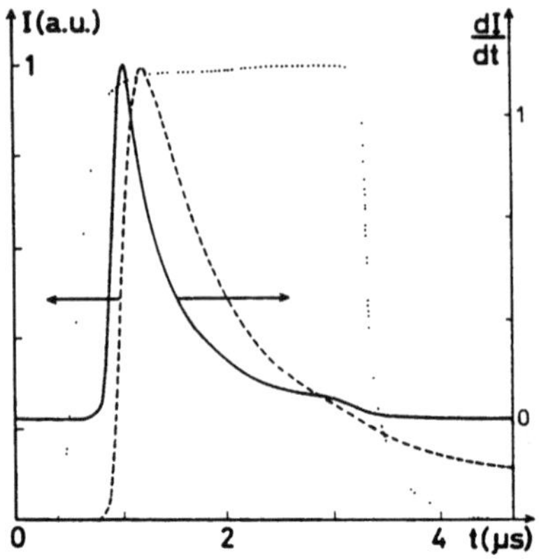

Fig. 8 : Characteristics of the time responses.
- - - I ($^5D_3$)
——— dI ($^5D_4$)/dt

time of the emissions are proportional to the number of the excited centers at each time. We report on Fig. 7 the dI (ZnS)/dt, dI ($^5D_3$)/dt and dI ($^5D_4$)/dt curves. Their beginnings don't show any delay and their maxima are reached at the same time t_1 of the order of 0.3 µs. Therefore, from t = 0 until t = t_1, we can conclude that the excitation rates of the three recorded emissions are proportional. After t_1, the dI (ZnS)/dt and dI ($^5D_3$)/dt curves decrease very quickly to zero, then are negative whereas the dI ($^5D_4$)/dt curve decreases slowly during more than 2 µs, but it is still positive. During this time, i(t) decreases quickly (see the ZnS emission curve) but is still present. Therefore, the direct excitation of $^5D_4$ take place. But the $^5D_4$ level is populated by another process ; it is by cross relaxation between 2 Tb^{3+} neighbour ions. This relaxation is very efficient ; the critical distance is of the order of 13 Å [29].

$$Tb^{3+}\ (^5D_3) + Tb^{3+}\ (^7F_6) \longrightarrow Tb^{3+}\ (^5D_4) + Tb^{3+}\ (^7F_{0,1})$$

This model can be supported by another argument : if $^5D_4$ level was populated only by $^5D_3$ transfer, the maximum of dI ($^5D_4$)/dt would be reached after the I($^5D_3$) one. We can see on Fig. 8 that it is not the case.

3. $^5D_3$ and $^5D_4$ emission shapes : this process we describe just above can explain the $^5D_3$ and $^5D_4$ emission shapes in Fig. 6. The maximum of the $^5D_3$ emission is reached about 0.1 µs after the maximum of the ZnS one. This delay can be explained by the fact that the $^5D_3$ emission has, for this first part of the curve, a decay we can characterize by a value of τ_e of the order of 1.1 µs, not much longer than the ZnS which is 0.5 µs due to the i(t) shape. Therefore, if there is an energy transfer from the ZnS to $^5D_3$ level, it is not significant.

The $^5D_4$ emission increases when the $^5D_3$ emission decreases ; its maximum is reached about 3 µs after the $^5D_3$ one. This delay and the shape of the rise part of the $^5D_4$ emission curve are functions of the dopant concentration for a same level of excitation [26].

5. CONCLUSION

We have studied P.L. and E.L. emissions versus time of ZnS:TbF_3 active layers. The data we have presented allow us to conclude :

1. The dominant mechanism of excitation of E.L. ZnS:TbF_3 devices is a direct impact excitation by energetic electrons.

2. Energy of these electrons is high enough to impact excite Tb^{3+} ions in the state $^5D_3$ or $^5D_4$; the $^5D_4$ level is also populated by cross relaxation between

Tb^{3+} neighbour ions from the $^5D_3$ level. This second process is the most important for the optimal 2.5 to 3 mol % concentration of E.L. devices. These luminescence properties of Tb^{3+} ions explain the efficiency of the E.L. Tb devices : if a Tb^{3+} ion is excited in its $^5D_3$ level, by cross relaxation a green "photon" is emitted. Therefore the emission is mainly in the green part of the spectrum ; this solves the need of an efficient green E.L. device.

Acknowledgements

We would like to express our thanks to J. Benoit, E. Bringuier and A. Geoffroy of the E.L. team of the laboratory for valuable discussions, C. Barthou for his collaboration in T.R.S. measurements, and Mrs Gouy and A. Nappey for their assistance in the sample preparation.

REFERENCES

1. T. Inoguchi and S. Mito, Topics in Appl. Phys. vol. 17, Springer Verlag (1977), 197
2. G.O. Müller, Phys. Stat. Sol.(a) 81, 597 (1984)
3. J. Benoit, P. Benalloul and B. Blanzat, J. of Lumines. 23, 175 (1981)
4. K. Okamoto and K. Watanabe, Appl. Phys. Lett. 49, 578 (1986)
5. A. Mikami, M. Ogura, K. Tanaka, K. Taniguchi, M. Yoshida and S. Nakajima, J. Appl. Phys. 61, 3028 (1987)
6. J. Mita, M. Koizumi, H. Kanno, T. Hayashi, Y. Sekido, I. Abiko and K. Nikei, Jap. J. of Appl. Phys. 26, L1205 (1987)
7. K. Hirabayashi, H. Kozawaguchi and B. Tsujiyamia, Jap. J. of Appl. Phys. 27, 587 (1988)
8. P. Benalloul, Thesis, Université P. et M. Curie, Paris, France (1986)
9. K. Okamoto and S. Muira, Appl. Phys. Lett. 49, 1596 (1986)
10. A. Mikami, M. Ogura, K. Taniguchi, M. Yoshida and S. Nakajima, J. of Lumines. 40-41, 784 (1986)
11. P. Benalloul, J. Benoit and A. Geoffroy, J. of Crystal Growth 72, 553 (1985)
12. D.C. Krupka, J. Appl. Phys. 43, 476 (1972)
13. H. Kobayashi, S. Tanaka and H. Sasakura, Jap. J. of Appl. Phys. 12, 1637 (1973)
14. V. Marello, L. Samuelson, A. Onton and W. Reuter, J. of Appl. Phys. 52, 3590 (1981)
15. I. Szczurek, H.J. Lozykowski and T. Szczurek, J. of Lumines. 23, 315 (1981)
16. Li-jian Meng, Chang-hua Li and Guo-zhu Zhong, J. of Lumines. 39, 11 (1987)
17. I.F. Chang and P.Y. Yu, Appl. Phys. Let. 37, 432 (1978)
18. H. Venghaus, J. of Crystal Growth 59, 403 (1982)
19. G.O. Müller and R. Mach, Phys. Stat. Sol.(a) 77, K179 (1983)
20. G.O. Müller, R. Mach, R. Reetz and G.U. Reinsperger, S.I.D. 88 Digest, p.23
21. W.W. Anderson, Phys. Rev. A 136, 556 (1964)
22. W.W. Anderson and S. Razi Proc. of the I.C.L. 1966, p. 1662
23. S. Shionoya in II-VI Semiconducting Conf. 1967, p. 1, ed. by D.G. Thomas
24. D.C. Krupka and D.M. Mahoney, J. Appl. Phys. 43, 2314 (1972)
25. Y.S. Chen, J.C. Burgiel and D. Kahng, J. Electrochem. Soc. Solid State Science 177, 794 (1970)
26. J.I. Pankove, M.A. Lampert, J.J. Hanak and J.E. Berkeyheiser, J. of Lumines. 15, 349 (1977)
27. P. Thioulouse, J. of Crystal Growth 72, 545 (1985)
28. E. Bringuier and A. Geoffroy, Appl. Phys. lett. 48, 1780 (1986)
29. D.L. Robbins, B. Cockayne, B. Lent and J.L. Glapser, Solid State Comm. 20, 673 (1976)
30. P. Benalloul, to be published

Time Resolved Emission Spectra in ZnS Thin Film Electroluminescent Devices

R. Nakano, H. Matsumoto, N. Miura, N. Sakagami, J. Shimada, and T. Endo

Faculty of Engineering, Meiji University,
1-1-1, Higashi-Mita, Tama-ku, Kawasaki 214, Japan

1. Introduction

The excitation mechanism in AC-electroluminescent (EL) devices has been the subject of much interest and controversy. Some EL properties have been successfully interpreted on the basis of a direct impact excitation model. However, a resonant energy transfer model as a dominant excitation process has been presented [1,2,3]. There are only a few reports on studies of the relation between broad-band emission intrinsic to ZnS host and emission of doped ion [2,4,5,6]. This paper reports the experimental results of transient behavior between broad-band and Tb^{3+} emission in ZnS thin-film EL devices excited by an alternating pulse voltage of very narrow width.

2. Experimental

The EL device was fabricated on a quartz glass substrate coated with In_2O_3:SnO_2 (ITO). The device structure is shown schematically in Fig. 1. As an emission layer, undoped ZnS or ZnS:TbF_x thin-films were prepared by electron-beam-evaporation at a substrate temperature of approximately 150℃, and subsequently subjected to heat treatment at as deposit, 100, 200, 400 or 600℃ for 1 hour. The TbF_3 concentration in the starting materials was 5 wt%. This film was sandwiched between electron-beam-evaporated Y_2O_3 insulating layers.

The transient behavior of emission under an alternating pulse voltage of 500 ns width at 12.5 Hz were taken by a system having a monochromator (Jobin YVON H-20), photomultiplier (HAMAMATSU PHOTONICS R1104), wide band preamplifier (NF BX-31) and digital boxcar integrator (NF BX-531). The gate widths of the boxcar integrator were 50 ns and 10 ns for the measurement of time-resolved spectra and time response, respectively. Particular attention was paid to the selection of photomutiplier output resistance, because the rise time of the emission

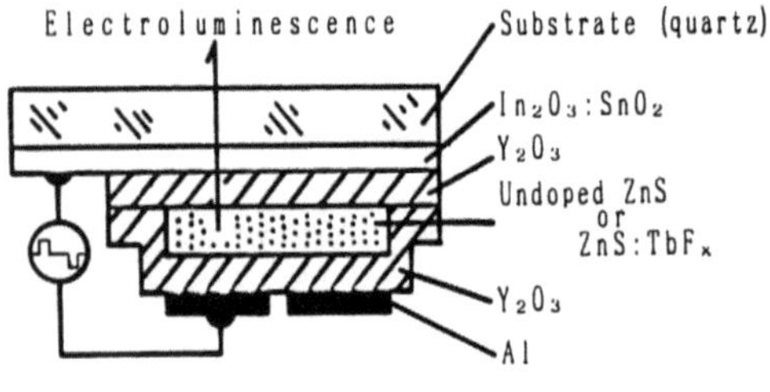

Fig. 1 Schematic structure of EL device

Springer Proceedings in Physics, Vol. 38 Electroluminescence
Editors: S. Shionoya and H. Kobayashi

strongly depends on this value. The spectral sensitivity of the detection system was not calibrated . All measurements were performed at room temperature.

3 . Results and Discussion

The emission spectra in $ZnS:TbF_x$ and undoped ZnS EL devices subjected to 400 ℃ heat treatment are shown in Fig. 2. In the $ZnS:TbF_x$ EL device, the strong emission peaks at 490, 545, 590 and 625nm correspond to the transition from $^5D_4$ to ^7F and weak emission peaks at 380, 420, 440 and 460nm correspond to the transition from $^5D_3$ to ^7F of Tb^{3+} ion respectively [7]. In undoped ZnS EL , very broad band emission with several peaks was observed in the wavelength region from 320nm to 700nm.

The time-resolved emission spectra in $ZnS:TbF_x$ and undoped ZnS EL subjected to 400℃ heat treatment are shown in Fig. 3 . In undoped ZnS, the emission

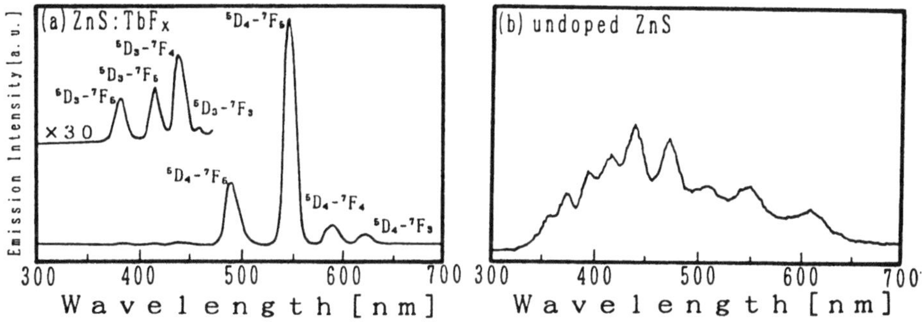

Fig. 2 Electroluminescent spectra for (a) $ZnS:TbF_x$ and (b) undoped ZnS EL devices subjected to heat treatment at 400℃

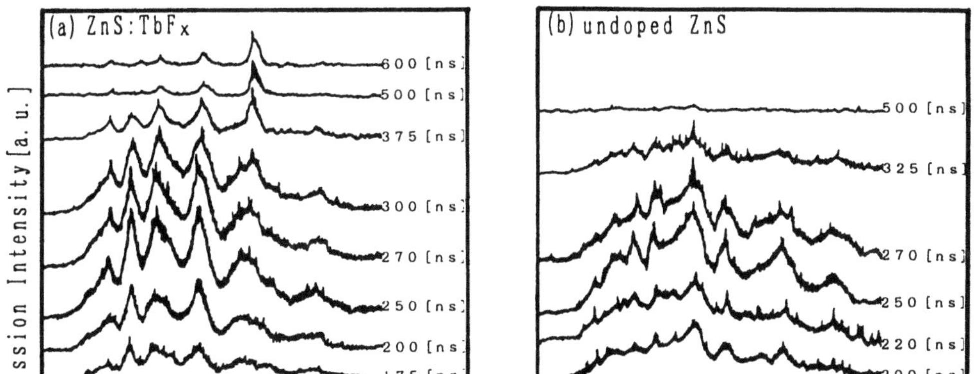

Fig. 3 Time-resolved emission spectra for (a) $ZnS:TbF_x$ and (b) undoped EL devices subjected to heat treatment at 400℃

extends to a shorter wavelength than with the ZnS:TbF_x structure. The emission related to the Tb^{3+} ion is delayed due to broad-band emission from the ZnS host.

The time responses for 545nm and 350nm emission in ZnS:TbF_x EL devices subjected to heat treatment at some temperature are shown in Fig. 4 (a) ~ (d) and (a') ~ (d'). Figure 4 (e) and (e') show the time responses in undoped ZnS EL device subjected to heat treatment at 400 ℃. The emissions of ZnS:TbF_x EL de-

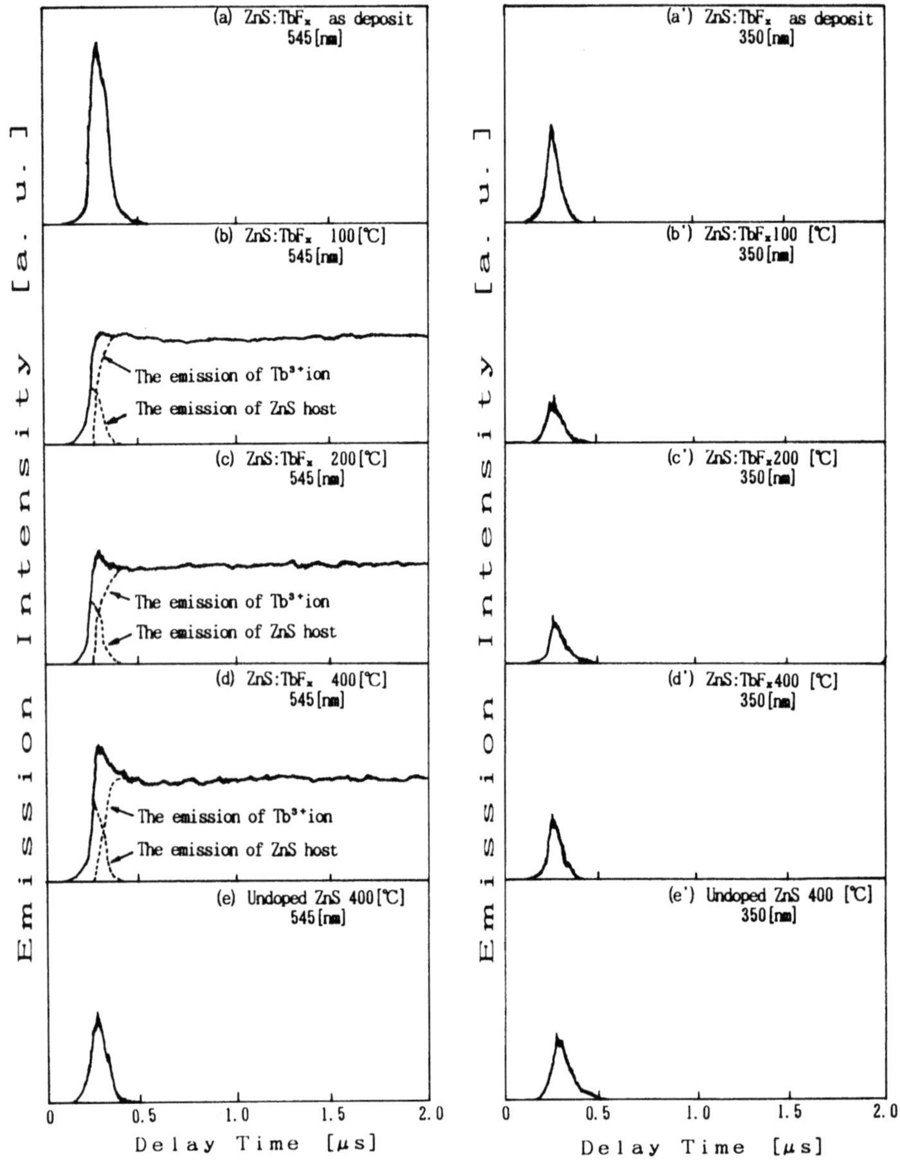

Fig. 4 Time responses for 545 and 350 [nm] in ZnS:TbF_x EL devices (a, a' ~ d, d') and undoped ZnS EL device (e, e'). Dotted lines indicate the separated 545 [nm] emission of ZnS host and Tb^{3+} ion

vices at 545nm shown in Fig. 4 (b) ~ (d) are separated into the emission intrinsic to the ZnS host and that of the Tb^{3+} ion as shown by dotted lines, considering the characteristics of an undoped ZnS EL device for 545nm emission (Fig. 4(e)). In the case "as-deposited" shown in Fig. 4(a), the emission of the ZnS host is observed but that of the Tb^{3+} ion is not observed. The time responses for 350nm as shown in Fig. 4 (a') ~ (d') give the same characteristics with undoped ZnS (Fig. 4(e')). Therefore it is considered that this emission is from the ZnS host. The dependence on heat treatment temperature of the 545nm emission from the Tb^{3+} ion and 350nm emission of the ZnS host are shown in Fig. 5. The emission of the Tb^{3+} ion and the ZnS host show a similar tendency except when "as-deposited". In this case, the emission of Tb^{3+} ion is scarcely observed in spite of the existence of 350nm emission intrinsic to the ZnS host.

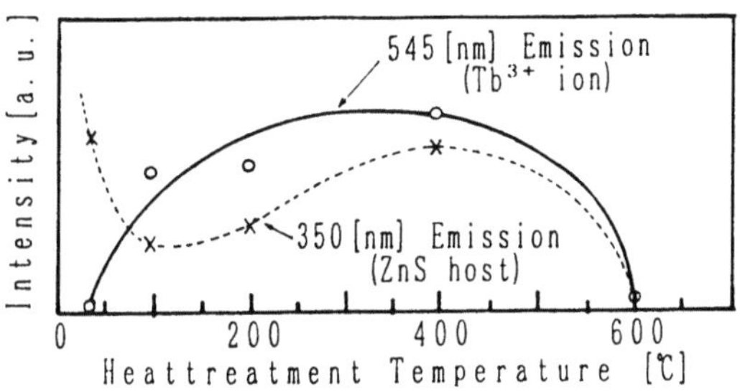

Fig. 5 Dependence on heat treatment temperature of 350nm emission of ZnS host and 545nm emission of Tb^{3+} ion in ZnS:TbF_x EL devices

4. Summary

1) The 545nm emission in ZnS:TbF_x EL devices is separated into the emission from the ZnS host and Tb^{3+} ion.
2) The emission of Tb^{3+} ion rises with decreasing the ultraviolet emission.
3) The dependence of the 350nm emission of the ZnS host and the 545nm emission of Tb^{3+} ion on heat treatment temperature show a similar tendency.
4) From these results, it is considered that energy transfer from the ZnS host participates in the excitation of the Tb^{3+} ion. The 350nm emission of the as-deposited device is observed, but the emission of Tb^{3+} scarcely observed. It seems that the efficiency of energy transfer from ZnS host to Tb^{3+} ion is low.

5. References

1) W.W.Anderson : Phys.Rev. A136, 556 (1964)
2) V.Marrello, L.Samuelson and A.Onton : J.Appl.Phys. 52, 3590 (1981)
3) K.Okamoto and S.Miura : Appl.Phys.Lett. 49, 1596 (1986)
4) D.H.Smith : J.Luminescence 23, 209 (1981)
5) W.E.Howard : O.Sahni and P.M.Alt : J.Appl.Phys. 53, 639 (1982)
6) P.Thioulouse : J.Cryst.Growth 72, 545 (1985)
7) D.C.Krupka and D.M.Mahoney : J.Appl.Phys. 43, 2314 (1972)

Excitation Mechanism Based on Field-Induced Delocalization of Luminescent Centers in $CaS:Eu^{2+}$ and $SrS:Ce^{3+}$ Thin-Film Electroluminescent Devices

H. Yoshiyama[1], *S.H. Sohn*[2], *S. Tanaka*[1], *and H. Kobayashi*[1]

[1]Department of Electronics, Tottori University, Koyama, Tottori 680, Japan
[2]Department of Physics, Kyung-Pook National University, Sankuk, Taegu 635, Korea

1. Introduction

Thin-film electroluminescent (TFEL) devices with rare-earth doped alkaline-earth sulfides such as $CaS:Eu^{2+}$ and $SrS:Ce^{3+}$ have recently received much attention, because they are promising candidates for full-color TFEL display panels [1-3]. Very recently, several experimental investigation on EL excitation mechanism of these TFEL devices have been reported [4-7]. From experimental data, it is argued that the EL excitation mechanism of Eu or Ce doped CaS and SrS TFEL devices is different from that of Mn or rare-earth doped ZnS TFEL devices. On the theoretical side, very few works on EL mechanism, especially in Eu or Ce doped CaS and SrS TFEL devices, have been reported.

In this paper, we discuss EL excitation mechanism of Eu^{2+} or Ce^{3+} doped alkaline-earth sulfides, in which the f-d transition of Eu^{2+} or Ce^{3+} centers is responsible for luminescence. Effects of a strong electric field on localized luminescent centers are theoretically investigated by introducing a model Hamiltonian relevant to Eu^{2+} or Ce^{3+} doped CaS and SrS.

2. Summary of the EL characteristics of $CaS:Eu^{2+}$ and $SrS:Ce^{3+}$ TFEL devices

The EL characteristics of $CaS:Eu^{2+}$ and $SrS:Ce^{3+}$ TFEL devices are summarized as follows: (1)The ratio τ_{EL}/τ_{PL} has a large value of 100, where τ_{EL} and τ_{PL} denote the EL decay time and the photoluminescent (PL) decay time, respectively. (2)Two emission peaks appear, one (L_{on}) at the leading edge and the other (L_{off}) at the trailing edge of the applied voltage pulse. The intensity of L_{off} is as large as that of L_{on}, even though only a small amount of conduction current density exists in the former case. (3)EL emission from Eu^{2+} and Ce^{3+} luminescent centers is strong near the anodic interfaces, while the dominant emission from ZnS:Mn occurs in the cathodic region.

3. Band model for $CaS:Eu^{2+}$ and $SrS:Ce^{3+}$

First, a model for $CaS:Eu^{2+}$ and $SrS:Ce^{3+}$ electroluminescent active layers is briefly described. Electronic energy band structure of CaS and SrS is schematically shown in Fig. 1(a): CaS (SrS) has an indirect band gap and the conduction band minimum at the X point comes from the 3d (4d) wave function of Ca (Sr) atoms [8]. The energy E_B denotes the band width of the lower conduction band with a d-electron character, and the value for E_B is approximately 2 eV. Also shown in the figure is a schematic diagram of the localized electronic states of Eu^{2+} (Ce^{3+}) in CaS (SrS). These energy levels relative to the band energy of CaS (SrS) were determined from the absorption measurements [9]. Luminescence from Ce^{3+} (Eu^{2+}) arises from the allowed transition due to the dipole-dipole interaction: $5d[T_{2g}] \rightarrow 4f$ ($5d[T_{2g}]4f^6 \rightarrow 4f^7$). The 5d electronic state is split into

Springer Proceedings in Physics, Vol. 38 **Electroluminescence**
Editors: S. Shionoya and H. Kobayashi

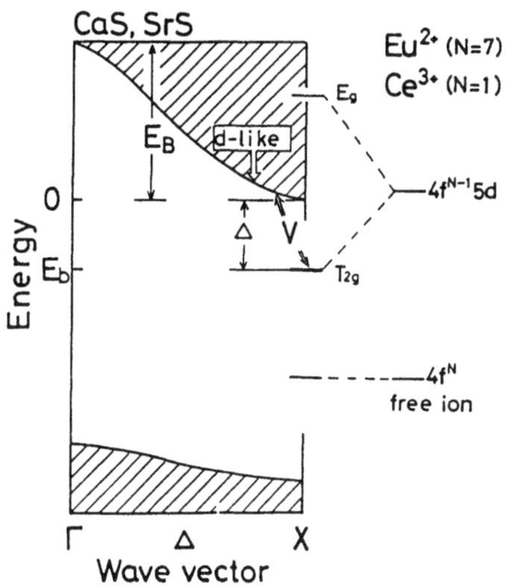

Fig. 1(a) Electronic energy structure of CaS:Eu^{2+} and SrS:Ce^{3+}. The band structure of CaS (SrS) is shown. The energy levels of Eu^{2+} (Ce^{3+}) relative to the band energy are also shown. The energy E_B denotes the band width of the lower conduction band with a d-electron character. The energy depth of the 5d(T_{2g}) level is denoted by Δ. The energy strength of the mixing interaction is denoted by V.

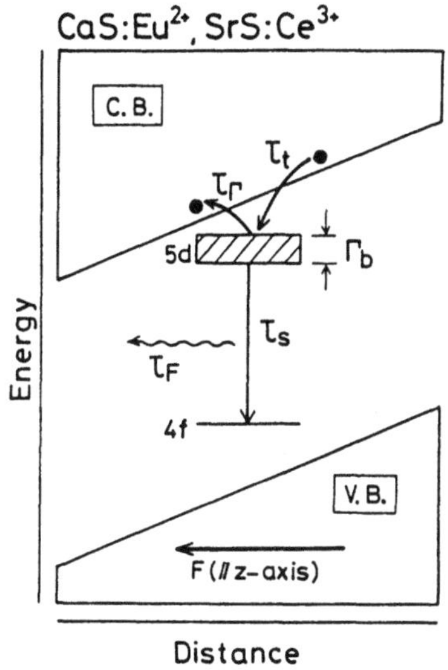

Fig. 1(b) Band model of the active layer for CaS:Eu^{2+} and SrS:Ce^{3+}, when a strong electric field is applied parallel to the z-axis. The luminescent process is also shown. The time constants of τ_t, τ_s, τ_Γ and τ_F correspond to the electron trapping, the transition from the 5d(T_{2g}) state to the 4f ground state, the field-induced delocalization and the luminescent decay time, respectively.

two states, E_g and T_{2g} states, with the crystal field split-off energy being approximately 1 eV. The higher energy state E_g lies in the conduction band, while the lower state T_{2g} and the 4f ground state lie in the forbidden band gap of CaS (SrS). As shown in the figure the energy of T_{2g} state is denoted by E_b, measured from the conduction band minimum at the X point, which is taken to be the ground level. Then the energy depth of T_{2g} measured from the ground level, Δ, is equal to $-E_b$, and the value for Δ is considered to be between 0.1 eV to 1 eV.

In this system it is expected that there exists a strong mixing interaction. The mixing interaction V comes from overlap between the d-like conduction electron wave function of Ca^{2+} (Sr^{2+}) and the 5d-electron wave function of Eu^{2+} (Ce^{3+}). This mixing interaction causes charge transfer from the conduction band to the 5d-excited T_{2g} state of the luminescent centers. The charge transfer is a kind of energy transfer from the host lattice to the luminescent centers. Contrary to this, only when a high electric field is applied, the 5d-excited T_{2g} electrons of the luminescent centers can be released into the conduction band through the mixing interaction. Here, the transition from the localized T_{2g} state to the conduction band state is called the delocalization.

Figure 1(b) shows a schematic band diagram of the model system when a strong electric field F is applied parallel to the z-axis. The strength of the electric field in energy units is given by aeF, where a is the lattice constant of CaS or SrS ($a \simeq 5\times10^{-8}$ cm for both CaS and SrS) and e is the electron charge ($e=1.6\times10^{-19}$C).

In the figure, luminescent process expected experimentally is shown: (1) Ionized luminescent centers, caused by the electric field excitation, capture conduction electrons with the time constant τ_t. (2) The captured electrons give rise to luminescence when they decay from the localized excited state T_{2g} to the ground state with the time constant τ_s. Under the strong mixing interaction, the high electric field causes delocalization of the excited state with the time constant τ_Γ. Due to the uncertainty principle, the excited energy level broadens with the energy band width Γ_b which is equal to $\hbar/2\tau_\Gamma$. Here, the energy band width Γ_b means the delocalization probability in energy units. It should be noted that the field-induced delocalization has a strong effect on the luminescent process; especially luminescent decay time τ_F is strongly influenced, when τ_Γ becomes shorter than τ_s. In the following, this strong effect is called the mixing effect.

4. Theory of field-induced delocalization of luminescent centers in CaS:Eu^{2+} and SrS:Ce^{3+}

4.1 Model Hamiltonian

From the discussion in the previous section, the model Hamiltonian of the system with electric field F parallel to the z-axis can be written as follows:

$$H = H_C(F) + H_{LC} + H_{MIX}(F) \tag{1}$$

$$H_C(F) = \sum_n \sum_{\mathbf{k}\perp} E_n(\mathbf{k}\perp;F)a_n^+(\mathbf{k}\perp)a_n(\mathbf{k}\perp) \tag{2}$$

$$H_{LC} = E_b b^+ b \tag{3}$$

$$H_{MIX}(F) = V \sum_n \sum_{\mathbf{k}} \{\phi_n^*(\mathbf{k};F)a_n^+(\mathbf{k}\perp)b + \phi_n(\mathbf{k};F)b^+a_n(\mathbf{k}\perp)\}, \tag{4}$$

with $\mathbf{k}=(k_x, k_y, k_z)$ and $\mathbf{k}\perp=(k_x, k_y)$, where $H_C(F)$ describes the d-like conduction electrons near the conduction band minimum of SrS (CaS), and H_{LC} describes localized electrons in the T_{2g} state of Ce^{3+} (Eu^{2+}). On the other hand, $H_{MIX}(F)$ represents the mixing interaction between the d-like conduction electrons and the localized electrons in the T_{2g} state.

In eq.(2), creation operator $a_n^+(\mathbf{k}\perp)$ and annihilation operator $a_n(\mathbf{k}\perp)$ are for the electrons in the nth Stark level with the wave vector $\mathbf{k}\perp$, where $\mathbf{k}\perp$ indicates the perpendicular components of the wave vector $\mathbf{k}$ with respect to the direction of the electric field [10]. The energy of this state is denoted by $E_n(\mathbf{k}\perp;F)$, which is given by [11]

$$E_n(\mathbf{k}\perp;F) = \frac{2\pi eF}{\kappa} n + \frac{1}{\kappa}\int_{-\kappa/2}^{+\kappa/2} \varepsilon(\mathbf{k})dk_z = aeF\cdot n + \frac{a}{2\pi}\int_{-\pi/a}^{+\pi/a} \varepsilon(\mathbf{k})dk_z , \tag{5}$$

where F is the electric field, n is an integer describing the Stark level, and κ is the length of the reciprocal lattice vector $\boldsymbol{\kappa}$, given by $2\pi/a$. $\varepsilon(\mathbf{k})$ represents the non-interacting conduction band energy in the absence of the electric field:

$$\varepsilon(\mathbf{k}) = \frac{\hbar^2}{2m^*}(k_x^2 + k_y^2 + k_z^2) , \tag{6}$$

In eq.(3), E_b is the electron energy of the localized T_{2g} state of Ce^{3+} (Eu^{2+}), and b^+ and b are creation and annihilation operators of the localized electrons in the T_{2g} state, respectively.

In eq.(4), V is the mixing interaction energy in the absence of the electric field. When the electric field is applied along the z direction, the mixing parameter should be replaced by an effective mixing parameter $V\cdot\phi_n(\mathbf{k};F)$, where $\phi_n(\mathbf{k};F)$ is given by the following equation [14]:

$$\phi_n(\mathbf{k};F) = \frac{1}{\sqrt{K}} \exp\left[-\frac{i}{F}\int_0^{k_z}[E_n(\mathbf{k}_\perp;F) - \varepsilon(k_x,k_y,k_z')]dk_z'\right] . \qquad (7)$$

4.2 Formulation of the delocalization probability Γ_b and the time constant of delocalization τ_Γ

In order to calculate the delocalization probability Γ_b in energy units and the time constant of the delocalization τ_Γ, thermal Green's function method is employed [12]. Using the Hamiltonian H, the Green's function for the localized electrons in the T_{2g} state is calculated. After lengthy calculation, we obtain the self-energy of the Green's function as follows:

$$\Sigma_{LC}(i\omega_\ell) = V^2 \sum_n \sum_{\mathbf{k}_\perp} \frac{\left|\sum_{k_z} \phi_n(\mathbf{k};F)\right|^2}{i\omega_\ell - E_n(\mathbf{k}_\perp;F)} , \quad \omega_\ell=(2\ell+1)kT \ (\ell=\text{integer}) \qquad (8)$$

The delocalization probability Γ_b is given by the imaginary part of the self-energy as follows:

$$\Gamma_b = -\operatorname{Im}\left[\Sigma_{LC}(i\omega_\ell \to E_b + i\delta)\right], \qquad (\delta = 0^+) \qquad (9)$$

$$= \pi V^2 \sum_n \sum_{\mathbf{k}_\perp} \left|\sum_{k_z} \phi_n(\mathbf{k};F)\right|^2 \delta(E_b - E_n(\mathbf{k}_\perp;F)). \qquad (10)$$

Substituting eqs.(5), (6) and (7) into eq.(10), we finally obtain the explicit formula of the delocalization probability Γ_b in the following form:

$$\Gamma_b = \frac{\pi E_B}{32}\left(\frac{V}{E_B}\right)^2 \frac{aeF}{\Delta} \exp\left[-\frac{4\pi}{3}\cdot\frac{E_B}{aeF}\left(\frac{\Delta}{E_B}\right)^{3/2}\right], \qquad (11)$$

where $\Delta= -E_b$ and $E_B = \hbar^2\pi^2/(2m^*a^2)$.

From the uncertainty principle, the time constant of the delocalization τ_Γ is given by

$$\tau_\Gamma = \hbar/(2\Gamma_b) . \qquad (12)$$

5. Results and discussion

5.1 Calculated results of the time constant of delocalization τ_Γ

First, we calculate the time constant of the delocalization, τ_Γ, as a function of the electric field aeF by using eqs.(11) and (12) for several sets of fixed values of E_B, V and Δ. Figure 2 shows the electric-field dependence of τ_Γ for the fixed value of E_B=2 eV. As seen from the figure, τ_Γ shows remarkable aeF-dependence, when V is large and Δ is small. In the case of the T_{2g} state with Δ=0.6 eV and V=1-2 eV, τ_Γ becomes shorter than 10 ns, when the electric field F (aeF) becomes stronger than 10^6 V/cm (0.05 eV). Therefore, assuming that τ_s is on the order of 10 ns, the luminescent decay time τ_F can be strongly influenced by the electric field of $F\simeq10^6$ V/cm (aeF$\simeq$0.05 eV) owing to the field-induced delocalization. The CaS:Eu^{2+} and the SrS:Ce^{3+} TFEL devices seem to be typical examples in this case.

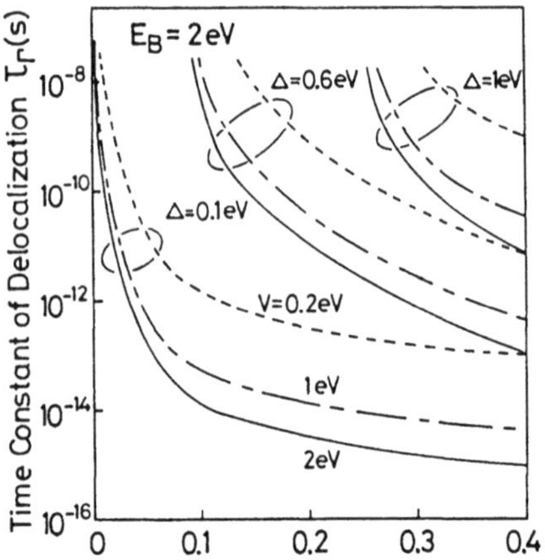

Fig. 2 The aeF-dependence of the time constant of the delocalization τ_Γ in the case where E_B=2 eV. The parameters Δ and V denote the energy depth of the $5d(T_{2g})$ state and the mixing interaction, respectively.

5.2 Effects of the delocalization on the EL decay time

In order to calculate temporal variation of EL intensity L(t), we assume the EL process as shown in Fig.1(b). Then, the EL intensity L(t) is determined by the following equations:

$$L(t) = -\frac{d}{dt}(n_C + n_{LC}), \tag{13}$$

$$\frac{d}{dt}n_C = -\frac{n_C}{\tau_t} + \frac{n_{LC}}{\tau_\Gamma}, \qquad \frac{d}{dt}n_{LC} = \frac{n_C}{\tau_t} - \frac{n_{LC}}{\tau_\Gamma} - \frac{n_{LC}}{\tau_s}, \tag{14}$$

where n_C is the number of the conduction electrons and n_{LC} is the number of the localized electrons in the T_{2g} state. Using the following initial condition;

$$n_C(0) = N_0, \text{ and } n_{LC}(0) = 0, \tag{15}$$

where N_0 is the total number of electrons in our considering system, we calculate the temporal variation of the EL intensity L(t) for several sets of aeF, E_B, V, Δ, τ_t and τ_s. From the calculated results, we can evaluate the luminescent decay time by using the following definition: $\tau_F=t_2-t_1$, where t_1 is the time corresponding to the maximum value of L(t), and t_2 is the time corresponding to the value of 1/e with respect to the maximum value of L(t) in the decay.

Figure 3 shows the field dependences of the relative luminescent decay time $\tau_F/\tau_{F=0}$ (= $\tau(F)/\tau(0)$) for the fixed values of E_B=2 eV, τ_t/τ_s=0.1 and τ_s=10 ns. The luminescent decay time τ_F increases with increasing electric field. The field dependence of τ_F for the shallow T_{2g} level of Δ=0.1 eV is stronger than that for the deep T_{2g} level of Δ=1 eV, or that for the intermediate energy depth of Δ= 0.6 eV. When the value of Δ is fixed, the field dependence of τ_F becomes remarkable with increase of the mixing interaction V.

We compare theoretical results of the ratio $\tau(F)/\tau(0)$ with experimental ones of the ratio τ_{EL}/τ_{PL} in the CaS:Eu^{2+} and SrS:Ce^{3+} TFEL devices. In the CaS:Eu^{2+} and the SrS:Ce^{3+} TFEL devices, a high field of the order of 10^6 V/cm is applied to the active layer in order to excite the luminescent centers or the host lattice. Therefore, luminescent decay time $\tau(F)$ at F=(1-2)x10^6 V/cm corresponds to the EL decay time τ_{EL}. On the other hand, luminescent decay time $\tau(F)$ at F=0 corresponds to the PL decay time τ_{PL}. As seen from Fig. 3, in the case of the shallow T_{2g} level of Δ=0.1 eV, the ratio $\tau(F=1\text{-}2\times10^6 \text{ V/cm})/\tau(F=0)$ becomes much larger than 100. In this case, luminescence does not arise, because the increase of the

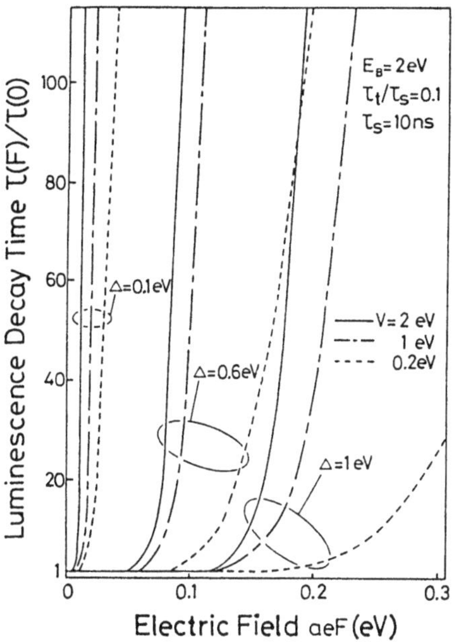

Fig. 3 The aeF-dependence of the relative luminescence decay time $\tau(F)/\tau(0)$ in the case where E_B=2 eV, τ_t/τ_s=0.1 and τ_s=10 ns. The parameters Δ and V denote the energy depth of the $5d(T_{2g})$ state and the mixing interaction, respectively.

luminescent decay time causes the decrease of the luminescent intensity L(t). When the T_{2g} level is Δ=0.6 eV, the ratio $\tau(F=1\text{-}2\times10^6\ \text{V/cm})/\tau(F=0)$ can become the order of 100, if the mixing interaction V has the order of 1 eV. In contrast to these, $\tau(F=1\text{-}2\times10^6\ \text{V/cm})$ nearly equals $\tau(F=0)$ in the case of the deep level of Δ=1 eV. In the CaS:Eu^{2+} and the SrS:Ce^{3+} TFEL devices, the experimental ratio τ_{EL}/τ_{PL} is on the order of 100. Therefore, the experimental result agrees well with the theoretical one for E_B=2 eV, Δ=0.6 eV and V=1-2 eV, and also these values of the parameters, E_B, Δ and V, are reasonable for the CaS:Eu^{2+} and the SrS:Ce^{3+} TFEL devices.

5.3 Energy-band diagram of EL excitation process

Figures 4(a) and (b) show the energy-band diagrams corresponding to the leading edge peak (L_{on}) and the trailing edge peak (L_{off}), respectively, of the CaS:Eu^{2+} or SrS:Ce^{3+} TFEL devices.

As shown in Fig. 4(a), in the period of L_{on}, a large band-bending occurs. In this case, the excitation process near the cathodic interface and that near the anodic interface should be discussed separately. Near the cathodic interface, the electric field is sufficiently high, so that impact ionization of the host lattice and the luminescent centers occurs efficiently. However, electron trapping is inefficient because of the field-induced delocalization. Contrary to this, near the anodic interface, the electric field is lower than 10^6 V/cm, so that impact ionization of the host lattice and the luminescent centers are inefficient, while conduction electrons are captured efficiently by the ionized luminescent centers which are generated at the previous stage of the pulse excitation. This is becaue the field-induced delocalization is much suppressed. The captured electrons, finally, give rise to luminescence. As a result, the dominant emission of Ce^{3+} and Eu^{2+} occurs near the anodic interfaces.

As shown in Fig. 4(b), in the period of L_{off}, the opposite internal electric field due to accumulated charges in interface states ejects trapped electrons from shallow levels into the conduction band. These conduction electrons are captured very efficiently by the ionized luminescent centers, which are generated in the

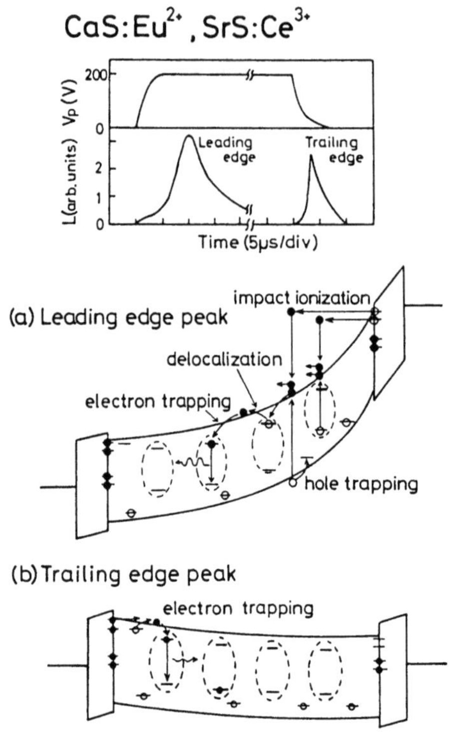

Fig. 4 Energy-band diagram of EL excitation process corresponding to (a) the leading edge peak (L_{on}) and (b) the trailing edge peak (L_{off}). Typical EL wave form observed for the CaS:Eu^{2+} and the SrS:Ce^{3+} TFEL devices is also shown together with a excitation voltage pulse (V_p).

previous period. This is because the internal electric field is low in the whole region of the active layer and then the field-induced delocalization is much suppressed. From this consideration, we can expect that the EL efficiency of L_{off} is higher than that of L_{on}.

6. Summary

We have theoretically investigated effects of the electric field on the localized luminescent centers, Eu^{2+} and Ce^{3+}, in the CaS:Eu^{2+} and the SrS:Ce^{3+} TFEL devices. We have shown that the mixing interaction between the conduction electrons in the host materials, CaS and SrS, and the electrons in the first excited state of the luminescent centers, Eu^{2+} and Ce^{3+}, plays an essential role in the EL excitation process.

References

1. W.A. Barrow, R.E. Coovert, C.N. King: 1984 SID Int. Symp., Digest of Technical Papers, p.249
2. S. Tanaka, V. Shanker, M. Shiiki, H. Deguchi, H. Kobayashi: 1985 SID Int. Symp., Digest of Technical Papers, p.218
3. S. Tanaka, Y. Mikami, J. Nishiura, S. Ohshio, H. Yoshiyama, H. Kobayashi: Proceedings of the SID, 28, 357 (1987)
4. S. Tanaka, H. Yoshiyama, Y. Mikami, J. Nishiura, S. Ohshio, H. Kobayashi: Proceedings of the 6th Int. Display Research Conf. (Japan Display '86)
5. R.S. Crandall: Appl. Phys. Lett. 50, 551 (1987)
6. R.S. Crandall: Appl. Phys. Lett. 50, 641 (1987)
7. S. Tanaka: J. Luminescence 40&41, 20 (1988)
8. A. Hasegawa, A. Yanase: J. Phys. C 13, 1995 (1980)

9. S. Asano, N. Yamashita, Y. Ogawa: phys. stat. sol. (b)118, 89 (1983)
10. G.H. Wannier: Phys. Rev. 117, 432 (1960)
11. E.N. Korol: Sov. Phys. Solid State 19, 1327 (1977)
12. For detailed description of the thermal Green's function method, see for instance A.L. Fetter, J.D. Walecka: In Quantum Theory of Many-Particle Systems (McGraw-Hill 1971)

Excitation Mechanism in White-Light Emitting SrS:Pr, K and SrS:Ce, K, Eu Thin-Film Electroluminescent Devices

S. Tanaka, H. Yoshiyama, J. Nishiura, S. Ohshio, H. Kawakami, K. Nakamura, and H. Kobayashi

Department of Electronics, Tottori University, Koyama Tottori 680, Japan

1. Introduction

Recently, thin-film electroluminescent (EL) devices with rare-earth activated CaS and SrS phosphors have been investigated extensively to realize multi- and full-color thin film EL displays. Very recently, we have developed two types of bright white-light emitting thin film EL devices with new phosphors, SrS:Pr,K and SrS:Ce,K,Eu [1-3]. White-light EL devices have the following important characteristics; White-light EL displays with a reverse mode are superior from an ergonomics point of view, because characters and/or figures are drawn in black on a white screen in this mode, just as in printed matter. In addition, full-color displays can be realized by combining a white thin-film EL with color filters.

In this paper, we report on the excitation mechanism in the novel white-light emitting SrS:Pr,K and SrS:Ce,K,Eu TFEL devices.

2. Experimental

The EL devices used in this experiment have a conventional double insulating structure of a glass substrate (HOYA NA-40), an indium tin oxide (ITO) transparent electrode, a first dielectric layer, a 1.0 μm thick SrS:Pr,K or SrS:Ce,K,Eu phosphor layer, a second dielectric layer, and an Al rear electrode. The SrS phosphor thin films were prepared at a substrate temperature of 400-500 °C by the electron beam evaporation method. Simultaneous sulfur coevaporation was employed. A heat treatment of the phosphor films after the deposition was not made.

The EL characteristics of these devices are summarized in Table 1, together with those of other white-light emitting thin film EL devices [4,5].

Table 1. EL characteristics of the white-light emitting thin film EL devices

Phosphors	Color coordinate x	y	Luminance (cd/m^2) 5k	1k	60Hz	η (lm/W)	Color control	Color variance due to V	
ZnS:Pr,K	0.33	0.48	300	70	5	0.02	impossible	none	[4]
SrS:Pr,K	0.38	0.40	1500	500	30	0.1	impossible	none	[2]
SrS:Ce,K,Eu	0.28-0.40	0.42-0.40	1500	500	30	0.1	possible	none	[1]
SrS:Ce,K/SrS:Eu	0.38-0.46	0.38-0.39	520	200	12	0.03	possible	arise	[3]
SrS:Ce,Cl/ZnS:Mn	0.30-0.42	0.38-0.42	1200	350	12	0.06	possible	arise	
SrS:Ce/CaS:Eu	0.25-0.35	0.40-0.35	1200	320	16		possible	arise	[5]

Springer Proceedings in Physics, Vol. 38 **Electroluminescence**
Editors: S. Shionoya and H. Kobayashi

3. Experimental Results and Discussion

3.1 SrS:Pr,K

As summarized in Table 1, the SrS:Pr,K device shows the saturated luminance of 500 cd/m^2 (1500 cd/m^2 at 5 kHz) and the EL efficiency of 0.1 lm/W at 1 KHz. It should be noted that the luminance level and the EL efficiency of the SrS:Pr,K device are five times higher compared with the ZnS:Pr,F device. To investigate the EL excitation mechanism of Pr^{3+} centers in the SrS lattice responsible for a higher EL efficiency, we have carried out photoluminescence (PL) and PL-excitation spectra measurements. The PL spectra and the PL-excitation spectra of the SrS:Pr,K thin films with various Pr concentrations are shown in Fig.1. The excitation spectrum consists of two main bands: one with a peak around 280 nm is due to SrS host excitation and the other with a peak around 320 nm is due to $(4f)^2$-(4f)(5d) transition of Pr^{3+} ions. Both PL and EL are due to the relaxation of the $(4f)^2$ electron configuration of Pr^{3+} ions. We also found that the PL intensities of the SrS:Pr,K films are a factor of 100 higher than those of the ZnS:Pr,F films.

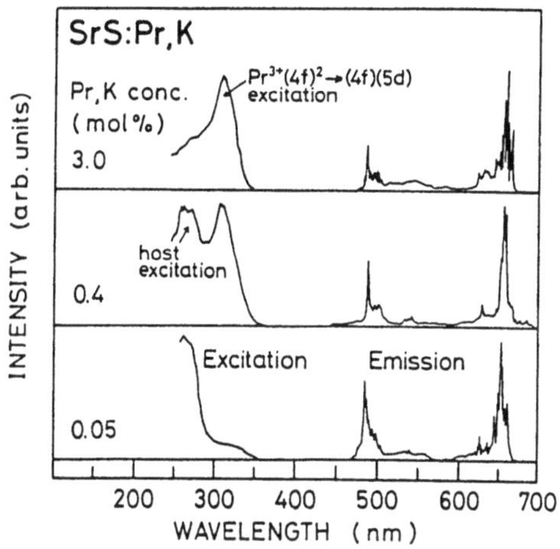

Fig. 1 PL spectra and PL-excitation spectra of the SrS:Pr,K thin-films.

It is known that SrS has an indirect band structure and the conduction-band minimum at X point comes from the wave function of unoccupied 4d levels of Sr atoms. Therefore this state has d-like characteristics. When the Pr^{3+} luminescent centers are doped in the SrS lattice, the $(4f)^2$ ground-state level of the Pr^{3+} centers lies in the forbidden band and the (4f)(5d) excited state lies fairly close to the conduction-band minimum at the X point of SrS. These features of the band structure of the SrS host and the energy states of the Pr^{3+} centers result in a highly efficient PL excitation mechanism, in which host excitation followed by an energy transfer to Pr^{3+} centers through the (4f)(5d) excited state takes place.

Taking these facts into account, it can be concluded that the dominant EL excitation mechanism responsible for the higher EL efficiency in the SrS:Pr,K devices results from the following processes. Ionization of the Pr^{3+} luminescent centers due to impact ionization and/or hole trapping occurs first. Subsequent recombination with electrons resulting in the Pr^{3+} centers in a $(4f)^2$ excited state through the (4f)(5d) excited state follows. Finally, the Pr^{3+} centers give rise to luminescence due to the transition in the $(4f)^2$ electron configuration of Pr^{3+} ions.

3.2 SrS:Ce,K,Eu

The EL spectrum of the SrS:Ce,K,Eu device is shown in Fig.2 (a). The PL spectra of SrS:Ce,K and SrS:Eu phosphors and the absorption spectrum of SrS:Eu phosphors

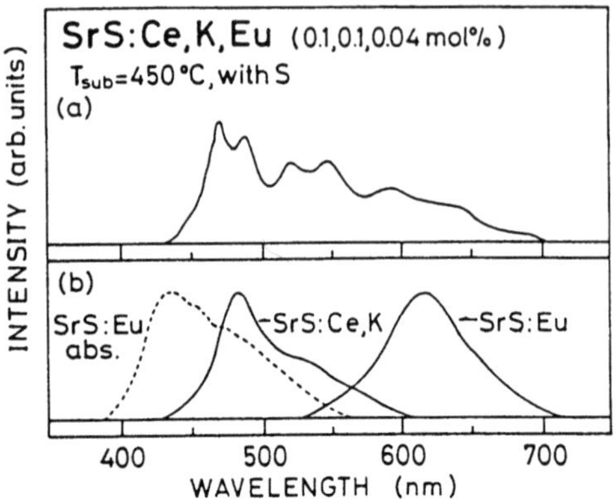

Fig. 2 (a) EL spectrum of the SrS:Ce,K,Eu EL devices.
(b) PL spectra of SrS:Ce,K and SrS:Eu phosphors and an absorption spectrum of SrS:Eu phosphors.

are shown in Fig.2 (b). It is expected that, in the SrS:Ce,K,Eu phosphors, a nonradiative energy transfer from Ce^{3+} to Eu^{2+} centers takes place efficiently, because the overlap between the PL spectrum of the SrS:Ce,K and the absorption spectrum of the SrS:Eu is remarkably large [4]. According to Dexter's theory, the probability of the nonradiative energy transfer from an energy donor (Ce^{3+} centers) to an acceptor (Eu^{2+} centers) due to a dipole-dipole interaction P^{dd}_{Ce-Eu} is given by [9]

$$P^{dd}_{Ce-Eu} = \frac{4\pi e^4 S^{em}_{Ce} S^{ab}_{Eu}}{3\hbar R^6 \varepsilon_\infty^2} \times \int f_{Ce}(E)\, f_{Eu}(E)\, dE, \qquad (1)$$

where $f_{Ce}(E)$ is the normalized emission spectrum of Ce^{3+}, $f_{Eu}(E)$ is the normalized absorption spectrum of Eu^{2+}, S^{em}_{Ce} is the (5d) - (4f) line strength of the Ce^{3+} ions for emission, S^{ab}_{Eu} is the $(4f)^7 - (4f)^6(5d)$ line strength of the Eu ions for absorption, ε_∞ is the optical dielectric constant, and R is the distance between Ce^{3+} and Eu^{2+} ions. The Eu concentration dependence of the relative intensity of the Ce emission can be calculated by using the critical transfer concentration of the Eu ions, defined by $C_0 = 3/(4\pi R_0^3)$ in terms of critical transfer distance R_0. The value of R_0 can be calculated by putting $P^{dd}_{Ce-Eu} = 1/\tau_D$, where τ_D is the radiative lifetime of the energy donor (Ce^{3+} centers). Fairly good agreement between the experimental and the calculated results have been obtained, although no fitting parameter was used [1]. Taking these facts into account, we can conclude that the nonradiative energy transfer from Ce^{3+} to Eu^{2+} centers play an important role in the SrS:Ce,K,Eu thin film phosphors.

The problem to be discussed is why Ce centers are likely to be excited rather than Eu centers at the same electric field. Recently, it has been reported that the EL excitation mechanism of Ce- or Eu-doped SrS thin films is significantly different from that of Mn or rare-earth doped ZnS thin-films [8,9]. The processes play an important role in that the ionization of luminescent centers first occurs, subsequent recombination with electrons follows, and finally luminescence arises [9]. To confirm this excitation process, we have measured the spectra of the photo-induced transferred charge of the SrS:Ce,K and SrS:Eu thin film EL devices. The results are shown in Fig. 3 together with the excitation spectra of PL. It is clearly shown that the Ce^{3+} or Eu^{2+} luminescent centers are ionized when the centers are excited to the $(4f)^{n-1}(5d)$ excited state. This process probably gives rise to the difference in the threshold electric field for the Ce^{3+} or Eu^{2+} excitation in SrS thin films, because the electric field required to ionize the luminescent centers are strongly dependent on the energy depth of $(4f)^{n-1}(5d)$ (T_{2g}) levels measured from the conduction band minimum at the X point of SrS lattice, which comes from the 4d wave function of Sr atoms [10]. We have found that the observed threshold field of 0.7-0.8×10^6 V/cm for SrS:Ce,K thin-films is lower than

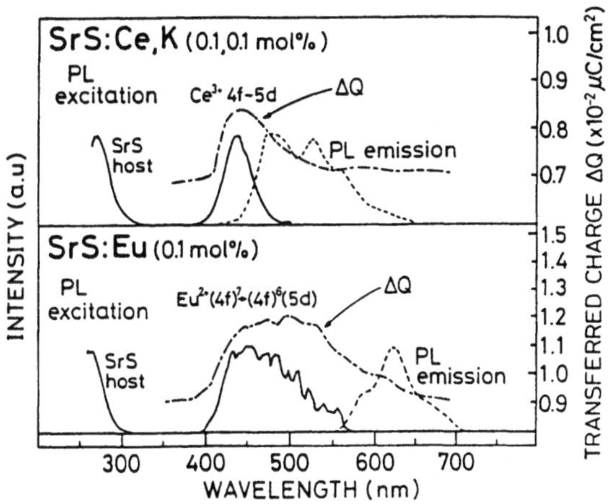

Fig. 3 Spectra of the photo-induced transferred charge of the SrS:Ce,K and SrS:Eu thin film EL devices. PL and PL-excitation spectra are also shown.

that of 1.0-1.1x10^6 V/cm for the SrS:Eu thin-films. This result probably comes from the shallow energy depth of the Ce^{3+} centers compared with the Eu^{2+} centers.

4. Summary

In the SrS:Pr,K and SrS:Ce,K,Eu phosphors, a strong mixing interaction between the conduction band state and the $(4f)^{n-1}(5d)$ excited state of luminescent centers (Pr^{3+}, Ce^{3+} and Eu^{2+}) plays an important role. The following EL excitation process is expected: Ionized luminescent centers capture conduction electrons efficiently, and consequently they give rise to luminescence. In the SrS:Pr,K films, the Pr^{3+} centers capture conduction electrons through the (4f)(5d) excited state. In the SrS:Ce,K,Eu films, the Ce^{3+} centers are excited rather than the Eu^{2+} centers, and then excitation of the Eu^{2+} centers arises from the nonradiative energy transfer from the Ce^{3+} to Eu^{2+} centers.

References

1. S. Tanaka, H. Yoshiyama, J. Nishiura, S. Ohshio, H. Kawakami, H. Kobayashi: Appl. Phys. Lett., 51, 1661 (1987)
2. S. Tanaka, S. Ohshio, J. Nishiura, H. Kawakami, H. Yoshiyama, H. Kobayashi: Appl. Phys. Lett., 52, 2102 (1988)
3. S. Tanaka, H. Yoshiyama, J. Nishiura, S. Ohshio, H. Kawakami, H. Kobayashi: 1988 SID Int. Symp., Digest of Technical Papers, p.293.
4. T. Suyama, N. Sawara, K. Okamoto, Y. Hamakawa: Jpn. J. Appl. Phys., Suppl. 21-1, 383 (1982)
5. Y.A. Ono, M. Fuyama, K. Onisawa, K. Taguchi, H. Kawakami: J. Luminescence, 40/41, 796 (1988)
6. D.L. Dexter: J. Chem. Phys., 21, 826 (1953)
7. K.B. Eisenthal, S. Siegel: J. Chem. Phys., 41, 652 (1964)
8. R.S. Crandall: Appl. Phys. Lett., 50, 551 (1987)
9. S. Tanaka: J. Luminescence, 40/41, 20 (1988)
10. H. Yoshiyama, S.H. Sohn, S. Tanaka, H. Kobayashi: to be submitted.

Novel Step Impact Electroluminescent Devices

H.J. Lozykowski

Ohio University, College of Engineering and Technology,
Department of Electrical and Computer Engineering, Stocker Center,
Athens, Ohio 45701-2979, USA

The optical properties of rare earth ions in solids have been of interest for many years. However, the luminescence properties of 4f ions in III-V semiconductors and silicon have remained unexplored, despite their obvious potential for light emitting and laser diodes. The characteristic photoluminescence and electroluminescence spectra of rare earth ions in III-V and silicon semiconductors were reported recently [1-4].

The excitation mechanisms of electroluminescence (EL) and photoluminescence (PL) from forward biased p-n junction diodes and when the photon excitation energy exceeds the band gap are not clearly understood.

The photoluminescence excitation experiment has shown that the intensity of neodymium emission from GaP : Nd^{3+} is about five times higher for above band-gap excitation than for resonant 4f excitation. This fact indicates that the energy transfer from the host semiconductor to the Nd^{3+} ions is efficient and strongly argued for resonant transfer mechanisms as a excitation mechanism of RE^{3+} ions in both photoluminescence and electroluminescence.

The alternative method of excitation of luminescent centers like RE^{3+}, Mn^{2+} and other transition metals in semiconducting hosts is a direct impact excitation mechanism which should have a higher efficiency. The excitation mechanism is achieved when electrons highly accelerated by the electric field impact luminescent centers. The direct impact excitation mechanism was experimentally proved for II-VI semiconductors doped with RE^{3+} and Mn^{2+} [5-9]. To improve the efficiency of the electroluminescence we propose new type devices in which the acceleration and collision excitation processes can be spatially separated [1]. The separation of the acceleration and collision excitation processes in different layers permit independent optimization of each function in different materials. The conventional thin film EL devices have low efficiencies. Present evidence suggests that only a small fraction of the conduction electrons have sufficient energy to impact excite the rare earth or Mn^{2+} ions. Electroluminescent devices based on spatial separation of the electron acceleration and collision excitation processes corresponds to cathodoluminescence where the acceleration occurs in vacuum. The power efficiency of ZnS : Tm^{3+} [10] under cathode ray excitation was determined as 0.216 W/W at 475 nm, the highest reported for any rare earth activated phosphor especially in view of the very narrow emission bandwidth (6 nm). The band centered at 775-800 nm has power efficiency of 0.59 W/W. This says that if the electrons can be accelerated to ballistic energies more efficient devices can be fabricated which use direct impact excitation.

The proposed novel devices [1] are a multi-layered heterojunction structures in which the acceleration and collision excitation process are spatially separated. Fig. 1 and Fig. 2 shows the energy band diagram with bias of the graded gap and graded doping single layer step impact electroluminescent device (SIED) and a step photon amplifier converter (SPAC) respectively. Each stage (assumed intrinsic) is linearly graded in composition from a low E_{g1} to a high E_{g2} band gap (region B) with an abrupt step back to low-band gap material (interface B-C). The graded doped (SIED) structure with the band gap E_{g2} are heavily doped in region A, and graded doped in region B from n type to intrinsic at the interface between regions B and C respectively. The conduction band

Springer Proceedings in Physics, Vol. 38 Electroluminescence
Editors: S. Shionoya and H. Kobayashi

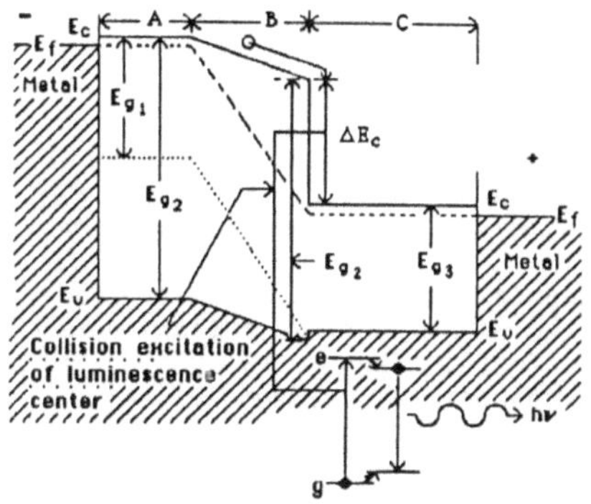

Fig. 1: Energy band diagram of a SIED devices under bias.

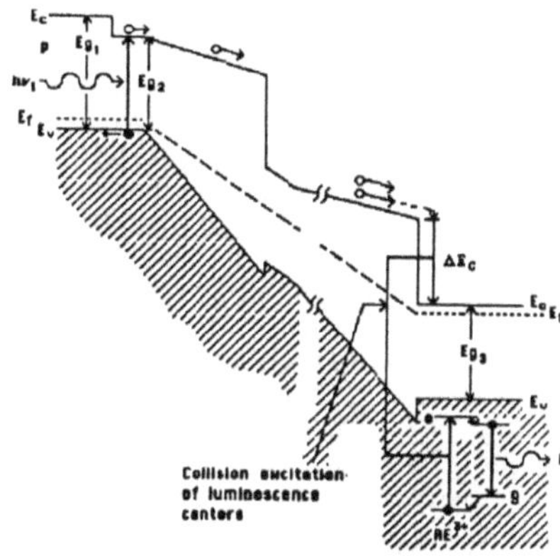

Fig. 2: Energy band diagram of a SPAC device under dias.

discontinuity ΔE_c for graded doped semiconductors at the interface B and C, is controlled by the difference in electron affinity of the semiconductors E_{g2} and E_{g3}, the doping levels and the interface states. For the SPAC Fig. 2, the abrupt steps are followed by ungraded layers with a thickness equal to a few ionization mean free paths. The conduction band steps ΔE_c are chosen to be greater than the electron ionization energy E_i of the smallest band gap. The last step has ΔE_c tailored to optimize the impact excitation process of the luminescence center.

For electron transport across the graded region (Fig. 1, 2) the external field must cancel the sum of the quasi-field term arising from the band-edge gradient and effective mass gradient term. This field must be high enough to assure electron transport by drift rather than diffusion in the graded regions [1].

Consider now an electron introduced near the left contact of either device. In SPAC the incident photons generate electron hole pairs next to the p^+ wide band gap semiconductor window contact. This photo electron is input to the SPAC. The electron is accelerated, however does not impact ionize in the graded region before the conduction band step, because the effective electric field is too low. After the step it abruptly gains energy equal to the conduction band edge discontinuity. If $\Delta E_c \geq E_i$ it creates electron hole pairs. This process is repeated at every stage of SPAC; note the ballistic nature of this energy gain. In order to achieve a high ionization yield per stage, electrons passing the step should have an energy well in excess of the phonon energy. This energy the electron gains from effective fields in the graded regions between steps.

At the last stage the electron enters the collision excitation region. In this region ballistic electrons excite the rare earth ions by direct impact. The energy of the ballistic electrons participating in the collision excitation must be at least equal to the threshold energy ϵ_T and not greater than ionization energy of deep centers and the band gap of collision excitation region. Because, the excitation efficiency of RE^{3+} centers would be reduced by deep center ionixation and electron-hole pairs production. In classic thin film electroluminescent devices the excitation process is more random, because electrons can excite RE^{3+} ions everywhere in the high field region, while in the staircase SIED and SPAC, electrons excite them in a well defined spatial region (i.e., the excitation process is more deterministic). Also exciting electrons are more nearly mono-energetic because immediately after passing the conduction band step, their energy is equal to ΔE_c plus the average kinetic energy before the step.

To maximize the excitation process the collision excitation region must be heavily doped with RE^{3+} ions (or other luminescence centers) and the luminescence centers should have large excitation cross sections. Because of the finite collision excitation length, $\lambda_{ex} = (Q_{ex}N)^{-1}$ (Q_{ex} is impact excitation cross section and N the concentration of centers), electrons with energy $E_e \geq \epsilon_\Gamma$, will impact after a distance of order λ_{ex}. Thus, the rare earth collision excitation region doped with RE^{3+} (or Mn^{2+}) must have a thickness of a few λ_{ex} to optimize the excitation process.

The literature to date has contained only limited information on the electron impact excitation cross sections Q_{ex} of the free atoms and ions of the rare earth elements and ions in semiconductors (see references in [1]). Experimental study of the collision excitation cross-sections have yet to be performed for III-V compounds.

The impact excitation probabilities of luminescence centers can be calculated if we know the electron energy distribution function and impact excitation cross section Q_{ex} [1]. The electron energy distribution function can be calculated if we know the electric field and the electron mean free path λ_e between collisions. The electric field causing impact excitation has two sources: the effective external field, discussed above, and the conduction band edge discontinuity. The first one is taken into account in the Baraff distribution function [1]. The band-edge discontinuity has to be treated in a different way because the abrupt energy changes occur over a distance which is much shorter than the mean free path λ_e between collisions. To take into consideration the effect of the conduction band-edge step on the excitation rate we deduct the discontinuity energy ΔE_c from the impact excitation threshold for a given luminescence center. By changing ΔE_c we can tune up the energy of excited electrons to the resonance condition of a particular excited state Γ of RE^{3+} ions.

The first approximated expression for cross section function near the threshold was given by Allen [11]. Recently Yu and Shen [12] derive the analytical expression and calculated the inelastic scattering cross sections based on the Born approximation considering the coulomb interaction between hot free electron and the 4f electrons of the rare earth ion. The impact cross section corresponding to the transition from i state to f state is given by:

$$Q = [18\pi^2 me^2h^2c^3/n(n^2 + 2)^2\epsilon_{fi}{}^3]\epsilon^{-1}\ln[(\sqrt{\epsilon} + (\epsilon - \epsilon_{fi})^{1/2}) \div (\sqrt{\epsilon} - (\epsilon - \epsilon_{fi})^{1/2})](g_f/g_i)A_{fi} \quad (1)$$

where: m - the mass of a electron, $g_{i,f}$ - degeneracy of initial and final states, ϵ_{fi} is the threshold energy, ϵ is the energy of the incident electrons, n is the refractive index, A_{fi} the electric dipole transition probability. Consider the excitation of the level from the ground state i. Radiation emitted during the downward transition f→i, is measured and its intensity is therefore determined by the population of the f-state, but this state may also be populated by downward transitions from a higher state j.

The rate equation then is:

$$dn_f/dt = nN \int f(E, \epsilon)\, \epsilon^{1/2} Q(\epsilon, \epsilon_\Gamma)\, d\epsilon - n_f A_{fi} + \sum A_{jf} n_j \quad (2)$$

where: n, N are the density of hot electrons and luminescence centers respectively, A_{fi} is the probability of the luminescence center making a radiative f→i transition in one second. If we assume that there are no other processes that populate the upper level f, (transition from the higher level j or energy transfer from other center) the last term $\sum A_{jf}n_j = 0$. At equilibrium, the luminescence intensity I is:

$$I = \eta_q r_e = \eta_q n_f A_{fi} = \eta_q\, nN \int f\,(E, \epsilon)\epsilon^{1/2} Q(\epsilon, \epsilon_\Gamma) d\epsilon \quad (3)$$

Figure 3 shows the f(ε, E) distribution function two excitation functions for threshold energy ϵ_Γ and reduced excitation threshold $\epsilon_\Gamma' = \epsilon_\Gamma - \Delta E_c$, and the integrands. The distribution function f(ε, E) (insert Fig. 3) and excitation function (2) were computed for parameters shown in Fig. 3 and normalized to unity at maximum. In our computation example we assume $\Delta E_c = 0.55$ eV which can be realized using $Al_xGa_{1-x}As$-GaAs graded band gap structure. The excitation threshold $\epsilon_\Gamma = 0.805$ eV corresponds to transition between spin orbit levels of

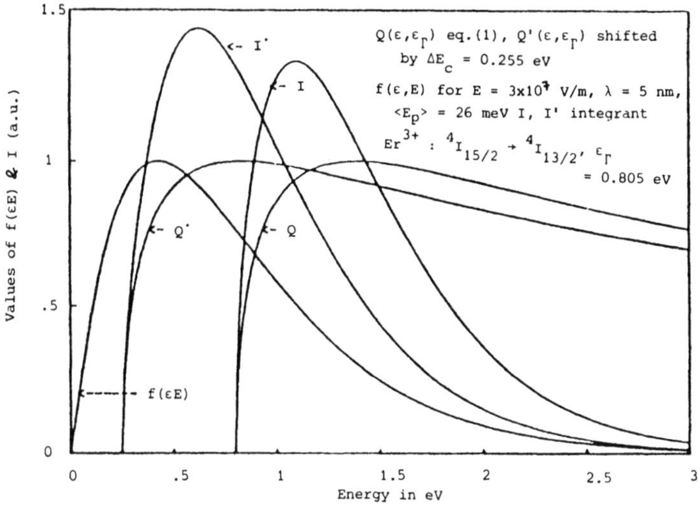

Fig. 3: The energy distribution function $f(\epsilon,E)$, two excitation functions Q and the integrands I (plotted for parameters given in insert).

Er^{3+}, $^4I_{15/2} \rightarrow {}^4I_{13/2}$. The graph shows that for the integral only the energy region above threshold is important.

The multistep photon amplifier converter (SPAC) shown in Figure 2 is a multistage graded gap heterostructure. The basic impact excitation mechanism in the SIED and SPAC devices are the same. In SPAC the incident photons generate electron hole pairs in narrow band gap semiconductor next to the p^+ wide band gap semiconductor window contact. This photo electron is input to the multistage graded gap heterostructure. If the total avalanche gain of the structure is high the emission from the device will exceed the input intensity. By proper choice of the ΔE_C at last stage, the band gap E_g of the collision excitation stage and the rare earth impurity, we can obtain emission at different wavelengths (radiation converter amplifier). The SIED structure is a very promising structure for studying the direct impact excitation mechanism for a localized impurity in semiconductors. It can be used as a general tool for studying the excitation function of RE^{3+} ions and other impurities with internal transitions like some transition metal impurities in III-V compounds.

Both proposed devices: SIED and SPAC, can be built using the possible combinations of direct or indirect band gap semiconductors containing III-V compounds and II-VI compounds (or other semiconductors) doped with rare earths, or transition metals. The above discussion has been concentrated on narrow band gap materials however, the possibility of using wide band gap semiconductors is not excluded.

Laser action should also be obtainable by properly designed SIED structure doped with RE or other impurities with internal transitions like some transition metal impurities [13]. The laser devices will attain lasing at the rare earth transitions with single longitudinal mode operations with reproducible precise lasing wavelength insensitive to temperature variations. Thus, the SIED structure with the unipolar injection is feasible to carry semiconductor lasers into the blue and ultraviolet region.

I. Acknowledgements

I wish to thank Vasant Shastri at Ohio University for help in computer programming and technical assistance.

II. References

1. H. J. Lozykowski, Proc. SPIE Vol. 836, 88 (1987)
   H. J. Lozykowski, Solid State Comm. 66, 755 (1988) and reference therein
2. J.P.vanderZiel, M.G.Oberg and R.A.Logan, Appl.Phys.Lett. 50, 1313 (1987)
3. K. Uwai, H. Nakagome, and K. Takahei, Appl. Phys. Lett. 50, 977 (1987)
4. K. Uwai, H. Nakagome, and K. Takahei, Appl. Phys. Lett. 51, 1010 (1987)
5. D. C. Krupka, J. Appl. Phys 43, 476 (1972)
6. I. Szczurek, H. J. Lozykowski, J. Lumin. 23. 315 (1981)
7. I. Szczurek and H. J. Lozykowski, J. Lumin. 14, 389 (1976)
8. J. W. Allen, J. Lumin. 23, 127 (1981)
9. R. Mach and G. O. Muller, Phys. Status Solidi (a) 81, 609 (1984)
10. R.E.Shrader, S.Larach, and P.N.Yocom, J. Appl. Phys. 42, 4529 (1971)
11. J.W.Allen and S.G.Ayling, J. Phys. C: Solid State Phys. 19 L 369 (1986)
12. J. Yu and Y. Shen J. Lumin. 40 and 41, 769 (1988)
13. P.B.Klein, J.E.Furneaux, and R.L.Henry, Appl. Phys. Lett. 42, 638 (1983)

Preparation of a Low Voltage ZnS Thin Film Electroluminescent Device Using Injection of Hot Electrons into the Emitting Layer

Y. Nakanishi, Guixi Zhou, T. Ando, and G. Shimaoka

Research Institute of Electronics, Shizuoka University,
3-5-1 Johoku, Hamamatsu 432, Japan

A ZnS:Mn thin film electroluminescent device has been prepared on a p-type Si substrate in order to lower the device driving voltage. Hot electrons that excite luminescent centers are injected into the ZnS emitting layer from Si as a result of the band bending in Si at the interface between SiO_2 and p-Si. When the device was prepared on a p-layer formed by ion implantation of a n-type Si wafer, the driving voltage for electroluminescence became lower and in addition the luminance is improved.

1. INTRODUCTION

Both the acceleration of electrons and the excitation of luminescent centers in an emitting layer are essential for obtaining light emission from a standard thin-film electroluminescent (TFEL) device in which the emitting layer is sandwiched between two dielectrics. Therefore, a high driving voltage is necessary for obtaining the light emission. Moreover, for EL of donor-acceptor (D-A) pair centers, it is very difficult because of a large gradient of the energy band in the emitting layer under a field of 10^6 V/cm. A method to overcome these problems is to inject hot electrons into the emitting layer [1].

In this study, a ZnS emitting layer was deposited on an SiO_2 insulating layer formed on a Si wafer. One of the advantages of this arrangement is that a thin SiO_2 insulating layer of good quality can be grown on the Si wafer by thermal oxidation. Also, when SiO_2 is formed on p type Si, the energy band is bent at the interface as shown in Fig.1 [2]. Therefore, electrons in the conduction band of Si can be injected into the ZnS emitting layer via the SiO_2 layer at a lower driving voltage. The injected electrons are sufficiently energetic to impact-excite a luminescent center. As a result, EL of D-A pair centers is expected in an EL panel based on this principle.

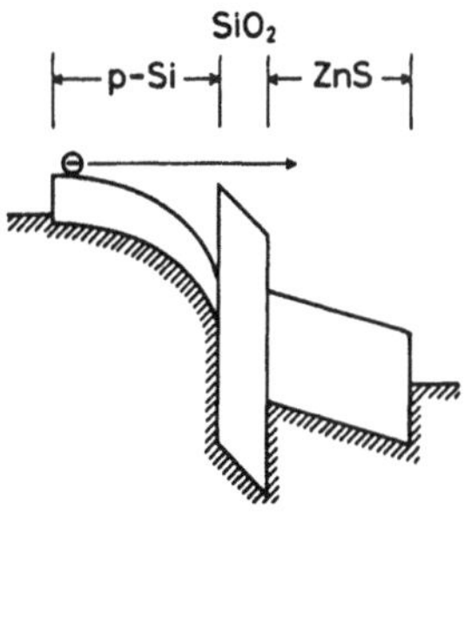

Fig.1 Schematic band diagram of $ZnS/SiO_2/p$-Si EL device.

2. EXPERIMENTAL

2.1 Formation of insulating SiO_2 layer

The structure of the TFEL device used in this work is Au/ZnS:Mn/SiO_2/p-Si. An insulating SiO_2 layer was formed by thermal oxidation on p-type Si(100) wafers with resistivities of 0.025 and 5 Ω-cm, and a p-layer formed by the ion implantation of boron into a n-type Si wafer (1 Ω-cm). The oxidation process was as follows :(1) the formation of a thick oxide (600 nm) on the Si wafer in wet oxygen, (2) the patterning of the thick oxide for the EL device by photolithographic technology and (3) the formation of a thin oxide as the insulating layer in the EL device in dry oxygen. The thickness of the thin oxide was 80 nm.

2.2 Preparation of EL devices

The ZnS emitting layer was deposited by electron beam evaporation. Vacuum pressure during the deposition was maintained in the order of 10^{-5} Pa using a turbo-molecular pump. The substrate temperature during the deposition was 220°C [3]. ZnS films were heat treated at 500°C for 1 hour in vacuum after the deposition. Mn was used as the luminescent center in the ZnS film. The emission was detected through a semi-transparent Au electrode.

2.3 Characterization and measurements

A depth profile measurement was carried out by Auger electron spectroscopy (AES) and Ar sputtering. EL properties of the TFEL devices were measured under excitation with a rectangular wave (a duty of 50 %) or sine wave, at room temperature using a photomultiplier (R374) or a photometer.

3. RESULTS AND DISCUSSION

3.1 AES-measured depth profiles

The depth profile results from the ZnS:Mn/SiO_2/p-Si structure are shown in Fig.2. Two different energies for LMM electrons from Si were detected. One corresponds to SiO_2 and the other to Si. They were detected at 76 eV and 92 eV respectively although the tail of their AES spectra did overlap slightly. The thickness of the ZnS:Mn film was made thin for AES measurement. It can be seen from Fig.3 that the ZnS is nearly stoichiometric throughout the whole of the film. Moreover, it is considered from the Auger results that diffusion between layers was slight.

3.2 EL properties of Au/ZnS:Mn/SiO_2/Si EL devices

Figure 3(A) shows the dependence of L-V characteristics on the thickness of ZnS:Mn films. The resistivity of the Si wafer was 0.05 Ω-cm. The excitation was

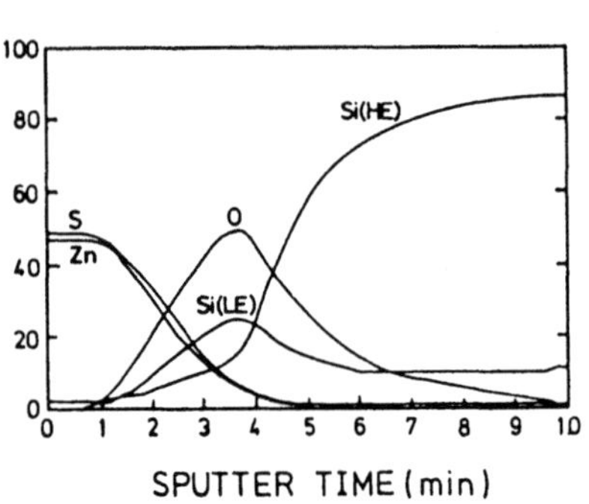

Fig.2 AES-measured depth profile of ZnS/SiO_2/Si multilayer.

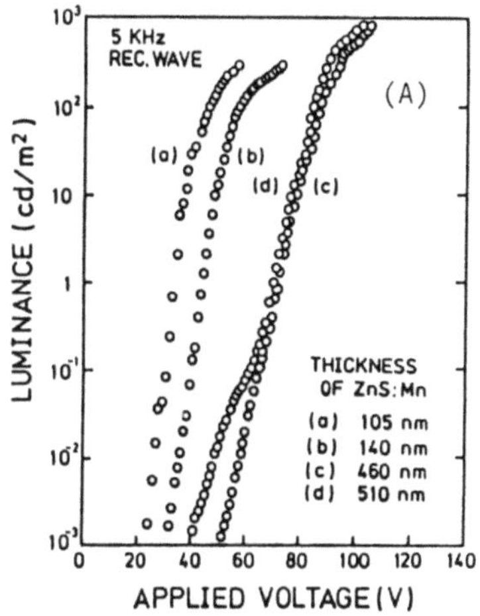

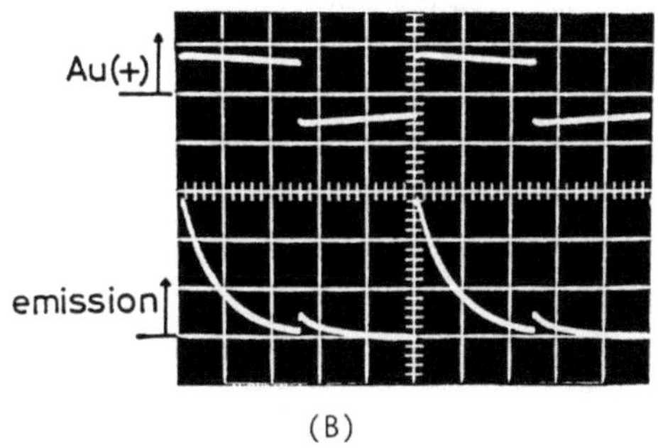

Fig.3 (A) Dependence of L-V characteristics on the thickness of ZnS:Mn and (B) EL waveform of Au/ZnS:Mn/SiO_2/p-Si EL devices.

done with a 5-kHz rectangular wave. It can be seen from this figure that the threshold voltage at which a EL device shows a luminance of 1 cd/m^2 is about 70 volts at 510 nm thickness. This driving voltage is considerably lower than for a standard TFEL device. It is also seen from this figure that the driving voltage lowers, but maximum luminance drops somewhat when the thickness of the ZnS:Mn films is reduced.The cause of the drop of the luminance is considered to be the deterioration of the crystallinity of the ZnS:Mn films as observed by RHEED.

Figure 3(B) shows the EL waveform for a ZnS:Mn film thickness of 140 nm. The excitation was done with a 100-Hz rectangular wave. The EL waveform at other thicknesses of ZnS:Mn films was nearly the same as this waveform. It can be seen from this figure that the light emission is obtained when the polarity of the applied voltage is positive at the Au electrode. This is consistent with the band configuration shown in Fig.1. The slow decrease of the light emission is considered to be due to the decrease in electrons injected into the emitting layer, because the barrier height of SiO_2 increases gradually due to the accumulation of electrons at the interface between Si and SiO_2.

Figure 4 shows L-V characteristics for various p-type Si substrates for a ZnS:Mn thickness of 500 nm. In this figure, p-Si and p^+-Si correspond to Si substrates with resistivities of 5 and 0.05 Ω-cm, respectively, and [p-n]-Si corresponds to Si substrates with p-layer formed by boron implantation into a 1 Ω-cm n-type Si wafer followed by thermal diffusion at 1200°C. It can be seen from this figure that nearly the same L-V characteristics were obtained for both p- and p^+-Si substrates. The high luminance obtained from electron injection, which are minority carriers in p-type Si may come about from two effects. One is the increase of electrons by avalanche in depletion region[2], another is the increase of electrons in the conduction band of Si by photo-excitation of Si by the emission from the Mn center in the ZnS:Mn emitting layer [4,5]. To what extent these effects act in our EL devices must be confirmed in future investigation .

What should be emphasized in Fig.4 is the luminance of the device prepared on a [p-n]-Si substrate. As shown in this figure, not only the applied voltage for EL is shifted toward lower but also the luminance is improved in contrast with the results using the p- and p^+-Si substrates. As shown in Fig.3(b), the light emission is obtained when the polarity of the applied voltage is positive at the Au electrode. During this period, the p-n junction in the Si substrate is forward biased. As a result, a number of electrons will be injected into the conduction

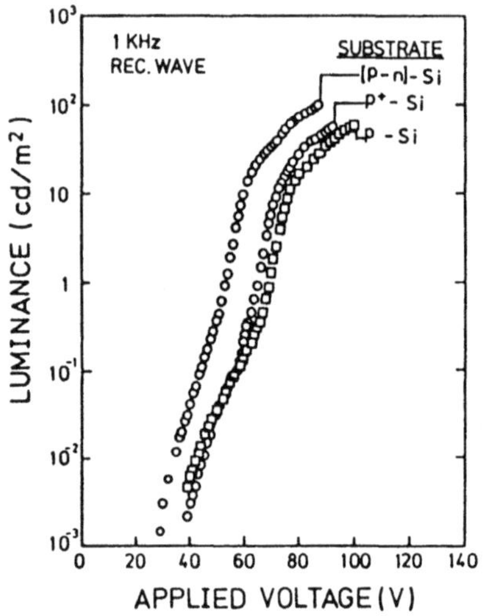

Fig.4 L-V characteristics of Au/ZnS:Mn/ SiO_2/p-Si EL devices for various p-type Si substrates.

band of the p-layer from the n-layer. Therefore, more hot electrons than in the case of p- and p^+-Si substrates are injected into emitting layer across SiO_2 layer. As a result the improved luminance shown in Fig.4 is obtained. As the thermal diffusion after the ion implantaion was carried out at 1200°C for 170 min in this experiment, the thickness of p-layer is estimated to be 4 to 5 µm. Therefore, the ratio of the number of the electrons injected into the emitting layer to the number of the electrons injected into the p-layer from the n-layer must be small. In other words, further improvement of the EL properties is expected by making the p-layer thinner.

4. SUMMARY

ZnS:Mn thin film EL devices were prepared on SiO_2/p-Si. It was shown by AES-measured depth profiling that the ZnS film was nearly stoichiometric throughout the whole of the film.

An EL device with a threshold voltage of about 70 volts was obtained for a ZnS:Mn film thickness of 510 nm, and it was shown that the driving voltage decreased the thickness of the ZnS:Mn films became less.

It was seen from the observation of EL waveform that light emission was obtained when the polarity of the applied voltage at the Au electrode was positive. This is consistent with the band configuration of the EL device in this work.

When the device was prepared on the p-layer formed on a n-type Si wafer by ion implantation, not only the driving voltage is shifted toward lower values but also the luminance is improved. The improvement is concluded to result from the injection of hot electrons into the emitting layer from the n-Si region.

Finally, the authors are very grateful to Mr. S. Hibino of YAMAHA CORPORATION for providing the ion-implanted Si wafers.

LITERATURE REFERENCES

1. J.I. Pankove: 172nd Meeting of The Electrochem. Soc., Inc. cosponsored by The Electrochem. Soc. of Japan with the cooperation of The Japan Soc. Appl. Phys., Extended Abstract No.1226, Honolulu, Hawaii, U.S.A.(1987).

2. T.H. Ning: Solid-State Electron. 21, 273 (1978).
3. Y. Nakanishi and G. Shimaoka: J. Vac. Sci. Technal. A 5, 2092 (1987).
4. Y. Komiya, T. Sakamoto, E. Suzuki and Y. Tarui: Jpn. J. Appl. Phys. 43, Suppl., 356 (1974).
5. T.H. Ning and H.N. Yu: J. Appl. Phys. 45, 5273 (1974).

Part III

Luminescence Characteristics and Materials

Secondary Light Output from ZnS:Mn Thin Film Electroluminescent Devices

H. Schade[a] *and M. Ling*[b]

RCA Laboratories, Princeton, New Jersey, USA

1. INTRODUCTION

Substantial insights into the processes responsible for light emission from ac thin-film electroluminescent (EL) devices can be deduced from the time dependences of the EL light output (emission waveforms). Under steady-state ac drive, differences in phosphor materials are exhibited not only by different emission decay times, but also by the number of emission peaks per period. Specifically under bipolar square pulse drive, ZnS:Mn devices normally exhibit light output only at the leading edges of the voltage pulses, but alkaline earth sulfide devices show both leading and trailing edge (secondary) light outputs. These differences stem from different light generation mechanisms [1,2]. Secondary light output is generally attributed to trap-controlled recombination of the transferred charge with excited activators. The absence of secondary light output in ZnS:Mn, which was also noted for ZnS:Mn EL powders, in contrast to ZnS phosphors with other activators [3], is ascribed to the specific nature of the Mn activator that allows only localized transitions within the luminescence center [3]. However, the specific light generation mechanism in ZnS:Mn does not rule out the appearance of secondary (trailing edge) light. Here we will both report its observation, and discuss the underlying cause and implications, going beyond the explanation of a previous observation [4].

Measurements of the threshold voltages for leading and trailing edge light output (LE and TE), as well as for UV-generated EL [5] in the same sample, will be shown to yield quantitative agreement between the polarization-induced reverse threshold field for TE, and the UV-generated threshold field. Both fields represent a hot-electron acceleration threshold for impact ionization of the Mn centers. The measured transferred charges, as well as the field dependences of the three types of light output, provide further evidence for TE to be caused by renewed impact ionization at the end of the pulses.

2. EXPERIMENTS

The experiments were done on conventional insulator/ZnS:Mn/insulator device structures with a typical "pixel" area of 7×10^{-2} cm^2. The insulator layers, about 200 nm thick on either side of the phosphor film, were SiO_xN_y prepared by plasma deposition, or Ta_2O_5 prepared by sputtering. The ZnS:Mn layer containing about 0.5 at.% Mn was deposited by evaporation, about 500 nm thick. The following measurements were performed:

* Voltage dependences of LE and TE under bipolar square pulse drive. In order to distinguish TE from the typically much larger, slowly decaying LE, long pulse lengths (8 msec) and low repetition rates (25 Hz) were chosen.

---

a) Present Address: Chronar Corporation, P. O. Box 177, Princeton, NJ 08542, USA
b) Present Address: David Sarnoff Research Center, Subsidiary of SRI International, Princeton, NJ 08543, USA

Springer Proceedings in Physics, Vol. 38 Electroluminescence
Editors: S. Shionoya and H. Kobayashi

* For the same pixels, charge vs. voltage characteristics were recorded [6] to determine the transferred charge as a function of the voltage.

* For pixels with semitransparent metallization, adjacent to the above-mentioned, the voltage dependence of UV-generated EL [5] was measured.

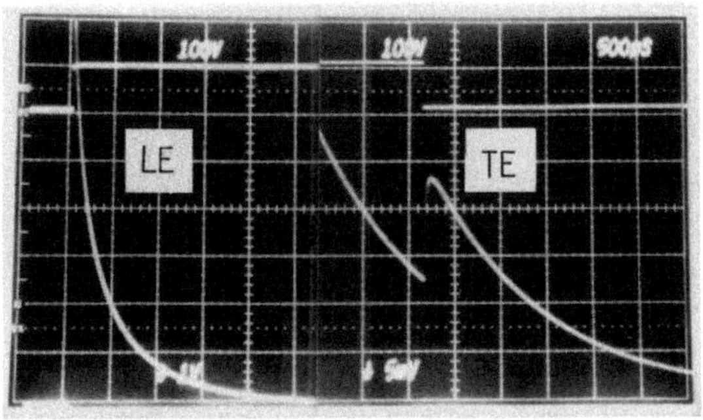

Fig. 1: Leading and trailing edge light output (lower traces) due to bipolar square pulse drive (shown for a positive pulse; upper trace). For illustrative reasons, TE is shown at a slightly higher drive voltage than LE; note that even then TE is considerably smaller than LE (5 mV/div vs. 1 V/div)

3. RESULTS

TE was observed for three devices obtained from different sources. For a fourth device having similar properties as the others, no TE could be detected; a possible reason is suggested within the context of our model. Figure 1 shows LE and TE during steady-state bipolar square pulse drive. We have also verified that both LE and TE are virtually identical for the positive and negative pulses.

With regard to the magnitude of TE, it is important to point out:

* TE is only detected at a substantially higher threshold voltage compared to that for LE (the conventional threshold voltage).

* TE remains always considerably smaller than LE.

In contrast, this behavior does not typically apply to secondary light output in other phosphor materials, like alkaline earth sulfides, for which we have measured almost identical threshold voltages for LE and TE [7], and moreover, frequently have observed even larger TE than LE.

4. EVALUATION

The appearance and the properties of the secondary light output in ZnS:Mn is explained as follows: The charge transfer between the insulator/phosphor interfaces leads to a polarization of the phosphor layer. Upon removal of the applied field (at the trailing edges of the voltage pulses) this polarization gives rise to a reverse field in the phosphor layer. For sufficiently high applied fields, the reverse field may exceed the threshold field for light generation also at the trailing edges. Since the electron supply for this emission consists of the charge already transferred during the pulse, the reverse threshold field is determined, just like the threshold field for UV-generated EL [5], only by the hot-electron acceleration requirement for impact ionization of the Mn centers.

4.1 Phosphor Threshold Fields

Based on capacitive voltage distribution at the onset of the pulses, field clamping towards the end of the pulses, and uniform fields, the polarization-induced reverse phosphor threshold field for TE is given by

$$F_{oT} = F_{pTTE} - F_{pT} \tag{1}$$

where F_{pTTE} is the phosphor threshold field for TE determined at the pulse onset,

$$F_{pTTE} = V_{TTE}/(d_p + 2d_iK_p/K_i), \tag{2}$$

F_{pT} is the phosphor threshold field for LE determined from (2) for the threshold voltage of LE, V_{TTE} is the threshold voltage of TE, and K_p, K_i and d_p, d_i are the dielectric constants and thicknesses of the phosphor and insulator layers, respectively. In Table I, the polarization-induced reverse fields at the TE threshold are compared with the UV-generated EL thresholds, also deduced from (2). The generally good agreement between these values lends quantitative support to the described origin of TE.

Table I: Phosphor threshold fields for LE, TE, and UV-generated EL

Sample	F_{pT} [MV/cm]	F_{pTTE} [MV/cm]	F_{oT} [MV/cm]	F_{TUV} [MV/cm]
P 5	1.22	1.69	0.47	0.51
K 7	1.15	1.76	0.61	0.73
B 1	1.31	2.09	0.78	0.83
O 11	1.72	(2.47)		0.75

In the case of sample O-11, the F_{pT} value is considerably higher than for the other three samples, which in combination with F_{TUV} would lead also to a higher F_{pTTE} value. The corresponding applied voltage would have to exceed 270 V, which, within the uncertainties of this estimate, is considered to fall beyond the voltage range of our supply.

4.2 Transferred Charge

Since the amount of the transferred charge must also be related to the field difference at the onset and the end of each pulse, we would expect, based on the assumptions above,

$$F_{oT} = Q_{TTE}/K_o(K_p + K_i d_p/2d_i) \tag{3}$$

where Q_{TTE} is the transferred charge measured at the TE threshold voltage, and $K_o = 8.854 \times 10^{-14}$ C/Vcm.

Table II shows that the threshold fields derived from (1) are mostly higher, compared to the values deduced from the measured transferred charge (3). This discrepancy may stem from the fact that during steady-state drive conditions a portion of the initially transferred charge stays trapped at the insulator/phosphor interfaces. This explanation is supported by earlier studies that attribute relaxation times on the order of 10 msec to the electron release from interfacial traps [8]. Due to the remaining charge, the phosphor field is no

Table II: Reverse phosphor threshold fields for TE, transferred charges, and trapped interface charges

Sample	$K_p+K_i d_p/2d_i$	F_{oT} [MV/cm]	$F_{oT}(Q_{TTE})$ [MV/cm]	Q_{TTE} [C/cm^2]	Q_{trap} [C/cm^2]
P 5	16	0.47	0.16	2.3×10^{-7}	2.2×10^{-7}
K 7	17	0.61	0.19	2.9×10^{-7}	3.0×10^{-7}
B 1	19	0.78	0.70	11.8×10^{-7}	-

longer uniform; the average field stays higher at the end of the pulses, which is consistent with the measured charge, and results in a lower average reverse field. However, due to the trapping, the reverse field at the interface remains larger. To account for the reverse phosphor threshold field, the remaining trapped charges were estimated (see Table II) under the assumption that space charges are set up to balance them. These amounts are consistent with relaxation times given in [8].

4.3 Brightness vs. Field Characteristics

The field dependence of the brightness reflects

* for UV-generated EL, the energy dependence of impact ionization of the Mn centers by hot electrons, and

* for conventional EL (LE), the combined energy dependence of both the electron supply (injection) and impact ionization [5].

We have evaluated the brightness of TE as a function of the reverse phosphor field, and compared it with the two other field dependences. Figure 2 illustrates all three field dependences obtained on the same sample. Note particularly that the curves for both LE and TE exhibit about the same steepness, while that for UV-generated EL has a considerably smaller slope. These differences in the field

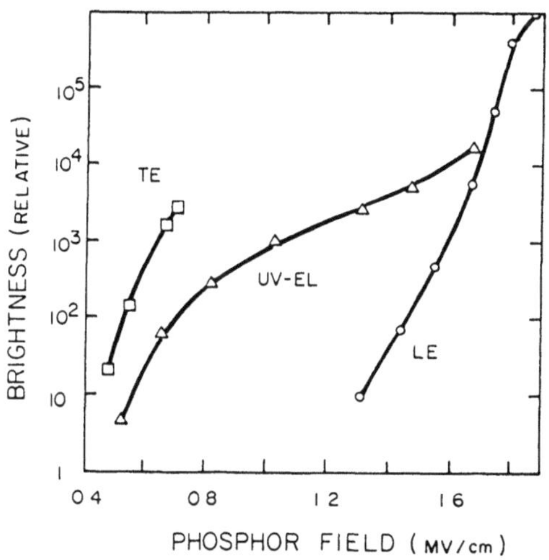

Fig. 2: Field dependences of the leading and trailing edge light outputs (LE and TE), and of UV-generated EL. All three light intensities are shown on the same relative scale; TE is plotted as a function of the reverse phosphor field (Eq. 1)

dependences are explainable by differences in the electron supply: For TE, just as for LE, the electron supply strongly increases with the field, while for UV-generated EL the electron supply, below the normal threshold field ($<F_{pT}$), is determined by the UV light intensity, and thus is constant.

5. CONCLUSION

Secondary light output in ZnS:Mn is governed by polarization-induced reverse phosphor fields that exceed the hot-electron acceleration limit for impact ionization of the Mn centers.

ACKNOWLEDGMENT

The depositions of ZnS:Mn by F. J. Tams, and of insulator layers by G. Kaganowicz and H. L. Pinch are gratefully acknowledged. We also would like to thank R. Tuenge of Planar Systems, Inc., for providing us with sample devices.

REFERENCES

1. S. Tanaka, H. Yoshiyama, Y. Mikami, J. Nishiura, S. Oshio, H. Kobayashi, Japan Display '86, 242 (1986)
2. R. S. Crandall, M. Ling, J. Appl. Phys. 62, 3074 (1987)
3. J. Mattler, J. de Phys. et le Rad. 17, 725 (1956)
4. E. Bringuier, A. Geoffroy, Appl. Phys. Lett. 48, 1780 (1986)
5. H. Schade, Proc. Intern. Electroluminescence Workshop, Warm Springs, Oregon (1986)
6. D. H. Smith, J. Luminescence 23, 209 (1981)
7. H. Schade, M. Ling, unpublished (1986)
8. H. Kobayashi, S. Tanaka, H. Sasakura, Y. Hamakawa, Japan. J. Appl. Phys. 12, 1854 (1973)

Thermally Stimulated Currents in Thin Film Electroluminescent Devices

Y. Sano and K. Nunomura

Opto-electronics Research Laboratories, NEC Corporation,
4-1-1, Miyazaki, Miyamae-ku, Kawasaki 213, Japan

Abstract

New signals from thermally stimulated currents were observed on thin film electroluminescent devices. The signals originate from polarization charge at the interface between a ZnS layer and an insulating layer. The interface trap depths ranged from 1.8 to 1.9 eV, which is about half the ZnS energy gap.

1. Introduction

Thin film electrolumimescent (TFEL) device characteristics are thought to be affected from traps around the interface between a light emitting layer and an insulating layer [1-3]. Thermally stimulated current (TSC) or light induced detrapping charge measurements are useful to determine the trap characteristics. These measurements were attempted on TFEL devices [4-7]. However, TSC signals caused by depolarization have not yet been observed. This paper describes TSC signals successfully measured at a higher temperature range than that for previous measurement. These newly observed signals seem to originate from interface traps.

2. Experimental

Formerly, TSC measurements on TFEL devices have been carried out at up to 150 °C. However, interface trap signals have not been identified. Therefore, the measurement range was chosen from room temperature to higher temperature, 300 °C. Figure 1 shows the TSC experimental setup. A Keithley model 619 electrometer was used to detect a small current with high precision. The device was kept in N_2 atmosphere. Sample device constructions are shown in Table 1. The device area is 4 mm^2. ZnS:Mn layers were thermally evaporated. Device No. 1 insulators were composed of a sputtered Y_2O_3 layer and a sputtered TaSiO high-dielectric-constant layer. Y_2O_3 was used as an adhesive layer between TaSiO and ZnS:Mn. In device No. 2, sputtered SiO_2 was used as an ion barrier between electrodes and electron-beam evaporated Y_2O_3. Device No. 3 insulators are sputtered Si_3N_4.

Table 1 TFEL Device Construction

No.	Structure
1	Al / TaSiO / Y_2O_3 / ZnS:Mn / Y_2O_3 / TaSiO / ITO /Glass
2	Al / SiO_2/ Y_2O_3/ZnS:Mn/ Y_2O_3/ SiO_2/ ITO/Glass
3	Al / Si_3N_4 / ZnS:Mn / Si_3N_4 / ITO/ Glass

In device No. 1, a large current appeared at above 200 °C, even without electric excitation (A in Fig. 2). This current was found to be effectively suppressed (B in Fig. 2) by annealing the device with the electrodes short circuited at 460 °C in N_2 atmosphere for more than 10 minutes before TSC measurement. Device No. 2 exhibited the same tendency. On the other hand, device No. 3, when used with Si_3N_4 insulators, did not show this no-excitation current. Therefore, the currents depend on the insulating layer materials.

Then, the device was excited by applying voltage at room temperature. This voltage was DC, or AC pulse voltage (120 Hz, 30 microsec. long), the same as that required for the device to emit light. After excitation, the device was heated at a constant temperature rising rate, and TSC spectra were recorded.

3. Results and Discussions

3-1 TSC from Device No. 1

After an AC pulse voltage train was applied to the device (inset in Fig. 3), TSC spectra were measured. Final pulse voltage polarity is denoted as "ITO(+) Exc." or "ITO(-) Exc." TSC signals were observed above the threshold voltage required to emit light. TSC spectra from device No. 1 consisted of two peaks (A and B in Fig. 3). Their polarity was changed when finally applied voltage was reversed. Peak A has the same charging direction as was previously reported [5,6]. Peak B, appearing at higher temperature and newly observed this time, has the discharging direction. A simple model to interpret these signals is introduced (Fig. 4). At first, applying voltage to the device, electrons move toward the anode-side interface between a ZnS:Mn layer and an insulating layer (Fig. 4 (a)). After electrons finished moving (Fig. 4(b)), the voltage was removed. Thereafter, the internal field, caused by electrons and residual holes, was established (Fig. 4(c)). Some electrons are also trapped in the insulating layers. From this stage, electrons are released from traps (alpha, beta or gamma in Fig. 4(c)) when the device is

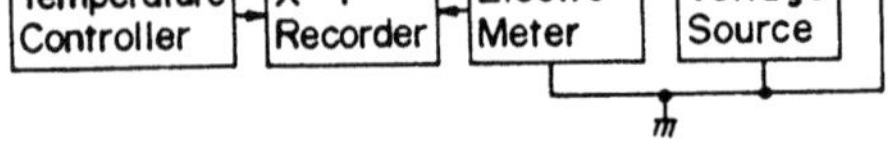

Fig. 1 TSC experimental set up

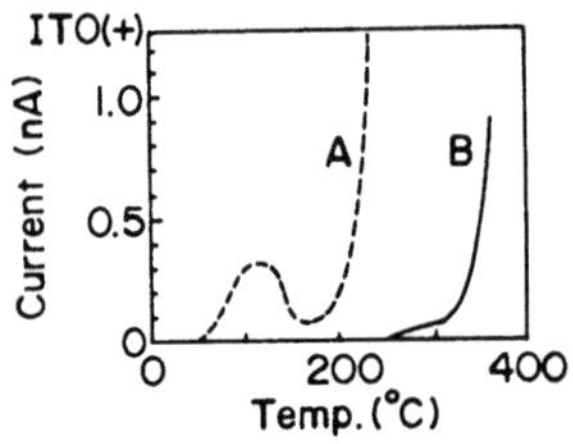

Fig. 2 Current without electric excitation for device No. 1

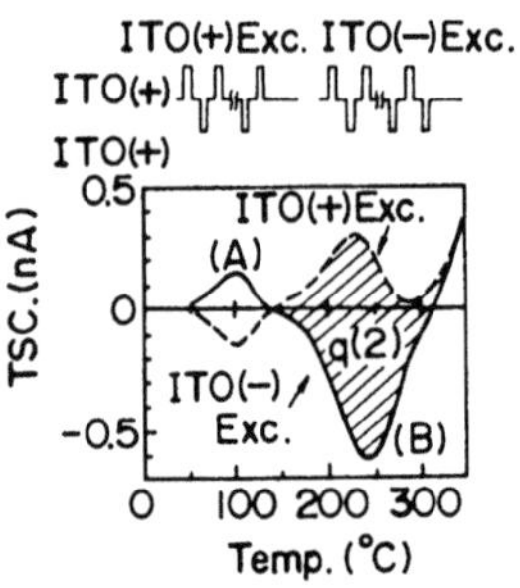

Fig. 3 TSC spectra for device No. 1

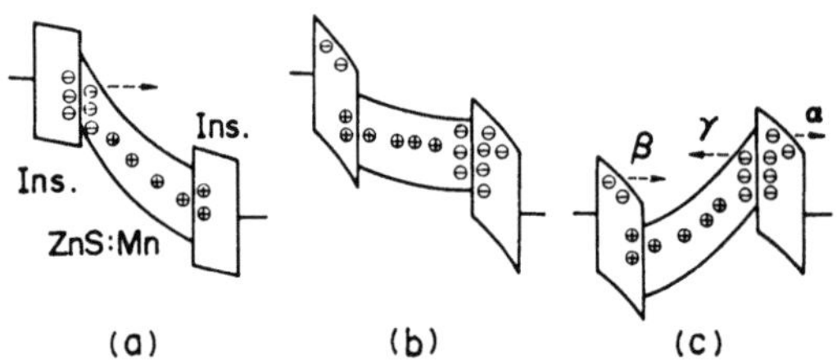

Fig. 4 Charge distribution model. (a) right after voltage applied. (b) after charge redistributed. (c) after voltage removed

heated up. The charging current, peak A in Fig. 3, appears to be the current for electrons from traps in insulating layers (alpha and/or beta in Fig. 4). The discharging current, peak B, appears to be the current from interface traps (gamma in Fig. 4). In addition, peak B is asymmetric between ITO(+) and ITO(-) excitation. When excited with an AC pulse train of gradually decreasing voltage, weak peak B appears. It seems that asymmetry of peak B is caused by the different nature for interface traps at both ZnS layer faces.

3-2 Peak B Identification

On the simple equivalent circuit model for TFEL devices, the light emitting layer could be regarded as ideal back to back Zener diodes [2,3] (inset in Fig. 5). The total charge transfer (Q) versus applied voltage (V) diagram is shown in Fig. 5. In the figure, "q" is the residual charge at zero applied voltage. Charge transfer, 2q, can be obtained by integrating the device conduction current (assigned to q(1)). On the other hand, if peak B in Fig. 3 comes from interface-trapped charge, q(2) in Fig. 3 must coincide with q(1). This relation was examined while varying the excitation pulse voltage. It was found that q(2) was actually proportional to q(1) at every voltage, as shown in Fig. 6. Furthermore, after heating up the device to more than 300 $^{\circ}$C to erase the TSC, a single pulse was applied whose polarity was the reverse from the polarity for the final pulse before heating up. The voltage was chosen as the threshold voltage. However, there was no light emission. This suggests that the residual charge, q, were erased by heating up the device to above 300 $^{\circ}$C. From these results, peak B is considered to result from interface-trapped charge.

3-3 Trap Depth Evaluation

In the next evaluation, excitation voltage was DC, rather than AC pulses. This was to make the TSC peak sharper, so that peak temperature was easily determined (Fig. 7). Device No. 1, excited with ITO(-) polarity, shows the variation in inverse peak-temperature versus inverse heating-rate in Fig. 8. From the line gradient in Fig. 8, trap depth was calculated [8] as being 1.9 eV. For ITO(+) excitation, trap depth was 1.8 eV.

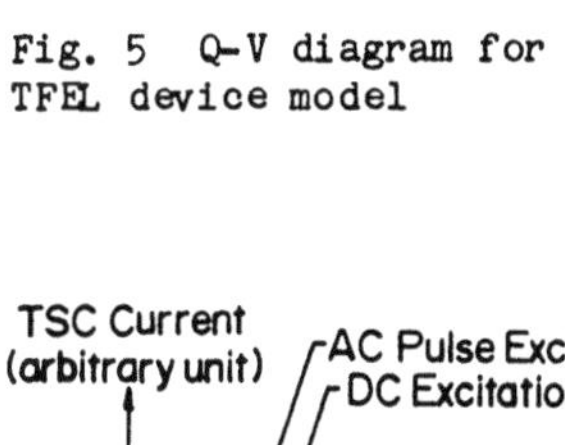

Fig. 5 Q-V diagram for a simple TFEL device model

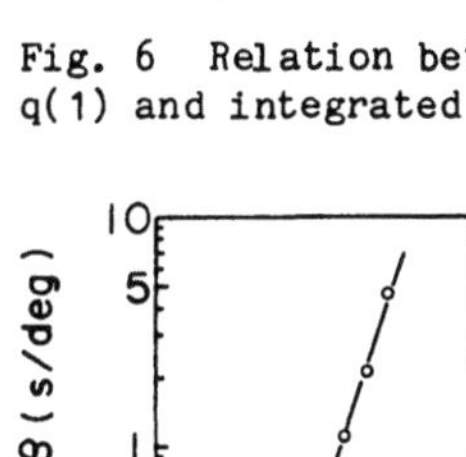

Fig. 6 Relation between conduction charge q(1) and integrated charge for TSC q(2)

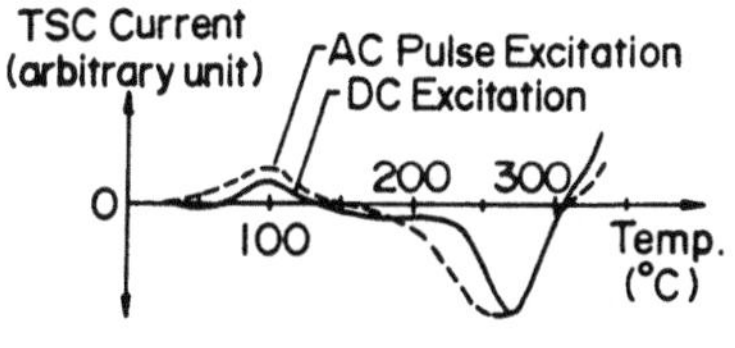

Fig. 7 Device No. 1 TSC spectra for DC and AC pulse excitation

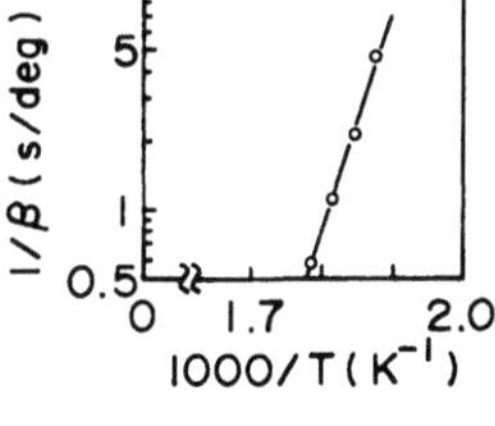

Fig. 8 Reverse TSC peak temperature versus reverse heating rate

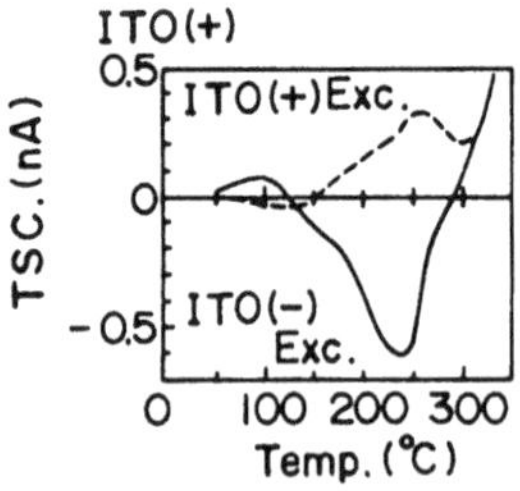

Fig. 9 TSC spectra for device No. 2

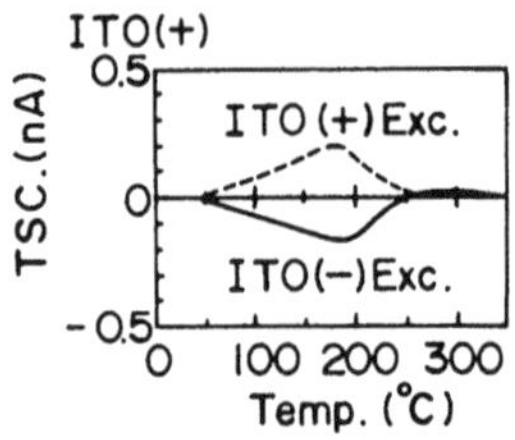

Fig. 10 TSC spectra for device No.3

3-4 TSC Spectra from Other Devices and Electric Field Strength in ZnS:Mn

TSC spectra from device No. 2 resembles those from device No. 1 (Fig. 9). This seems to be due to the fact that Y_2O_3 layers were in contact with the ZnS:Mn layer for both devices. Trap depth was calculated to 1.8 eV for ITO(+) excitation. However, trap depth for ITO(-) excitation could not be measured, because of the device electrical instability. Device No. 3 TSC spectra, shown in Fig. 10, differ greatly from those for device No. 1 or No. 2. First, there is no charging direction TSC. Second, discharging direction TSC peaks at under 200 °C. However, trap depth, corresponding to the peak, was 1.9 eV for both excitation polarities. In addition, the integrated charge for this TSC peak, q(2), was about half the conduction charge, q(1). This suggests that other not measured traps exist.

The electric field in ZnS:Mn was calculated as 2 MV/cm for all devices, No. 1 to No. 3, at the threshold voltage required to emit light for 1 cd/m^2 luminance. On the other hand, as mentioned above, trap depths corresponding to the TSC spectra peaks spread in a rather narrow range, 1.8 to 1.9 eV. Therefore, there seems to be a correlation between electric field in ZnS:Mn and trap depths.

4. Conclusion

TSC signals from interface traps were observed. The TSC spectra shape varied when insulating-layer materials, in contact with ZnS:Mn layers, were changed. However, trap depths, corresponding to the TSC spectra peaks, spread in a rather narrow range, 1.8 to 1.9 eV, which is about half the ZnS energy gap. On the other hand, the electric field in ZnS:Mn was almost the same for these devices. Therefore, there seems to be a correlation between trap depths and device characteristics. Further research on TSC spectra shape is expected to provide more information.

Acknowledgements

The encouragement and support from C. Tani and T. Saito are greatly appreciated. Special thanks are due to M. Ishiko and T. Yoshioka for helpful discussions, to N. Koyama for sample fabrication, and to A. Ueki for measurement.

References

1. R. Mach and G. O. Muller, Physica Status Solidi (a), 69 (1982) p.11
2. W. E. Howard, Journal of Luminescence, 23 (1981) p.155
3. D. H. Smith, Journal of Luminescence, 23 (1981) p.209
4. T. Tatsumi et al., The Institute of Electronics, Information and Communication Engineers Technical Report, 87, No. 407 (1987) p.49 (in Japanese)
5. K. Miyashita and M. Shibata, Digest of Japan Display '83, p.100
6. T. Shibata et al., Extended Abstracts (The 34th Spring Meeting, 1987); The Japan Soc. of Appl. Phys. and Related Societies, No. 3, p.879 (in Japanese)
7. T. Inoguchi and S. Mito : In Electroluminescence, ed. by J.I.Pankove, Topics in Applied Physics, Vol.17 (Springer, Berlin, Heidelberg, New York,1977) p.197
8. J. G. Simons and G. W. Taylor, Physical Review B, 5 (1972) p.1619

Measurement of Trap Levels in Electroluminescent Devices by Photon-Released Residual Charges

H. Uchiike, M. Noborio, T. Tatsumi, S. Hirao, and Y. Fukushima

Faculty of Engineering, Hiroshima University,
Shitami, Saijou-chou, Higashi-Hiroshima 724, Japan

1. Introduction

Thin-film ac electroluminescent (EL) devices, consisting of a ZnS:Mn active layer sandwiched by a pair of insulating layers, have trap levels in the active layer as well as in the active layer–insulating layer interfaces [1]. Carriers trapped in these levels generate polarized charges and the memory function of a ZnS:Mn EL display is considered to be caused by these polarized charges [2]. Therefore, identification of these trap levels is necessary in order to clarify the operation mechanism of the EL display.

Residual charges Q_0, which are proportional to the polarized charges, are important parameters in EL devices. In this study, these residual charges and the energies of the trap levels were measured via the photon-released residual charge. At the same time, EL measurements using short pulse voltages and photoluminescent (PL) measurements using a nitrogen laser were performed to confirm that these trap levels are relevant to EL emission.

2. Experimental

Figure 1 shows an experimental setup for photon-released residual charge measurement. A halogen lamp was used as the light source. Monochromatic light was obtained using a monochromator and color filters. Charges were detected with the sense capacitor in the Sawyer-Tower circuit. Output signals were stored in a digital boxcar integrator.

In this measurement, eight successive voltage pulses with a pulse width t_a of 800 μs were applied. This pulse width is large enough to saturate the residual charge Q_0, as shown in Fig. 2. Just after the applied voltage is turned off, light with a monochromatic wavelength is irradiated onto the EL cell to release residual charges in trap levels. By changing the photon energy of the incident light, it is possible to estimate the density of states of trap levels from the measurement of photon-released residual charge.

Springer Proceedings in Physics, Vol. 38 Electroluminescence
Editors: S. Shionoya and H. Kobayashi

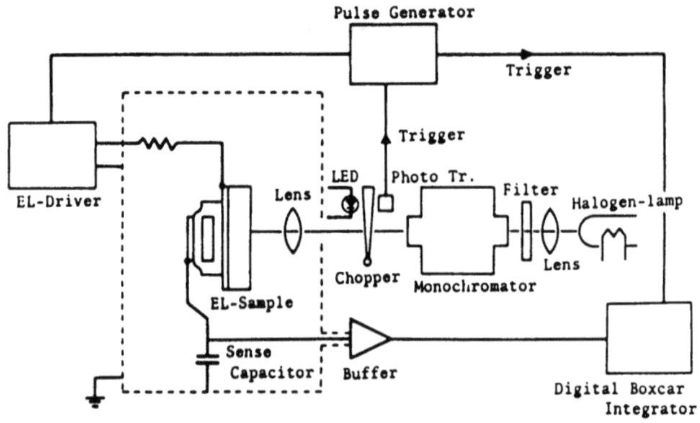

Fig. 1. Experimental setup for photon-released residual-charge measurement

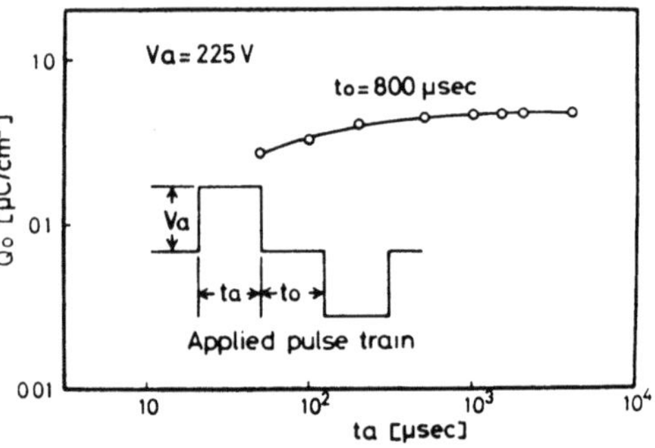

Fig. 2. Pulse width dependence of residual charge

3. Results and Discussion

Figure 3 illustrates a schematic time sequence of the charge Q. Residual charges were released by irradiating monochromatic light for 10 s. Figure 4 shows an example of a time sequence of residual charge Q_0. In this figure, the time sequence of residual charge for several photon energies, A(3.65 eV), C(3.05 eV), D(2.85 eV), E(2.55 eV), and F(2.25 eV) is shown. Curve H corresponds to no irradiation.

Figure 5 shows photon-released residual charges ΔQ_0 after ten second irradiation as a function of photon energy. We observe several peaks; A(3.65 eV), B(3.25 eV), C(3.05 eV), D(2.8 eV), E(2.55 eV), F(2.25 eV) and G(1.95 eV). The trap level indi-

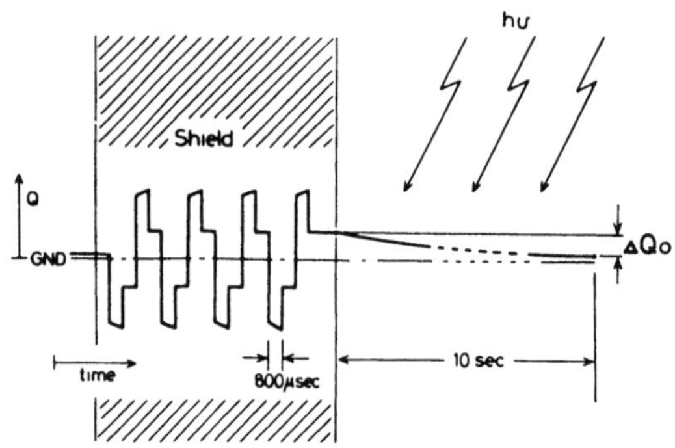

Fig. 3. Schematic time sequence of charge Q

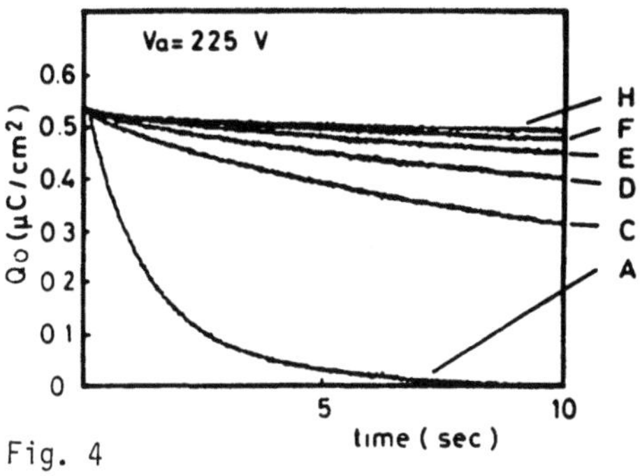

Fig. 4

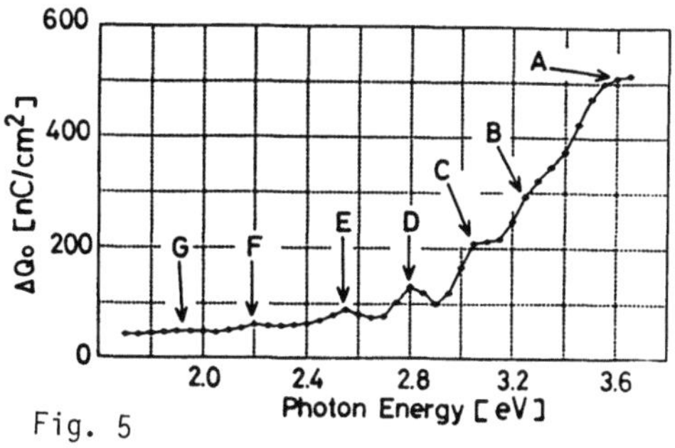

Fig. 5

Fig. 4. Example of time sequence of residual charge Q_0 under light irradiation

Fig. 5. Photon-released residual charges ΔQ_0 as a function of photon energy

cated by A is due to the ZnS bandgap, because the photon energy is equal to the band gap energy, 3.6 eV. Other peaks were considered to be due to trap levels.

In order to confirm that the energies of several trap levels were observed by measuring photon-released residual charges, the following measurements were carried out: an EL measurement by the application of short pulse voltage and a PL measurement using a nitrogen laser. Results are shown in Figs. 6 and 7: the PL emission spectrum is shown in Fig. 6 and the EL emission spectrum is shown in Fig. 7. Here the arrows B through E indicate the peaks shown in Fig. 5 and I indicates emission from Mn. In both PL and EL measurements, emissions B(3.25 eV), C(3.05 eV), D(2.85 eV), and E(2.55 eV) were observed. These emissions were assigned to ZnS bulk material rather than Mn from the measurements of temporal light response characteristics [3]. In both EL and PL spectra, we could not distinguish the emissions from ZnS and that from Mn, because the main emission from Mn is at more than 500 nm ($E < 2.5$ eV).

From the above experiments it was not possible to determine where these trap levels are in the band. However, the energies of two of the levels, 2.8 eV and 2.55 eV are equal to those reported using TSC (Thermally Stimulated Current) measurements [4].

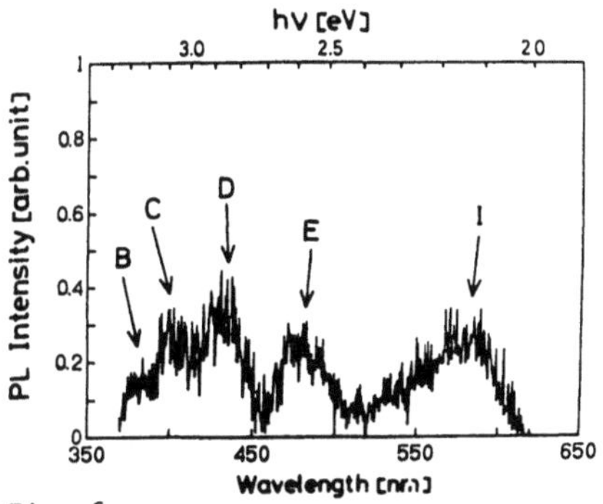

Fig. 6

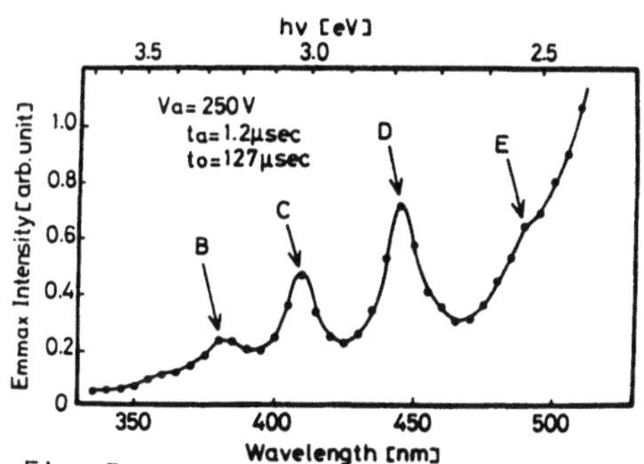

Fig. 7

Fig. 6. PL emission spectrum of ZnS using a nitrogen laser

Fig. 7. EL emission spectrum of ZnS by the application of voltage pulses of 1.2 μs

4. Summary

In order to study energy states of trap levels, we considered polarized charges and implemented a new method: photon-released residual charge measurement. The results show the existence of trap levels with 3.25 eV, 3.05 eV, 2.8 eV, 2.55 eV, 2.25 eV and 1.95 eV in one of our samples. It is not clear where these trap levels are situated in the band. However, this measurement technique is expected to present more information in future.

References

1. D.H. Smith: J. Lumin. **23**, 209 (1981)
2. H. Kobayashi, S. Tanaka, H. Sasakura, K. Tsuji, R. Tueta: Appl. Phys. Lett. **40**, 1024 (1982)
3. K. Okamoto, S. Miura: Appl. Phys. Lett. **49**, 1596 (1986)
4. T. Shibata, K. Hirabayashi, H. Kozawaguchi, B. Tsujiyama: Ext. Abstract of 34th Spr. Mtg., J. Soc. Appl. Phys. 879 (1987)

Influence of the Mn Concentration and the Level of Excitation on Efficiency of ZnS:Mn Devices

J. Benoit, P. Benalloul, and A. Geoffroy

Laboratoire d'Optique de la Matière Condensée, Université P. et M. Curie, Tour 13, 4 place Jussieu, F-75252 Paris Cedex 05, France

1. INTRODUCTION

Several papers dealing with the mechanism of $ZnS:Mn^{2+}$ emission in TFEL devices have been reported and two principal phenomena which govern the decay of the yellow Mn^{2+} emission have been registered [1,2,3] :

- Diffusion mechanism and energy trapping : an excited Mn^* transfers its energy via non excited Mn to killers that reduces the emitted light by the introduction of non radiative channels.

- Interaction between excited Mn^* : This phenomena occurs only at high excitation level ; the higher the excitation level, the lower the decay time.

In previous work [2] we have clarified the conditions in order to separate these two phenomena : we have demonstrated that interactions between Mn^* do not exist at low level of excitation near the electroluminescence threshold. Then the decay curve was fitted by the function $B = B_0 \exp - (t/\tau)^{\alpha}$ where α and τ depend on the Mn concentration. This law describes the decay curves in overall four decades in intensity and takes the diffusion limited relaxation process into account.

2. EXPERIMENTAL DETAILS

An acquisition system using a 12 bit Nicolet digital memory scope interfaced to a microcomputer allows to digitize and accumulate the decay curves. A power supply gives bipolar pulses the width of which range between 3 and 5μs. We have studied two kinds of devices evaporated with an electron beam (EB) and devices made by atomic layer epitaxy (ALE) by the french society CIMSA-SINTRA. The mean value of the dopant concentration has been measured by electron probe microanalysis (EPMA).

3. THE INTERNAL EFFICIENCY FORMULATION

The internal efficiency of an EL device can be written as :

$$\eta(c) = K\ \eta_{exc} \times \eta_{lum}\quad [4]\ ;\ c \text{ being the Mn concentration}$$

η_{exc} : including the impact cross section excitation of the dopant $\eta_{exc} = \sigma . c$ [4].

η_{lum} : relates to the competition between radiative and non radiative processes. Taking into account both mechanisms outlined in the introduction, we can introduce a new expression for η_{lum} : $\eta_{lum} = D_{diff} \cdot I_{int}$

D_{diff} : including the diffusion and trapping mechanism defined by the experimental law $B = B_0 \exp - (t/\tau)^{\alpha}$

I_{int} : taking into account Mn^* interactions : the value of this term will be 1 at low level of excitation.

Editors: S. Shionoya and H. Kobayashi

4. THE INTERNAL EFFICIENCY : LOW LEVEL OF EXCITATION (E.L. THRESHOLD)

At low level, if we increase the intensity of light emission by a factor 10, the decay rate does not vary, which implies that there are no interaction between Mn^* (Int = 1). If the decay is a pure exponential decay we could express the luminous efficiency as :

$$\eta_{lum} = \frac{\tau(c)}{\tau(o)}$$

$\tau(o)$ decay time of isolated ion in ZnS

$\tau(c)$ decay time of diluted ion in ZnS depending on c.

We have shown it is not the case for ZnS:Mn active layers, the decay time can be fitted by the function $B = B_0 \exp - (t/\tau)^{\alpha}$; therefore η_{lum} must be expressed as :

$$\eta_{lum} = S(c)/S(o)$$

$S(c) = \int_0^{\infty} \exp - (t/\tau)^{\alpha} dt$ is the normalized total emitted light when a pulse excitation is applied to the device.

$S(o) = \int_0^{\infty} \exp - (t/\tau_0)\, dt = \tau_0$ ($\tau_0 = 1770\ \mu s$ in cubic symmetry).

We have measured the value of α and τ for Mn concentration between 0.2 and 2 mole % (Fig. 1). The values have been obtained for ALE and EL devices.

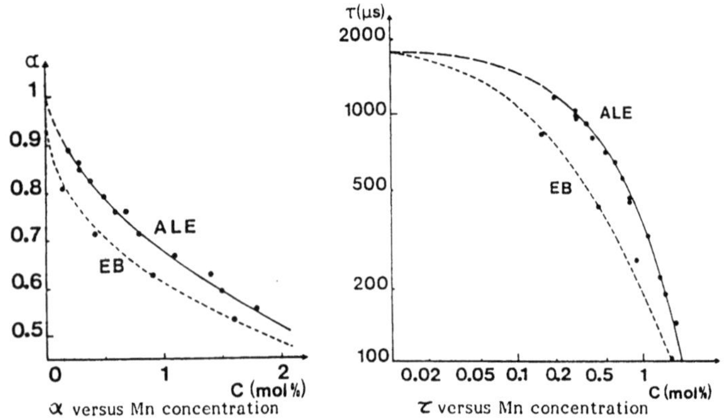

Figure 1

These values of α and τ allow us to calculate the function S(c) (Fig. 2). We also report on this figure the experimental values of the normalized total emitted light obtained by numerical integration which are in good agreement with calculated values. Consequently we can obtain the value of $\eta(c)$ as :

$$\eta_{low\ level}(c) = K\,\sigma\, c\ \ S(c)/S(o)$$

As it can be seen, the efficiency is higher for ALE devices compared with our EB devices by a factor 2. This result shows that the emission of ZnS:Mn EL devices at low level of excitation allows study of the quality of the active layer : the best value of $\eta_{low\ level}$, the best quality of ZnS:Mn layer. We report on Fig. 3 the curve $\eta_{low\ level}$ corresponding to ALE devices.

The optimal concentration is about 1 mole % : it is not in agreement with experimental values due to the fact that we do not take into account the Mn^* interactions as the device works at low level of emission.

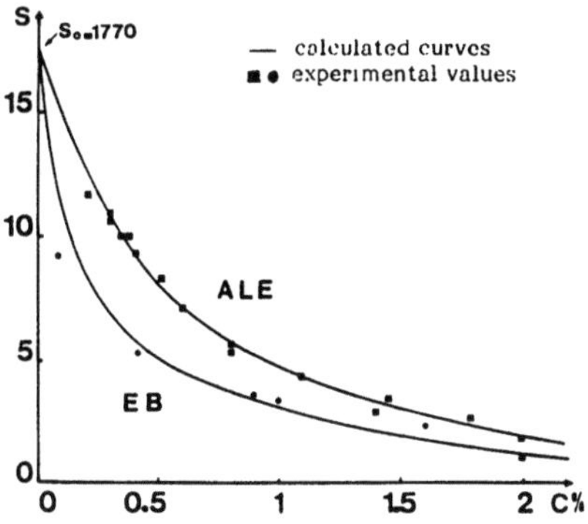

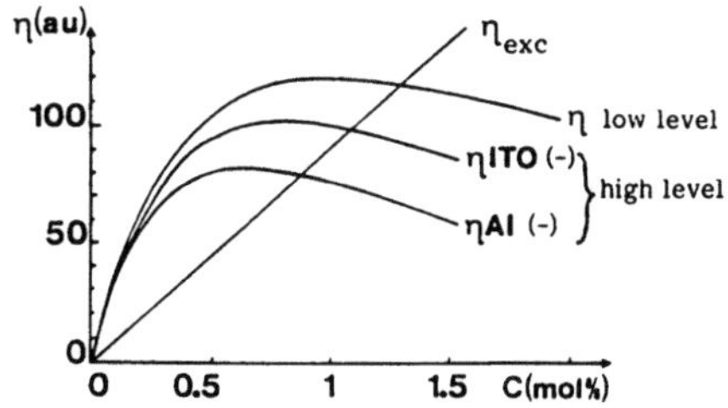

△ Fig. 3. Efficiency η versus Mn concentration

◁ Fig. 2. Normalized total emitted light S(c) versus Mn concentration

5. THE INTERNAL EFFICIENCY : HIGH LEVEL EXCITATION (E.L. SATURATION)

In this domain, when the level of emission increases, an increase of the decay rate of Mn^{2+} is observed at the beginning of the decay ; this phenomenon can be explained by interactions between Mn^* which interfere with the diffusion mechanism.

In this domain, until now, the complexity of the process forbids the use of any approximative function. However it is possible to measure the normalized total emitted light $S^*(c)$ obtained in the EL saturation domain. So the interaction term can be written as : $I_{int} = S^*(c)/S(c)$ and the internal efficiency is given by :

$$\eta_{high\ level}(c) = K\,\eta_{exc}\,D_{diff}\,I_{int} = K\sigma c\ S(c)/S(o).S^*(c)/S(c) = K\sigma c\ S^*(c)/S(o)$$

$S^*(c) = S(c)$ at low level ; $S^*(c)$ decreases with increasing level of excitation.

In order to compare several devices made by a same technique at different Mn concentration, we must have a very good reproducibility in the fabrication, which is the case for ALE devices.

Therefore we only perform I_{int} measurements with ALE devices. We have observed I_{int} varies linearly with the concentration and depends on the nature of the interface. With negative polarity on Al we observe :

$I_{int} \simeq 1$ for 0.1 mol %

$I_{int} \simeq 0.5$ for 1.5 mol %

for ALE device.

We have reported on Fig. 3 the efficiency curve at high level of excitation ($\eta_{high\ level}(c)$) with respect to the influence of interface.

As expected, the optimal concentration values are in agreement with experimental values (0.6% - 0.8%). In a more detailed paper [5] we will show that I_{int} term depends directly on the brightness level and indirectly on the Mn concentration. The dependence on the nature of the interface will also be discussed.

6. CONCLUSION

In this paper we have used the possibility to separate the diffusion phenomena from the interactions between Mn^* [2]. This possibility has been fruitful.

At low level of excitation where there is only diffusion mechanism, our method for studying the diffusion limited relaxation process let us compare accurately EL devices elaborated by two different methods (ALE, EB). Now it is possible to test the effect of a modification in the fabrication process on the diffusion process and consequently on the efficiency of the active layer.

At high level of excitation we have estimated the influence of interactions between Mn^* on the efficiency and we have shown that the optimal concentration depends on the excitation level which confirms Muller's hypothesis [3]. So that, for ALE devices, the optimal concentrations range from 1 mol % (when low level excitation is applied on the device) to 0.6 mol % when high level excitation is applied. On the other hand the optimal concentration at high level of excitation depends on the nature of the interfaces.

REFERENCES

1. R. Tornquist, J. Appl. Phys. **54**, 4110 (1983)
2. J. Benoit, P. Benalloul, A. Geoffroy and C. Barthou, Phys. Stat. Sol. (a) **105**, 637 (1988)
3. G.O. Müller, J. Neugebauer, R. Mach and U. Reinsperger, J. Crystal Growth **86**, 890 (1988)
4. R. Mach and G.O. Müller, Phys. Stat. Sol. (a) **81**, 609 (1984)
5. J. Benoit et al., to be published

Characterization of Isolated Mn^{2+} Ions in ZnS:Mn Thin Film

H. Uchiike, S. Hirao, M. Noborio, and Y. Fukushima

Faculty of Engineering, Hiroshima University,
Shitami, Saijyo-cho, Higashi-Hiroshima 724, Japan

In the fabrication of AC thin film electroluminescent devices, annealing in a vacuum after the deposition of an emission layer influences the device characteristics. The deviation of composition and a quantity of isolated Mn^{2+} ions before and after the annealing have been investigated in ZnS:Mn thin film. Especially, it is clarified by electron spin resonance analysis that annealing increased isolated Mn^{2+} ions, which contributed electroluminescense. Longer, and high-temperature annealing results in increase of isolated Mn^{2+} ions. Effects of annealing on the composition of ZnS:Mn thin film is also discussed.

1. Introduction

AC thin film electroluminescent(ACTFEL) devices have high luminance and efficiency, and good display quality. However, there are many challenges, such as clarifying the operating mechanism, full color realization , and low voltage drive. Optoelectric characteristics, such as luminous efficiency, threshold voltage, and memory margin of ACTFEL devices depend on the fabrication conditions[1]. Therefore it is very important to clarify the relationship between the fabrication conditions and the film quality, and EL characteristics.

Until now we have investigated film quality of ZnS:Mn concerning the sinter process of evaporation sources and the evaporation process of emission layer. From the results, it became clear that in the sinter and evaporation processes, ZnS reacts with Mn in the evaporation source and MnS and Zn were produced. The composition of ZnS:Mn thin film was not equal to that of evaporation source caused by the evaporation temperature difference of MnS and ZnS[2].

In particular, the vacuum annealing after the emission layer deposition gives the most dominant effects on the performance of the ACTFEL device, such as the decrease of threshold voltage, stable memory margin[3]. Therefore it is necessary to clarify annealing effects on the film quality in order to fabricate the device with good performance.

In the present work, we study annealing effects on the composition and the quantity of isolated Mn^{2+} ions of electron beam evaporated ZnS:Mn thin films.

2. Experimental Method

The composition and the quantity of isolated Mn^{2+} ions of ZnS:Mn thin film were measured by using particle induced X-ray emission(PIXE) method[4] and electron spin resonance(ESR) method. ZnS:Mn thin films were prepared by electron beam evaporation on glass substrate for ESR measurements and carbon substrate for PIXE measurements by electron beam evaporation. They were annealed in a vacuum under various conditions in the heat treatment apparatus illustrated in Fig.1.

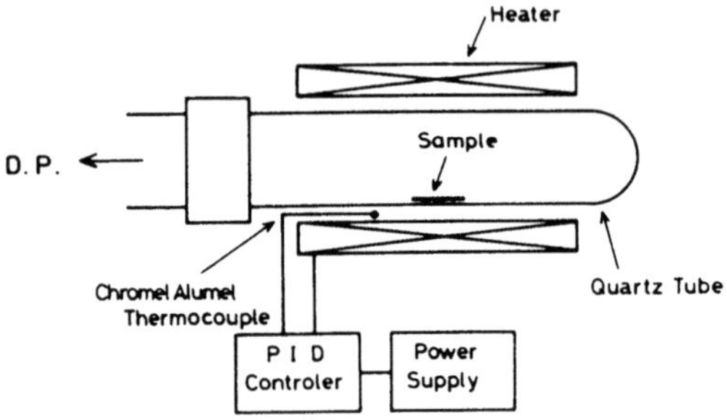

Fig.1 Heat treatment apparatus

3. Result and Discussion

3.1 Composition

In order to clarify the deviation of composition due to annealing, the quantitative analysis was carried out for ZnS:Mn thin film by PIXE. Figure 2 shows the composition ratio of ZnS:Mn thin film after 2 hour annealing. Horizontal and vertical axes indicate the annealing temperature and the composition ratio, respectively. Mn concentration is not shown because it is lower than 0.1mol%. From the figure we find that the composition ratio of ZnS:Mn thin film deviates from the stoichiometry condition at the temperature lower than 600°C and approaches the stoichiometry condition only at 650°C.

A possible explanation of the composition deviation is as follows. Generally, electron beam evaporated sulfide films made

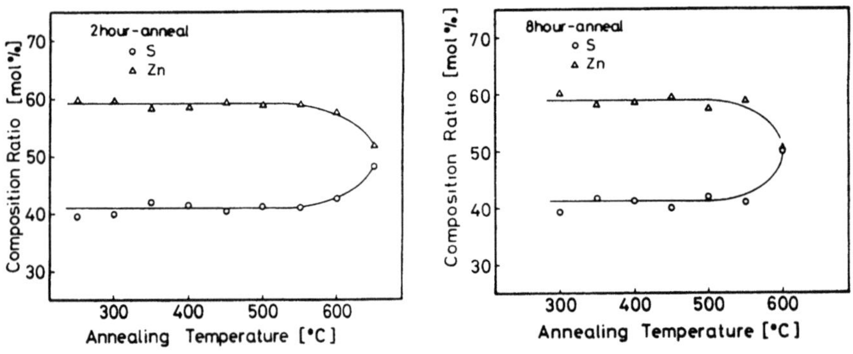

(a) After 2 hours annealing (b) After 8 hours annealing

Fig.2 Composition of ZnS:Mn thin film

of zinc sulfide, strontium sulfide, and calcium sulfide are to be sulfur poor films. Furthermore ZnS reacts with Mn in the evaporation sources due to electron beam bombardment and Zn which is produced in this reaction evaporates. Therefore, the composition ratio of Zn is rich. The deviated composition is constant at temperature under 600°C. Figure 2(b) shows the composition ratio after 8 hours annealing. The stoichiometry condition is realized at 600°C. On the other hand, PIXE analysis showed that a quantity of ZnS:Mn decreased at higher temperature than 600°C. From these results we conclude that in this temperature range, ZnS begins to evaporate from ZnS:Mn thin film and an excess of Zn evaporates at same time.

3.2 Isolated Mn^{2+} Ions

We analyzed ZnS:Mn thin film by using ESR method, where the absorption of microwaves by unpaired electrons with Zeeman-splitting under magnetic field was measured. From the ESR measurements, information of interaction of nucleus and electrons and extent of unpaired electron can be elucidated. Mn^{2+} in ZnS is a subject of ESR measurement, since Mn^{2+} ions doped as emission centers have unpaired 3d electrons.

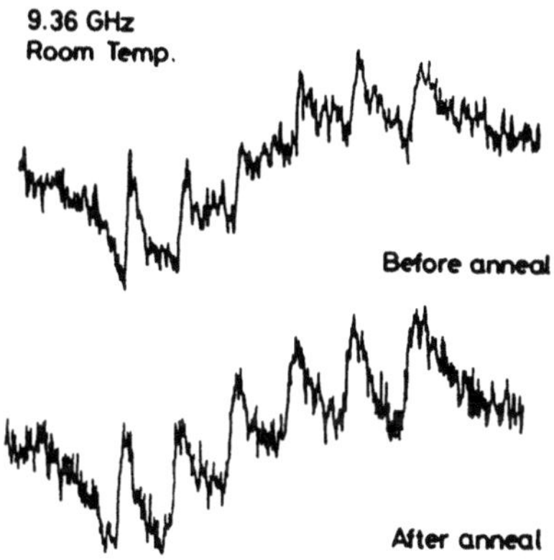

Fig.3 Differential ESR spectra of ZnS:Mn

Figure 4 shows differential ESR spectra of ZnS:Mn thin film. The upper spectra is before annealing, the lower is after annealing. In the figure, six sharp peaks are distinctly shown overlapping a broad peak. These six peaks are the shape specific to Mn^{2+} ion called hyperfine structure and caused by the interaction of electron and nuclear spin. Broad background peak is due to electron exchange between Mn^{2+} ions when Mn^{2+} ions are close to each other. There should exist isolated and clustered Mn ions in ZnS:Mn thin film.

We calculated a relative quantity of isolated Mn^{2+} ions by integrating six peaks. Figure 4 shows the calculated result. In Fig.4(a) a horizontal axis indicates annealing temperature and vertical axis indicates the increase rate of an integrated intensity, Ia/Ib where Ia and Ib are integrated intensity after and before annealing, respectively. Here EL samples were annealed for ten hours. We observe that the increase rate of integral intensity Ia/Ib increases with annealing temperature.

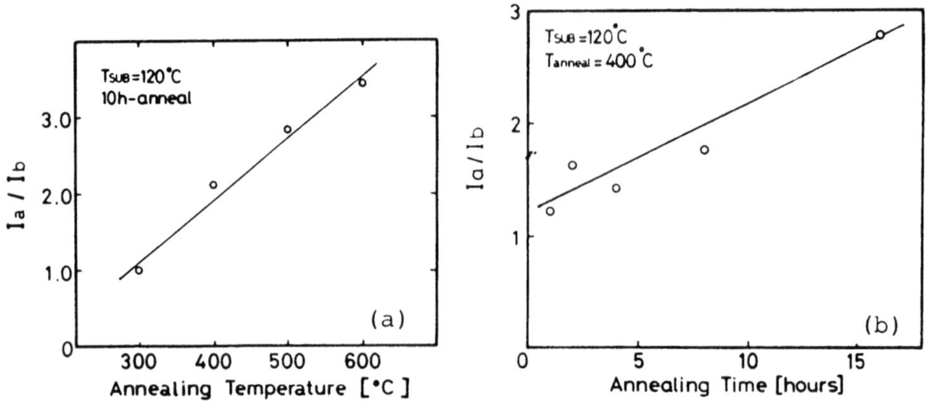

Fig.4 Integrated intensity of six peaks in ESR spectra
(a) Dependence on annealing temperature for 10 hours
(b) Dependence on annealing time at 400°C

Figure 4(b) shows the dependence of Ia/Ib on annealing time with annealing temperature of 400°C. Here Ia/Ib increases with annealing time. From these results we conclude that number of isolated Mn^{2+} ions increases by annealing ZnS:Mn thin film. Longer time and higher temperature anneal cause the increase of Ia/Ib . We consider this effect due to the diffuse of Mn ions from clustered state. Generally the isolated Mn^{2+} ions are more efficient for EL emission than clustered Mn^{2+} ions. So the luminance should increase due to increase in number of isolated Mn ions.

4. Conclusion

From the measurements of PIXE and ESR of ZnS:Mn EL thin films, the following findings of annealing effect are obtained:
(1) With the annealing above 600°C, composition of ZnS reaches the stoichiometry condition.
(2) Post deposition annealing of ZnS:Mn increases the number of isolated Mn^{2+} ions, which is expected to increase EL emission.

References

1. H.Sasakura et al.: J.Appl.Phys,53,5186(1982)
2. H.Uchiike et al.: Applied Surface Science 33/34,661(1988)
3. P.T.Alt et al.: J.Appl.Phys,53,5186(1982)
4. K.Chiba: Radioisotopes,34,648(1985)

Bound-Excitonic Emissions in Undoped and Mn-Doped ZnS Single Crystals

T. Taguchi

Faculty of Engineering, Osaka University, Suita, Osaka 565, Japan

Excitonic emissions observed in undoped and Mn-doped ZnS single crystals with cubic modification have been investigated by means of the temperature dependence of photoluminescence intensity and the doublet structure. Reflectance spectrum at 4.2 K indicates that an exciton dip appears at 3.803 eV and the longitudinal-transverse splitting energy of free exciton is about 4 meV when $4\pi\beta$ is 1.05×10^{-2} and ϵ is 8.1, respectively. Thermal dissociation energy of the free exciton is estimated to be about 40 meV. Assuming that the zero-phonon line associated with an Mn emission is ascribed to an isoelectronic-trap bound exciton, the doublet structure and thermal dissociation process are reasonably interpreted.

1. INTRODUCTION

Studies on excitonic luminescence of high-purity ZnS bulk single crystals and MO-CVD-grown thin films, under the optical pumping condition capable of band-to-band excitation, have been extensively carried out by my group at OSAKA University since 1983 /1,2/. The sharp-emission lines due to acceptor or donor bound-excitons are well-known features of the radiative recombination luminescence; they may be used to characterise extrinsic impurities and native defects due to deviations from stoichiometry in the crystals /2/. However, little has been known about the excitonic features in Mn-doped ZnS crystals so far.

This paper describes the characteristics of excitonic emissions observed in undoped and Mn-doped ZnS single crystals excited by Xe-Cl (308 nm) excimer or He-Cd (325 nm) laser. A possibility of the isoelectronic-trap Mn bound-exciton (BE) emission is proposed on the basis of the experimental evidence .

2. EXCITONIC EMISSIONS IN UNDOPED ZnS

In order to understand the behaviour of excitons bound to several kinds of impurities, we have recently studied the reflectance spectrum at 4.2 K for determining the free-exciton binding energy and its position, and have compared it with a 4.2 K photoluminescence spectrum. By and large, there appear free-exciton lines at 3.803 and 3.800 eV, a donor-bound-exciton (I_2) line at 3.789 eV and a Na acceptor-bound-exciton line (I_1^a) /2/ at 3.782 eV at 4.2 K as shown in Fig. 1 (a). The free-exciton line corresponds to a minimum dip in the reflectance spectrum (shown in Fig. (b)), which indicates the longitudinal (L) exciton energy. The broken curve shows the theoretical reflectance spectrum calculated for the 1s exciton using a dipole function ($4\pi\beta=1.05\times10^{-2}$, $\epsilon=8.1$ and $\hbar\Gamma=3.1$ meV), where ϵ is a dielectric constant and Γ is a damping factor. Therefore, the transverse (T) exciton energy E_T is equal to 3.799 eV and then the LT splitting energy is calculated to be about 4 meV. This analysis can yield a dielectric constant $\epsilon_\infty=5.1$ which is a reasonable value.

From the temperature dependence of the emission intensity of the I_{ex} line as shown in Fig. 2, an activation energy for the decrease in intensity is estimated to be about 0.04 eV, suggesting that this value is equal to the dissociation energy in

Springer Proceedings in Physics, Vol. 38 Electroluminescence
Editors: S. Shionoya and H. Kobayashi

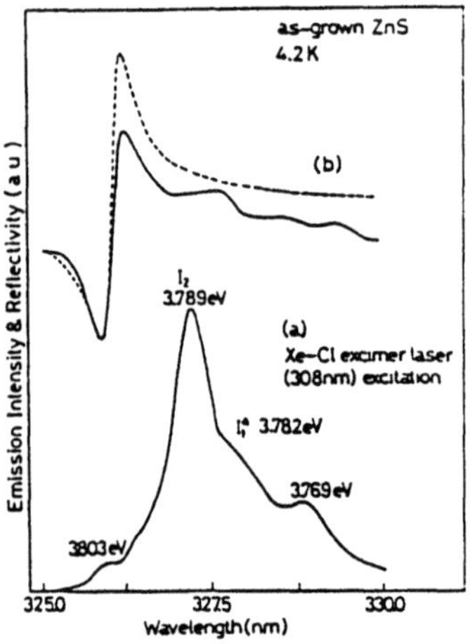

Fig. 1 Excitonic emissions (a) and reflectance (b) spectra of undoped ZnS crystal at 4.2 K. The broken curve is drawn using a dielectric function.

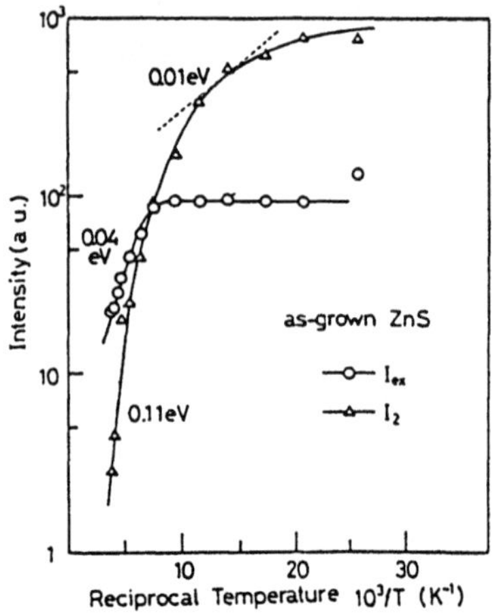

Fig. 2 Thermal quenching of the free-exciton (I_{ex}) and neutral-donor BE (I_2) lines.

which excitons are thermally liberated to electrons and holes. On the other hand, the I_2 line indicates two thermal dissociation processes connected with (i) the thermal release of free excitons from the neutral donor and (ii) with the ionization of the neutral donor which bind excitons. These activation energies are estimated to be 0.01 and 0.11 eV, respectively.

3. Mn EMISSION BAND

It is interesting to know about the energy level of the ground-state of Mn 3d core electrons. We have measured the X-ray photoelectron spectroscopy (XPS) of Mn 3d level as a function of Mn concentrations in the range of 1 to 50 wt%. Fig. 3 shows the XPS spectrum of a ZnS crystal doped with Mn at the concentration of 1 wt%. It is noted that at 1 wt% the Mn zero-phonon line can be barely detected. It is therefore understood that the Mn 3d level is located at about 5 eV below the Fermi level at entire concentrations of Mn impurity. Since the as-grown undoped ZnS:Mn crystal from the iodine-transport method is usually high-resistive, it is reasonable to assume that the Fermi level is situated at about 1.8 eV ($\sim E_g/2$). So, we expect that the Mn 3d level is located at about 3 eV below the valence band.

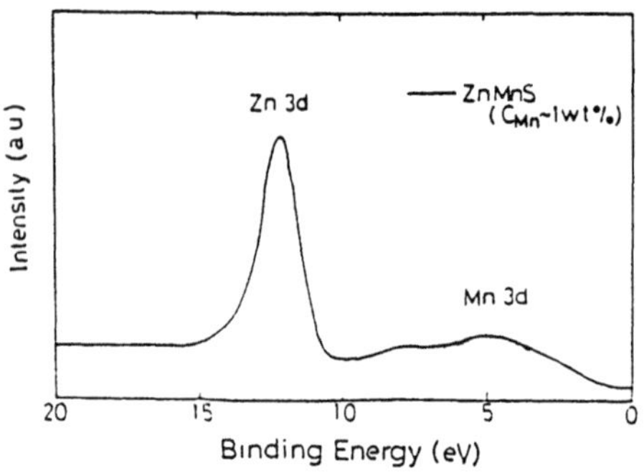

Fig. 3 XPS spectrum of ZnS doped with Mn at the concentration of 1 wt% using Mg K_α (1253.6 eV) X-ray source.

3.1 A Possibility of Isoelectronic-trap Bound Excitons

ZnS single crystal doped with Mn (the concentration of 0.2 wt%) does not give rise to the excitonic-emission lines as observed in Fig. 1, but is dominated by the characteristic Mn orange band associated with TO-phonon replicas as shown in Fig. 4. Kawai and Hoshina /3/ have recently suggested the importance of the ls exciton formation on the Mn emission in the excitation spectra. It is expected here that a neutral Mn should result in a localized level acting as an acceptor-like impurity when considering the electronegativity in its substitutional replacement of Zn and also as a result of an ionization potential smaller than that of Zn.

In Fig. 5, a strong zero-phonon line (a) in the vicinity of 559.0 nm can be seen at 4.2 K. In particular, with increasing temperature the doublet structure (b) clearly appears. The higher and lower energy components are designated as the J_1 and J_2 line, respectively. The energy separation between the J_1 and J_2 lines is then estimated to be about 1.2 meV. Assuming the Gaussianline shape for each spectrum, we can analyze the intensity ratio between them as a function of reciprocal temperature.

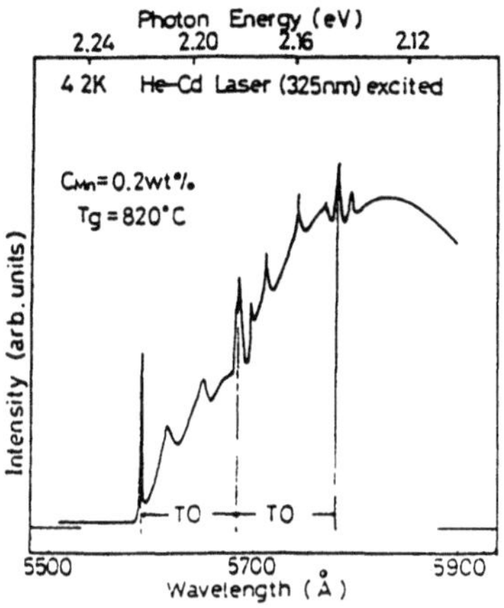

Δ
Fig. 4 Mn emission band in ZnS:Mn cubic crystal at 4.2 K.

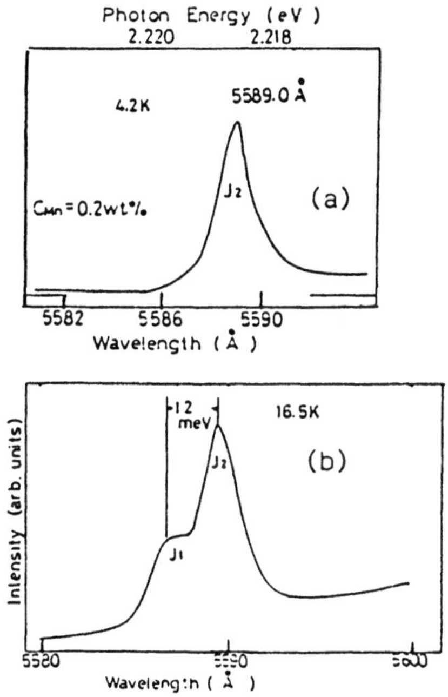

▷
Fig. 5 Enlarged zero-phonon line observed at 4.2 K (a) and 16.5 K (b).

Fig. 6 shows the temperature dependence of the intensity ratio(J_2/J_1). From the slope of the straight line, an activation energy is calculated to be about 1 meV which corresponds to the energy separation between the J_1 and J_2 components. With increasing temperature the J_1 component becomes dominant and then a thermal quenching of the emission intensity (takes place). From the result described in Fig.7 the activation energy for the decrease in intensity is estimated to be about 44 meV, being nearly equal to the dissociation energy of the ls free exciton which was found in Fig. 2.

3.2 Proposed Model

The doublet structure may originate from J-J coupling scheme in which Γ_6 electron and Γ_8 hole of angular momentum j=1/2 and j=3/2, respectively, are combined to

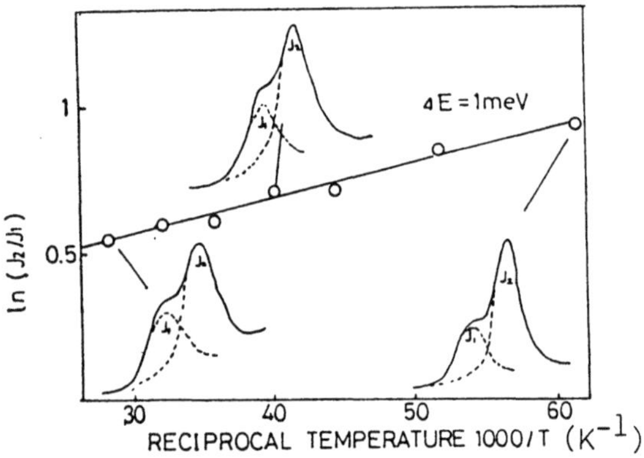

Fig. 6 J_2/J_1 ratio as a function of reciprocal temperature.

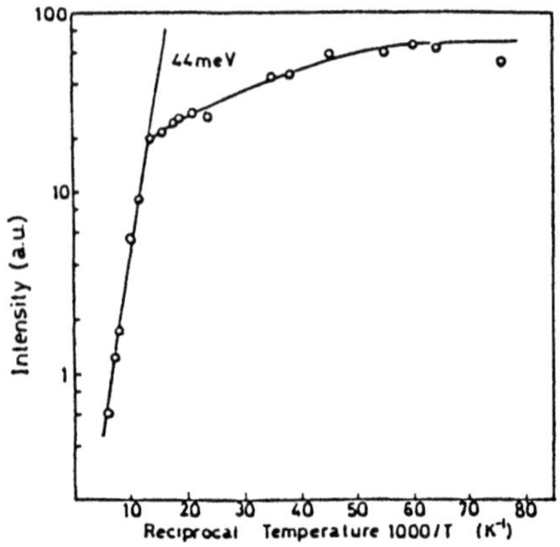

Fig. 7 Thermal quenching of the Mn zero-phonon line.

form J=1 and J=2 states. The optical transition from J=2 to J=0 is dipole-allowed by a perturbation due to the crystal field. The doublet splitting energy of about 1.2 meV can be explained by the spin-exchange interaction between electrons and holes trapped at the Mn impurity, as suggested in the isoelectronic-trap BE model of Fig. 8. This energy separation can be theoretically calculated /4/ and is in fairly good agreement with the value of about 2 meV obtained by Bir et al /4/. The temperature dependent behaviour seems to be very similar to those observed in ZnTe: O and GaP:N /5/.

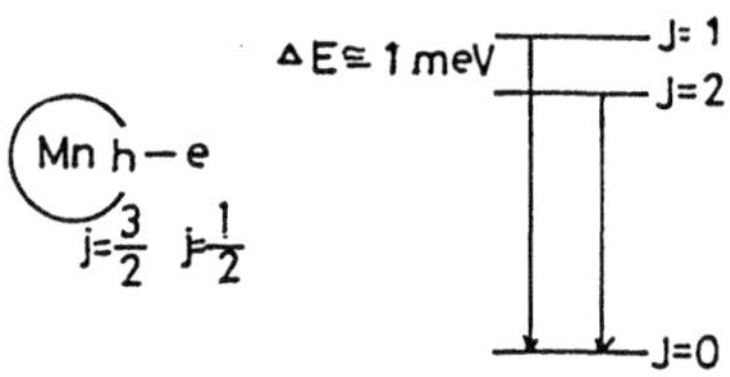

Fig. 8 The proposed model of the Mn emission.

The Mn 3 d ground-state energy level is located at about 3 eV below the valence band in ZnS and may correspond to the $^6A_1$ level. We believe that the doublet structure is originated from the excited state of excitons bound to Mn impurities, but does not derive from the splitting of $^4T_1$ excited state /6/.

4. CONCLUSION

It is tentatively suggested that there is the possibility of excitons bound to Mn impurities, namely the formation of the isoelectronic bound excitons, and as a result the 1s exciton plays an important role for the Mn emission.

I am indebeted to my former students A.Sawada and T.Ueda for measuring photoluminescence spectra and for growing ZnS crystals. This work was partly supported by Mitsubishi Kasei Industry and also supported by a Grant-in-Aid for Scientific Research on Priority Areas, New Functionality Material-Design, Preparation and Control NO. 62604583, from the Ministry of Education, Science and Culture of Japan.

REFERENCES

/1/. T.Taguchi, T.Yokogawa and M.Yamashita, Solid State Communication 49 550 (1984).

/2/. T.Taguchi et al. The 172 nd Electrochemical Society Meeting, Extended Abstract, vol. 87-2 p. 1748 (1987).

/3/. H.Kawai and T.Hoshina, Solid State Communication 22 391 (1977).

/4/. G.L.Bir, G.E.Pikus, L.G.Suslina and D.L.Fedrov, Sov. Phys. Solid St. 12 2602 (1971).

/5/. T.Taguchi and B.Ray, Progress and characterisation of crystal growth, ed.B. Pamplin (Pergamon Press, Oxford 1983) p.136.

/6/. J.Schulze and H.E.Gumlich, J. Crystal Growth 59 347 (1982).

About the Microstructure of Luminescent Centers in SiO_x and ZnS Films Doped with Tb and Mn Fluorides

N.A. Vlasenko, I.N. Geifman, A.B. Goncharenko, Ya.F. Kononetz, and V.S. Khomchenko

Institute of Semiconductors, Academy of Sciences of Ukrainian SSR, P.O. Box 252650, Kiev, Prospekt Nauki 45, USSR

Rare earth and manganese ions are generally used for the doping of electroluminescent (EL) films such as ZnS, ZnSe, CdF_2 and SiO_x. These ions are introduced commonly in the form of fluorides. However, the partial dissociation of fluorides as well as exchange reactions between the fluorides and basic materials take place during the film preparation. Therefore, besides molecular centers [1], centers with different crystal field symmetry around activator ions are formed. It is essential to establish the model of these luminescent centers because different centers may be unlike in EL characteristics.

Most suitable techniques for the investigation of center microstructures are X-ray photoelectron spectroscopy (XPS), Auger electron spectroscopy (AES), secondary ion mass spectroscopy (SIMS) and electron spin resonance (ESR) spectroscopy. However, these techniques were seldom used for EL films up to now. XPS was used to clarify the microstructure of Tb centers in sputtered ZnS:Tb,F films [2]. Insufficient sensitivity resulting from some difficulties measuring XPS spectra of shallow-energy levels of Tb was found. SIMS including measurements of yields of secondary single-atomic and cluster ions was employed for studies of the center microstructure in EL amorphous $SiO_x:TbF_3$ films [3,4]. However, it is desirable to check the conclusions obtained by SIMS by other techniques.

The purpose of this paper is to use other techniques as mentioned above, namely AES and ESR spectroscopy, for the investigation of the microstructure of luminescent centers in EL films doped by Tb and Mn fluorides. The results obtained by AES for $SiO_x:TbF_3$ films and by ESR spectroscopy for $ZnS:MnF_2$ films are reported.

1. Experimental Methods

Films of 500 nm thickness were deposited by thermal coevaporation of basic material (SiO, ZnS) and TbF_3 or MnF_2 on silicon and glass substrates heated to 150°C. The film annealing was carried out at T_a=350–600°C. The concentration of Tb and Mn in the films measured by EPMA was about 1 mol%.

Auger spectra and the composition (x) of $SiO_x:TbF_3$ films were studied by JAMP-10S. The electron energy and the current density in the beam were 10 keV and 10^{-6} A/cm^2, respectively. Films were etched by Ar^+ beam. Its energy was reasonably low to avoid some structural disturbance in the films. It was found that the Auger spectra of the films etched at the Ar^+ beam energy of ≤2 keV have not any distinctions. Si, SiO_2, Tb and TbO_y (y=1–1.5) were used as etalons.

ESR spectra of $ZnS:MnF_2$ films were investigated by a radiospectrometer RE-1307 of 3-cm range with high frequency modulation (100 kHz) of magnetic field.

Springer Proceedings in Physics, Vol. 38 Electroluminescence
Editors: S. Shionoya and H. Kobayashi

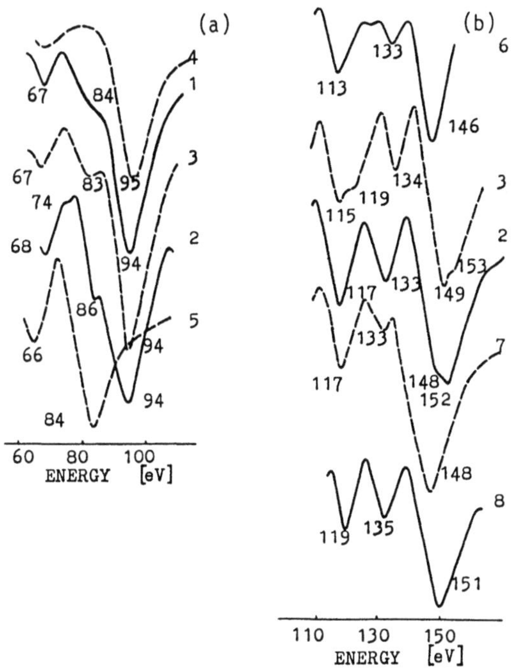

Fig. 1 Peaks of (a) Si and (b) Tb in Auger spectra from SiO_x(1) and $SiO_x:TbF_3$(2,3) films as-deposited (1,2) and annealed at 550°C (3) and from etalons: Si(4), SiO_2(5), Tb(6), TbF_3(7) and TbO_y(8).

2. Results and Discussion

The figure shows typical Auger spectra obtained from SiO_x and $SiO_x:TbF_3$ films and from etalons for the $L_3M_{2,3}M_{2,3}$, $L_3M_1M_{2,3}$ and $L_2M_1M_2$ transitions in Si and for the $N_{4,5}O_{2,3}O_{2,3}$, $N_{4,1}O_{2,3}O_{2,3}$ and $N_{4,3}N_{6,7}N_{6,7}$ transitions in Tb. There are some differences in the shape and energy of the Si and Tb peaks for the SiO_x and $SiO_x:TbF_3$ films as-deposited and annealed. It follows that microstructural rearrangements proceed during the doping and annealing both in the matrix itself and in the neighbourhood of Tb ions. The comparison of the Auger spectra of the $SiO_x:TbF_3$ films and etalons shows that in the former there are the peaks characteristic not only for TbF_3 (116 and 148 eV) but also for Tb-O complexes (119 and 152 eV). The Tb-O peaks are intensified after annealing at $T_a \geq 500$ °C. These results confirm the conclusions made in [3,4]:

i) TbF_3 molecules are dissociated during the film preparation;

ii) F atoms are incorporated in the matrix forming Si-F bonds (see new peak at 74 eV on curve 2) and Tb atoms substitute Si atoms in $SiO_{4/2}$ tetrahedra;

iii) the annealing brings about some decrease of the O content (x decreases from 1.0 to 0.9) and the formation of $Tb^{3+}-O_y$ complexes which are luminescent centers.

It was found from Auger spectra that the composition of SiO_x and $SiO_x:TbF_3$ films is the same and the depth profile of x is almost uniform except the thin surface, i.e. ∿10 nm-thick layer, where x is higher (1.6-1.8). X was equal to ∿1 and ∿0.9 for the films as-deposited and annealed at 500-600°C. This is in good agreement with x measured by EPMA.

The investigation of ESR spectra of $ZnS:MnF_2$ films has shown that for the films annealed at ≥500°C the spectra are the same as those of ZnS:Mn films, if the Mn concentration in the films is equal. The ESR spectrum in this case consists of five lines with the fine splitting constant $A= -(64.2\pm0.8)\cdot10^{-4}$ cm^{-1} and $g=2.005\pm0.002$. Hence most of MnF_2 molecules are dissociated at such annealing temperatures, and Mn centers with the same crystal field symmetry as in ZnS:Mn (T_d) are formed. It is consistent with the fact that EL characteristics of such

$ZnS:MnF_2$ films and ZnS:Mn films are identical. Their EL spectra consist of one orange band (λ_{max}=583 nm), and EL decay time (τ) is ∿1.3 ms.

However, ESR spectra and EL characteristics of $ZnS:MnF_2$ films as-deposited and annealed at $T_a \leq 350°C$ have certain features. Thus, new peaks appears in ESR spectra. The red band in a range of 650–750 nm arises in EL spectra simultaneously. The decay time in these bands is more smaller than in the orange band (τ_r=0.5–0.6 ms). It follows that molecular MnF_2 centers predominate in such films. The crystal field symmetry around Mn^{2+} in these centers is lower (C_{2v}) than that around Mn^{2+} in ZnS:Mn.

References

1. D. Kahng: Appl. Phys. Lett. 13, 210 (1968)
2. J. Mita et al.: Jap. J. Appl. Phys. 26, 1558 (1987)
3. P.I. Didenko et al.: Phys. stat. sol. (a)100, 501 (1987)
4. N.A. Vlasenko et al.: J. Luminescence 40/41, 792 (1988)

On the Stability of Rare Earth Centers in II–VI Compounds

D. Hommel[1], *H. Hartmann*[1], *F.J. Bryant*[2], *M.J.R. Swift*[2], *W. Busse*[3], *and H.-E. Gumlich*[3]

[1]Zentralinstitut für Elektronenphysik, Hausvogteiplatz 5–7, 1086 Berlin, GDR
[2]University of Hull, Department of Physics, Hull HU6 7RX, Great Britain
[3]Technische Universität Berlin, Institut für Festkörperphysik, Hardenbergstr. 36, D-1000 Berlin 12

After a brief review of the results obtained on ZnS:Sm samples by site selective spectroscopy, different excitation channels and the non-linear optical behaviour as a function of the excitation density will be described as a background for the following discussion of the unexpected of rare earth (RE) sites in respect to host lattice damage. It will be shown that after an Ar implantation the ZnS emission bands are reduced drastically, whereas the RE luminescence is nearly unchanged. From cathodoluminescence (CL) and dye laser spectroscopy it will be concluded that the dominant RE sites in a II-VI lattice form stable complexes.

1. INTRODUCTION

RE ions incorporated into various II-VI compounds are well studied luminescence centers emitting in the whole visible spectral region /1,2/. Unfortunately, until now only ZnS:Tb thin film structures are promising for real device applications due to their brightness and long time stability /3/. One of the key questions is, whether the RE-sites are stable in high field electroluminescence (EL) and how they are incorporated into a II-VI lattice. The picture still common in literature is that of an isolated RE-site (mostly on a lattice site) with some charge compensation. Our results obtained on Sm doped ZnS crystals, made such interpretations questionable /4/. In that paper it was shown that at least 10 different samarium sites are present. Further the RE incorporation and concentration strongly depends on the sample preparation /5/.

2. EXPERIMENTAL and SAMPLE PREPARATION

The results were obtained on ZnS bulk crystals, containing 0.1 mol% Sm and co-doped with Li, grown by the iodine transport method /5/. The site selective spectroscopy was performed using a dye laser (FL 2001) pumped by an excimer laser (EMG 53 MSC) with 18 ns pulses at 308 nm. The pulse energy of the excimer laser was 50 mJ per pulse, resulting in an average excitation power at the sample of 30 mW. The laser power has been varied by a factor of 200.

One half of the sample surface was then Ar^+-implanted with a beam energy of 300 keV, a dose of $2*10^{15}$ cm^{-2}, using a flux of 1 μAcm^{-2}, in order to achieve a serious lattice damage within the protection range of 400 nm. At 300 keV the Ar^+-ions clearly have a sufficient energy to produce displacements within each of the sub-lattices of the ZnS. It has been proved that, using an electron beam energy of 5 keV in CL, the emission came exclusivly from the damaged part of the sample.

3. RESULTS and DISCUSSION

From site selective spectroscopy the following can be concluded:

- All of the 10 separated Sm^{3+} sites are excitable with similar efficiencies directly (4f-4f), whereas only those of them dominate the average emission spectrum (Fig.1), which are excitable via an energy transfer from recombining electrons and holes trapped by donor-acceptor pairs.
- Due to non-linear optical processes (two-photon-processes) even under selective 4f-4f excitation of a given RE-site with an energy <Eg, electrons can be excited to the conduction band. This gives the unique possibility to observe both excitation channels with increasing laser power at the same time and to compare their relative importance for each RE-site (For details see /4/).

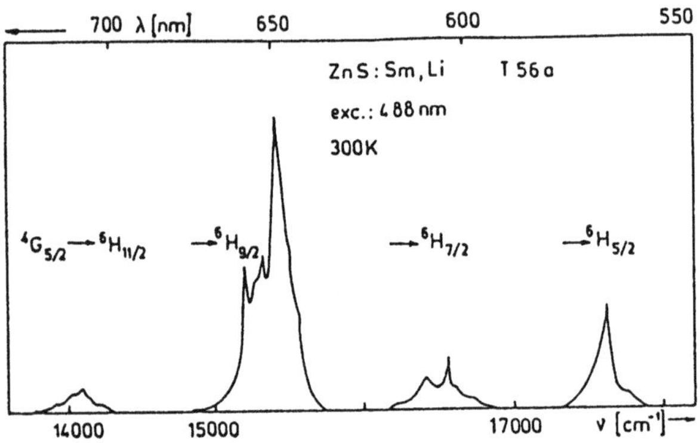

Figure 1: Sm emission spectrum in the ZnS crystal at 300 K.
The dominant transition at 650nm will be discussed below in detail.

Figure 2 illustrates the presence of both excitation channels. Site 1 is excited using a laser power, at which the two step processes also occur - therefore other sites also observed in the integral spectrum (a). Shortly after the laser pulse,the luminescence due to the directly inner shell excited site 1 is the more pronounced (b). In contrast to it , for longer time gates (2-12 ms), an "average" emission spectrum is seen (c). This emission is very similar in line structure and peak intensities to those observed under a band-to-band excitation.

This knowledge will be now used to understand the behaviour of the RE-sites after host lattice damage induced by argon implantation.

The ZnS:Sm crystal has been partly Ar-implanted in order to damage the lattice of the ZnS. The implanted and the as grown part of the sample were studied by cathodoluminescence with low beam energy to ensure that the emission came exclusively from the implanted layer /6/. The results can be summarized as follows:

- The lattice emission (near-band-edge- and blue self-activated-emission) dropped strongly after the implantation. It illustrates that the host lattice has been really damaged.
- In contrast to this, the Sm emission was only slightly reduced.
- No new sites have been created. This has been proved by site selective spectroscopy as well.
- No signs for a re-arrangement between different RE-sites are registered.

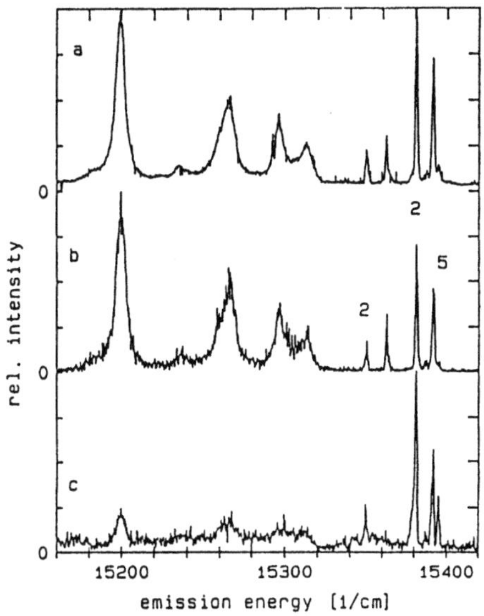

Figure 2:
Time resolved spectra of the Sm emission at a laser power, where both excitation channels are active.
Site 1 is excited selectively. (Sites,2 and 5, are labeled)
a) integral emission spectrum
b) first 10 μs after laser pulse
c) 2 - 12 ms after excitation

It can be therefore concluded that the RE-centers are very stable in respect to changes in the host lattice environment. Such a result is not understandable in a model treating the RE-ions as more or less isolated impurities with some charge compensation within the host lattice. They rather form stable complexes, which are embedded in the semiconductor matrix.

In Fig.3 the luminescence of the implanted and as-grown part of the ZnS:Sm crystal is compared for the most intense Sm transition in the red. Again site 1 is excited selectively. It turns out that the emission of the other sites (2 and 5), which are excited indirectly via the host crystal,is much weaker in the implanted part. In order to confirm this finding, the excitation spectra have been measured for each of the RE-sites. The result for site 1 is illustrated in Fig. 4.

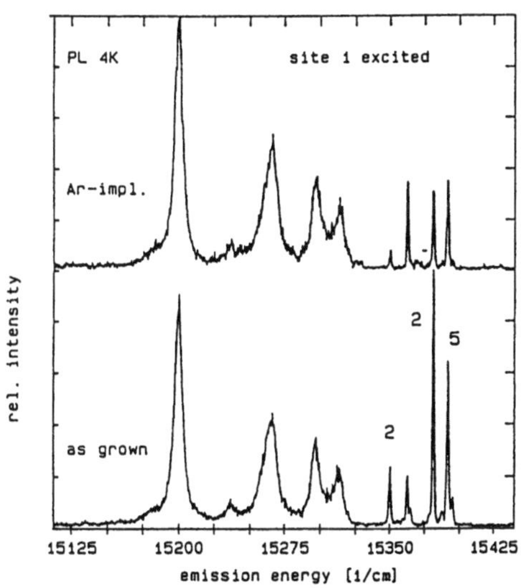

Fig. 3: Samarium luminescence at 4 K. Site 1 excited directly 4f-4f. The emission of other sites (labeled is strong in the as grown sample, but reduced drastically after implantation (upper part)

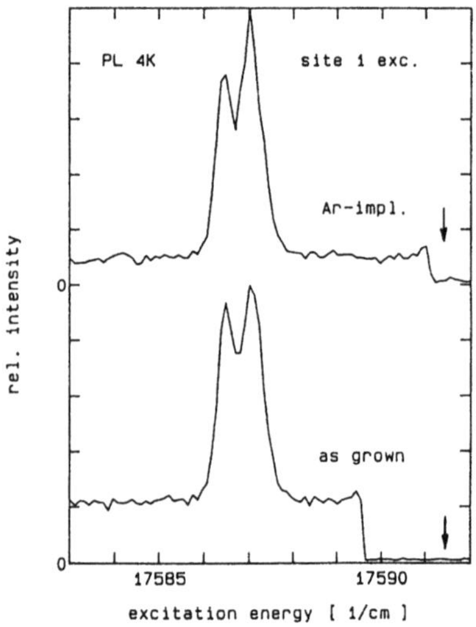

Figure 4:
Excitation spectra of the Sm before and after the ZnS lattice damage.
Arrows indicate the dark current level.

The peak is due to the 4f-4f inner shell excitation of samarium and the broad "background" is connected the band-to-band excitation due to non-linear optical processes described before.

It can be seen that the 4f-4f inner shell excitation of the RE ions is rather unaffected by the lattice damage, whereas the excitation channel via the ZnS lattice is reduced drastically. This confirms the results shown before in Fig.3.

4. Conclusions

Using a growth method, which is close to the thermodynamical equilibrium, RE-centers are very stable in a II-VI lattice und relatively unaffected by changes of the host lattice environment. The dominant RE-sites seem to be therefore stable complexes. Changes in the crystalline environment may vary the excitability of the rare earth ions but do not change the sites themselves. This unexpected high stability of the RE-sites is a promising feature for their application in thin film electroluminescence devices.

5. REFERENCES

1. H. Kobayashi, S. Tanaka, V. Shanker, M.Shiiki, T. Kunou, J. Mita and H. Sasakura: phys. stat. sol. (a) 88, 713 (1985)
2. T. Tohda, Y. Fujita, T. Matsuoka and A. Abe: Appl. Phys. Lett. 48, 95 (1986)
3. K. Okamoto and K. Watanabe: Appl. Phys. Lett. 49, 578 (1986)
4. D. Hommel, W. Busse, H.-E. Gumlich and D. Suisky: Submitted to Journal of Luminescence
5. D. Hommel, H. Hartmann, M. Godlewski, J.M. Langer and A. Stapor: J. Crystal Growth 72, 346 (1985)
6. D. Hommel, M.J.R. Swift and F.J. Bryant: Submitted to Journal of Luminescence

The Relation of Thin Film Electroluminescence and Photoluminescence Excitation Spectra

Zhilin Zhang, Zhuotong Li, Biao Mei, Xueyin Jiang, Peifang Wu, and Shaohong Xu

Department of Materials Science, Shanghai University of Science and Technology, Shanghai 201800, P.R. China

The optical excitation spectra of many materials which have been used in thin film electroluminescence (TFEL) have been measured. It is found that for more efficient TFEL materials they always have some impurity excitation peaks in the long wavelength side. This can be used to select the new material for TFEL. The Mn, Tb have very different behavior in ZnS and CaS TFEL; it can be explained by the lattice symmetry.

1. INTRODUCTION

The Zns:Mn thin film electroluminescence (TFEL) devices have been applied to display systems for several years. Now people pay more attention to multi-colour TFEL devices. Many materials such as ZnS:Tb, ZnS:Sm, CaS:Eu, SrS:Ce [1,2] have been investigated. But the brightness and chromaticity for red and blue color are not quite satisfactory. It is important and urgent to discover some new materials. In this work it is pointed out that there is a relation between the excitation spectra and TFEL properties. We propose that this can be used to select the new materials which might be a candidate for efficient TFEL. These results also can help us to understand the different behavior of Mn, Tb in ZnS and CaS TFEL.

2. EXPERIMENT AND RESULTS

The powder phosphors of ZnS doped by Mn, Tb, Ho, Er, Sm, Ce and CaS doped by Ce, Eu and SrS:Ce were prepared by the conventional method. The excitation and emission spectra were measured by the fluorescence spectrophotometer model 850. Some ac TFEL devices (the insulting layers are Y_2O_3) were made by electron beam evaporation.

The excitation and emission spectra of those phosphors are shown in Fig.1. The arrow above the emission spectra is the wavelength selected for measuring the exc. spectra. The wavelength, energy transition type of the lowest energy excitation peak and published brightness of the relevent TFEL devices are summarized in table 1. It is found that (1) For efficient TFEL materials such as ZnS:(Mn, Tb, Er, HO, Sm), CaS:(Ce,Eu) and SrS:Ce, in addition to the intrinsic excitation band there are some impurity excitation peaks on lower energy side corresponding to the d-d, f-f and f-d transition of the luminescent center. (2) On the contrary for inefficient TFEL materials such as ZnS:(Tm, Ag), CaS:(Mn, Tb) only the intrinsic excitation band can be observed. (3) Mn, Tb have very different excitation spectra in ZnS and CaS. In ZnS:(Mn, Tb) there are some strong impurity excitation peaks, the lowest excitation energy is about 2.5eV. But in CaS:(Mn, Tb) no impurity excitation peak can be observed. The excitation energy is the energy gap of CaS (4.4eV).

Based on these results we can assume that the phosphor which would be an efficient ac TFEL material should have some impurity excitation band, the lower the excitation energy the higher the efficiency of the TFEL. Lower excitation energy facilitates excitation by hot electrons.

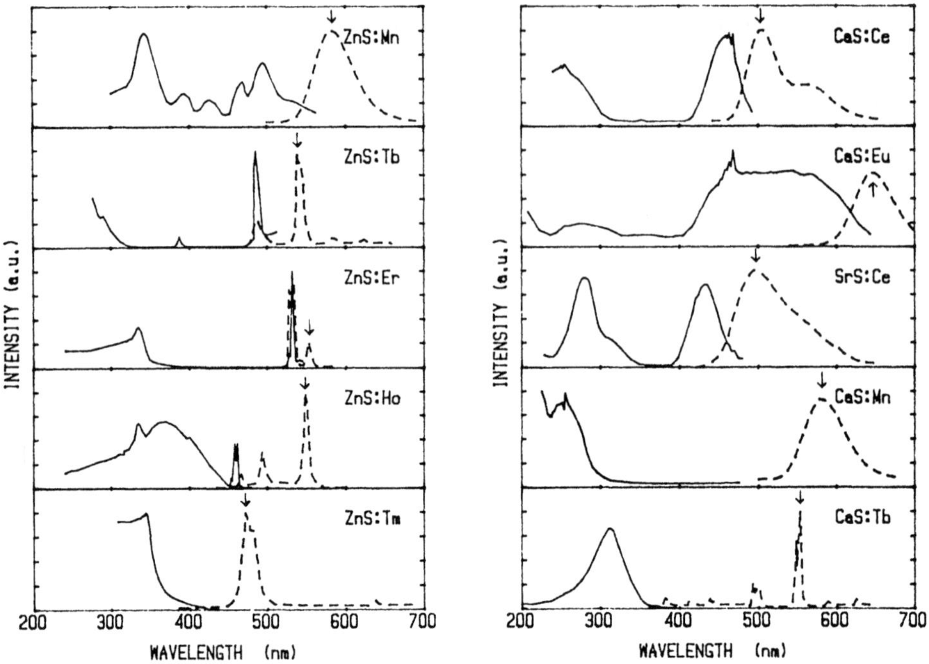

Fig.1. The excitation spectra (full line) and emission spectra (dash line) of different phosphors. The arrows are the monitor wavelengthes for exc. spectra

Table 1. The excitation peak parameters and TFEL device brightness

host material	lum. center	lowest excitation peak			brightness of TFEL (cd/m^2)
		wavelength(nm)	energy(eV)	transition	
ZnS	Mn^2	530	2.34	d--d	4000
ZnS	Tb^3	490	2.54	f--f	2300
ZnS	Er^3	530	2.34	f--f	1100
ZnS	Ho^3	460	2.70	f--f	500
CaS	Ce^3	470	2.64	f--d	650
CaS	Eu^2	450-550	2.25	f--d	170
SrS	Ce^3	440	2.82	f--d	650
ZnS	Tm^3	330	3.76	band edge	3
CaS	Mn^2	280	4.4	band edge	2
CaS	Tb^3	310	4		2

3. DISCUSSION

Even though the TFEL and photoluminescence (PL) are very different, the fundamental processes seem to be connected to each other. The fundamental processes in luminescence can be divided into three steps;the excitation, energy transfer and light emission. Among them the light emission processes are the same, but the excitation and energy transfer processes are different for TFEL and PL. As shown in Fig.2 there are two kinds of excitation processes in TFEL: the direct impact excitation of luminescent center and impact ionization of host lattice. We shall discuss the relation of TFEL and PL in these two cases as follows.

3.1 Direct Impact Excitation

As shown in Fig.3a when an electron with enough energy excites the luminescent center from initial state i to excitation state f, the excitation probability is pro-

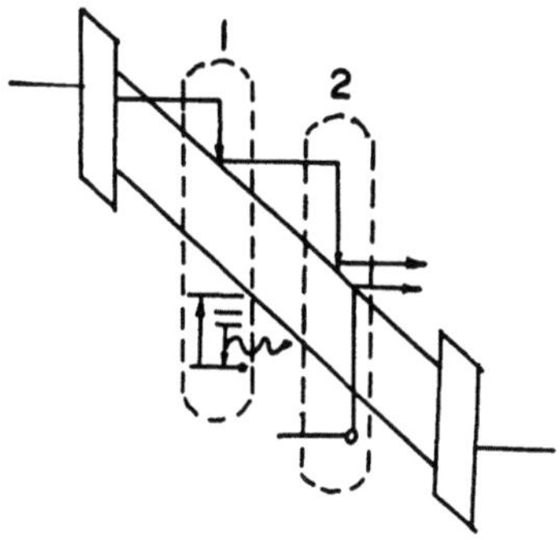

Fig.2 The excitation processes (1) impact excitation (2) impact ionization of the lattice

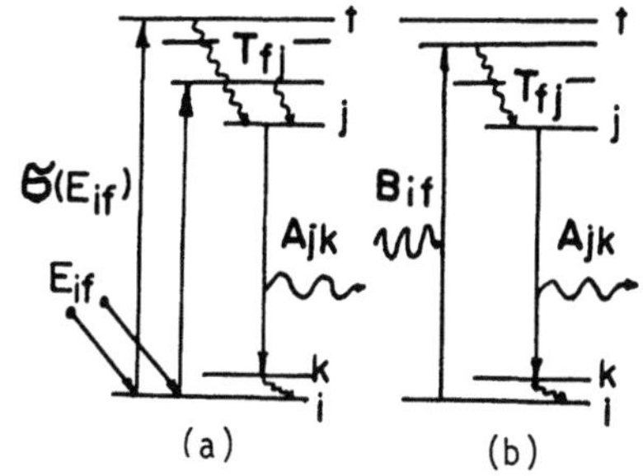

Fig.3 The luminescence processes (a) for TFEL (b) for PL

portional to the excitation cross section $\sigma(E_{if})$ of luminescenct center and energy distribution of electrons in the conduction band f(E). Then they relax from state f to j with probability T_{fj}, and the emission occur with probability A_{jk}, therefore the emission intensity of TFEL has the form

$$I_{jk} \sim \int_{E_{if}}^{\infty} \sigma(E_{if}) f(E) dE \; T_{fj} A_{jk} \tag{1}$$

here E_{if} is the energy difference between i and f states. E is the energy of electron. As deduced by Yu [3] $\sigma(E_{if}, E)$ has the form

$$\sigma(E_{if}, E) = \frac{144\pi^3 m^2 e^2}{n(n^2+2)^2} \frac{1}{E} \ln\left[\frac{E^{\frac{1}{2}} + (E-E_{if})^{\frac{1}{2}}}{E^{\frac{1}{2}} - (E-E_{if})^{\frac{1}{2}}}\right] B_{if} \tag{2}$$

here B_{if} is the absorption probability from state i to state f, n is the refractive index. Substituting (2) into (1) we obtain

$$I_{jk} \sim \int_{E_{if}}^{\infty} \frac{144\pi^3 m^2 e^2}{n(n^2+2)^2} \frac{1}{E} \ln\left[\frac{E^{\frac{1}{2}} + (E-E_{if})^{\frac{1}{2}}}{E^{\frac{1}{2}} - (E-E_{if})^{\frac{1}{2}}}\right] f(E) \, B_{if} dE \; T_{fj} A_{jk} \tag{3}$$

Since the absorption, relaxation and emission processes take place in the luminescent center, they are not influenced by electric field. It can be considered that the parameters B_{if}, T_{fj}, A_{jk} are the same in TFEL and PL.

As for PL the excitation and emission processes are shown in Fig.3b. The monochromatic light is absorbed by luminescent center with probability B_{if} to excite the luminescent center from state i to f, then the center relaxes to j state with probability T_{fj}, and then makes an emission transition to k state with probability A_{jk}. The emission intensity I_{jk} for PL is as follow

$$I_{jk} \sim B_{if} T_{fj} A_{jk} \tag{4}$$

By comparing (3) and (4) it can be seen that the emission intensity of both TFEL and PL are proportional to $B_{if} T_{fj} A_{jk}$. In other words the efficient TFEL material should have large product of these three parameters, and optical excitation spectra could provide very useful information.

3.2 Impact Ionization

In TFEL the electric field may be as high as 2×10^6 V/cm. In such a high field, impact ionization of the host lattice should occur. Although the number of electron-hole pairs created by the field may be very large, the efficiency of EL may not be high. Energy has to be transferred from the lattice to the luminescence centers. The transfer efficiency would decrease since the electron capture cross-section by hole at the luminescence centers becomes smaller due to the increase of electron velocity, and thus the recombination rate decreases.

Luminescence of ZnS:Tm shows that these considerations are correct. ZnS:Tm gives intense blue emission in PL. Its excitation spectrum is a band at 340nm, corresponding to host excitation. This implies that direct excitation of Tm in ZnS is not effective. Energy has to be obtained from the host. It has been shown by Ma Li et al [4] that in TFEL of ZnS:Tm the Tm is excited by energy transfer from the host, in this case ZnS:Tm gives very poor blue emission. Thus it may be concluded that such kind of energy transfer becomes inefficient in TFEL.

3.3 Different Behavior of Mn or Tb in ZnS and CaS

As shown by Tanaka's results [2] Mn and Tb are quite efficient center of TFEL in ZnS, but in CaS they are not. Our experiments also reveal the PL excitation spectra of Mn and Tb in ZnS have some strong impurity excitation peaks, but in CaS no impurity excitation peak was observed. This may be caused by difference of crystal field symmetry for ZnS and CaS. Since the Mn^2 and Tb^3 are d-d and f-f transition respectively. They are parity forbidden for dipole transition. In the case of ZnS (T_d group) the activator ion site has no inversion center, the parity forbiddeness can be more or less lifted. While in the case of CaS (O_h group) the activator ion site has inversion symmetry, the parity forbiddeness can not be lifted. This makes Mn and Tb very difficult to be excited in CaS.

4. CONCLUSIONS

From the relation of TFEL and optical excitation spectra we may say that effective TFEL materials always have some impurity excitation peaks on the long wavelength side. It appears that this may be used as a condition to select new materials for TFEL. Different behavior of Mn and Tb in ZnS and CaS can be explained by the difference of symmetry of ZnS and CaS lattice.

5. ACKNOWLEDGEMENT

This work is sponsored by The Chinese National Natural Science Fundation. Special thanks are due to Prof. Yu Jiagi and Zhong Guozhu for helpful discussions.

REFERENCES

1. H. Kobayashi, S. Tanaka, V. Shanker, M. Shiiki, T. Kunov, J. Mita and H.Sasakura; Phys. Stat. Sol. (a) 88 (1985) p.713
2. S. Tanaka, V. Shanker, M. Shiiki, H. Deguchi and H.KObayashi; SID (85) DIGEST p.278
3. J. Yu and Y. Shen; J. Lumin. 40/41 (1988) p.769
4. L. Ma, G. Zhong and S. Xu; Luminescence and display devices 5 (1985) p.192

The Dependence of Near Band Edge Electro- and Photoluminescence on Purity of Starting Materials in ZnSe Crystals

Xiwu Fan and Jiying Zhang

Changchun Institute of Physics, Academia Sinica, Changchun, P.R.China

The dependence of exciton emission in electroluminescence (EL) and photoluminescence (PL) on purity of starting materials in ZnSe crystals is studied. Three types of ZnSe single crystals with different purity of starting materials grown by sublimation are used in this work. Increasing the purity of starting material of ZnSe crystal, the EL emission bands related to the impurity, such as deep center and free to bound (FB) emissions decrease, but the EL emission bands related to the free exciton increase; and the P band in PL in ZnSe under high excitation density, which is ascribed to interaction between free excitons, appears and becomes intense.

1 INTRODUCTION

ZnSe, with its direct band gap around 2.7eV at room temperature, has the potential to become an important material for the fabrication of blue light emitting diodes[1,2]. In order to reach this purpose, high quality ZnSe single crystal with low concentration deep centers must be prepared for blue light emitting diode application, because the incorporation of such deep centers remarkably affects the luminescence properties, and the colour of luminescence in such crystal easily changes from blue to yellow-orange. Yoneda et al.[3] and Huang et al.[4] have studied the dependence of near band edge PL on crystal quality of ZnSe, which is improved by purification of Se materials used as starting materials. Some of our earlier work was concerned with the effect of crystal quality on its near band edge emission[5-8].

In this paper effort is devoted to the dependence of near band edge EL and PL on purity of starting materials in ZnSe crystals. For this purpose the starting material of ZnSe crystal is refined by sublimation once and twice. The results of the exciton emission spectra are described.

2 EXPERIMENTAL

Three types of ZnSe crystals with different purity of starting materials were grown by sublimation in our laboratory. These were 408, 413 and 420 ZnSe single crystals with (110) surface. The starting material of 408 crystal was ZnSe powder produced by direct synthesis from 6N-Zn and 5N-Se. The starting materials of 413 and 420 single crystals were also ZnSe powder produced by sublimation once and twice from the starting material of 408 ZnSe single crystal, respectively. The dice of ZnSe single crystal with (110) surface were cleaved and heated in molten zinc to reduce their resistivities to the range of 1 to 10 $\Omega\cdot$cm. They were then polished and etched in bromine in methanol, after which an ohmic contact was made to one of the large area faces with a pellet of indium. A thick (500--1000Å) insulating layer of ZnSe was deposited on the opposite large area face by electron beam deposition. Finally a circular gold electrode, 1mm in diameter, was evaporated on top of the insulating layer.

A nitrogen pulsed laser of Model QJD-9 was used as the excitation source with a peak power I_{max} of $3MW/cm^2$, when PL in ZnSe crystal was measured. The ZnSe MIS diodes were excited by direct current in forward bias, when EL in ZnSe diodes was measured. The spectral distributions of the EL and PL were measured using a grating monochromator of Model 44W with a RCA-C31034 cooled photomultiplier.

3 RESULTS

Fig.1 shows the EL emission spectra at 77K from 408, 413 and 420 ZnSe single crystals. It is found that by increasing the purity of ZnSe starting materials, both the deep center and FB emission bands decrease, but E_s band, which is associated with free exciton recombination following scattering from free electrons in the conduction band, increases. An important fact should be noticed that the free exciton emission band E_x with zero phonon at 4437 Å(2.794eV) can be resolved for 420 ZnSe single crystal, and E_x-2LO band at 4540 Å(2.731eV) can also be resolved for 413 and 420 ZnSe single crystals. Fig.2 shows the EL emission spectra at RT from 408, 420 ZnSe single and 420* twin crystals. It is obvious that no deep center emission can be observed in 420 single crystal, but there is deep center emission in both 420* twin and 408 single crystals.

Spatial distributions of EL at RT in 408, 420 single and 420* twin crystals were observed with Model XJZ-01 optical microscope. The densely packed spots in EL are observed for 408, 420 single and 420* twin crystals. All the EL spots are pure blue for 420 single crystal, but for 408 single and 420 twin crystals, a few orange-red EL spots disperse among the densely packed blue EL spots at random.

Fig.3 shows the PL spectra at 77K from 408, 413 and 420 ZnSe single crystals. The spectrum from 408 ZnSe single crystal shows a dominant emission band E_s. When the starting materials were purified, the P band appeared and became intense, as shown in Fig.3. Fig.4 shows the PL spectra at 77K from 420 ZnSe single crystal under different excitation densities. It is clear that by increasing the excitation density the P band increases.

Fig.5 shows the PL spectra at 77K in ZnSe single crystal, annealed in molten zinc at 900°C for 0, 80 and 120h, respectively. It is found that the P band increased with an increase in the annealing time.

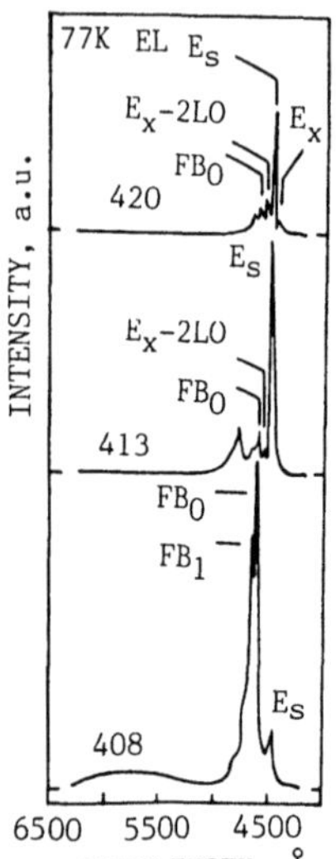

Fig.1

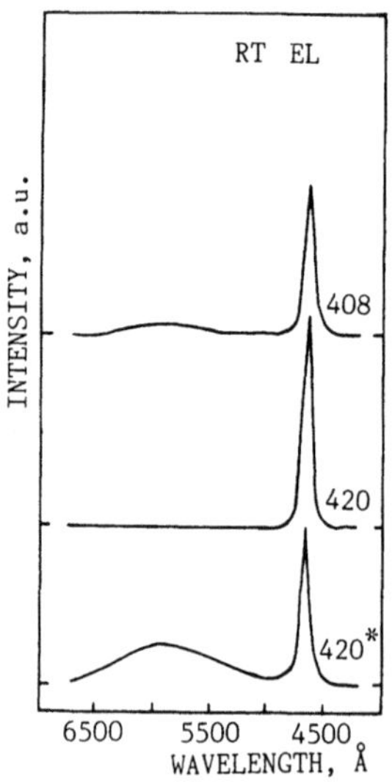

Fig.2

Fig.1 EL emission spectra of 408, 413 and 420 ZnSe single crystals at 77K

Fig.2 EL emission spectra of 408 and 420 ZnSe single crystals and 420* ZnSe twin crystal at room temperature

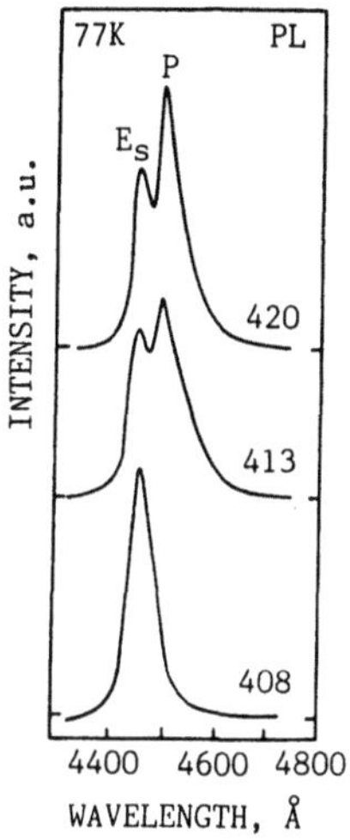

Fig.3 PL spectra in ZnSe single crystals at 77K

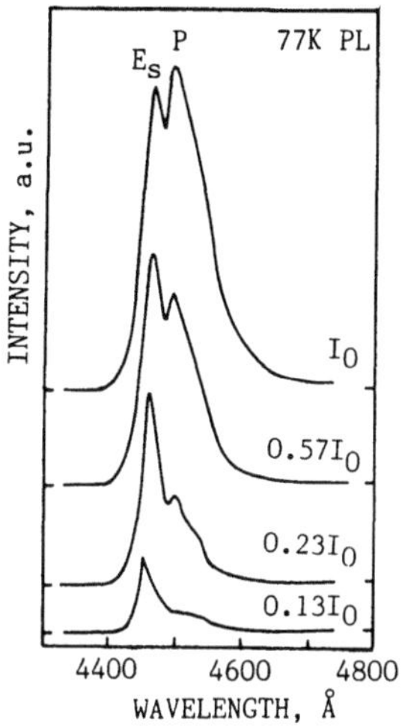

Fig.4 PL spectra in ZnSe under different excitation densities at 77K

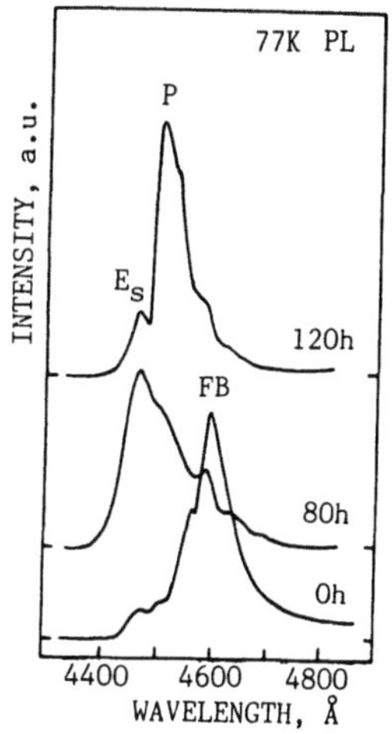

Fig.5 PL spectra in ZnSe for different annealing time at 77K

4 DISCUSSION

In describing the spectral distribution of the EL at 77K in Fig.1, it is convenient to refer to three regions of photon energy[5]. The first of these is the high energy range from 2.82 to 2.72eV, which is the region of exciton emission; the second is from 2.72 to 2.47eV, where the so-called edge emission occurs. A series of FB bands, which are equally spaced in energy by 31.5$\pm$0.5meV, are observed.The third is at lower energy where deep center luminescence and self-activated emission in particular would be expected. As mentioned above, when improving the purity of ZnSe crystal, either deep center or FB emssion bands decrease, but the emission bands related to the free exciton increase. On basis of the same consideration the result in Fig.2 can also be explained.

One of the interesting aspects of the work described here is concerned with the nature of the P band in PL in Fig.3,4. Saito et al.[9] attributed the P band, which they observed in PL in ZnSe crystal, to the interaction between free excitons (E_x-E_x). Comparing the results obtained in Fig.3,4 with that obtained by Saito et al. [9], it is reasonable to think that the P band in Fig.3,4 can be attributed to the E_x-E_x interaction. Considering that the high quality of ZnSe crystal favours the free exciton emission, it follows that by improving the purity of ZnSe starting materials, the P band of ZnSe crystal becomes intense under high excitation density. According to the improvment of ZnSe crystal quality with the increase inannealing time[10], the result in Fig.5 can also be explained.

In summary, therefore, it can be concluded that the quality of ZnSe crystal, as well as the free exciton emission in ZnSe can be improved by purification of starting materials.

This work was supported by the National Natural Science Foundation of China.

REFERENCES

1. M. Yamaquchi et al.: J. Appl. Phys. 48(1), 196(1977)
2. J. Iton et al.: J. Appl. Phys. 57(6), 2210(1985)
3. K. Yoneda et al.: Appl. Phys. Lett. 45(12), 1300(1984)

4. X.M. Huang et al.: J. Cryst. Growth 78, 24(1986)
5. X.W. Fan and J. Woods: IEEE Trans-ED ED-28(4), 428(1981)
6. X.W. Fan and J. Woods: J. Phys. C. 14(13), 1863(1981)
7. Z.K. Tang, X.W. Fan and W.Z. Li: Chinese J. Lumin. 7(4), 349(1986)
8. J.Y. Zhang, X.W. Fan and B.J. Yang: Chinese J. Lumin. 8(2), 100(1987)
9. H. Satio and S. Shionoya: J. Phys. Soc. Japan 37(2), 423(1974)
10. J.Y. Zhang and X.W. Fan: J. Lumin. 40 & 41, 798(1988)

Photoluminescence of Zinc-Sulfo-Selenide Single Crystals Grown by Sublimation Method

S.R. Tiong[1], *M. Hiramatsu*[1], *Y. Matsushima*[1], *M. Ohishi*[2], *K. Ohmori*[2], *and H. Saito*[2]

[1]College of Liberal Arts and Sciences, Okayama University, Okayama 700, Japan
[2]Department of Applied Physics, Okayama University of Science, Okayama 700, Japan

Zinc-sulfo-selenide ($ZnSe_xS_{1-x}$) single crystals which can be used in blue luminescent devices are prepared by sublimation method and their photoluminescence spectra are measured. Free exciton luminescence is very weak; but, the bound exciton emissions are very pronounced indicating the presence of substitutional impurities in the form of neutral donors and acceptors. The incorporation of Na or Li is observed to enhance donor-acceptor pair transitions where the DAP band of the Na-doped sample is at 12 meV higher energy than the DAP band of the Li-doped sample. The energies of the luminescence transitions decrease curvilinearly with the increasing ZnSe composition x ($0.80 \le x \le 1$).

1. EXPERIMENTAL PROCEDURE

The crystals were grown by sublimation method (modified Piper Polich method) and their crystal structure and consequently, their respective lattice constants, were determined by x-ray diffraction. Some crystals were annealed in Zn vapor for a day to several days at 900℃ and several were purposely doped with Na or Li during the annealing process. Some of the samples were doped during growth. Luminescence measurements were carried out under excitation by UV light from either an Hg or Xe lamp. The spectra were analyzed by a monochromator HR-1000 and a photomultiplier R562 was used as a detector. The experiments were performed at the liquid helium bath temperature of 4.2 K.

2. RESULTS and DISCUSSION

Figure 1(a) shows the photoluminescence spectrum of as-grown $ZnSe_{0.96}S_{0.04}$. I_2 and I_1^{deep} accompanied by LO phonon replicas are very prominent and no DAP band is present. However, as shown in Fig. 1(b), for the same crystal annealed in Zn vapor, the intensity of the I_1^{deep} line diminishes appreciatively with respect to the I_2 line. The I_1^{deep} line is known to be due to the recombination of excitons bound to neutral acceptors. The aforementioned acceptors are considered as Zn vacancies (V_{Zn}) or associated defects containing V_{Zn} (V_{Zn}-complex)/1/. The decrease in the intensity of the I_1^{deep} line therefore can be explained by the decrease in the density of the Zn vacancies after heat treatment in Zn vapor. A DAP luminescence band is also observed. The difference between the I_1^{deep} - LO phonon replicas is approximately 31.5±1 meV which can be assigned as the value of the LO phonon energy for $ZnSe_{0.96}S_{0.04}$.

Springer Proceedings in Physics, Vol. 38 **Electroluminescence**
Editors: S. Shionoya and H. Kobayashi

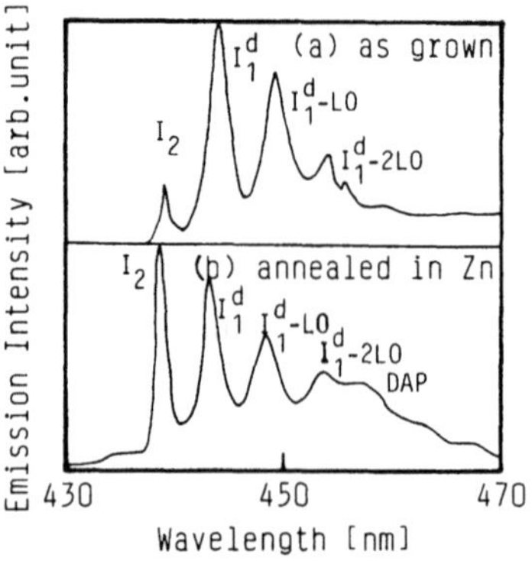

Fig.1 Photoluminescence spectra of $ZnSe_{0.96}S_{0.04}$

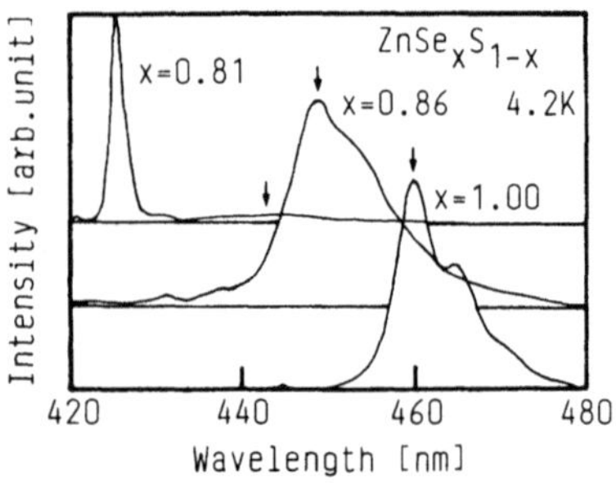

Fig.2 Photoluminescence spectra of Li-doped $ZnSe_xS_{1-x}$ $(0.81 \le x \le 1)$

The photoluminescence spectra of Li-doped $ZnSe_xS_{1-x}$ $(0.81 \le x \le 1)$ samples are shown in Fig.2. For the x = 1 sample, the observed band at 2.696 eV (Q_o) is attributed to the presence of Li, confirming earlier conclusions of MERZ et al./2/. A weak bound exciton line is also present with the 2.696 eV band. This bound exciton is associated with the Li acceptor. The peak therefore is ascribed to be I_1^{deep} which is the luminescence due to the recombination of an electron-hole pair captured by a neutral acceptor. It is very evident in the figure that the emission energy shifts to the lower energy side as the composition of x in $ZnSe_xS_{1-x}$ increases. For the x = 0.81 sample the bound exciton line becomes more prominent with an emission energy of 2.914 eV. A free exciton line is also observed at 2.946 eV. The Q_o-DAP band for x = 0.86 is 2.765 eV.

The effects of Na- as well as Li-doping are evaluated. As shown in Fig.3, no DAP band is present in the as-grown sample. A P_o-DAP band however, is observed in the Na-doped sample at 2.795 eV. For the same ZnSe composition of x = 0.81, a DAP of different emission energy of 2.807 eV is observed for the Li-doped sample. This band is assigned as Q_o which is due to the transition involving Al^+_{Zn} donor and Li^-_{Zn} acceptor which agrees with MERZ et

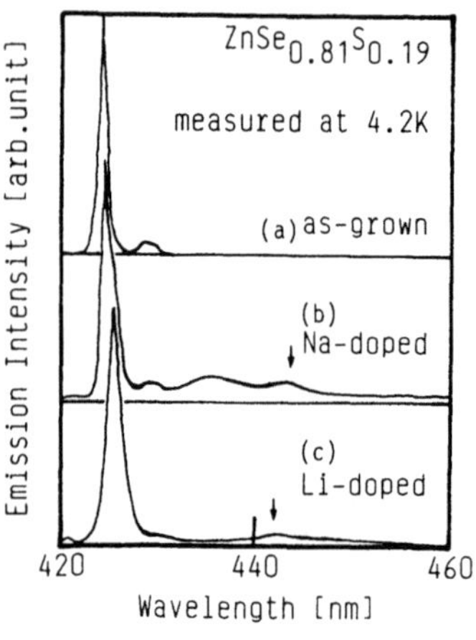

Fig.3 Photoluminescence spectra of $ZnSe_{0.81}S_{0.19}$
(a)as-grown (b)Na-doped (c)Li-doped

al. The presence of Al donor is determined by ion micro analysis. The Q_o-DAP band occurs at 12 meV higher energy than the P_o-DAP band which may be explained by the difference in the acceptor binding energies of Na and Li with Al as the common donor in both samples. I_1^{deep} is strong in all samples which is an indication that the concentration of neutral acceptors is high. E_x which is due to the radiative annihilation of free excitons is barely discernible in the Li-doped sample and its presence is not detected in the other samples. Since all samples have high concentrations of V_{Zn} (acceptors in the high-intensity I_1^{deep}), the excitons bind to the V_{Zn} or V_{Zn}-complex.

The photoluminescence spectra of as-grown $ZnSe_xS_{1-x}$ ($0.80 \leq x \leq 0.96$) is shown in Fig.4. As the ZnSe composition increases, the respective energies of the emission peaks of I_2 and I_1, the DAP band and its corresponding LO-phonon replicas shift to the lower energy side. E_x due free exciton radiative annihilation is not observed in the figure.However, bound exciton lines are remarkably present. For x = 0.93, the relative intensities of the peaks of the donor- as well as the acceptor-bound excitons are very strong with the emission of the acceptor-bound exciton accompanied by LO-phonon processes which result due to strong coupling to the lattice. The LO-phonon energy in this sample is 30±1 meV. For x = 0.91, I_2, I_1, and a DAP band are observed. The broadening of the bands are due to the merger of the LO-phonon replicas of both I_1 and the DAP band. The broadening of the I_1^{deep} in the samples x = 0.96, 0.93, 0.91, and 0.86 can be explained by the presence of a high concentration of acceptors.

The relationship between the respective emission energies and the ZnSe composition in $ZnSe_xS_{1-x}$ is shown in Fig.5. It is clearly seen that the corresponding emission energies decrease curvilinearly with the increasing ZnSe composition x. Strong DAP emission observed for $0.91 \leq x \leq 0.96$ can be explained by the fact that a different preparation procedure for the starting material is adapted; i.e., KCl is used as flux.

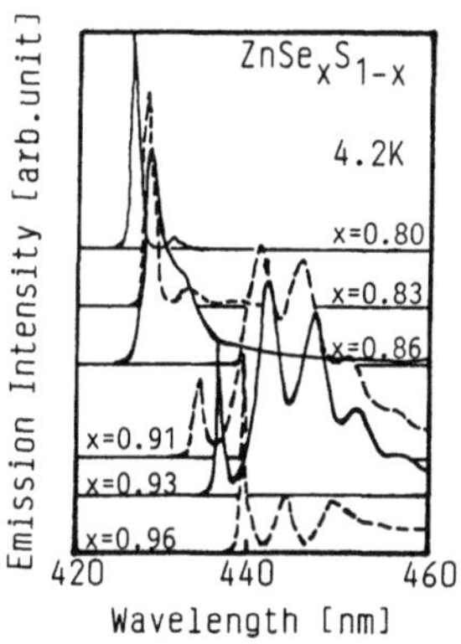

Fig.4 Photoluminescence spectra of $ZnSe_xS_{1-x}$ ($0.80 \leq x \leq 0.96$)

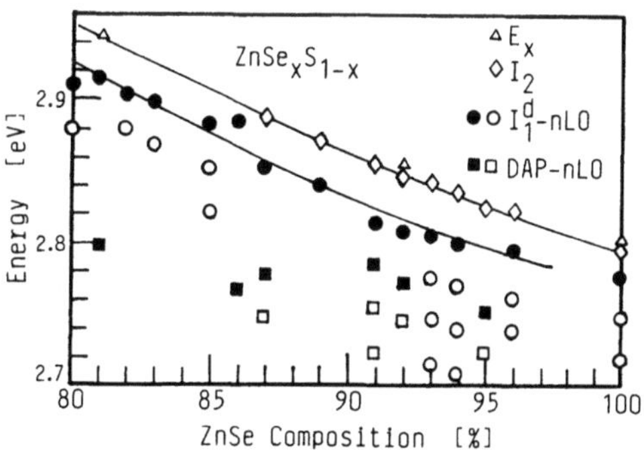

Fig.5 ZnSe composition dependence of emission energies

References

1. P.J. Dean and J.L. Merz: Phys. Rev. 178, 1310 (1969)
2. J.L. Merz, K. Nassau, and J.W. Shiever: Phys. Rev. B8, 1444 (1972)

CdS-ZnS Superlattice Electroluminescent Device Prepared by Hot Wall Epitaxy

H. Fujiyasu[1], *N. Katayama*[1], *H. Yang*[1], *K. Ishino*[1], *A. Ishida*[1], *M. Kaneko*[2], *and T. Ohiwa*[3]

[1]Department of Electronics, Faculty of Engineering, Shizuoka University, 3-5-1 Johoku, Hamamatsu 432, Japan
[2]Research Lab., Koito Manufacturing Co., Ltd., 500 Kitawaki, Shimizu 450, Japan
[3]Technical Research Lab., Hitachi Maxcell Ltd., 1-1-88 Ushitora, Ibaraki 567, Japan

Insulator-CdS-ZnS Superlattice-Insulator Electroluminescent Devices (ISUPIEL's) have for the first time been prepared by hot wall epitaxy and the electroluminescence of the photonenergy(2.0--3.0 eV) has been observed at room temperature. The luminances of the devices are 5 and 50 cd/m^2 for undoped and doped superlattice cases, respectively.

1. Introduction

II-VI wide gap semiconductors held promise for use as blue light emitting devices for a long time. However, this promise has not been forthcoming for the p-n junction type devices. On the other hand, the electroluminescent(EL) devices contacting insulating layers[1] have made steady progress, even though the blue light intensity is not yet intense enough. Compared with the properties of the light emitting III-V diodes, it is noted that the II-VI EL device has very long life(10^4 hrs) in spite of severe use under the high electric fields and photon densities, and the fact that the active layer is grown on an amorphous insulating layer. Many defects and grain boundaries must be introduced in the active layer. The type I superlattice structure where both electrons and holes are confined into a narrower gap layer(well layer) is considered to be a hopeful structure to develope the EL device. Recently we have succeeded in making CdS-ZnS superlattices(SL's) on GaAs(100) substrates by hot wall epitaxy(HWE) and observed strong photoluminescence(PL) associated with the band gap of the SL but weak PL associated with deep level defects at room temperature[2]. The band gap and the lattice mismatches between CdS(2.5 eV, 5.82 A) and ZnS (3.7 eV, 5.412 A) are 1.2 eV and 7.5 %, respectively. For the light emitting materials the larger the band gap difference or the band offset the better in type SL active layers. The large band offset confines many carriers into the well. This not only increases the probability of the direct recombination, but also decreases the probability of those through defect states there. The large lattice mismatch produces high compressive strain on the larger lattice(or narrower gap) layer and supresses the introduction of lattice defects. The ISUPIEL of CdS-ZnS SL was made and the properties will now be described.

2. Sample preparations

Figure 1 shows the HWE system for the CdS-ZnS SL's.First CdS and ZnS films were deposited on the GaAs(100) substrates and the growth rates were determined from optival reflectance measurements. The growth rates were 0.4 - 0.5μ m/h. The source temperature for the doped SL's was 600 C. For all film preparation a flip-flop (growth-interruption) process[3] was used

Springer Proceedings in Physics, Vol. 38 Electroluminescence
Editors: S. Shionoya and H. Kobayashi

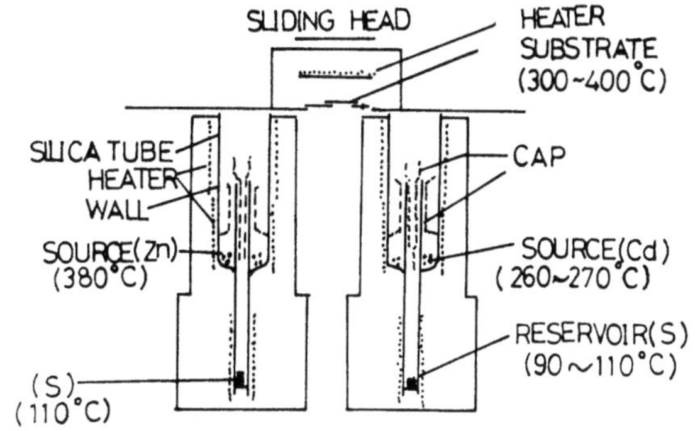

Fig.1 Hot wall epitaxy system for CdS-ZnS SL's. For the preparation of CdS-ZnS(TbF_3 doped) SL's CdS and ZnS(doped) compounds were used as the source materials, put into the source containers without using Caps.

to obtain stable growth rate and flat surfaces of each of the high quality layers of the SL. For EL devices the following two kinds of the EL structure were formed. One was glass/ITO/SiO_2/Si_3N_4/SL/Si_3N_4/Al_2O_3/Al(A) and the other was glass/ITO/Ta_2O_5/SL/Ta_2O_5/AL(B). TbF_3 was used for obtaining intense luminescence from the quantum well due to high excited carriers originated in the excited states of Tb^{3+}.

3. Optical properties

Figure 2 shows the experimental results of the optical characteristics for the ISUPIEL's(undoped,A); CdS(20A)-ZnS(60 A) SL(120 periods) for the sample 5 and CdS(35A)-ZnS(115) SL(120) for the sample 6. Comparing the EL, PL and the transmission spectra with each other, it is seen that the EL(PL) spectra and the optical absorption edge shift to shorter wavelength side for the sample 5 owing to the carrier confinement effect in the quantum well(CdS layer). The barrier layer(ZnS layer) thicknesses are considered to be enough to confine the carriers for both samples. The rather sharp decreases of the luminescences with decreasing the wavelength below 480 nm are due to the loss by the grating used in our instrument. The PL and EL spectra in the photon energy region less than the band gap of CdS(2.5 eV or 500 nm) seems to be due to the emissions through defects and impurities, some of which may be introduced into the SL layers during growth on the amorphous insulating layers. The EL device(B) of thick insulating layer (4000 A) was produced to examine the applied voltage dependence on the

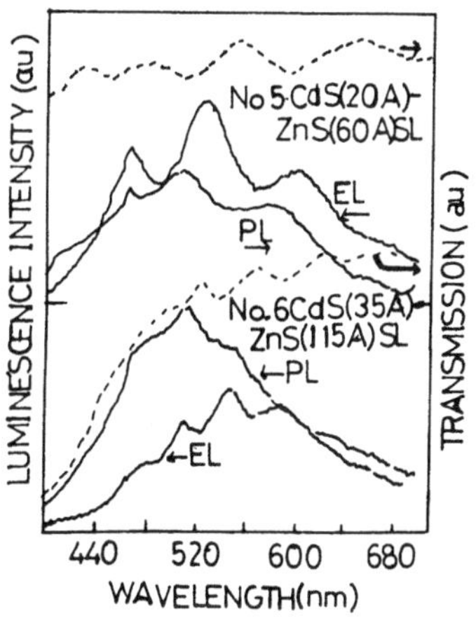

Fig. 2 EL, PL and transmission properties of ISUPIEL's (A). The insulating layer thicknesses are 500 A (SiO_2) and 2000 A (Si_3N_4). EL was measured under the conditions of rms voltage =150 V at 2 kHz and 170 V at 1 kHz for # 5 and # 6, respectively. For PL measurement Hg lamp was used as an excitor. Some humps seen in PL and EL spectra are considered to be due to the interference effect in the luminescent layers. The illuminance of the device(undoped) was 5 cd/m^2.

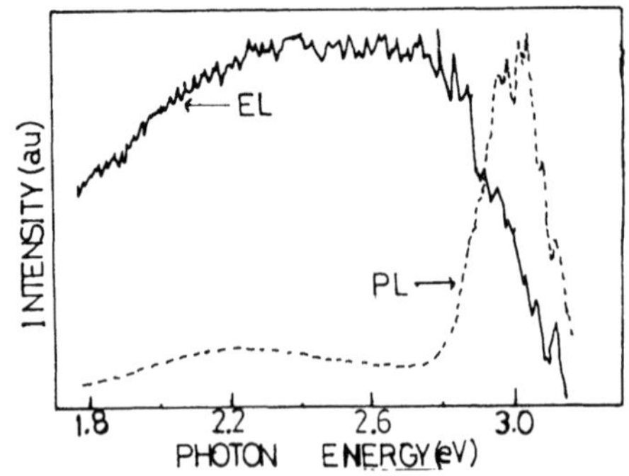

Fig. 3 EL and PL spectra of the ISUPIEL of CdS(10A)-ZnS(30A, TbF_3 doped)(350 periods). EL was measured unter the condition of rms voltage=170V at 5 kHz.

luminance. The increase with the voltage indicates that more excited carriers are required to emit more intense light.
The ISUPIEL of TbF3 doped(ZnS) SL was prepared and the PL and EL results are shown in Fig. 3. The concentration of TbF_3 in the source material was 1 %. In the PL spectra it is very interesting that a very strong emission peak appears in the energy region of 3.0 eV which is nearly equal to the band gap of the CdS(10 A)-ZnS(30 A) SL and that weak spectrum associated with emission from Tb^{3+} or/and with deep level defects appears in the lower energy region. Moreover, it is noted that one shoulder appears in the EL spectrum in the energy region of the strong PL spectrum. Luminescence must occur from the band gap of the quantum well layer of the SL. For EL intense luminescence in the energy region of green to blue is observed. The green(4) must be from Tb^{3+} and blue one may be from Tb^{3+} and the inhomogeneous well width. The illuminance of the device was 50 cd/m^2 and this is one order of magnitude larger than that of the undoped device. A more intense illuminance will be obtained for more highly doped device with protection from water.

3. Summary

Insulator-CdS-ZnS Superlattice-Insulator Electroluminescent Devices have for the first time been prepared by hot wall epitaxy and the electroluminescennes of the photon energy(2.0 - 3.0 eV) has been observed at room temperature. The luminescence from the quantum well has been observed. The luminances of the devices are 5 and 50 cd/m^2 for undoped and doped superlattice cases, respectively. The device will not only be promising for the EL devices which can be easily designed for desired spectrum characteristics but also they are useful for studying EL mechanisms.
Part of this work was supported by a Grant-in-Aid for Scientific Research areas, the Ministry of Education Science and Culture, JAPAN 6264007.

References

1. T. Inoguchi, M. Takeda, Y. Kakihara and M. Yoshida, SID Int. Symp. Dig. Tech. Papers, vol. V, 86, 1974.
2. H. Fujiyasu, T. Sasaya, M. Katayama, K. Ishino, A. Ishida, H.Kuwabara Y. Nakanishi and G. Shimaoka, Appl. Surface Science vol. 33/34, 854, 1988.
3. H. Fujiyasu, A. Ishida, T. Shimomura, S. Takaoka and K. Murase, Surface Science, vol. 142, 579, 1984.
4. D. Kahng, Appl. Phys. Letters, vol. 13, 210, 1968.

An Electroluminescent Device Using Sintered Manganese-Doped Zinc Sulfide Phosphor Ceramics

T. Minami, T. Nishiyama, S. Tojo, H. Nanto, and S. Takata

Electron Device System Laboratory, Kanazawa Institute of Technology, 7-1 Oogigaoka, Nonoichimachi, P.O.Kanazawa-South 921, Japan

1. Introduction

Recently, we have proposed new electroluminescent (EL) devices using a sintered manganese-doped zinc sulfide phosphor ceramic which acts not only as the active layer but also as the substrate [1]. The initial research clarified that the phosphor ceramic electroluminescent (PCEL) devices had advantages in regard to low cost, low dc voltage driving, breakdown-free operation and high legibility [1,2]. However, it has not been satisfactorily considered to optimize preparation conditions of the PCEL devices. In this paper, we will report the EL characteristics of the PCEL devices using ceramics fabricated by a newly developed method and will also describe some of the aging characteristics of the devices driven under various modes of excitation voltages.

2. Experiments

A cross sectional structure of the PCEL device is shown in Fig. 1. In the newly developed method, a luminescent grade ZnS:Mn phosphor powder was firstly copper-coated in copper chloride ($CuCl_2$) solution. A mixture of copper-coated ZnS:Mn phosphor powder and paraffin binder was moulded into a disk-plate with a pressure of 4 [ton/cm²] using a conventional cold press method. The pressed phosphor plate was heated at below 700 [℃] for 5 hours in vacuum to remove the paraffin binder, and then sintered at 900-1100 [℃] for an hour in the gas mixed argon with carbon disulfide. The thickness of as-sintered phosphor ceramic plate was about 0.7 [mm]. An aluminum-doped zinc oxide (ZnO:Al) transparent conducting film as a top electrode was deposited on the surface of the phosphor ceramic plate by the rf magnetron sputtering method [3]. An Al back electrode was evaporated in a vacuum. PCEL devices were driven by continuous dc, pulsed dc or triangle wave voltages in an atmosphere of nitrogen. The luminous efficiency was measured by using a Sawyer-Tower circuit. Before measurements, all PCEL devices were formed under the condition that ZnO:Al electrode was positively biased by continuous dc voltage.

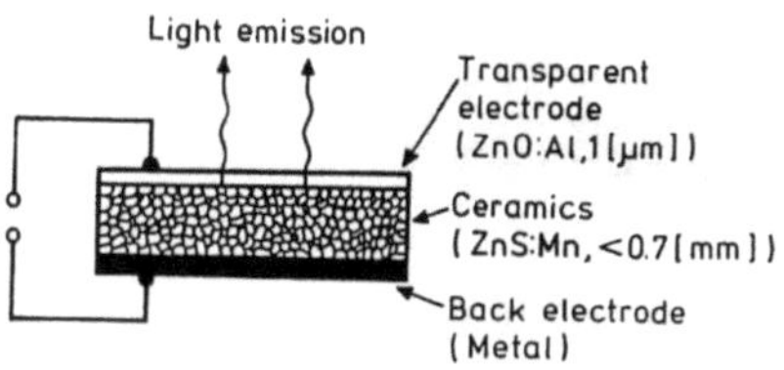

Fig. 1 Cross sectional structure of the PCEL device.

3. Results and Discussion

The newly developed method for the fabrication of PCEL devices had advantages in regard to the simplification of preparation process and the improvement of reproducibility in the device fabrication. However, EL characteristics of the devices were strongly dependent on the preparation conditions of ZnS:Mn phosphor ceramics. It was found that EL characteristics such as the applied voltage (Vth) at onset of a light emission and luminance (L)-applied voltage (V) characteristics were strongly dependent on the content of copper coated on ZnS:Mn phosphor powder in the $CuCl_2$ solution. In this experiment, the content of copper in ZnS:Mn was selected about 0.18 [wt%] in order to obtain devices operating under an applied voltage of about 100 V. On the other hand, EL characteristics were also dependent on the thickness (d) of phosphor ceramics, in the range of 0.2-0.7 [mm]. It was found that EL characteristics are gradually improved with decreasing the thickness and also the Vth is slightly lowered. The L-V and luminous efficiency (η)-V characteristics are shown in Fig. 2 for PCEL devices using ceramics with a thickness of 0.25 and 0.7 [mm]. The improvement of EL characteristics obtained for the PCEL devices using a thinner ceramic may be mainly attributed to the decrease in series resistance of the ZnS:Mn ceramics. In the PCEL devices fabricated by the new method, the maximum luminous efficiency (ηmax) of 2.7 [lm/W] was obtained by the optimization of preparation conditions. Figure 3 shows typical waveforms of the voltage and the corresponding current (I) and luminance of a PCEL device under the excitation of pulsed dc (50 [μs] width, 100 [V]). A capacitive transient current was observed at the beginning and the end of the pulse. As can be seen from Fig. 3, the voltage current luminance relationship observed in this device is similar to that in dc EL powder cells using ZnS:Mn,Cu phosphor powder, reported by Chadha et al. [4]. On the other hand, ac operation may help to understand the mechanism of PCEL devices [5]. The PCEL devices also indicated good maintenance when they were operated under the ac excitation. The L was observed for both polarities of the applied voltage as shown in Fig. 4, whereas the waveform of L changed with the increase in V. It should be noted that a very large dc current flowed into the device when ZnO:Al electrode was negatively biased. The dc current flowing at higher applied voltages regardless of the polarity may be related to V-I characteristics of the formed PCEL devices [1].

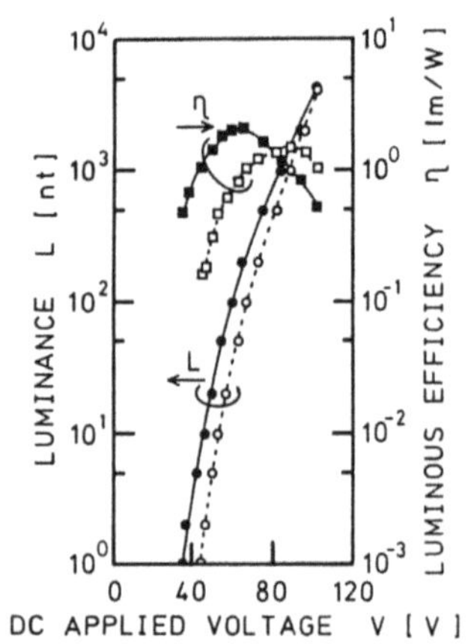

Fig. 2 L-V and η-V characteristics. ●,■;d=0.25[mm], ○,□;d=0.7[mm].

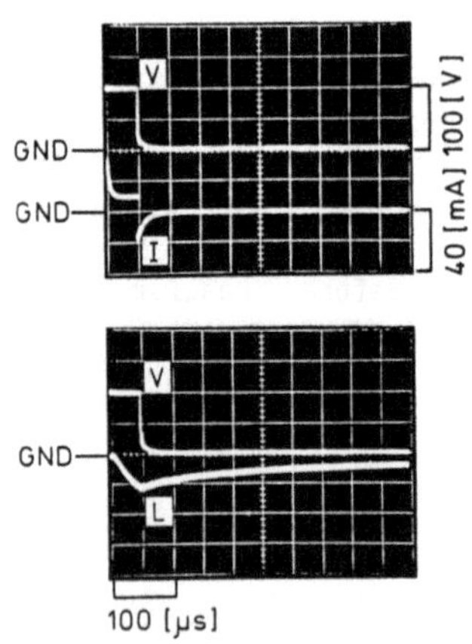

Fig. 3 Waveforms of V, I and L under pulsed dc voltage driving.

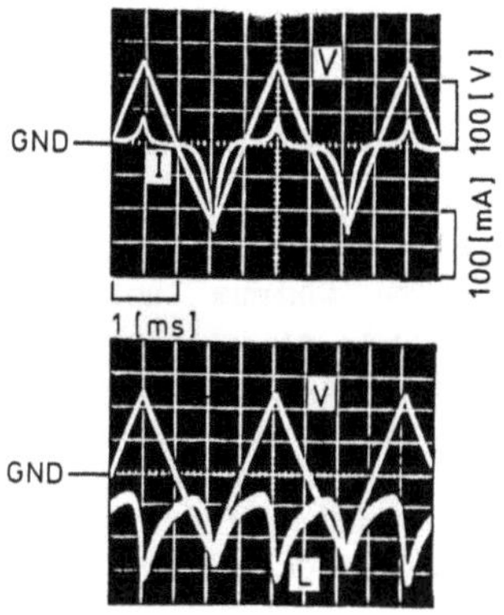

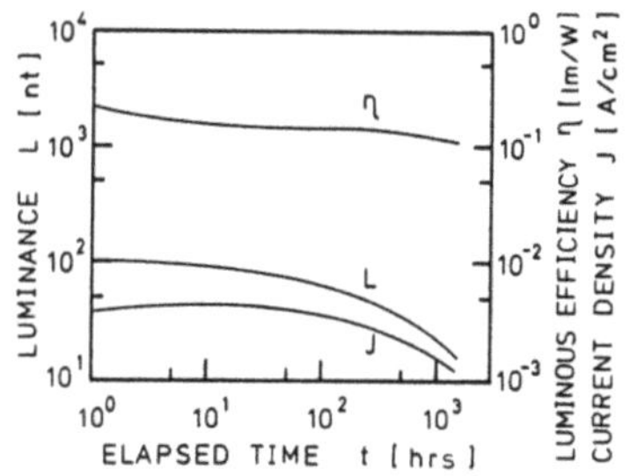

Fig. 5 L, J and η vs. aging time under a continuous dc voltage.

Fig. 4 Waveforms of the V, I and L under ac triangle voltage driving.

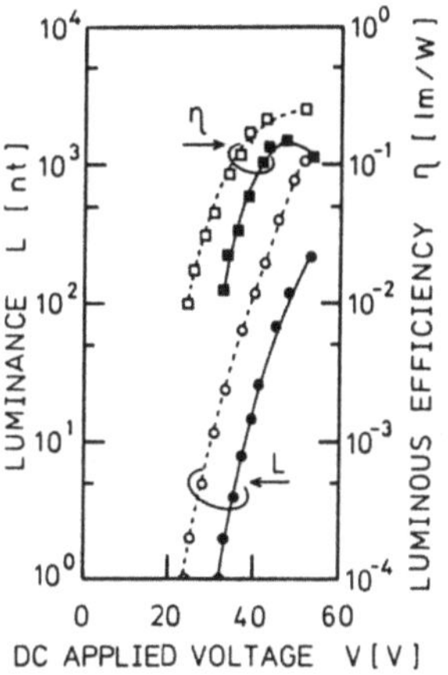

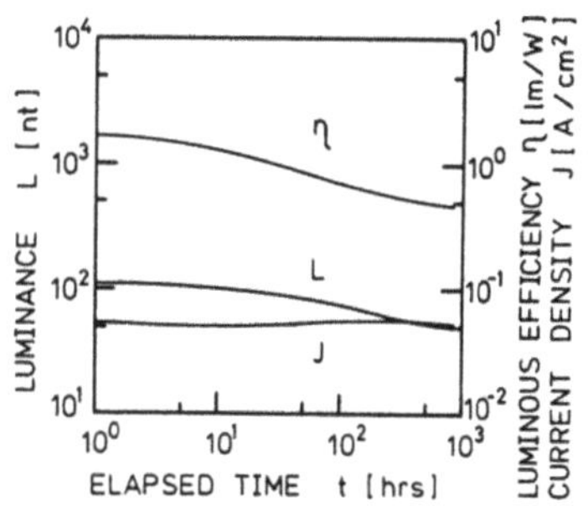

Fig. 7 L, J and η vs. aging time under a pulsed dc voltage driving.

Fig. 6 L-V and η-V characteristics before (○,□) and after (●,■) aging.

Thus, these results suggest that the V-I characteristics of PCEL devices is dominated by the ZnS/Cu_2S heterojunction rather than the bulk effect [6]. Figure 5 shows aging characteristics of L, J and η in a PCEL device driven by a continuous dc voltage of 39 [V]. The L falls from 100 [nt] to 20 [nt] during one thousand hours. However, the η was gradually lowered rather than the L. Figure 6 shows the L-V and η-V characteristics of the device before and after the aging as shown in Fig. 5. It was found that the L-V and J-V characteristics after the aging are shifted to higher voltage sides as compared with those before the aging. This result suggests that the degradation of the L may be caused by the further forming described by Alder et al. [6]. It may be, therefore, expected that the degradation in the PCEL devices is improved by driving under a pulsed dc voltage. It was also observed that the η of PCEL devices driven by a pulsed dc voltage was always larger than one of the devices driven by a continuous dc voltage. Figure 7 shows aging characteristics of L, J and η in a PCEL device under the excitation of pulsed dc (10 [μs] width, 1.0 [%] duty, 118 [V]). It can be seen that the EL characteristics gradually degrade during aging. However, the degradation in the PCEL devices driven by a pulsed dc voltage can not be related to the further forming. In all of PCEL devices driven by a pulsed dc voltage, half the initial luminance was still maintained when operated for 800 [h].

4. Conclusions

The EL characteristics of PCEL devices using ZnS:Mn phosphor ceramics fabricated by the newly developed method have been investigated. The maximum luminous efficiency of 2.7 [lm/W] was obtained by the optimization of preparation conditions of the ceramics. Some of EL characteristics and the aging process of the devices driven under various modes of excitation voltage have been also described. It was found that the V-I characteristics of PCEL devices are dominated by the ZnS/Cu_2S heterojunction rather than the bulk effect. The lifetime of PCEL devices when driven by a continuous dc and pulsed dc voltages was above 300 and 800 [h], respectively.

Acknowledgements

The authors wish to acknowledge T. Kasai, S. Kida, Y. Kosaka, A. Dannoue and H. Morimoto for their technical assistance in the experiments. This work was partially supported by a Grant-in-Aid for Scientific Research No.61550024 from the Ministry of Education, Science and Culture of Japan.

References

1. T. Minami, M. Komano, H. Nanto and S. Takata: Jpn. J. Appl. Phys., 25 (1986) L961.
2. T. Minami, M. Komano, H. Nanto and S. Takata: Proc. of the SID 29/1 (1988) 71.
3. T. Minami, H. Nanto and S. Takata: Jpn. J. Appl. Phys., 23 (1984) L280.
4. S. S. Chadha, C. V. Haynes and A. Vecht: SID International Symposium Digest of Technical Papers, 21 (1988) 35.
5. Y. Fujii and T. Hoshina: Proc. Intern. Display Research Conf., Kobe, 1983 (Society for Information Display, Kobe, 1983) p. 96.
6. C. J. Adler, A. F. Cattel, K. Dexter, J. Kirton and H. S. Skoluick: I.E.E.E. Trans. Electron Devices ED 28 (1981) 680.

Role of Sulfur Vacancies in Luminescence of Pure CaS

P.K. Ghosh and V. Shanker

Division of Materials, National Physical Laboratory,
Dr. K.S. Krishnan Road, 110012-New Delhi, India

Results on EPR and luminescence of pure CaS are discussed in a new light to provide an interpretation that the same sulfur vacancies give rise to both EPR signals and fluorescence emission depending upon the relaxation times of trapped electrons.

1. INTRODUCTON

The alkaline earth sulfides, particularly CaS and SrS, have come into prominence due to the new multicolored thin film electroluminescent (EL) devices using these materials /1,2/. A long series of investigation carried out on CaS, by the present authors, had revealed some interesting aspects of sulfur vacancies. These had begun with the detection of EPR due to a photoexcited electron trapped at a V_S^{2+} and of F^{1+} (or V_S^{1+}) needing no photoexcitation /3,4/. A trap has been revealed in thermoluminescence (TL) also, at a temperature of 226 K corresponding to an electron trap at an energy E_1 of 0.32 eV below the conduction band. This TL peak also corresponds to a photo-electroluminescent peak indicating that probably electrons are more mobile than the holes which are trapped at acceptor levels in pure CaS /5/. TL glow peak analysis revealed Ca vacancies and interstitials in CaS /6/.

The EPR signal has a characteristic activation energy between room temperature and about 90 K of 0.26 eV. This was later correlated with the TL trap, which was estimated to have an escape frequency S of 10^5 sec^{-1}. These values of E_t and S, when used to estimate lifetimes, yielded values close to the decay lifetimes of EPR state, which are very long, of the order of 10-100 secs, justifying the correlation of TL trap and EPR state.

In the present paper, we shall present some results on a 490 nm fluorescence emission of the same pure CaS samples in which indications are strong that this emission also is connected with the same sulfur vacancy centres, which give rise to the EPR signals. A 580 nm fluorescence emission due to F^{1+}, an electron permanently trapped at a Vs^{1+}, is also discussed.

2. EXPERIMENTAL

For details of sample preparation and designation, we would refer to our earlier work /3,4/. It should suffice here to mention that pure CaS was obtained by reduction of CaSO by flowing H_2 or H_2S. CaS H-60, for example denotes a sample which was reduced in H_2 for 60 mts. Any subsequent treatment, such as firing in N_2 with 4 mole% NaCl, X-ray irradiated for 2 hrs or CaS mixed with CaO and fired, are designated as CaS:NaCl, CaS:X-ray or CaS:CaO respectively.

Luminescenece emission was excited by an electron beam using a demountable system described elsewhere /7/. An EHT of 15 KV and beam current in the region of 10-100uA/cm^2 have been used. These beam currents are higher than normally used in CRT's. Nevertheless, the results deserve a close analysis, keeping these limitations in mind.

3. RESULTS

The main result is the thermal activation of fluorescence at 490 nm and its similarity with that of the EPR signal a_0 . The 490 nm emission exhibits an activation energy in the range of 0.26 in a partially reduced CaS to 0.06 in a fully reduced or treated CaS. EPR of all samples, however, exhibits a single value of 0.26 eV. This thermal activation of intensity of both signals, could be expressed in terms of an efficiency as,

$$\eta = I/I_0 = \exp(E_2/kT)/(1 + C^* \exp(-E_1/kT)) \quad (1)$$

Where I_0 is the saturation intensity at low temperatures, E_2 is Boltzmann energy for distribution of electrons amongst the EPR centres and E_1 is the trap depth. A curve fitting procedure reveals that these temperature variation curves, log (I) vs (1/T) plots, for both 490 nm emission and EPR signal a_0 probably belong to the same family of curves of the type of Eq.(1), and therefore results from the same centre. In case of EPR, as shown in Fig. 1, the experimental results (points) for different samples fall on a single curve (solid line)corresponding to $C=2*10^6$, whereas luminesecnce results correspond to values of C ranging from $2*10^6$ to 10^3 .

Further, in CaS:X-ray, the EPR signal a_0 is replaced by a F^{1+} signal associated with a strong luminescence band emission at 580 nm. This emission is nearly temperature independent /7/.

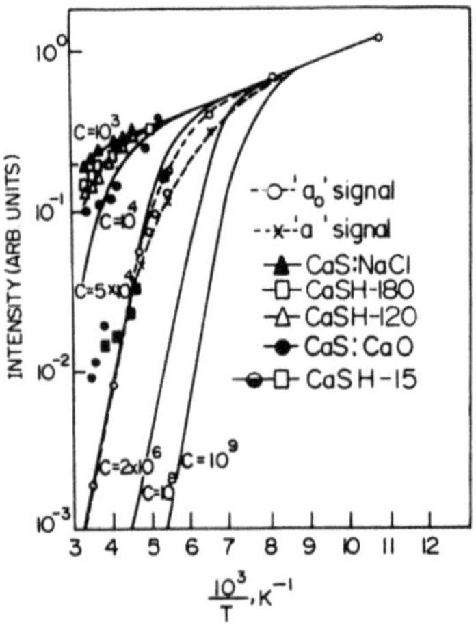

Fig. 1. Temperature variation of EPR signal a_0 and 490 nm emission shown with theoretical plot of Eq. (1) for different values of C for pure CaS

3.1 The 490 nm emission

As shown in Fig.1, both 490 nm emission and EPR signal a_0 exhibit similar thermal activation pattern. There has always been a strong temptation to explain these in similar terms. The EPR centre has been attributed, without any reasonable doubt, to an electron trap by virtue of the g value of 1.9998. This centre was correlated with a TL glow peak at 226 /K/. The difficulty arises, however, when attempt is made to explain the origin of 490 nm fluorescence from the same set of traps, primarily due to the fact that the decay constant of this emission is in us. The decay is exponential and the decay constant, varies with temperature, between RT and 190 K, according to the well known expression for escape probability P of charge carriers from traps or trap lifetime τ ,

$$P = 1/\tau = S \exp(-E_1/kT) \quad (2)$$

E_1 ,is estimated to be about 0.337 eV and the escape frequency S about 10^{11} sec^{-1}. Therefore, it appears at first that, this emission may be originating from a different centre, although the trap depth is nearly the same as for the a_0 centre.

3.2 The 580 nm emission

The EPR signal at a g value of 2.0018, and a corresponding fluorescence at 580 nm , both appearing at the expense of a_0 and 490 nm emission respectively, have been strong evidence in favour of our interpretation that the same centre, i.e., sulphur vacancies V_S^{2+} are responsible for the EPR signals as well as the luminescence emissions. The emission at 580 nm does not show any temperature dependence just like the EPR signal of F^{1+} . The decay constant of the emission is about 1.0 us.

4. DISCUSSION

A simple analysis, following CURIE /8/, for ionic lattices and using recent values of static and optical dielectric constants /9/, yields a thermal trap depth of 0.32 eV for a hydrogen-like trap in CaS. This agrees with the trap depth values obtained for the EPR centre a_0 and 490 nm emission. To explain the different values of escape frequencies, the following model is proposed, which we feel explains most of our experimental observations fairly well.

It is hereby proposed that only one trap is involved both in EPR and luminescence. This is situated at a depth of 0.32 eV below the conduction band of CaS. This trap depth values may be slightly modified for different charged states of V_S , but the analytical methods used here are not sufficiently accurate to distinguish these differences. Signal a_0 is detected when a V_S^{2+}traps one electron excited from the valence band by UV . This electron has a very long relaxation time in this trap, a requirement for an EPR signal to be observed. In the singly charged state, $V_S^{1+}$is now able to trap another electron converting the trap to a neutral vacancy. The second electron, however, has a very short residence time, corresponding to a large escape frequency. The electrons recombine directly to a ground state with a radiative lifetime of about 0.5 ms or more, the decay constant in the saturation region. The efficiency of fluorescence can be expressed as,

$$\eta = \frac{P_r}{P_r + P_{nr}} \qquad (3)$$

where P_r is the probability of radiative transition and P_{nr} is nonradiative transition of the thermally released electrons, which can be expressed by Eq. (2) and is temperature dependent. Eqs.(1) and (3) are essentially the same with the preexponential factor $C= S/P_r$. The increase in the number of neutral traps does not manifest in a reduction of the a_0 signal (EPR) strength due to the very short relaxation time of the second electron in the trap. The first electron is also released, and should recombine with a ground state giving rise to an emission, when the excitation is cut off. But the decay would be long and hyperbolic, like that for a_0 and contribute to the tail portion of phosphorescence. Any luminescence due to the first electron, is likely to be due to transition from the conduction band and in a different spectral region and not in the 490 nm band, which has an exponential decay.

The 580 nm emission is also likely to be a direct transition to a ground state. Since no thermal activation is involved, thermal escape of electrons to conduction band is negligible. In this case also the F-centre is most certainly involved in the emission process as the excited state. Here again, the trapped electron converts the F -centre to a neutral centre, for a very brief duration, making it undetectable in EPR. Further, it seems that X-rays, the primary cause for creation of this centre, are reponsible for inducing other changes, which include. the removal of electrons from surrounding ions thereby forming a permanent trapping site at sulfur vacancies for elctrons. The bleaching temperature of about 600 /K/ of F- centres provides some indications of the charging process induced by X-rays. Fig. 2 provides a model for the mechanism proposed and indicates the different transitions involved in excitation, trapping, thermal release of electrons and recombination. The position of the 580 nm emission centre at a depth of 0.7 eV is based on the energy difference between the 490 nm and 580 nm emissions, which is 0.4 eV. The recombination transitions are of Prener-Williams and Schon-Klasens types.

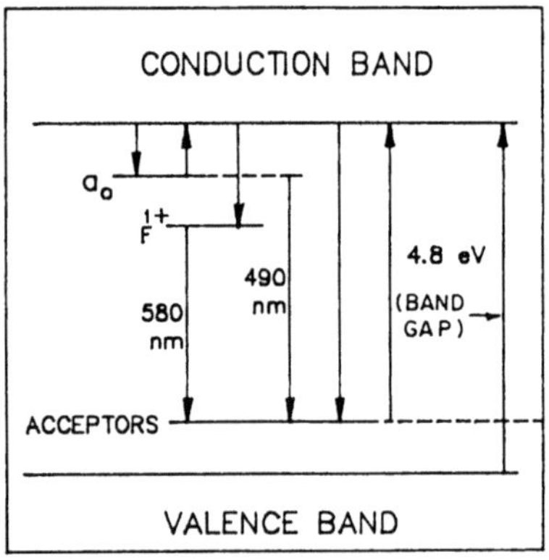

Fig. 2. Proposed model for V_S^{2+} traps in pure CaS acting as a_0 EPR centre, 490 nm and 580 nm emission centres

5. REFERENCES

1. V. Shanker, S. Tanaka, M. Shiki, H. Deguchi, H. Kobayashi and H. Sasakura : Appl. Phys. Lett., 45, 960 (1984)
2. S. Tanaka : J. Luminescence 40 &41, 20 (1988)
3. V. Shanker, P.K. Ghosh & T.R.S. Reddy : Ind. J. Pure & Appl. Phys., 14 193 (1976)
4. P.K. Ghosh and V. Shanker : J. Luminescence 20, 139 (1979)
5. R. Pandey and P.K. Ghosh: Phys. Stat. Sol., 93, 2 (1986)
6. P.K. Ghosh and R. Pandey : J. Physics (C) 15, 5875 (1982)
7. V. Shanker :Thesis, Delhi University (1978)
8. D. Curie : In Luminescence in Crystals (Methuen, London 1963)
9. R. Pandey and J.H. Harding : Phil. Mag.(B) 49, 135 (1983)

Structural Disorders in Gd_2O_2S:Tb Phosphors

V. Shanker, P.K. Ghosh, H.P. Narang, and H. Chander

Division of Materials, National Physical Laboratory,
Dr. K.S. Krishnan Road, 110012-New Delhi, India

The luminescence characteristics of Gd_2O_2S:Tb phosphors prepared under different conditions are described and the possibility of electroluminescence discussed. X-ray diffraction and cathodoluminescence results are used to optimise synthesis parameters. The analysis of diffraction data for reflection intensity distribution from different (hkl) plane shows the presence of a disordered crystal structure. Brightness of luminescence emission is found to depend upon the extent of this disorder. The green cathodoluminescence is primarily due to the characteristic line emission $^5D_4$-$^7F_4$ of Tb^{3+} in the spectral region of 540 nm. Under ac field excitation, the voltage dependence of electroluminescnece (EL) brightness follows the Destriau relation, $B = B_0 \exp(-b/V^{1/2})$.

1. Introduction

Recent investigations of new EL materials have revealed CaS and SrS to be very promising thin film EL materials /1,2/. However, not much has been reported on oxide and oxysulfide phosphors. The rare earths(RE) are known to form a class of highly efficient oxysulfide phosphors. Of this, Gd_2O_2S:Tb is of particular interest because of its widespread applications in CRT displays, solid state lasers and x-ray fluorescent screens /3/. Therefore, synthesis of efficient GOS phosphors was undertaken by us using indigenous raw materials.. We have used sulfurisation flux method which has been frequently used by others for the preparation of Yttrium oxysulfide phosphors /4/. High temperature treatment usually controls the crystal structure along with stoichiometry and defect properties etc.

The chemical composition and structural defects are analysed by powder x-ray diffraction method. To assess the phosphor under different synthesis parameters, luminescence properties are also studied. In this paper, we report a dependence of luminescence of GOS:Tb phosphor on synthesis parameters and the possibility of using it as an EL material.

2. EXPERIMENTAL

For sample preparation, RE oxides of Gd and Tb are mixed with sulfur, sodium carbonate and sodium or potassium phosphate, and fired at 1100 C in flowing nitrogen in a covered crucible. By varying the different conditions such as, temperature, atmosphere, fluxes and Tb concentration etc., a number of samples have been made.

For cathodoluminescence (CL) measurements, the sample is excited by a 6 KV electron beam at 1.0 microamp beam current. A thin phosphor layer, 3-4 mg/sq.cm, is used to reduce charging effects. A pulsed electron beam excitation is used to measure the decay characteristics of Tb luminescence.

For EL, a conventional EL cell is used. The phosphor is embedded in an epoxy resin acting as dielectric medium, between tin oxide coated glass and silver paste electrodes. A variable frequency sine-wave generator and a wide band amplifier formed the ac excitation source for EL and the EL brightness is detected by a photomultiplier (EMI 9658 B).

3. RESULTS

The XRD pattern of each sample is taken using CuK_{α} radiation at 35 KV tube voltage and 10 mA tube current. Two types of XRD patterns of samples are obtained. The samples which have mixed oxides, Na_2CO_3, and S in the ratio 1:1:3 with varying Tb concentrations (0.15 - 3.0 mole %) and fired in nitrogen for one hour, give XRD patterns exhibiting a single phase, corresponding to Gd_2O_2S:Tb (Fig. a). The same samples, when fired in air for extended periods in nitrogen exhibit XRD patterns due to a mixed phase (Fig.1c). XRD of a sample from GTE Sylvania, is given in Fig.1(b).

It is further observed that, in the first type of XRD patterns, there is a variation in the relative intensities of diffraction lines related to planes (101) and (100, 002). The reflection intensity ratios formed three groups with mean values, 2.68, 2.43 and 2.14.

The CL spectrum showing the characteristic emission of TB^{3+} is observed in our GOS phosphors also. In the samples with identical XRD patterns and having reflection intensity ratio I_x/I_y (x denotes the plane (101) and y denotes the planes (100, 002) around 2.14, Tb concentration is varied from 0.15 to 3.0 mole %. The effect of Tb concentration on CL output is shown in Fig. 2. This shows a maximum brightness for an optimum Tb of 2.0 mole %. This brightness is about 60% of that of a GOS:Tb phosphor obtained from GTE Sylvania (type 2611), which has a I_x/I_y ratio of about 2.68. Samples containing mixed phases exhibit low CL intensities.

In the frequency range of 10 to 14 KHz, a bluish EL is observed when 800 V and above is applied. With increasing voltage the brightness increases rapidly until breakdown occurs. The voltage dependence of EL brightness is shown in Fig. 3.

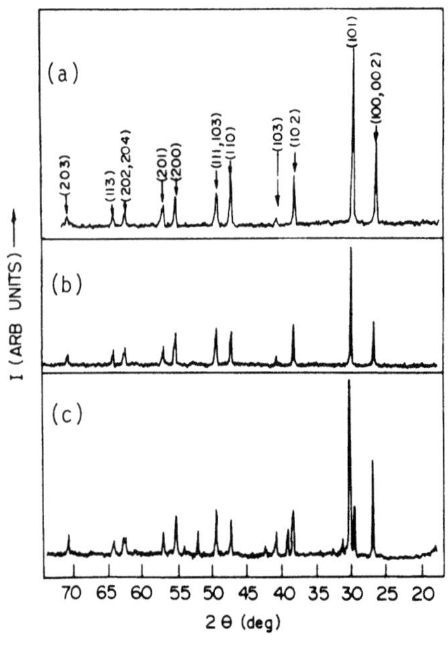

Fig. 1(a). XRD patterns of single phase Gd_2O_2S are obtained for samples prepared by sulfurisation flux method (b). Sample supplied by GTE, Sylvania (c). Sample as in (a) with extended time of firings in N_2 or air

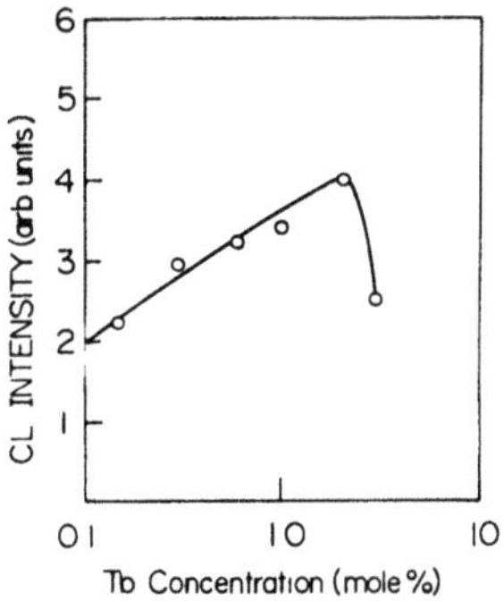

Fig. 2. Variation of CL output with Tb concentration

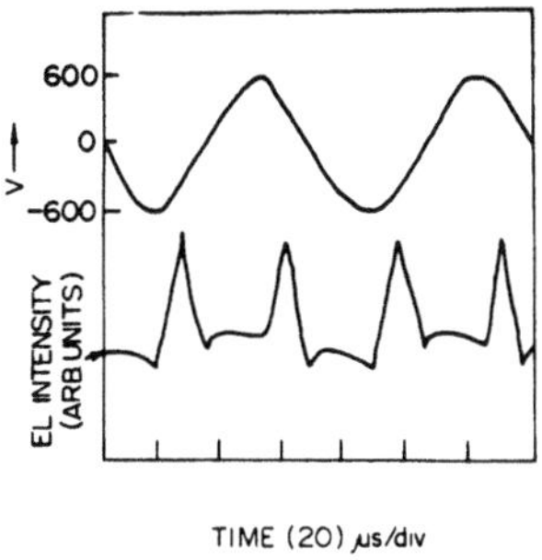

Fig. 3. The applied voltage dependence of EL brightness

4. DISCUSSION

All samples exhibit a line emission spectrum corresponding to transitions $^5D_3$ to ^7F_n and $^5D_4$ to ^7F_n where n = 0 to 6. Decay constants of luminescence are 200 microsecs and 1.0 msec respectively. The green emission is mainly due to the $^5D_4$ to $^7F_4$ transitions in the region of 540 nm.

CL brightness is found to depend directly on the reflection intensity ratio (XRD). Further, in samples where I_x/I_y ratio is constant, CL output depends on Tb concentration. On the other hand, in some samples where Tb concentration is not optimum, a comparable value of CL output is obtained if the ratio is 2.46. These observations indicate the following.
(i) The relative XRD intensity distribution is not related to Tb concentration, but the synthesis parameters.
(ii) CL brightness improves with the reflection intensity ratio approaching the standard value of 2.86 (ASTM 26.1422).
(iii) CL brightness also depends on Tb concentration.

To account for the dependence of CL brightness on x-ray reflection intensity ratio, a consideration of the atomic configuration of GOS unit cell and diffraction intensity dependence on atomic structure factor is made in the following. It is shown that even our pure samples have a disordered structure with Gd^{3+} and O^{2-} vacant sites.

It is well known that the intensity of a diffraction line depends on atomic structure factor or form factor which describes the scattering at a given wavelength and angle Θ. Also, the relative intensities of the various reflections depend on the number, type and distribution of atoms in the unit cell /5/. The XRD intensity is given as,

$$I \propto |F(hkl)|^2 \qquad (1)$$

where F(hkl) is the structure amplitude. Further for similar atoms,

$$F(hkl) = fS \qquad (2)$$

where f is the form factor and S is the geometrical structure factor which accounts for the presence and absence of a particular plane in a given lattice. Therefore, variation in the reflection intensity of a single plane will entirely depend on the form factor f, all other factors remaining constant. Before analysing the form factor for variations in XRD intensities, let us consider the crystal structure of Gd_2O_2S /6/.

Gd_2O_2S has hexagonal crystal structure with space group D^2_{3d} - $P\bar{3}m$. The atomic positions for two metal atoms are $\pm(1/3, 2/3, u_1)$, where u_1=0.29, and two oxygens are

$\pm$(1/3, 2/3, u_2) where u_2=0.64. The sulfur atom is at (0,0,0). Each metal atom is bonded to four O atoms and to three S atoms. A unit cell contains two atoms of Gd, two atoms of O and one atom of S. Gd_2O_2S lattice contains a plane consisting of S perpendicular to c-axis of the hexagonal structure. The arrangement of S ions and the S-S distance of 3.8 A in this plane are similar to those in the (111) plane consisting of S in cubic ZnS /7/. This suggests a common behaviour of holes generated by sulfur ions that constitute the valence band in these two types of most efficient phosphor hosts.

In the XRD pattern, the reflections from two planes (100) and (002) are superimposed having d=3.33 A. In this, (200) plane belongs to S only. The scattering powers of S and O are very small. Therefore, the intensity variations of (101) plane , which has mostly Gd ions, and maximum reflection intensity (ASTM 26.1422), have been considered against that of (100, 002) plane.

For simplicity of calculations, f is approximated to be equal to Z, the atomic number for an atom at θ =0. The variation in intensity ratio can therefore be written,

$$I_x/I_y \propto (Z-a)^2/Z^2 \tag{3}$$

where a represents missing ions. The value of a can be calculated for various observed ratios I_x/I_y. It is estimated that, on an average, one atom of Gd is missing for every twenty atoms or four unit cells corresponding to a ratio of 2.43. Corresponding to a ratio of 2.14, one atom of Gd and an atom of O are missing for twenty atoms. This shows that, even our pure GOS samples have a disordered structure with Gd and O vacant sites.

4.2 Electroluminescence

The EL brightness wave pattern shows that the frequency is twice that of the applied ac field, indicating the existence of potential barriers /8/. Also, for every period the brightness wave shows a small secondary peak which is attributed to an inversion of the local internal electric field in the emission zone caused by repulsion of the electrons making up a space charge. The relation $B = B_0 \exp(-b/V^{\frac{1}{2}})$ also proves that EL is originating from Mott-Scottky barriers. Impact processes are possibly involved in the excitation mechanism of luminescence centres.

5. REFERENCES

1. V. Shanker, S. Tanaka, M. Shiki, M. Deguchi, H. Kobayashi and H. Sasakura: Appl. Phys. Lett. 45, 960 (1984)
2. S. Tanaka: J. Luminescence, 40&41, 20 (1988)
3. R.A. Buchmann: IEEE Trans. Nuclear Sc., NS-19, 81 (1972)
4. Lyuji Ozawa, P.H. Jaffe: J. Electrochem. Soc. 117, 1297 (1970)
5. C. Kittel: In Introduction to Solid State Phys., (Wiley & Sons, London 1971)
6. H.A. Eick: J. Am. Chem. Soc. 80, 43 (1958)
7. T. Kano: J. Luminescence, 29, 177 (1984)
8. A. Cingolani, A. Levialdi: Phys. Lett., 17, 271 (1965)

Part IV

Color Electroluminescence

Thin Film Electroluminescent Phosphors for Patterned Full-Color Displays

R.T. Tuenge

Planar Systems, Inc., 1400 NW Compton Drive, Beaverton, Oregon 97006, USA

1. INTRODUCTION

Although there has been a strong interest in color EL flat panels, the first report of a full-color TFEL matrix display panel has only recently appeared [1]. This panel utilized a patterned phosphor structural approach. Advantages of the patterned side-by-side structure over the stacked approach that has been previously reported [2] include better reliability and need for only a single substrate. The design and performance of a matrix addressed 320 (x3) x 240 full-color display panel was recently described [3]. Although this report clearly demonstrates the viability of the patterned approach for fabricating high quality full-color TFEL displays, further improvement is still needed. The average panel "white" luminance reported was only 1.72 fL. The efficiency and chromaticity of the blue phosphor were the main limitations to achieving better performance.

In this paper, I will describe improvements that we have achieved in the luminance and efficiency for the red, green and blue thin film phosphors. Results will also be presented on a comparison of the stability and aging behavior for color phosphors and the effect of activator composition and the deposition processes. Still another important requirement of the phosphor materials for application in the patterned device structure is etchability. Reactive ion etching processes have been developed for each of the green, red and blue-emitting phosphors. The improved phosphor deposition and etching processes have been incorporated into the fabrication of the 320 x 240 full-color panel.

2. PHOSPHOR IMPROVEMENTS

Green Phosphor

Bright green ZnS:Tb,F monochrome EL devices have been reported by researchers using several deposition techniques including sputtering [4,5], coevaporation [6] and MOCVD [7]. The highest reported luminance value for this phosphor was obtained using a sputtering process by Ehime University [8]. On small research samples they have achieved a luminance of more than 30 fL at 60 Hz, V_{th}+30 V. However, some concern has been expressed [6] regarding the viability of the sputtering method as a manufacturing process for this material. Uniformity over large areas, brightness stability and reliability were factors of particular concern.

We have evaluated both coevaporated and sputtered ZnS:Tb,F for application to multi-color and full-color TFEL panels. Figure 1 compares the luminance and efficiency versus voltage characteristics at 60 Hz drive frequency for the two processes. A factor of two increase in both efficiency and luminance is achieved with the sputtering process. The ZnS:Tb,F film thickness was approximately 6000 A in both deposition methods. Threshold voltages of 170 and

Springer Proceedings in Physics, Vol. 38 Electroluminescence
Editors: S. Shionoya and H. Kobayashi

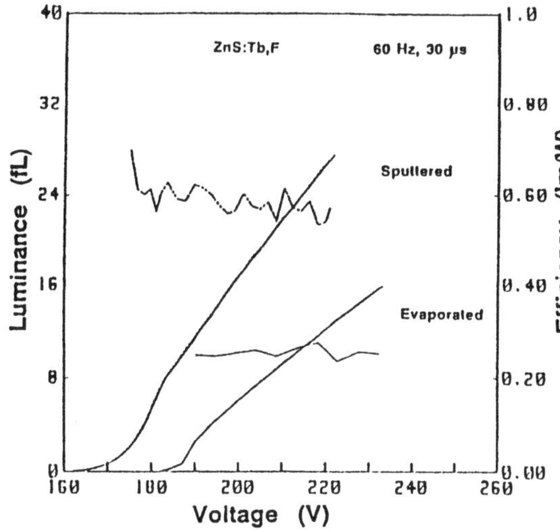

Fig. 1. Luminance and efficiency versus voltage characteristics for evaporated and sputtered ZnS:Tb,F

185 V are measured respectively for the sputtered and evaporated devices. A luminance of 22 fL is obtained at 30 volts above threshold on the sputtered device.

The stability of brightness and threshold was measured for sputtered and evaporated ZnS:Tb,F. Figure 2 shows the variation in threshold voltage and luminance with operation time at 1 kHz, V_{th}+40 V for devices prepared using the two deposition processes. Both samples show a 10 volt increase in the threshold voltage during the first 50 hours. After the initial aging period the threshold is stable for both devices. A luminance decrease of about 8 percent is measured after about 3000 hours of operation for the sputtered device. While this is not as stable as the evaporated device, it is still acceptable for most display applications.

Another major concern for the sputtered ZnS:Tb,F phosphor is uniformity. For a patterned phosphor multi-color device structure, it is even more important since one phosphor layer must be completely etched without overetching other areas on the panel. Using an RF magnetron sputtering process for the ZnS:Tb,F layer, we measure a thickness uniformity of ± 3% over an 8 inch diameter. A brightness nonuniformity of less than 20% can then be realized for the sputtered green phosphor on a half page-size display panel.

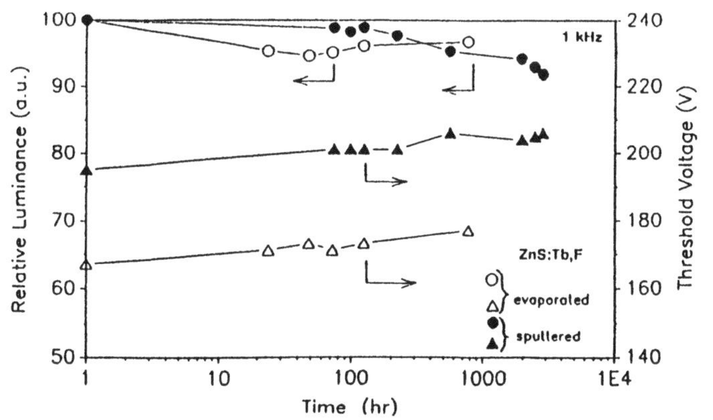

Fig. 2. Threshold voltage and luminance as a function aging time at 1 kHz for evaporated and sputtered ZnS:Tb,F

Red Phosphor

There has been considerable effort to develop red TFEL phosphors using both zinc sulfide and alkaline earth sulfide hosts. Table I lists the CIE color coordinates and 60 Hz luminance values for the various phosphors that have been reported. Samarium fluoride-doped zinc sulfide is not an ideal red phosphor due to the orange-red emission color. Improvements in the chromaticity for this phosphor have been reported by employing different co-activators and deposition techniques [10,11]. We have used a coevaporation process for ZnS:Sm,Cl and have achieved a chromaticity of x = 0.65, y = 0.34. Although this red is not as deep as the CaS:Eu phosphor, it has essentially the same coordinates as the NTSC red CRT phosphor.

Table I. Luminance and chromaticity of TFEL Red Phosphors

Material	Deposition Process	Luminance at 60 Hz	Chromaticity CIE x	y	Ref.
ZnS:Sm,F	e-beam	3.5 fL	0.60	0.38	9
ZnS:Sm,P	sputter	3.5 fL	0.63	0.36	10
ZnS:Sm,Cl	MOCVD	3.5 fL	0.64	0.35	7
ZnS:Sm,Cl	Coevaporation	2.5 fL	0.65	0.34	this work
ZnS:Mn with filter	Coevaporation	9.0 fL	0.65	0.35	this work
SrS:Eu	CVD	6.5 fL	0.59	0.40	11
CaS:Eu	e-beam	3.0 fL	0.69	0.31	12
CaS:Eu,F,Cu,Br	sputter	6.0 fL	0.66	0.34	13

The emission spectrum of the coevaporated ZnS:Sm,Cl phosphor is shown in Fig. 3. A shift in the emission peak wavelengths as well as different relative intensities compared with ZnS:Sm,F is observed. Table II summarizes the emission spectral data for the two phosphors. The transitions with $\Delta J = 0,1$ have decreased relative intensities for the chlorine co-activated phosphor. Since the chlorine ion has about the same ionic radius as sulfur, it can substitute on an S site with less lattice distortion than would result by replacement with the smaller F ion.

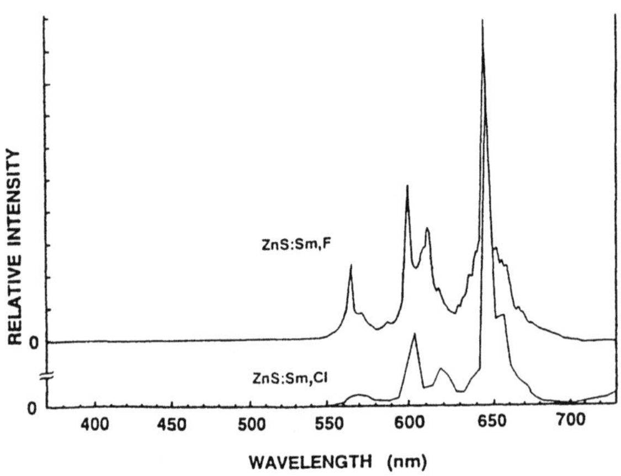

Fig. 3. Electroluminescent emission spectra of evaporated ZnS:Sm,F and ZnS:Sm,Cl

Table II ZnS:Sm,X Emission Spectral Analysis

Peak Wavelength X = Cl	X = F	Transition	ΔJ	Relative Intensity X = Cl	X = F
649	645	$^4G_{5/2}-^6H_{9/2}$	2	100	100
620	613			12	38
606	602	$^4G_{5/2}-^6H_{7/2}$	1	23	50
572	565	$^4G_{5/2}-^6H_{5/2}$	0	4	23

Using the coevaporation process we have achieved a luminance of 2.5 fL at 60 Hz, V_{th}+40 V for the ZnS:Sm,Cl phosphor. The aging behavior of this phosphor is very similar to that measured for the terbium-doped green phosphor. After 3000 hours of operation at 1 kHz the threshold voltage shows good stability. A luminance decrease of 10 % is measured at the end of this accelerated test for ZnS:Sm,Cl.

Red color EL can also be produced by filtering the broad ZnS:Mn emission. Figure 4 shows the CIE x and y-values that are obtained as a function of filtered luminance by varying the concentration of an organic dye. A ZnS:Mn device with a luminance of 50 fL at 60 Hz was used as the emission source. A bright red color measuring 9 fL at V_{th}+30 V with a chromaticity of x = 0.65, y = 0.35 can be achieved. A disadvantage, however, of the filter approach is that a two substrate device structure must be employed.

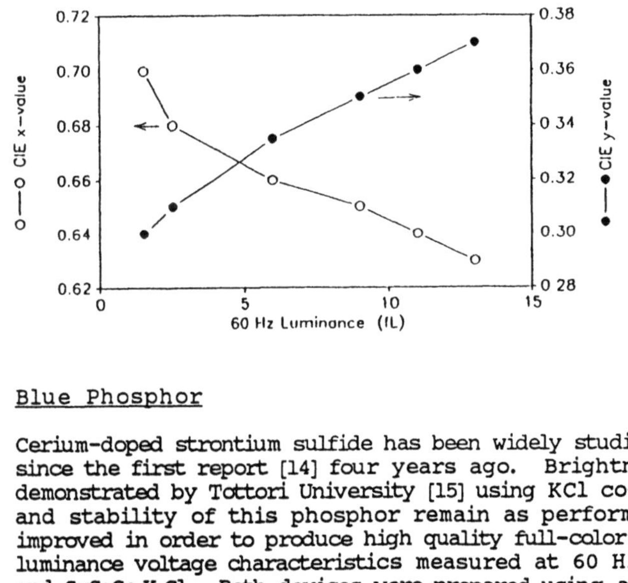

Fig. 4. CIE X and Y values as a function of percent transmission for red filtered ZnS:Mn

Blue Phosphor

Cerium-doped strontium sulfide has been widely studied as a blue TFEL phosphor since the first report [14] four years ago. Brightness improvement has been demonstrated by Tottori University [15] using KCl co-activation. Chromaticity and stability of this phosphor remain as performance areas that must be improved in order to produce high quality full-color displays. Figure 5 shows luminance voltage characteristics measured at 60 Hz for evaporated SrS:CeF_3 and SrS:Ce,K,Cl. Both devices were prepared using a substrate temperature of 350 C and a post-deposition anneal temperature between 650 and 700 C. The potassium and chlorine co-activated sample achieves more than twice the luminance compared with the CeF_3-doped device. However, the chromaticity was shifted more to the green when K and Cl are used. CIE coordinates measured x = 0.24, y = 0.46 for SrS:Ce,K Cl and x = 0.21 y = 0.37 for SrS:CeF_3. The reason for the bluer chromaticity for SrS:CeF_3 can be seen from the emission spectra shown in Fig. 6. The emission peaks are sharper when fluorine is used as the co-activator whereas a broad tail on the low energy side is observed for

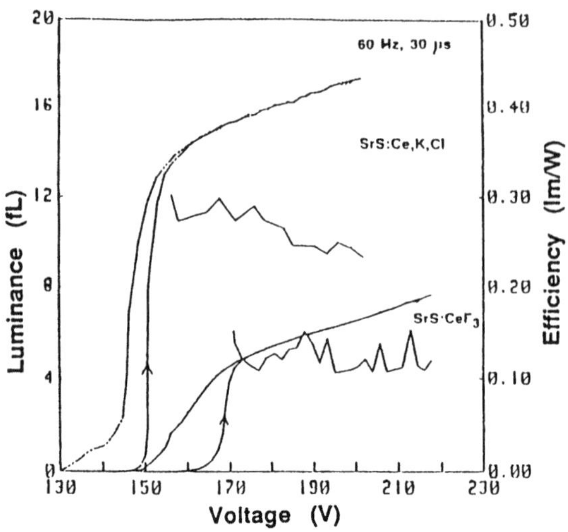

Fig. 5. Luminance and efficiency versus voltage characteristics for evaporated $SrS:CeF_3$ and SrS:Ce,K,Cl

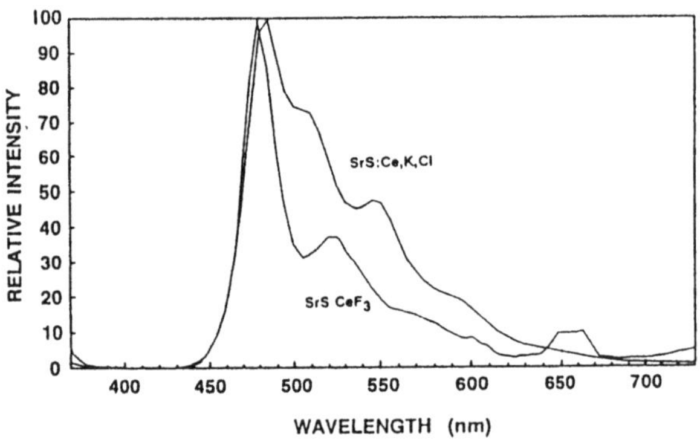

Fig. 6. Electroluminescent emission spectra of evaporated $SrS:CeF_3$ and SrS:Ce,K,Cl

SrS:Ce,K,Cl. The luminance of coevaporated $SrS:CeF_3$ is strongly dependant on the substrate temperature during deposition. Increasing the substrate temperature from 250 to 400 C results in a luminance improvement by nearly a factor of three. An enhancement in the relative intensity of the 530 nm emission band is also observed when higher temperatures are used.

As demonstrated by Tanaka [15], the chromaticity of SrS:Ce can be improved by filtering. Using an organic dye to filter the emission of $SrS:CeF_3$ results in a deep blue having CIE coordinates x = 0.12, y = 0.17. Even though the brightness is reduced to about 15 % of the unfiltered value, a luminance of 1.2 fL at 60 Hz, V_{th}+30 V can be obtained for a filtered $SrS:CeF_3$ blue device. However, a dual substrate device structure is required to utilize this improved blue in a multi-color panel.

Aging characteristics were measured at 1 kHz, V_{th}+30 V for $SrS:CeF_3$ and SrS:Ce,K,Cl devices. The threshold voltage of the CeF_3-doped film decreases about 10 V in the first 20 hours of operation and remains relatively constant after that. The luminance of this device has an initial large decrease and then assumes a much slower aging rate reaching the half-brightness level in about 2000 hours. The SrS:Ce,K,Cl device shows more severe aging behavior. The luminance of the K and Cl co-activated device decreases by about 70 % in the first 24 hours and continues to decline after that. Although both of these cerium-doped strontium sulfide devices show luminance degradation during this life test, better maintenance is obtained with the CeF_3 activator. Since a better unfiltered blue chromaticity was also obtained using $SrS:CeF_3$, this phosphor was selected for use in the patterned color panels.

3. DEVICE STRUCTURE AND FABRICATION

The full-color patterned phosphor device structure has been previously described [3]. All three primary phosphors reside on a single substrate and are arranged in parallel stripes aligned with the column electrodes. In the 320 x 240 display panel that has been reported, the phosphor stripes are of equal width measuring 0.003 inch. Compensation for differences in the pixel brightness of the red, green and blue phosphors to achieve a balanced white color can be accomplished by independently adjusting the stripe width for each phosphor.

The transparent ITO electrode processing and bottom insulator deposition are same as for fabrication of a monochrome panel. Each phosphor film is then sequentially patterned by a dry etching process. The blue $SrS:CeF_3$ film is the first to be deposited and etched since a higher annealing temperature than employed for the zinc sulfide phosphors is used in order to achieve improved luminance. Because this material is hygroscopic, it must be protected from moisture during all photolithographic and etching processes. A parallel plate reactive ion etching (RIE) system operating at 850 watts RF power was used to etch each of the phosphors. A mixture of BCl_3 and Cl_2 gases was introduced to a total pressure of 0.01 torr. No adverse effect on phosphor performance or device stability was encountered using this process.

Table III lists the etching rates measured for the SrS and ZnS active layers and some insulator layers. Using the same reactive gases, the etch rate for SrS is more than a factor of four less than that for the ZnS phosphors. The reason for the lower SrS rate is probably due to the very low volatility of a strontium chloride reaction product compared with that of zinc chloride. Since barium tantalate was the only dielectric with a slower etch rate than the $SrS:CeF_3$ phosphor it was selected as the bottom insulator. Although the etch rates for the zinc sulfide-based red and green phosphors are sufficient for panel processing, the rate for $SrS:CeF_3$ needs to be improved in order to develop a viable manufacturing process.

Table III. Reactive ion etching results using $BCl_3 + Cl_2$ gases

Material	Etch Rate (A/min)	Selectivity ZnS etch	SrS etch
ZnS:(Mn, Sm or Tb)	300	1:1	0.2:1
SrS:Ce	67	4.5:1	1:1
SiON	320	0.9:1	0.2:1
Ta_2O_5	280	1.5:1	0.3:1
$BaTa_2O_6$	23	13:1	3:1

4. DEVICE PERFORMANCE

When the phosphor brightness improvements described above for ZnS:Tb,F, ZnS:Sm,Cl and $SrS:CeF_3$ were incorporated in the 6 inch full-color panel, the following panel pixel luminances were measured at 60 Hz frame rate: green: 13 fL, red: 2 fL, and blue: 6 fL. The fill factor for each color is only about 15 % for the reported 320 (x3) x 240 display panel. When all pixels are fully activated an average brightness of 8 fL can be achieved when driven at 180 Hz frame rate. Stability of threshold voltage and brightness is reasonably good after an initial burn-in period of about 100 hours. The stability of the SrS:CeF_3 blue phosphor is still inferior to the red and green phosphors. Further improvement in unfiltered chromaticity and stability of the blue phosphor will lead to better performance for the EL full-color display panel.

REFERENCES

1. C. N. King, R. E. Coovert, and W. A. Barrow: Conf. Record 1987 Int. Display Research Conf. (Euro Display 87), 14 (1987).

2. W. A. Barrow, R. T. Tuenge and M. J. Ziuchkovski: Digest 1986 SID Int. Symp.,25 (1986).

3. W. A. Barrow, R. E. Coovert, C. N. King and M. J. Ziuchkovski: Digest 1988 SID Int. Symp., 284 (1988).

4. H. Ohnishi, K. Yamamoto, and Y. Katayama: Conf. Record 1985 Int. Display Research Conf., 159 (1985).

5. T. Ogura, A. Mikami, K. Tanaka, K. Taniguchi, M. Yoshida, and S. Nakajima: Appl. Phys. Lett. 48 (23), 1570 (1986).

6. S.K. Tiku, M.A. Mazed, and M.P.R. Panicker: Extended Abstract 1234, 172nd Meeting Electrochemical Society, Honolulu, HA (1987).

7. K. Hirabayashi, H. Kozawaguchi and B. Tsujiyama: Conf. Record 1986 Int. Display Research Conf., 4254 (1986).

8. H. Ohnishi, Y. Yamasaki, and R. Iwase: Digest 1987 SID Int. Symp., 238 (1987).

9. J. Ohwaki, B. Tsujiyama, and H. Kozawaguchi: Jpn. J. Appl. Phys. 22, 699 (1984).

10. T. Tohda, Y. Fujita, T. Matsuoka, and A. Abe: Appl. Phys. Lett. 48(2), 95 (1986).

11. J. Kane, W.E. Harty, M. Ling, and P.N. Yocom: Conf. Record 1985 Int. Research Conf., 163 (1985).

12. M. Yoshida, A. Mikami, T. Ogura, K. Tanaka, K. Taniguchi, and S. Nakajima: Digest 1986 SID Int. Symp., 41 (1986).

13. H. Ohnishi, R. Iwase, and Y. Yamasaki: Digest 1988 SID Int. Symp., 289 (1988).

14. W.A. Barrow, R.E. Coovert, and C.N. King: Digest 1984 SID Int. Symp., 128 (1984).

15. S. Tanaka, H. Deguchi, Y. Mikami, M. Shiiki, and H. Kobayashi: Digest 1986 SID Int. Symp., 29 (1986).

The TbOF Complex Center and the Brightness of ZnS Thin-Film Green Electroluminescent Devices

K. Okamoto, T. Yoshimi, and S. Miura

Fujitsu Limited, 10-1 Morinosato-Wakamiya, Atsugi, Japan 243-01

1. Introduction

Tb- doped ZnS thin-film is a promising material for practical green electroluminescent (EL) panels. Recently, the brightness of the green EL panels has been increased considerably [1-4]. This is mainly due to sputtering. At initial stage of development, the main advantage of sputtering is thought to be a capability of uniform doping of the emission centers of TbF_3 molecules into the ZnS host [1]. Further research has revealed that the advantage of the sputtering is not only uniform doping of the emission centers but also creating efficient TbF emission centers in the ZnS host [5,6]. The EL device having TbF complex centers, which are formed during the sputtering, exhibits bright green EL. The TbF complex center consists of a substitutional Tb atom in the Zn site with a F atom in the nearest interstitial site [5]. Furthermore, we have found that oxygen doping plays an important role in obtaining high brightness [7]. In this paper, we will report our experimental results indicating that F/Tb atomic ratio is very important in obtaining high brightness, and additional doping of oxygen to the TbF complex center is very effective in further improvement. The mechanisms of these improvementsare also discussed.

2. Experiment

ZnS:TbF thin-films were prepared by rf sputtering with a mixture of ZnS and an appropriate amount of TbF_3 powders. A typical concentration of Tb atom in the ZnS thin-film was 4 mol%. The post annealing was done at 500°Cfor 1h to maintain a F/Tb unity [5]. The Tb, F, and O concentrations were measured by electron probe microanalysis (EPMA). The 750-nm-thick ZnS:TbF films were sandwiched between sputtered SiON insulating layers of 200-nm in thickness and electrodes of Al and indium tin oxide were placed on each side of the insulating layers to complete EL devices. The luminance was measured at 30V above threshold voltage with a 60 Hz, 25 μs alternating pulse voltage excitation.

3. Results and discussions

Fig.1 shows the relationship between F/Tb concentration ratio and luminance. The F/Tb decreased at lower gas pressures or higher powers. The luminance increases as F/Tb decreases, implying that the F dissociation from TbF_3 is essential for higher efficiency. We analyzed the EL spectra of F/Tb=1 and 3 films to investigate why the F/Tb=1 gives higher luminance.

Fig.2 shows the EL spectra of $^5D_4$-$^7F_5$ transitions for F/Tb=3 and 1 films. The discrepancy between these EL spectra suggests a different crystal field around Tb ions. To identify the doping conditions of Tb and F, we measured the PL spectra of standard samples. TbF_3 powder and a ZnS:Tb single crystal grown from the melt under high pressure (no F atoms included) [8] were used as standard

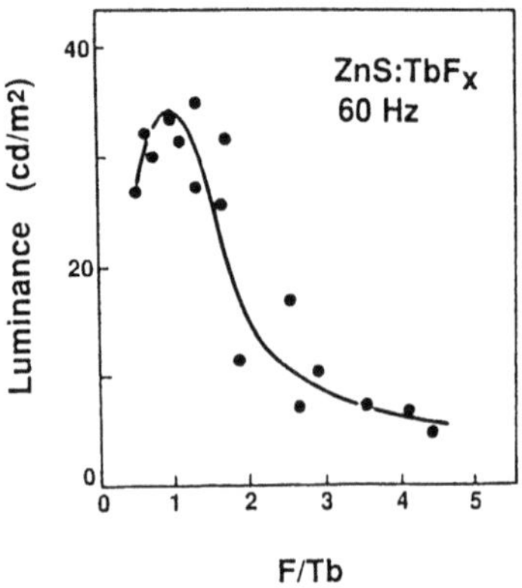

Fig.1 Dependence of EL luminance on the F/Tb ratio in ZnS film. Tb concentration is 4 mol.% for all samples.

samples for the TbF_3 molecular center and isolated Tb center substituted for the Zn site, respectively. The results are shown in Fig. 2. The envelope of the PL spectrum for TbF_3 is quite similar to the EL spectrum of F/Tb=3 film. This indicates that the Tb exists as a TbF_3 molecular center in the ZnS film. On the other hand, the EL spectrum for F/Tb=1 film looks like a mixture of the PL spectra of these two standard samples. This suggests that the Tb is not in the same state as the TbF_3 powder and the ZnS:Tb single crystal. We think that the F remaining after the dissociation of the TbF_3 molecule modulates the crystal field around the Tb ions. The lattice constant of the F/Tb=1 film is about 5.44 Å. This is considerably wider than the 5.42 Å of non-doped films. Therefore, We believe that Tb substitutes for the Zn site, and F occupies the nearest interstitial site to the Tb. This type of emission center is called the "TbF complex center".

To reproducibly form TbF complex centers in the ZnS film, we tried two methods. One is to sputter in an H_2S atmosphere, and the other uses a TbSF compound (sintered TbF_3 and Tb_2S_3 powder mixture) as the emission center. Fig. 3 shows the results. F/Tb=1 can be achieved by sputtering in an H_2S atmosphere. However, the luminance increases only slightly. Because the EL spectra of these samples are very narrow, we concluded that Tb and F are doped separately in these films. F does not occupy the nearest site to the Tb atom. On the other hand, the TbSF compound gives F/Tb=1 reproducibly and yields high luminance. However, the luminance fluctuates as shown in Fig. 3 even if the F/Tb ratio was

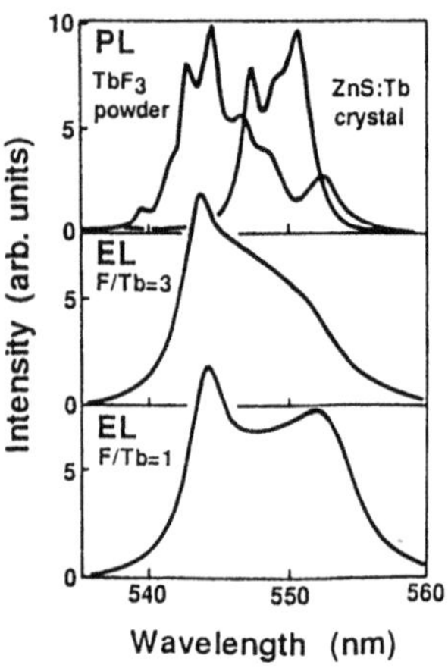

Fig. 2 PL (photoluminescence) spectra for TbF_3 powder and ZnS:Tb crystal excited by 325 nm He-Cd laser, and EL spectra for F/Tb= 3 and 1 films.

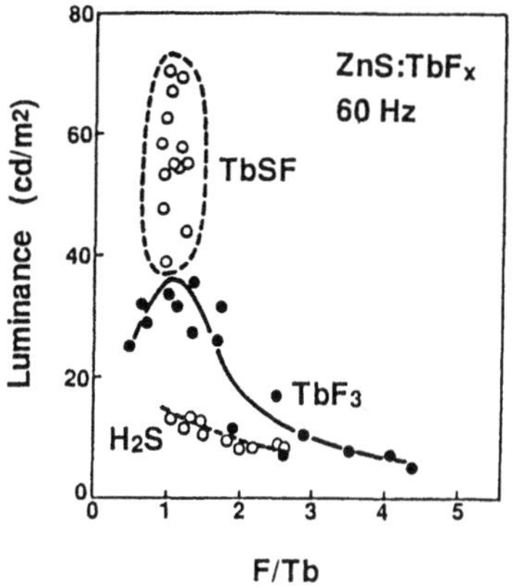

Fig. 3 Luminances obtained by sputtering in an H_2S atmosphere, and by TbSF powder as the emission center.

kept near unity. We analyzed impurities in the ZnS film since we suspected the cause of the fluctuation might be contamination, and detect no impurities other than oxygen by our EPMA.

Fig. 4 shows the relationship between the oxygen concentration and the luminance. The maximum luminance is obtained when the oxygen concentration is equal to the Tb concentration. This means that the oxygen doping is essential to improving luminance. The XPS and EPMA analysis showed that the TbF_3 did not contain any detectable oxygen, but Tb_2S_3 was considerably oxydized. The oxygen concentration in the ZnS film decreased as the presputtering time increased, but the luminance also decreased. Therefore, we can conclude that the source of the oxygen is the oxidized Tb_2S_3. Because the highest luminance is obtained when the O/Tb ratio is around unity, we speculate that the TbOF complex center contributes to high efficiency.

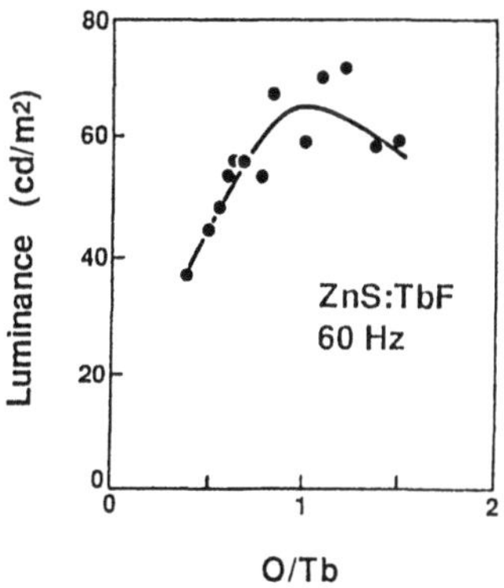

Fig. 4 Dependence of EL luminance on the O/Tb ratio in ZnS film. F/Tb ratio was kept near unity.

To obtain stable oxygen doping characteristics, we evaluated two methods. One is sputtering in an Ar-O_2 atmosphere to oxidize the Tb. The other uses a TbSF compound for the emission center (sintered TbF_3 and Tb_4O_7 powder mixture). Fig. 5 shows the relationship between the S/Zn and F/Tb atomic ratios and oxygen partial pressure in the sputtering gas. Although the oxygen concentration in the film can be changed by varying the oxygen partial pressure, the S/Zn and F/Tb decreased with increasing oxygen partial pressure. We think that the oxygen is doped as a substitutional impurity in the S sites, and some of it replaces F in the TbF complex centers. The luminance obtained by this method is shown by triangles in Fig. 6. Since the oxygen not bonded to Tb acts only as scattering

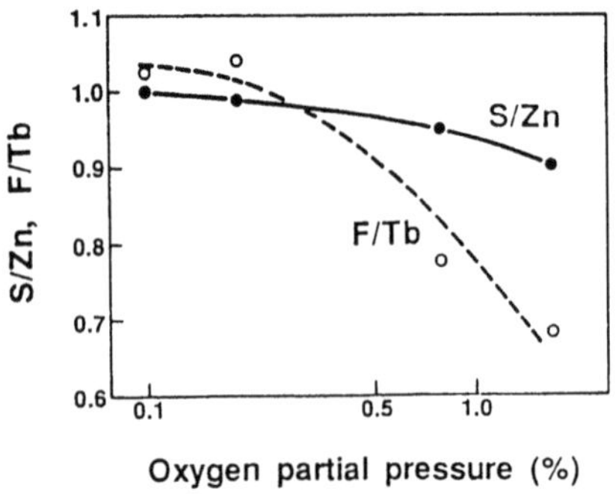

Fig. 5 Variations of S/Zn and F/Tb ratios with oxygen partial pressure in the sputtering gas.

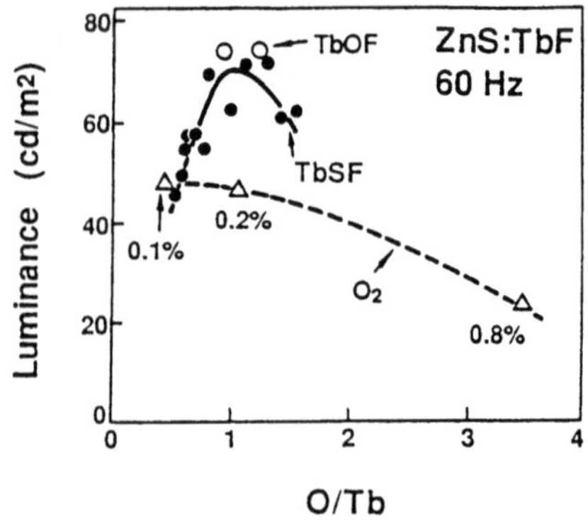

Fig. 6 Dependence of luminance on the oxygen concentration in ZnS film. Triangles are obtained by sputtering in $Ar-O_2$ atmosphere. Open circles are from TbOF powder as the emission center.

centers for hot-electrons, high luminance is not obtained by this method. However, high luminance was obtained with TbOF powder as the emission center as indicated in Fig. 6. S/Zn, F/Tb and O/Tb atomic ratios are kept almost unity by this method.

To clarify oxygen doping conditions in the ZnS:TbOF film, the chemical states of oxygen were analyzed by EPMA. Fig. 7 (a) shows the results. ZnS:O film prepared by sputtering under an oxygen partial pressure of 0.1% is also analyzed, in which the oxygen is believed to be a substitutional impurity occupying S sites and bonding with Zn but not with Tb. ZnO and TbOF powders were also measured as standards for the oxygen bonding with Zn (but not with Tb) and for the oxygen that is bonded to Tb in the TbOF complex center. A big chemical shift is seen for I_2 peaks. The peak position for the ZnS:O film is obviously shorter than that for the ZnS:TbOF thin-film. The peak positions and difference

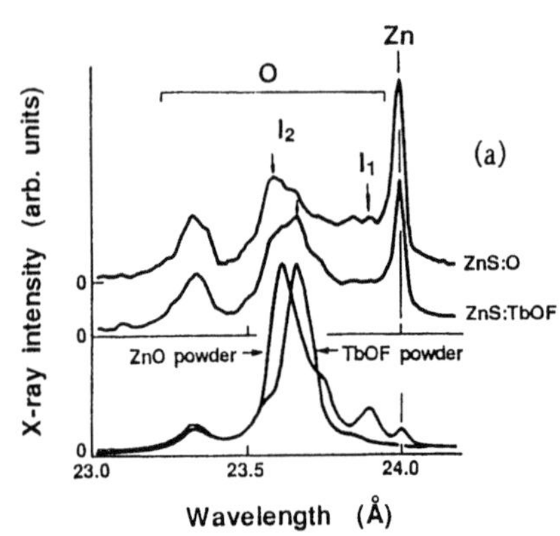

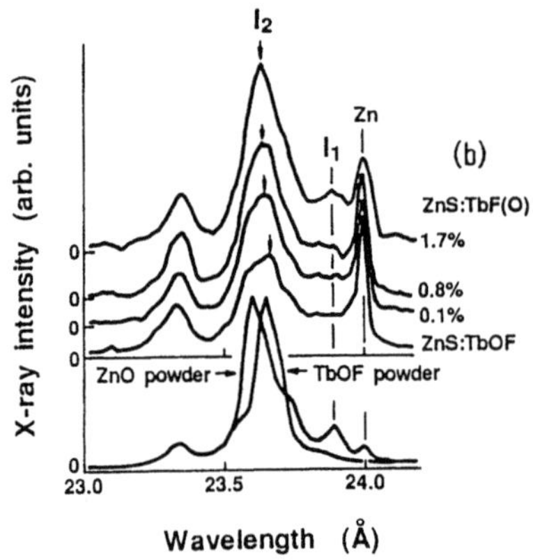

Fig. 7 EPMA spectra for oxygen K_a in ZnS films. I_1 is the characteristic peak for the oxygen not bonded with Tb. The I_2 peak is influenced by the presence of Tb in the nearest site. (a): ZnS:TbOF and ZnS:O films (b): ZnS:TbF(O) films prepared by sputtering in $Ar-O_2$ atmosphere.

are very similar to those observed for ZnO and TbOF powders. In addition, I_1 peak in the ZnS:TbOF film, which was identified as a characteristic peak for oxygen bonded with Zn only, is definitely smaller than that in the ZnS:O film. Therefore, it is clear that the oxygen in the ZnS:TbOF film occupies the S site nearest to the substitutional Tb to form the TbOF complex center with an interstitial F. The oxygen spectra of ZnS:TbF(O) films prepared by sputtering in an oxygen atmosphere are shown in Fig. 7 (b). The I_2 peak moves to a wavelength shorter than that for the ZnS:TbOF film. Also, the I_1 peak increases with oxygen partial pressure. These results indicate that sputtering in oxygen introduces oxygen bonded with Zn only as a substitutional impurity in S site in the ZnS:TbF film, although some of the oxygen is thought to form the TbOF complex center. Since the hot electrons loose energy by scattering with this kind of oxygen, improved brightness is not expected from such films. We concluded that the O occupies one of the S sites which bond to Tb atom in the TbOF complex center according to EPMA analysis.

Our model of the efficiency improvement by the TbOF complex center is illustrated in Fig. 8. The oxygen doping in the site nearest the Tb enlarges the host lattice distortions. This means an increased impact cross section of the host lattice for hot-electrons, and efficient generation of electron-hole pairs at the Tb site. If the Tb is excited by the energy transfer from the electron-hole pair's recombination at the Tb doping site, enlarged lattice distortions at Tb site could improve the excitation efficiency of the Tb center [7, 9, 10]. If the wave functions of the $4f^8$ electronic system of the Tb center and the host lattice are mixed well, the direct impact cross section of the $4f^8$ electronic system might be increased by adding O to the TbF complex center [11]. Further research is required to determine the excitation mechanism of the Tb center.

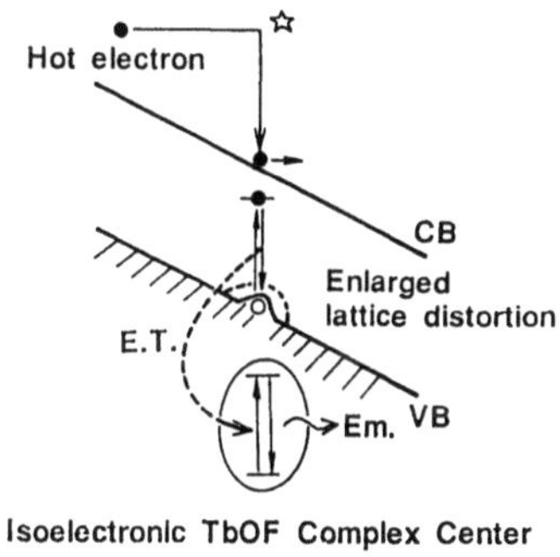

Fig. 8 Model for efficiency improvement by the TbOF complex center. The oxygen enlarges the host lattice distortions around Tb ions.

4 Summary

We have investigated the relationship between the brightness and the state of the Tb emission centers in the ZnS host. It is concluded that the TbF complex center, which consists of a substitutional Tb atom in the Zn site with a F atom in the nearest interstitial site, is more efficient than the TbF_3 molecular centers. Additional doping of oxygen to the nearest S site to Tb to form TbOF complex center is very effective in further improvement. This improvement is interpreted by increased impact cross section of the host lattice for hot electrons at Tb doping site attributed to the enlarged host lattice distortions by the TbOF complex center.

Acknowledgment

The authors would like to thank Dr. S. Shionoya, Professor Emeritus of The University of Tokyo, for kindly supplying the ZnS:Tb crystal. The authors would also like to thank Dr. Y. Toyama of Fujitsu Laboratories for his encouragement. Thanks are also given to S. Sato and M. Sigeno for their useful discussions. This work was carried out as part of a research program supported by JRDC (Research Development Corporation of Japan).

References

1. S. Miura, K. Okamoto, S. Sato, S. Andoh, H. Ohnishi, and Y. Hamakawa: Proceedings of the 3rd International Display Research Conference, Kobe, 1983, p. 84.
2. T. Tohda, Y. Fujita, T. Matsuoka, and A. Abe: Appl. Phys. Lett. 48, 95 (1986)
3. H. Ohnishi, K. Yamamoto, and Y. Katayama: Proceedings of the 1985 International Display Research Conference, San Diego, 1985, p. 159.
4. T. Ogura, A. Mikami, K. Tanaka, K. Taniguchi, M. Yoshida, and S. Nakajima: Appl. Phys. Lett. 48, 1570 (1986)
5. K. Okamoto and K. Watanabe: Appl. Phys. Lett. 49, 578 (1986)
6. A. Mikami, T. Ogura, K. Tanaka, K. Taniguchi, M. Yoshida, and S. Nakajima: J. Appl. Phys. 61, 3028 (1987)
7. K. Okamoto, T. Yoshimi, and S. Miura: Appl. Phys. Lett. 53, 678 (1988)
8. H. Kukimoto, S. Shionoya, T. Koda, and R. Hioki: J. Phys. Chem. Solids 29, 935 (1968)
9. K. Okamoto and S. Miura: Appl. Phys. Lett. 49, 1596 (1986)
10. K. Okamoto, T. Yoshimi, and S. Miura: submitted to J. Appl. Phys.
11. D. Kahng: Appl. Phys. Lett: 13, 210 (1968)

Doping Conditions of Tb, F Luminescent Centers in ZnS: Tb, F Films - Effects of Fabrication Methods on Doping Conditions

J. Mita, T. Hayashi, Y. Sekido, and I. Abiko

Research Laboratory, OKI Electric Industry Co., Ltd.,
550-5, Higashiasakawa-cho, Hachiohji-shi, Tokyo 193, Japan

1. Introduction

Following a practical application of ZnS:Mn electroluminescent (EL) devices /1/, the multi-coloring of EL devices has been researched intensively /2-8/. Especially, the EL devices with a ZnS:Tb,F film have been developed to show a practical luminance level /2/. The ZnS:Tb,F films were fabricated by coevaporation from a crucible and a boat /3/ or electron beam evaporation /4/ formerly. The Tb and F atoms are expected to form the TbF_3 molecular centers in the films fabricated by the coevaporation /5/. According to recent reports /6,7/, however, the maximum luminance for the rf sputtered films is attained when the F/Tb atomic ratio is unity. This difference seems to relate to the fabrication methods. This paper describes the doping conditions of the Tb and F atoms in the ZnS:Tb,F films fabricated by electron beam evaporation and rf magnetron sputtering.

2. Sample Preparation

The EL devices used in this experiment have a conventional doubly-insulating structure of indium tin oxide transparent electrode, SiO_2/Ta_2O_5 insulating layer, ZnS:Tb,F phosphor layer, Ta_2O_5/SiO_2 insulating layer and Al rear electrode. Each insulating layer consists of an insulating SiO_2 film (80 nm thick) and a semi-insulating Ta_2O_5 film (300 nm thick). The SiO_2 and Ta_2O_5 films were fabricated by rf magnetron sputtering and electron-beam evaporation, respectively. The ZnS:Tb,F phosphor films were fabricated by electron beam evaporation (EB) and rf magnetron sputtering (SP) to compare the doping condition of the Tb and F atoms. The thickness of these films was about 700 nm. The deposition sources of pellets for the EB and sintered targets for the SP were made from the mixture of ZnS and TbF_3 powder. Substrate temperature during deposition was kept at 200 °C. After the deposition, annealing was done in a vacuum for an hour at the temperature of 300 ~ 600 °C.

3. Results and Discussion

The dependence of the luminance L_{30} on the annealing temperature T_{an} is shown Fig.1 (a). The luminance was measured under 1 kHz sinusoidal voltage excitation. For the EL devices having the EB films, the luminance increases very slightly from 320 to 350 cd/m^2 by the annealing at 450 °C and decreases to 230 cd/m^2 at 600 °C. For the devices having the SP films, the luminance increases considerably from 240 to 330 cd/m^2 by the annealing at 450 °C and reaches a maximum luminance of 350 cd/m^2 at 600 °C. Thus, the L_{30}-T_{an} characteristics are quite different. This difference seems to originate in the fabrication methods.

Firstly, the F/Tb atomic ratio in these films was measured by electron probe micro-analysis (EPMA). Figure 1 (b) shows the dependence of F/Tb atomic ratio on

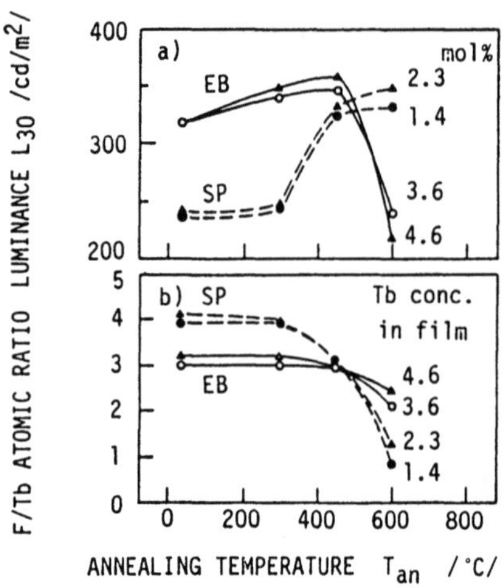

Fig.1 Dependences of (a) luminance L_{30} and (b) F/Tb atomic ratio on the annealing temperature T_{an} for the devices fabricated by electron beam deposition and rf magnetron sputtering

the annealing temperature T_{an}. The Tb concentration in the films shown in the figure remain unchanged by the annealing. The F/Tb ratio in the as-deposited EB film is around 3, and decreases to 2-2.5 after annealing at 600°C. The F/Tb ratio in the as-deposited SP film is around 4, and decreases to 1 after annealing at 600°C. Thus, the F/Tb ratio for the SP films decreases remarkably, compared with those for the EB films. The F atoms in the SP films migrate easily and escape through the surface of the ZnS film. The value of 4 clearly indicates a part of the F atoms do not combine with the Tb atoms, because the valency of Tb and F ions are +3 and -1, respectively.

To clarify the combination condition of the F and Tb ions directly, the chemical states of the Tb and F were measured by X-ray photo-electron spectroscopy (XPS). Figures 2 (a) and (b) show the XPS spectra of Tb and F atoms in the as-sputtered films and the annealed films, respectively. For the EB films, the XPS spectra was not obtained clearly. The spectra for the TbF_3 powder are also shown for reference. The spectra of Tb have two peaks at 1242.7 eV (Tb $3d_{5/2}$) and 1277.5 eV (Tb $3d_{3/2}$). The binding energy of the Tb is not affected by the annealing and are equal to those for the TbF_3 powder. This may result from the fact that the analysis was done on 3d-core-electron. Therefore, the chemical state of the Tb ions could not be clarified. For the F atoms, the core electrons

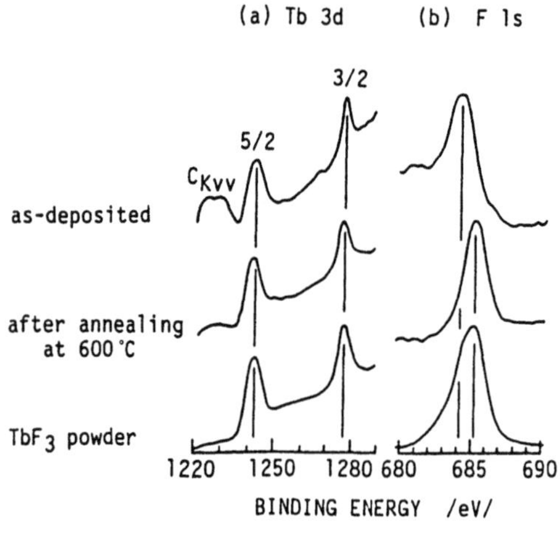

Fig.2 XPS spectra of (a) Tb 3d and (b) F 1s for the as-sputtered films and the films after annealing at 600°C. Those for TbF_3 powder are shown for reference

of 1s were measured. The peak of the F 1s spectra shifts to the higher energy side by the annealing at 600°C. The binding energy of the F 1s in the as-sputtered film is 684.2 eV, which is close to that of the ZnF_2 (684.2 eV) /11/ rather than that of the TbF . The binding energy shifts to 685.2 eV after annealing at 600°C, which is consistent with the value for TbF_3 powder. These results show that many of the F ions exist separately from the Tb ions and do not contribute to the formation of luminescence centers with the Tb ions. These F ions may act as scattering centers for hot-electrons. Taking the results of the F/Tb ratio into consideration, the luminance improvement caused by annealing is believed to be due to the decreased number of scattering centers. In addition, the F ions may combine with the isolated Tb ions by Coulomb's force when they migrate through the ZnS film. Efficient Tb-F complex centers formed in this process also contributes to the luminance improvement.

Next, the crystallinity of the ZnS host layer was evaluated by X-ray diffraction to investigate the effect of the doping of the Tb and F atoms. The change of the lattice spacing d of ZnS (111) by the annealing is shown in Fig. 3. The d of the as-sputtered SP film is considerably larger than that the film annealed at 600°C, which is close to the d of bulk ZnS. This result is consistent with the fact that interstitial F ions are released from ZnS film by annealing /7/. Beside this, the crystallinity of the ZnS film itself is improved by the annealing because the d of un-doped ZnS decreases. This improvement of crystallinity may also contribute to the enhancement of luminance. However, the d of the as-deposited EB film is close to that of bulk ZnS and hardly change by annealing. In the EB films, the Tb atoms accompanying three F atoms seem to be doped in a stable form.

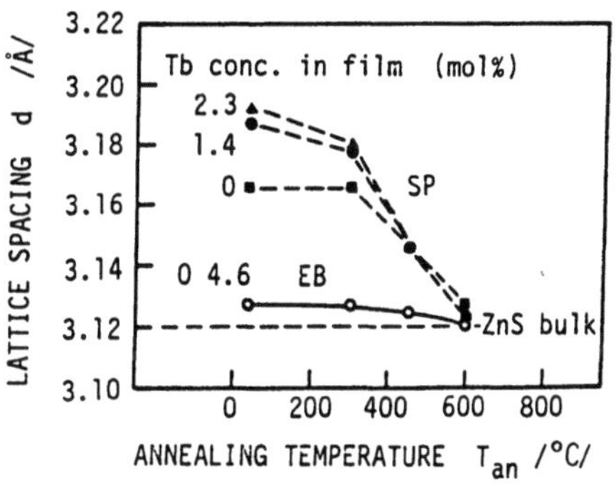

Fig.3 Dependences of lattice spacing d of ZnS (111) on the annealing temperature T_{an}

The reduction in the luminance for the EB film annealed at 600°C is not explained by the results mentioned above. We suspected the diffusion process which occurs by the annealing at high temperature and measured depth profiles by secondary ion mass spectroscopy (SIMS). Figure 4 shows the depth profiles of O, Zn, and S atoms for the EB films after annealed at (a) 450°C and (b) 600°C. The oxygen diffusion was not observed for the film annealed at 450°C. The films annealed at the temperature of less than 450°C show almost same profiles. Only the profile for the EB film annealed at 600°C shows the oxygen diffusion from the underlying Ta_2O_5 film. The peaks of the Zn and S at the ZnS/Ta_2O_5 interface result from the increase in the ionization probability of Zn and S due to the existence of diffused oxygen. This diffusion seems to relate with the difference in the property of ZnS itself, because the same diffusion was observed in un-doped ZnS/Ta_2O_5 film. As is well known, the first 200 nm of EB film is a dead layer and is of poorer crystallinity than the region above it. Since this dead layer is believed to be somewhat porous, the oxygen in the Ta_2O_5 dielectric layer can diffuse during annealing. The reduction of the luminance for the EB film after annealing at 600°C is considered to be due to the oxidation of ZnS phosphor and/or Tb ions.

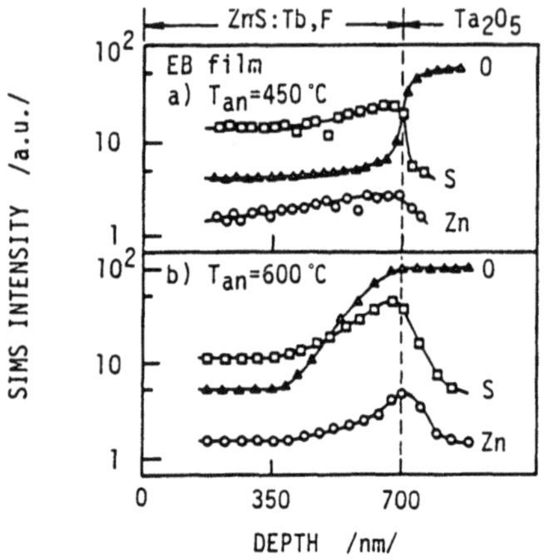

Fig.4 Depth profiles of Zn, S and O in the electron-beam evaporated films after annealing at (a) 450°C and (b) 600°C

4. Conclusion

The doping condition of the Tb and F atoms was investigated to clarify the difference of the ZnS:Tb,F films fabricated by rf magnetron sputtering and electron beam evaporation. For the EB films, the F/Tb ratio of 3 and the luminance are hardly affected by the annealing up to the temperature of 450°C. For the EB film annealed at 600°C, however, oxygen diffusion from underlying Ta_2O_5 insulating layer was observed, resulting in the reduction of the luminance. For the SP films, interstitial F ions are released from the ZnS film and Tb-F complex centers are formed by annealing. The enhancement of the luminance for the SP films may be attributed to not only the increase in the number of Tb-F complex centers but also the improvement of the crystallinity.

References

1. T.Inoguchi, S.Mito: In Topics in Applied Physics, ed. by J.I.Pankove Luminescence, Vol.17 (Springer, Berlin, Heidelberg, 1977) p.197
2. H.Ohnishi, K.Yamamoto, Y.Katayama: Conf.Record Int.Display Research Conf. (1985) p.159
3. E.Chase, R.T.Hepplewhite, D.C.Krupka, D.Kahng: J.Appl.Phys. 40 (1969) 2512
4. K.Okamoto, Y.Hamakawa: Appl.Phys.Lett. 13 (1979) 508
5. D.Kahng: Appl. Phys. Lett. 13 (1968) 210
6. T.Ogura, A.Mikami, K.Tanaka, K.Taniguchi, M.Yoshida, S.Nakajima: Appl.Phys. Lett. 48 (1986) 1570
7. K.Okamoto, K.Watanabe: Appl.Phys.Lett. 49 (1986) 578
8. J.Mita, M.Koizumi, H.Kanno, T.Hayashi, Y.Sekido, I.Abiko, K.Nihei: Jpn.J. Appl.Phys., 26, 5 (1987) L558
9. J.Mita, M.Koizumi, H.Kanno, T.Hayashi, Y.Sekido, I.Abiko, K.Nihei: Jpn.J. Appl.Phys., 26, 7 (1987) L1205
10. C.W.Wagner, W.M.Riggs, L.E.Davis, J.F.Moulder, G.E.Muilenberg: In Hand Book of X-ray Photoelectron Spectroscopy, (Perkin-Elmer Corp., Pairie, Minesota, (1979)

Effects of Preparation and Operation Conditions on Electroluminescence Spectra of $ZnS:TbF_3$ Film Structures

N.A. Vlasenko, V.S. Khomchenko, S.F. Terechova, M.M. Chumachkova, L.I. Veligura, S.I. Balyasnaya, and Yu.A. Tzircunov

Institute of Semiconductors, Academy of Sciences of Ukrainian SSR, P.O. Box 252650, Kiev, Prospekt Nauki 45, USSR

Among rare earth elements Tb is the most effective activator for ZnS thin-film electroluminescent structures (TFELS). It is usually introduced into the lattice as TbF_3. The microstructure of the centers formed and their dependence on film preparation conditions have not yet been established. The model for the TbF_3 molecular center has recently been reconsidered (see e.g. [1]). Essential information about the crystal field symmetry (CFS) around Tb^{3+} ions can be obtained from the emission spectra. Conclusions about the excitation mechanism of electroluminescence (EL) can be obtained from the field dependence of the relative intensity of the $^5D_4-^7F_j$ and $^5D_3-^7F_j$ bands. However $ZnS:TbF_3$ TFELS spectra have been poorly investigated with most results reported at room temperature without allowance for interference effects. In this paper such spectra have been studied at temperatures in the range 4.2 - 300 K and the dependence on film preparation conditions and applied voltage are reported. EL spectra obtained are compared with Tb^{3+} emission spectra for materials which have different CFS and the results of the group-theory analysis of the level energy scheme are given. Center models as well as the origin of temperature and voltage effects on EL spectra are discussed.

1. Experimental Method

Two types of conventional MISIM TFELS with standard 250 nm-thick insulating layers (e.g. Y_2O_3) were used. $ZnS:TbF_3$ films (600 nm thick) were deposited by thermal evaporation of ZnS and TbF_3 from separate sources (type I) and by electron beam evaporation of $ZnS:TbF_3$ pellets (type II). The substrate (T_s) and post-annealing (T_a) temperatures were over ranges of 100-250 °C and 350-600 °C respectively. The concentration (C) of Tb and F in the EL film were measured by EPMA. C_{Tb} was about 1 mol%, but C_F depended on the doping method and T_a. The ratio $x=C_F/C_{Tb}$ was equal to ∿3 and 3.2-3.6 for the films as-deposited of the type I and II respectively. However in the latter there were some precipitates in the the films with x∿1 and x>4. The required decrease of x was observed after annealing at T_a>400 °C for the both types. EL spectra were measured by a computer-controlled spectrometer with the resolution ≤ 0.2 nm. TFELS with the same layer thickness were selected to rule out interference effects.

2. Results and Discussion

Some differences have been found in the spectra for TFELS of type I and type II. They are illustrated by Fig. 1 for the most intensive green $^5D_4-^7F_5$ band. The emission bands are always narrower for type I than for the type II. The band structure is well-defined and reproducible, at 4.2 K, for the former, but it is less defined and observed rarely for the latter (Fig. 1a-c). There are some differences in the spectral distribution of the intensity in the same bands and in the position of peaks in them, especially in the case of the as-deposited films. Thus the long-wavelength part of green band are intensified in EL spectra of the

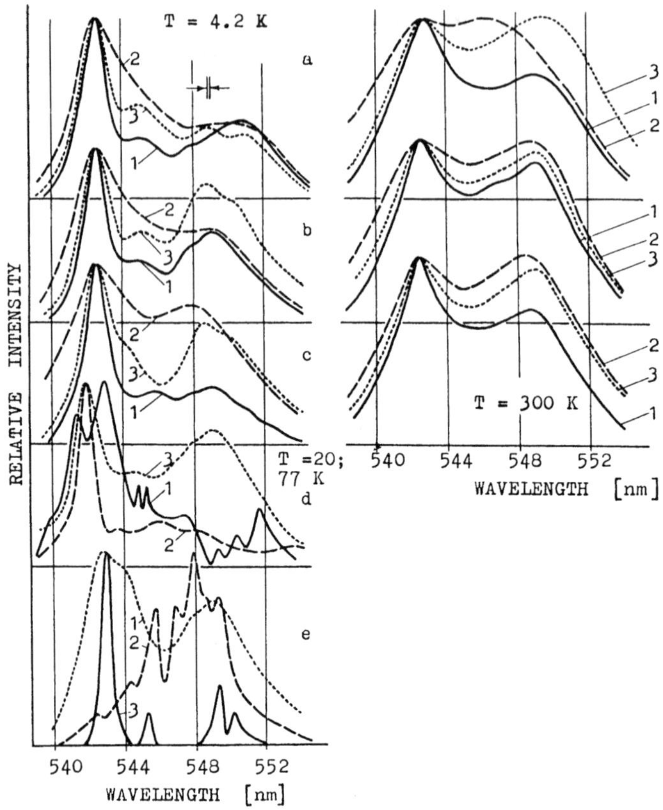

Fig. 1 $^5D_4-^7F_5$ band at 4.2 and 300 K for (a-c) $ZnS:TbF_3$ TFELS of type I (1) and type II(2,3) as deposited (c) and annealed at 350°C (b) and 500-600°C (a); (d) SiO_x:Tb,F (3), $CdF_2:TbF_3$ (2) TFELS at 77 K (our data), and TbF_3 film (1) at 20 K [4]; (e) ZnS:Tb film (1) [4] and crystals (2,3) [3] at 20 and 77 K, respectively.

type II (Fig. 1c). Hence there are more different Tb centers in the $ZnS:TbF_3$ film deposited by electron beam evaporation than in the type I. These are due to exchange reactions between ZnS and TbF_3, which proceed in the pellets during their annealing and evaporation and result in the formation of ZnF_2 and Tb_2S_3. ZnF_2 is evaporable at temperatures close to those of ZnS and TbF_3, but Tb_2S_3 remains in the pellets and only partly dissociates. Therefore C_F/C_{Tb} is higher in the films than in the pellets and Tb is transfered to a substrate as both molecules and the atoms. This causes centers with different CFS to be formed.

T_s did not affect EL spectra if $T_s<T_a$. Some narrowing of the bands occurs with increasing T_s up to 250°C as well as after the film annealing at 300-350°C. This effect is due to a crystal ordering in the film. Further increase of T_a results in some band broadening and the redistribution of the intensity (I). For example, I is intensified over 549-555 nm and a new peak at 550.5 nm appears, which is present also in the spectra of TFELS as-deposited of type II. It follows that new Tb centers are formed during the annealing at T_a>400°C.

The intensity in the ranges of 538-540 nm and 545-552 nm increases relative to I_{max} (λ=542.6 nm), with activation energy equal to 6-12 meV depending on λ, as temperature (T) rises from 20 to 300 K for the both types of TFELS. Such behaviour cannot be explained by the strengthening of lattice vibrations only and are attributed to the change in the population of $^5D_4$ level substrates.

The I_j/I_{max} ratio for the $^5D_4-^7F_j$ bands (j=3••6) did not depend on the applied voltage (U), whereas the $I(^5D_3-^7F_5)$ to $I(^5D_4-^7F_5)$ ratio increases as U rises. This confirms that the direct impact excitation of Tb^{3+} takes place in TFELS studied.

The green band in emission spectra of TFELS studied and of materials with different CFS around Tb^{3+} has been compared (Table 1, Fig. 1d, 1e). In all cases considered one can separate two groups of subbands, short-wavelength (S) and long-wavelength (L), which are due to transitions from the lowest substate of $^5D_4$ level to two groups of substates originating as a result of $^7F_5$ level splitting in the crystal field. A definite correlation of the energy interval (Δ) between the most intensive subbands in the S and L groups with CFS has been established. As it is seen from Table 1, the four values of Δ (8-9, 16-18, 25-27 and 33-34 meV) correspond with the following sequences of CFS: D_{3h} or T_d, C_{4v}, C_{3v} and C_{2v}. An exception is Δ for amorphic SiO_x:Tb,F films, in which CFS for Tb^{3+} is C_{3v} [5], which is just above that for other materials with the same CFS. This is due to a crystal disordering. It follows from above: 1) CFS around Tb^{3+} in TFELS of the type I is C_{3v} (Δ=26 meV); this is consistent with the model of the molecular center as far as TbF_3 molecule has such a point symmetry. 2) Centers with the lowest CFS among those considered (C_{2v}, Δ=33 meV) are in TFELS of the type II as-deposited and in the both types after the annealing at T_a>400°C. These centers are believed to involve Tb^{3+} ion associated with some acceptor defect such as F_i^- or V_{Zn}^-.

Table 1 Energy interval (Δ) between the most intensive peaks in S and L groups of subbands in green band of Tb^{3+} for the TFELS studied and for materials with different CFS

Material, sample and excitation			Crystal system	Possible CFS	Δ /meV/	T /K/	References
LaCl:Tb,	crystal,	PL	hex. a≃c	D_{3h}	9	4.2	[2]
ZnS:Tb,	crystal,	PL	cub.	T_d, C_{4v}, C_{3v}	8, 16, 26	77	[3]
TbF_3,	film,	CL	rombic	C_{2v}	34	20	[4]
	powder,	PL			33	300	[1]
CdF_2:TbF_3,	film,	EL	cub.	O_h, C_{4v}, C_{3v}	18, 25	77	This paper
SiO_x:Tb,F,	film	PL	amorph.	C_{3v}	29	77	- " -
ZnS:TbF_3,	film,	EL	cub.			4.2	- " -
type I,	as-depo			C_{3v}	26		
	T_a≤350°C			C_{3v}	26		
	T_a>400°C			C_{3v}, C_{2v}	26, 33		
type II				C_{3v}, C_{2v}	26, 33		

The group-theory analysis of $^5D_4$ and $^7F_5$ level splitting and selection rules for T_d and C_{3v} symmetry have been carried out. Also the splitting of the levels for the T_d symmetry has been estimated by the perturbation theory using the point charge model as a first approximation for the crystal field potential. According to this estimation, the $^7F_5$ level is split into the four substates forming two groups (T_1, E and T_1, T_2) with the splitting in the each one of ≤1 meV and between the groups about 9 meV. This is in good agreement with the lowest value of Δ observed for ZnS:Tb crystals. Some estimation of the level splitting for C_{3v} is impossible now due to lack of information about parameters required. However the comparison of the qualitative level scheme obtained from the group-theory analysis with experimental data for emission spectra of various materials, in particular of the type I TFELS, will aid in the valuation of them.

References

1. K. Okamoto, K. Watanabe: Appl. Phys. Lett. 49, 578 (1986)
2. K.S. Thomas, S. Singh, G.H. Dieke: J. Chem. Phys. 38, 2180 (1963)
3. W.W. Anderson, S. Razi, D.J. Walsh: J. Chem. Phys. 43, 1153 (1965)
4. J. Benoit, P. Benalloul: Phys. stat. sol. (a) 15, 67 (1973)
5. N.A. Vlasenko et al.: J. Luminescence 40/41, 792 (1988)

Green AC Electroluminescence in ZnS Thin Films Doped with Tb, Er and Ho Ions and Concentration Quenching Models

Guozhu Zhong, Changhua Li, Lijian Meng, and Hang Song

Changchun Institute of Physics, Academia Sinica, P.R. China

1. Introduction

ZnS thin films doped with rare earth ions in sandwich structure can give various color of ACEL [1,2]. Tb shows a bright green EL, other ions show weaker EL. The present work is to compare three green ACEL in ZnS thin films doped with Tb, Er and Ho and to discuss why they have different brightness.

There are three reasons for concentration quenching (CQ) in ZnS ACELTF doped with RE ions. When RE are introduced into ZnS, lattice periodic field will be destroyed. (1) Some killers will be formed and lead to nonradiative energy transfer from RE ions to them. Because the strength of main emission energy levels of three RE ions is similar, the action of killers will also be similar. (2) The increase of scattering center will result in the decrease of the mean free path of hot electron and then the decrease of emission intensity. Because the difference of ionic radius of three RE ions is very small, this effect will be the same. (3) The distance between RE ions will be decreased with increasing concentration. The interaction between them will result in energy transfer [3,4], i.e, cross-relaxation (CR) [5].

2. Experimental Results and Discussion

Figure 1 shows brightness B vs. concentration of these ACEL samples. At lower concentration, brightness increases almost linearly with increasing concentration. Brightness saturates and then falls down with increasing concentration. The optimum concentration values C_{opt} of Tb, Er and Ho in ACELTF are approximately 1.4×10^{-2} mol, 7×10^{-3} mol and 3×10^{-3} mol, respectively. Values for the maximum

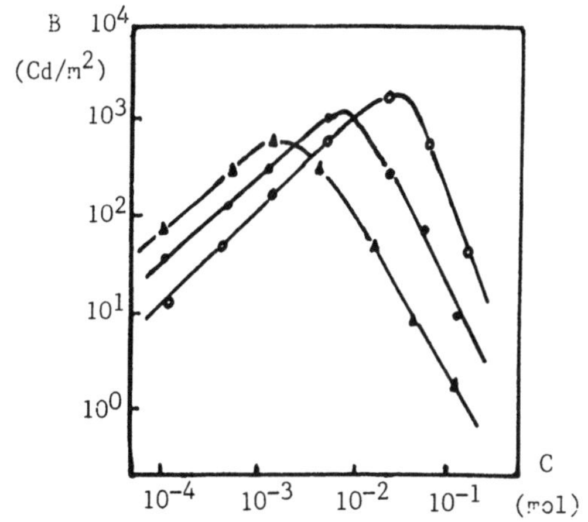

Fig. 1 ACEL brightness of ZnS:TbF(o), ZnS:ErF(o) and ZnS:HoF(Δ) TF vs. doping concentration

brightness B_{max} of these films are 2300 cd/m^2, 1100 cd/m^2 and 500 cd/m^2, respectively. From these results we find that the ACEL of ZnS thin films doped with RE are strongly affected by optimum concentration.

As shown in Fig. 2, main emission energy levels of Tb, Er and Ho are $^5D_4$, $^2H_{11/2}+^4S_{3/2}$ and $^5S_2$, respectively and emission peaks are located at 542 nm, 528 nm + 552 nm and 550 nm, respectively.

Figure 2(a) shows ACEL spectra of ZnS:TbF thin films with various concentrations. The emission intensity of $^5D_3-^7F_j$ decreases with increasing Tb concentration. When concentration is higher than 4.5×10^{-2} mol, no emission of $^5D_3-^7F_j$ can be observed. On the other hand, the emission intensity of $^5D_4-^7F_j$ increases with increasing concentration. This means that there is a CR process of $^5D_3-^5D_4 == ^7F_6-^7F_0$ [5]. This CR process will enhance the green emission of $^5D_4$ energy level. Therefore, ZnS:TbF thin film has the highest optimum concentration and brightness.

Figure 2(b) shows ACEL spectra of ZnS:ErF thin films with different Er concentrations. The green emission intensity of $(^2H_{11/2}+^4S_{3/2})-^4I_{15/2}$ decreases with increasing Er concentration, but red one of $^4F_{9/2}-^4I_{15/2}$ increases relatively with increasing Er concentration. It is obvious that CQ is different for different energy levels. From Fig. 2(b), the energy gap between $^2H_{11/2}+^4S_{3/2}$ and $^4F_{9/2}$ is smaller than that between $^4I_{15/2}$ and $^4I_{13/2}$. The overlap integration of them is very small [4], i.e., the probability of CR $^2H_{11/2}-^4F_{9/2} == ^4I_{15/2}-^4I_{13/2}$ is also very small. In experimental results, we find that emission intensities of $^4F_{7/2}-^4I_{15/2}$ and $(^4F_{5/2}+^4F_{3/2})-^4I_{15/2}$ decrease with increasing Er concentration,

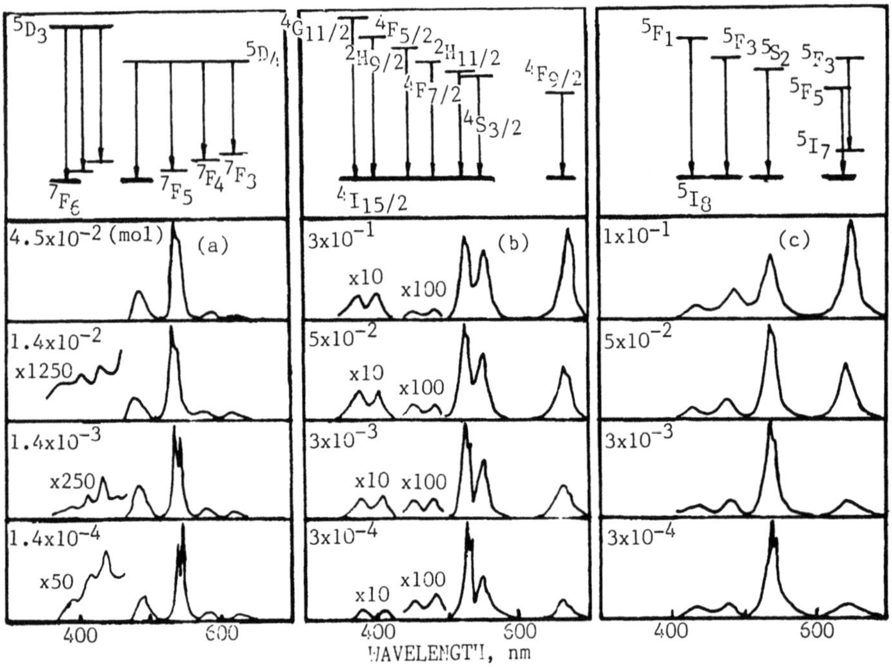

Fig. 2 ACEL spectra of ZnS:TbF(a), ZnS:ErF(b) and ZnS:HoF(c) TF vs. concentration and related transition

but the emission intensities of $^2H_{9/2}$-$^4I_{15/2}$ and $^4G_{11/2}$-$^4I_{15/2}$ increase with increasing Er concentration, as shown in Fig. 2(b). The energy difference of $^2H_{11/2}$ and $^4F_{9/2}$ is close to that of $^4F_{7/2}$ to $^2H_{9/2}$ and of ($^4F_{5/2}$+$^4F_{3/2}$) to $^4G_{11/2}$. Therefore, the concrete CR processes are $^2H_{11/2}$-$^4F_{9/2}$==$^4F_{7/2}$-$^2H_{9/2}$ or $^2H_{11/2}$-$^4F_{9/2}$==($^4F_{5/2}$+$^4F_{3/2}$)-$^4G_{11/2}$. These processes occur between two excited Er ions with energy migration mechanism which will be discussed elsewhere. Although there are also the CR processes in ZnS:ErF ACELTF for a main emission energy level, the probability of the CR is small. Therefore, the optimum concentration value of Er ion is higher and Er is a better green luminescent center.

Figure 2(c) shows the ACEL spectra of ZnS:HoF thin films vs. Ho concentration. Green emission is due to transition from $^5S_2$ to $^5I_8$ and red emission is due to transition from $^5F_3$ to $^5I_7$ or from $^5F_5$ to $^5I_8$. The green emission intensity is decreased relatively to red one at higher Ho concentration. From energy level of Ho ion, it is found that the energy gap between $^5S_2$ and $^5I_4$ is almost the same as the energy gap between $^5I_8$ and $^5I_7$. The CR process may take place between Ho ion in the excited state and Ho ions in the ground state of the nearest Ho site. The probability of the CR is larger because there are a large number of unexcited Ho ions near an excited Ho ion. Therefore, the brightness of ACEL in ZnS:HoF thin films is low and the optimum concentration is low.

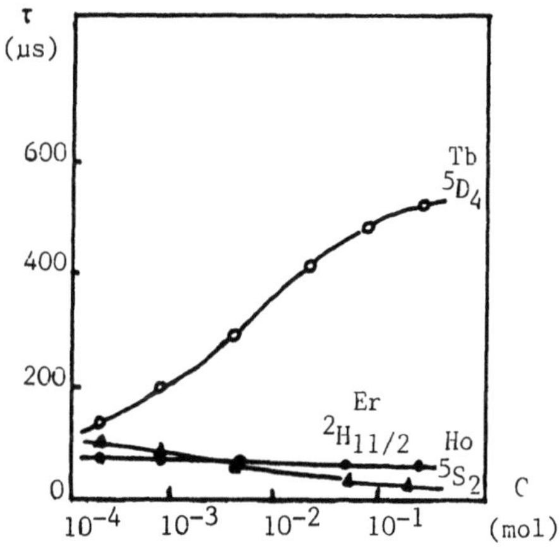

Fig. 3 The relation of lifetimes of excited states $^5D_4$, $^2H_{11/2}$ and $^5S_2$ vs. doping concentration

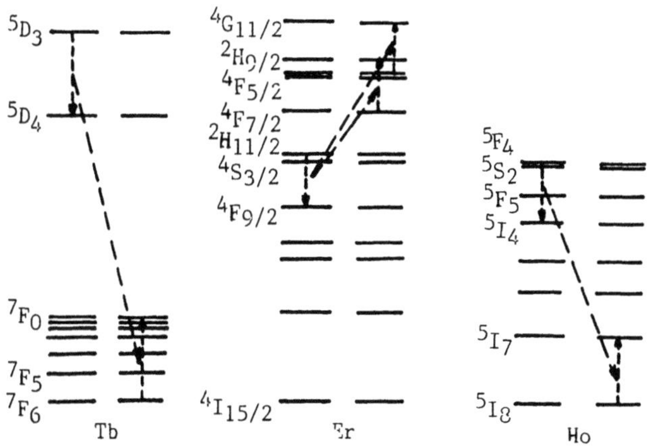

Fig. 4 Cross-relaxation processes of ACEL in ZnS TF doped with Tb, Er and Ho ions

The decay time of $^5D_4$ energy level of Tb ion increases with increasing Tb concentration, as shown in Fig. 3. The nonradiative CR process from $^5D_3$ to $^5D_4$ not only favours green emission but also increases the decay time of $^5D_4$ energy level. The decay times of $^2H_{11/2}$ and $^5S_2$ energy levels of Er and Ho ions decrease with increasing Er and Ho concentration. This result shows that the nonradiative transition of these energy levels will increase with increasing concentration. The decay time of $^5S_2$ energy level of Ho ion decreases more rapidly with increasing Ho concentration. It proves that the probability of the CR of Ho ion is large.

From the experimental results mentioned above, CR models for ZnS:TbF, ZnS:ErF and ZnS:HoF ACELTF are proposed, which are shown in Fig. 4.

3. Conclusion

ZnS:TbF, ZnS:ErF and ZnS:HoF thin films ACEL emit green light. Their ACEL spectra, EL decay and brightness all depend on concentration. Three RE ions have different CR models, therefore, have different optimum concentration values and different maximum brightness.

References

1. D. Kahng: Appl. Phys. Lett. 13, 210 (1968)
2. T. Inoguchi: SID'74, 84 (1974)
3. D.L. Dexter: J. Chem. Phys. 21, 836 (1953)
4. S.Y. Zhang: Luminescence and Display Devices (in Chinese) 4, 31 (1983)
5. H. Kobayashi: Phys. stat. sol. (a)88, 713 (1985)

Pulse-Excited Characteristics of Electroluminescent Device Based on ZnS:Tb, F Thin Films

H. Ohnishi

Faculty of Engineering, Ehime University, Matsuyama, Ehime 790, Japan

1. Introduction

Intensive attention has been paid to electroluminescence from the ZnS:Tb,F thin films because they have high feasibility to develop a green emitting device in a full color display panel /1/. A laboratory model of the device exhibits a luminance beyond 100 [cd/m^2] at 60 [Hz] drive /2/. This luminance level is acceptable for practical applications. In addition to the technological approaches, efforts have also been made to clarify luminescent mechanisms in thin film electroluminescent devices /3-11/. The unsolved problems, however, still remain in understanding the excitation mechanisms of luminescent centers, the carrier transport processes in high electric field, etc.

One of the difficulties in the theoretical treatment originates in the fact that the free electrons drift in a long range order through the phosphor in electroluminescence, which is essentially different from photoluminescence. This difficulty is eased on the assumption that the excitation probability is independent of the radiative transition probability in luminescent centers. If this tentative model is accepted, an expression for the electroluminescent intensity can be obtained by using the excitation probability P_{ex} and the radiative transition probability τ_r^{-1}. This work deals with the theoretical expression for the light output waveform and the experimental results on the pulse-excited characteristics of the device, then the obtained results are submitted for discussion.

2. Expression for Current and Light Output Waveforms

2.1 Current Waveform

A thin film ac electroluminescent device has the double insulator layer structure which is constructed with layers of transparent electrode, insulator, zincsulfide doped with terbium fluoride, insulator, and aluminium electrode. The device is driven in an ac pulse excitation mode. The carrier generation in the active layer is due to the electron emission from the traps at the interface between insulator and phosphor layers. The injected electrons gain the high kinetic energy during travel through high fields in the active layer. The energetic electrons may generate the electron hole pairs in addition to the impact excitation of luminescent centers, but in this work the electronic avalanching is neglected. These assumptions offer easy analyses for the current and light output waveforms.

Let n_0 be the trapped electron density at the interface and τ_t be the relaxation time for the electron emission. After the pulse is turned on, the rate of change in the trapped electron number is given by

$$dn/dt = -n/\tau_t . \tag{1}$$

Taking t=0 at the start of the voltage pulse, the solution for the rate equation is $n = n_0 \, exp(-t/\tau_t)$. Thus the injected electron number n_{in} is

$$n_{in} = n_0 \{1 - exp(-t/\tau_t)\} . \tag{2}$$

The conductive current through the active layer is obtained by differentiating (2). Thus, the current waveform is expressed as

$$J(t) = (qn_0/\tau_t)\, exp\,(-t/\tau_t). \tag{3}$$

2.2 Light Output Waveform

The phosphor layer under consideration is activated uniformly with terbium impurity in a concentration level N_i. The number of terbium centers is $N_i\,dx$ in a small region dx at a distance x from the cathodic interface. The injected electrons can arrive at the position x during the traveling time t_x given by $x/\mu F$, where μ is the electron mobility and F is the electric field strength. The electrons travel through the small region dx and excite the luminescent centers with the probability P_{ex}. The rate equation for the excited state luminescent centers N_{ex} is

$$\frac{dN_{ex}}{dt'} = P_{ex}\frac{J(t')}{q} N_g - \frac{N_{ex}}{\tau_r}, \tag{4}$$

where τ_r is the relaxation time for the light emission and $t' = t - t_x$.

The number of ground state luminescent centers approximates to $N_i\,dx$ when $N_i\,dx > N_{ex}$. Assuming this approximation and substituting (3) to (4), we can obtain the solution for N_{ex} as a function of t'. The light output in the small region dx is given by N_{ex}/τ_r, and the time that the injected electrons arrive at x is expressed as $t' = t - t_x = t - x/\mu F$. By using these relationships, the light output waveform can be expressed as follows.

$$L(t) = h\nu\,\tau_r^{-1}\int^{s} N_{ex}dx$$

$$= h\nu\, P_{ex}n_0 N_i \frac{\mu F}{\tau_r - \tau_t}\left[\tau_r\, exp\,(-\frac{t}{\tau_r})\left\{exp\,(\frac{s}{\tau_r\,\mu F}) - 1\right\}\right.$$

$$\left. - \tau_t\, exp\,(-\frac{t}{\tau_t})\left\{exp\,(\frac{s}{\tau_t\,\mu F}) - 1\right\}\right], \tag{5}$$

where $h\nu$ is the photon energy, s is the thickness of active layer. The expression (5) shows that the light output rises up when the pulse is turned on and has a maximum value at

$$t_{max} = \frac{\tau_r\,\tau_t}{\tau_r - \tau_t}\, ln\left[\frac{exp\,\{s/(\tau_t\,\mu F) - 1\}}{exp\,\{s/(\tau_r\,\mu F) - 1\}}\right]. \tag{6}$$

Thus it is shown that the light output waveform depends on the relaxation time for the electron emission and the electron transition in luminescent centers.

3. Experimental Results

3.1 Sample Preparation and Measurements

A sample used for the investigation is the ac electroluminescent device consisting of 300-nm thick Al_2O_3, 600-nm thick ZnS:Tb,F, and 300-nm thick Y_2O_3 layers. The insulator layers were prepared by electron beam evaporation. The phosphor layers were deposited by conventional type diode rf sputtering under the condition that: the sputtering gas is argon, the gas pressure during sputtering is 1.45 [Pa], the rf power density is 1.27 [W/cm²], the substrate temperature is 150−250 [℃], the deposition rate is about 20 [nm/min]. After the deposition, the phosphor films were thermally treated at 350−450 [℃] for 30 [min] in a vacuum.

The conductive current through the active layer was measured by a high speed capacitance bridge having the electrical configuration of electroluminescent device, variable capacitance C_2, and two resistors R_1 and R_2. Under the balanced conditions of $R_1=R_2=R$ and $C_2=(C_s\ C_{ox})/(C_s+C_{ox})$ in the electrical bridge, the voltage V_1 across R_1 and voltage V_2 across R_2 are detected by a differential amplifier with a voltage amplification of 10. The conductive current through the active layer is given by $\{(C_{ox}+C_s)/C_{ox}\}\times(V_1-V_2)/10R$, where C_{ox} is an equivalent capacitance for the insulator layers and C_s is an equivalent capacitance for the phosphor layer. The light output was measured by a calibrated photomultiplier tube (PMT). The rise time and fall time of the pulse used for measurements are about 1 [μs]. The measurement system of current and light output has a response speed of 0.5 [μs].

3.2 Current and Light Output Waveforms

Figure 1 is a trace of conductive current waveform in a half cycle with the aluminium electrode positively biased. The current rises from zero to a maximum value of 1.25 [A/cm^2] at 1 [μs] after the pulse is turned on. A time lag of 1 [μs] in the current waveform is mainly due to the rise time of pulse voltage. It is confirmed from the *log* J vs t plots that the decay is an exponential function. The decay time in the figure is estimated to be 2.15 [μs]. Figure 2 shows the light output waveform when the device is driven under the same condition in the current waveform. The light output rises and has a maximum value of 2 [mW/cm^2] at 15 [μs] after the pulse is turned on. The luminescent decay time is estimated to be 660 [μs] from the slope of the *log* L vs t characteristic.

Figure 3 shows the voltage dependence of decay time in current and luminescence. The luminescent decay time τ_r is slightly decreased from 700 [μs] to 650 [μs] and the relaxation time of electron emission is rapidly decreased from 4.7 [μs] to 1.65 [μs] while the applied voltage is elevated from 190 [V] (= V_{th}) to 262 [V]. Figure 4 shows

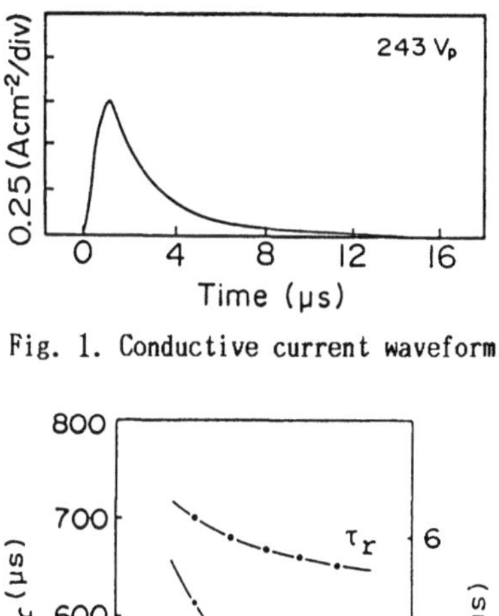

Fig. 1. Conductive current waveform.

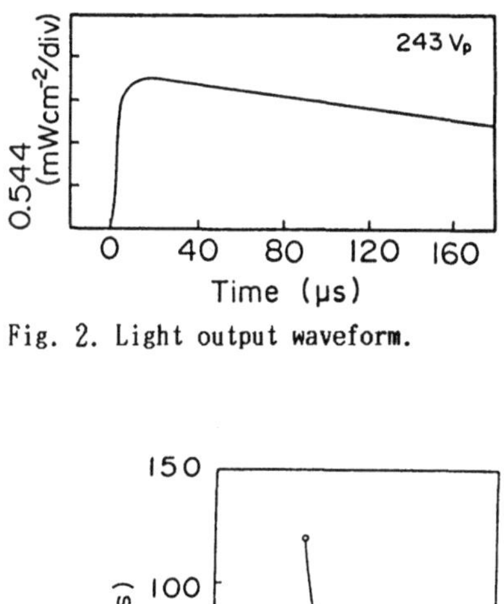

Fig. 2. Light output waveform.

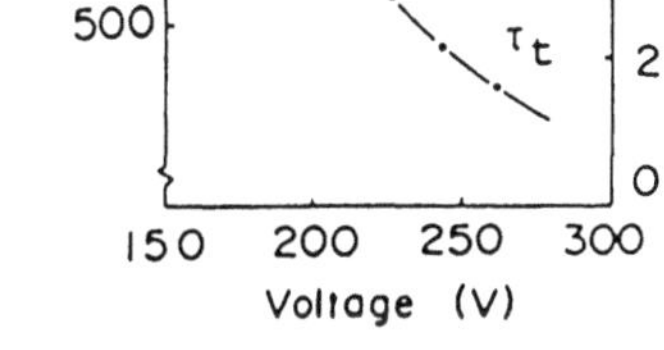

Fig. 3. Voltage dependences of τ_r and τ_t.

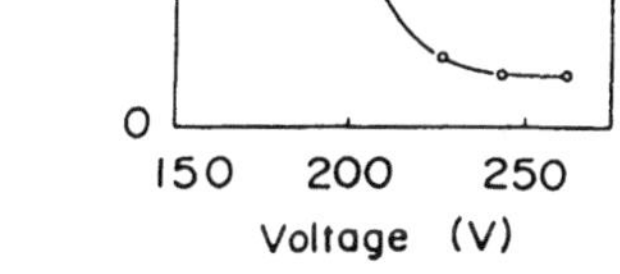

Fig. 4. Voltage dependence of t_{max}.

the voltage dependence of t_{max} which means time when the light output is maximum. As can be seen from the figure, a response speed of light output is slow in a voltage region just above the threshold voltage, and t_{max} is weakly dependent on the voltage in a region beyond 225 [V] .

4. Discussion

The fitting between the theoretical expressions and the experimental results will be examined in the case that the device is driven at 243 [V] . τ_r and τ_t are found to be 660 [μs] and 2.15 [μs] , respectively, from Fig. 3. The electric field strength in the phosphor film is estimated to be about 2 [MV/cm] considering the film thicknesses and the threshold voltage. When the electron drift mobility, which is not measured in the experiment, is assumed to be above 1 [cm^2/Vs] , the term $\tau \mu F$ is larger than s of the phosphor thickness. In this situation the expessions (4) and (5) are simplified by using the approximation of *exp* (y) $\fallingdotseq$ 1+y when y$\ll$1. The simplified expressions for L(t) and t_{max} do not contain the term μF so that it is easy to compare with the experimental results. It has been confirmed that a good fitting is obtained taking into account the time lag in conductive current. However, it is very difficult to calculate the drift velocity of electron from the expression (6) by using the measured values because the exponential terms in the expression (6) are close to unity.

This paper does not discuss the excitation probability of luminescent centers P_{ex} and the relaxation time τ_t , which are important parameters to understand the luminescent mechanisms. If the direct impact excitation is accepted to explain the luminescent mechanism, the excitation probability is obtained from the hot electron distribution functions (for example, Baraff's distribution function /6,13/) and the impact cross section of luminescent centers with hot electrons (for example, the assumption of step function /6/). An expression for the electron emission rate τ_t^{-1} has been derived in the case of the field emission of electrons trapped in a spherical well /13/. The knowledge of the excitation probability and the electron emission rate can give more details of the light output waveform. The investigation for this is in progress.

Acknowledgment

The author wish to express his gratitude to Mr. T. Okuda of Ehime University for his technical assistance during the course of this work, and to partial support from Fuji Xerox Co. (Scientific Relations).

References

1. C. N. King, R. E. Coovert, and W. A. Barrow : In Conf. Record 1987 Inter. Display Res. Conf., 14 (1987)
2. H. Ohnishi, Y. Yamasaki, and R. Iwase : Proc. Soc. Information Display 28, 345 (1987)
3. Y. S. Chen, M. V. DePaolis, Jr., and D. Kahng : Proc. IEEE 58, 184 (1970)
4. D. C. Krupka : J. Appl. Phys. 43, 475 (1972)
5. D. C. Krupka and D. M. Mahoney : J. Appl. Phys. 43, 2314 (1972)
6. Y. S. Chen and D. C. Kurupka : J. Appl. Phys. 43, 4089 (1972)
7. H. Kobayashi, S. Tanaka, H. Sasakura, and Y. Hamakawa : Jpn. J. Appl. Phys. 12, 1854 (1973)
8. H. Kobayashi, S. Tanaka, H. Sasakura, and Y. Hamakawa : Jpn. J. Appl. Phys. 13, 264 (1974)
9. R. Mach and G. O. Muller : phys. stat. sol. (a)69, 11 (1982)
10. K. Okamoto and K. Watanabe : Appl. Phys. Lett. 49, 578 (1986)
11. K. Okamoto and S. Miura : Appl. Phys. Lett. 49, 1596 (1986)
12. G. A. Baraff : Phys. Rev. 133, A27 (1964).
13. S. R. Pollack and J. A. Seitchik : In Applied Solid State Science , ed. by R. Wolfe, (Academic, New York, London 1969)

AC Electroluminescence of Ho-Implanted ZnS Thin Films

Lijian Meng, Changhua Li, and Guozhu Zhong

Changchun Institute of Physics, Academia Sinica, P.R. China

1. Introduction

Rare earth ions are considered to be advantageous in the production of a wide variety of emission colors. To develop multi-color displays, ZnS:RE electroluminescence thin film (ELTF) devices have been designed [1]. In order to get uniform doping and avoid the effect of fluorine ions we fabricate ZnS:Ho thin films by means of ion-implantation. This paper will discuss the effect of the annealing temperature on ACEL characteristics of these devices.

2. Device Preparation

The ELTF devices have a glass-ITO-Y_2O_3-ZnS:Ho-Al structure. The Y_2O_3 insulator layer is prepared by electron-beam evaporation. The ZnS layer was grown by the atomic layer epitaxy (ALE) method [2], into which Ho ions, with energy of 150 keV and dose of 1.7×10^{14} cm^{-2}, were implanted at room temperature. The samples implanted were of three types, designated A, B and C, respectively. Sample A was not annealed, however, samples B and C were annealed in N_2 atmosphere for two hours. The annealing temperatures were 300 C and 400 C, respectively. An Al electrode was evaporated onto the samples.

3. Results and Discussion

Figure 1 shows ACEL spectra of samples A, B and C at room temperature. The main emission peaks are located at 462.6 nm, 492.6 nm, 548.7 nm and 658.3 nm, corresponding to the transitions of $^5F_1$-$^5I_8$, $^5F_3$-$^5I_8$, $^5S_2$-$^5I_8$ and $^5F_5$-$^5I_8$. The relative emission intensities are defined as I_1, I_2, I_3 and I_4, respectively. It was found that I_3/I_4 increases with increasing annealing temperature, but that I_1/I_4 and I_2/I_4 at first increase and then decrease with increasing annealing

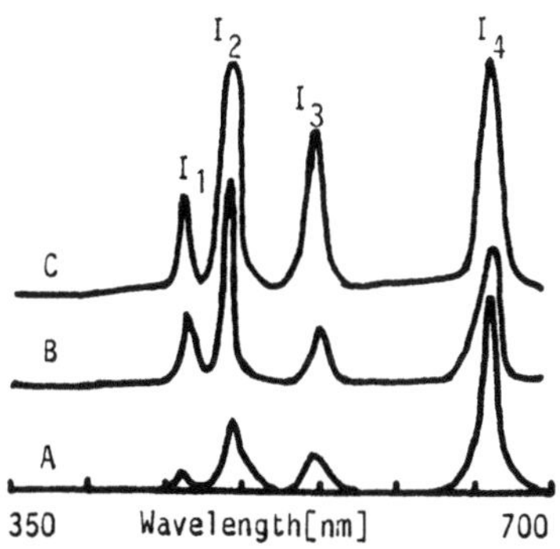

Fig. 1 ACEL spectra of samples A, B and C at RT (V=30 V, f=1 kHz)

Table 1 Ratio of intensities of spectra lines

	I_1/I_4	I_2/I_4	I_3/I_4
Sample A	0.13	0.36	0.23
Sample B	0.58	1.42	0.44
Sample C	0.46	0.95	0.70

temperature, as shown in Table 1. The ion implantation damages the periodicity of ZnS crystal lattice and will thus result in the increase of the probability of scattering of hot electrons. This process will strengthen emission intensities of lower energy levels. The damage to the crystal lattice was partly restored after annealing, and this should mean that the hot electrons can more easily obtain higher energies. This process will strengthen emission intensities of higher energy levels. It can be seen that the emission intensities of higher energy levels fall when annealing temperature is higher than 300°C. Figure 2 shows the B-V curves of samples A, B and C. The brightness increases when annealing temperature is lower than 300°C and decreases when annealing temperature is higher than 300°C. Brightness reaches maximum value when the annealing temperature is 300°C. Therefore we suppose that some new defect centers appear when the annealing temperature is higher than 300°C, the rare earth ions and these defect centers forming complex centers [3]. The emission intensities of the higher energy levels of these complex centers are weaker. Some experimental results have shown that I_1 and I_2 are weaker in $ZnS:HoF_3$ thin films than in ZnS:Ho thin films [1,4]. We suggest that F ions and Ho ions form complex centers and that these complex centers are unfavourable for emission of higher levels.

Figure 3 shows the EL decay curves of the emission lines, located at 492.6 nm and 658.3 nm, of sample A. Although their EL decay curves all obey the exponential law, their EL decay times are different. Therefore we suggest that the origin of emission lines located at 492.6 nm and 658.3 nm are transition from $^5F_3$ to $^5I_8$ and from $^5F_5$ to $^5I_8$, respectively [1,5].

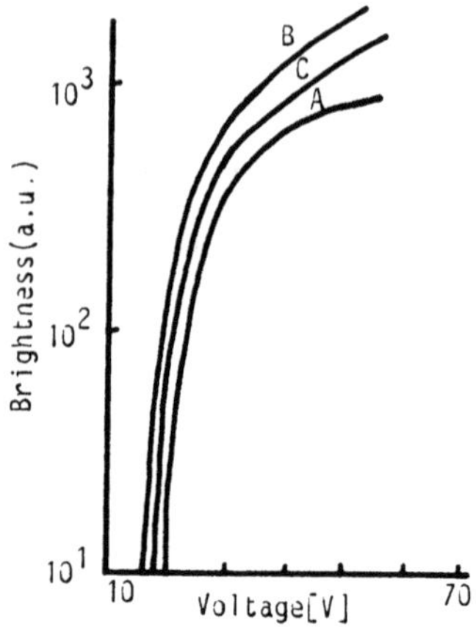

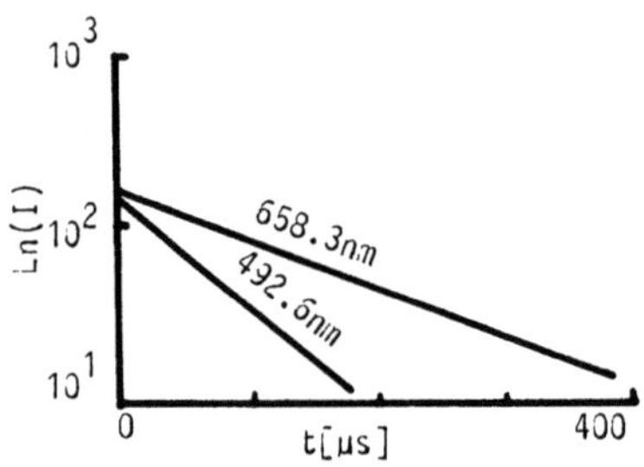

Δ Fig. 3 The EL decay curves of emission lines located at 658.3 nm and 492.6 nm of sample A (RT)

◁ Fig. 2 B-V curves of samples A, B and C

4. Conclusion

We have fabricated ZnS:Ho thin films by means of ion-implantation and obtained green ACEL. The properties of these devices may be improved by proper post-deposition annealing. Optimum annealing temperature was found to be about 300°C.

The complex centers consisting of RE ions with crystal defects are unfavourable for emission of higher levels.

References

1. E.W. Chase: J. Appl. Phys. 40, 2512 (1969)
2. T. Suntola: U. S. Patent, 4058430
3. J.Q. Yu: Chinese J. Semi. 1, 192 (1980)
4. G.Z. Zhong: SID 88 Digest, 287 (1988)
5. I. Szczurek: J. Luminescence 14, 389 (1976)

High-Luminance ZnS:Sm, F Thin-Film Electroluminescent Devices Using Ferroelectric $PbTiO_3$ Thin-Film

R. Fukao[1;*], *H. Fujikawa*[1], *M. Nakamura*[1], *Y. Hamakawa*[1], *and S. Ibuki*[2]

[1]Faculty of Engineering Science, Osaka University,
1-1 Machikaneyama-cho, Toyonaka-shi, Osaka 560, Japan
[2]Faculty of Engineering, Setsunan University,
17-8 Nakamachi, Ikeda, Neyagaea-shi, Osaka 572, Japan

ABSTRACT

Performances of ZnS:Sm,F thin-film electroluminescent(EL) devices using ferroelectric $PbTiO_3$ and silicon nitride(SiNx) thin films as insulating layers have been examined. The threshold voltage of the device(Vth) and the field strength in the phosphor layer(Eth) are 90V and 1.1MV/cm respectively and are low as compared with those of the conventional EL devices. Considerably high luminance of 400cd/m^2(at 1kHz) is obtained, and is due to the large amount of carriers injected from $PbTiO_3$ layer into the phosphor layer.

1 INTRODUCTION

Intensive studies have been made to improve the performances of EL devices, especially in the three primary-colors, for the purpose of the development of full color EL displays. To realize this potential one must address the important problem of reducing the driving voltage. This is because it exceeds 200V in current technology, resulting in high costs and requiring custom-made ICs. We have attempted to reduce the Vth and investigated the application of ferroelectric $PbTiO_3$ thin films, which have high dielectric constant, to insulating layers of EL devices. As a result a Vth of less than 60V was obtained in ZnS:Mn EL device using $PbTiO_3$ thin film as the only insulating layer/1/. Further investigation into a EL double insulating structure has been carried out using $PbTiO_3$ to improve emissive characteristics while lowering the driving voltage. Consequently a high luminance of 5300cd/m^2 was obtained in a ZnS:Tb,F green color EL device with a low threshold voltage of 100V by using $PbTiO_3$ and SiNx prepared by plasma CVD(PCVD-SiNx) as the insulating layers/2/. This structure was extended to ZnS:Sm,F red/orange EL devices and an improvement was obtained in their device performances.

In this paper, the preparation method and the characteristics of the ZnS:Sm,F red/orange EL devices are reported.

2 FABRICATION of EL DEVICES

The structure of the fabricated EL device is shown in Fig.1. The first insulating layer, $PbTiO_3$ of thickness 7000Å, was deposited on an ITO-coated glass substrate by rf magnetron sputtering. The sputtering conditions are shown

*on leave of absence from Hithchi Maxell Co.,Ltd.

Springer Proceedings in Physics, Vol. 38 **Electroluminescence**
Editors: S. Shionoya and H. Kobayashi

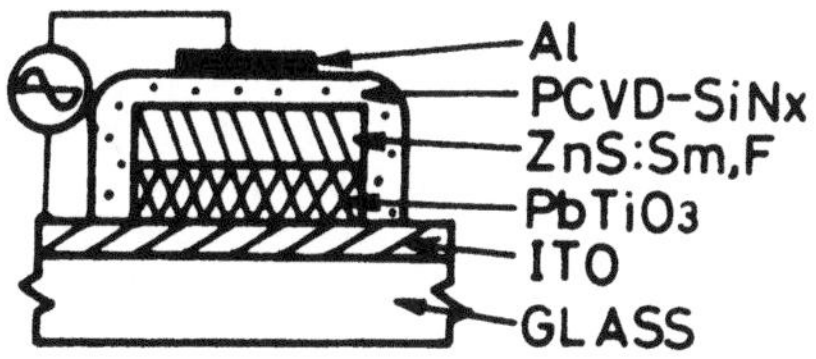

Fig.1. Schematic diagram of a ZnS:Sm,F EL device.

Table 1. Sputtering conditions

RF input power	150W
Target-substrate distance	40mm
Sputtering gas	$Ar(90\%)+O_2(10\%)$
Gas pressure	26.6Pa
Deposition rate	55Å/min
Substrate temperature	640°C

in Table 1. The dielectric constant and the dielectric breakdown strength of the film were 120 and 0.2MV/cm respectively. The phosphor layer, ZnS:Sm,F of thickness 5000Å, was deposited on the $PbTiO_3$ layer at 200°C by electron beam evaporation. The second insulating layer, PCVD-SiNx of thickness 1000Å, was deposited on the ZnS:Sm,F layer. The dielectric constant and the dielectric breakdown strength of the SiNx film were 4.5 and 7MV/cm respectively.

3 EL CHARACTERISTICS

Luminance and luminous efficiency of the ZnS:Sm,F EL device as a function of applied voltage are shown in Fig.2. The device was driven under an 1kHz sinusoidal wave. The characteristics of the ZnS:Sm,F EL device using Ta_2O_5 instead of $PbTiO_3$ as insulating layers are also shown for comparison. For the EL devices using $PbTiO_3$, Vth is 90V, and is 50V lower than that of the EL device using Ta_2O_5. Eth is low, 1.1MV/cm, while that of EL device using Ta_2O_5 is 1.85MV/cm. The luminance at the voltage of Vth+60V(L_{60}) in the EL device is 220cd/m^2, and is three times higher than that of the EL device using Ta_2O_5. The maximum luminance(Lmax) is 400cd/m^2, and is about four times higher than that of the EL device using Ta_2O_5. The maximum luminous efficiency(ηmax) is 0.083 lm/W, and is about twice as high as that of the EL device using Ta_2O_5. However, its luminous efficiency is still lower than those of EL devices using other kind of phosphors such as ZnS:Mn or ZnS:Tb,F.

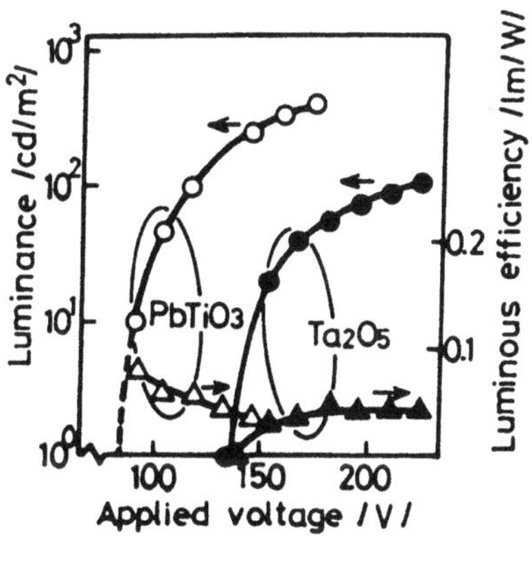

Fig.2. Luminance and luminous efficiency of ZnS:Sm,F EL devices as a function of applied voltage.

Figure 3 shows the density of transferred charge plotted against applied voltage for the EL devices using $PbTiO_3$ and Ta_2O_5. For the EL device using $PbTiO_3$, the transferred charge increases abruptly with increase of applied voltage. The density of transferred charge is very large in comparison with that of the EL device using Ta_2O_5 and the large magnitude of transferred charge is interpreted as carrier generation from the shallow electron trap state in the $PbTiO_3$ layer. The large number of carriers is considered to be the cause of the EL device's high luminance and low clamped electric field.

Luminance is plotted against the density of transferred charge in Fig.4. The luminance of the EL device using $PbTiO_3$ is lower than that of the EL device using Ta_2O_5. The results indicate that a large amount of carriers injected into the phosphor layer is not effective for impact excitation of the Sm centers because their kinetic energies are insufficient.

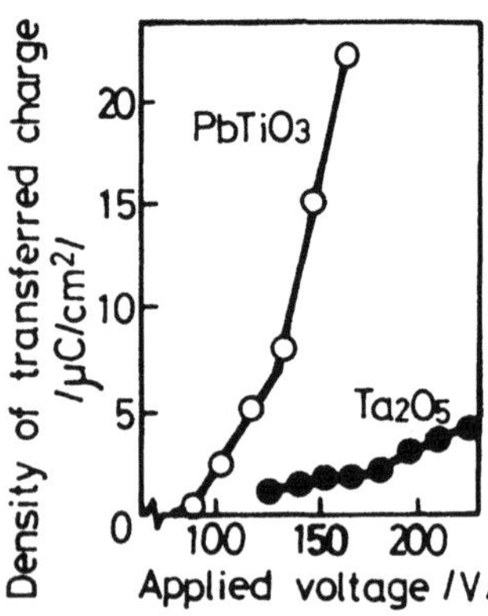

Fig.3. Density of transferred charge as a function of applied voltage in the ZnS:Sm,F EL device.

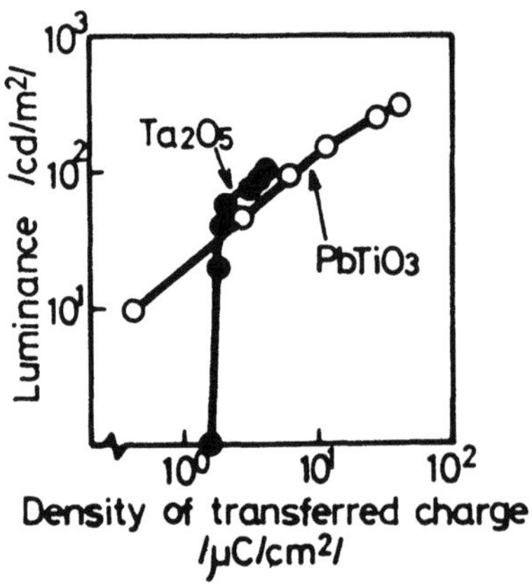

Fig.4. Luminance as a function of density of transferred charge in the ZnS:Sm,F EL devices.

4 SUMMARY

A low threshold voltage of 90V and a high maximum luminance of 400cd/m^2 have been attained in ZnS:Sm,F red/orange EL devices using ferroelectric $PbTiO_3$ thin films. These are attributed to a large number of carrier injections from the $PbTiO_3$ layer into the phosphor layer. Further investigation is desired to improve the performance of the EL device, especially its luminous efficiency.

5 LITERATURE

1. K.Okamoto, Y.Nasu and Y.Hamakawa, IEEE Trans. Electron Devices ED-28, 698(1981).
2. R.Fukao, T.Manabe and Y.Hamakawa, Extended abstracts of 172nd electrochemical society meeting, 87-2, 1719(1987).

Electroluminescent Devices With $CaS:Eu^{2+}$ Active Layer Grown by R.F. Reactive Sputtering

D. Yebdri, P. Benalloul, and J. Benoit

Laboratoire d'Optique de la Matière Condensée, Université P. et M. Curie, 4 place Jussieu, F-75252 Paris Cedex 05, France

1. INTRODUCTION

Over the past few years, intensive work has been made to achieve color thin film electroluminescent (E.L.) devices. Alkaline-earth sulphide CaS and SrS are the most promising materials for red and blue colors [1,2,3]. Nowadays, the active layer $CaS:Eu^{2+}$ is the material of choice which emits the fundamental red color. This layer is usually grown by electron-beam evaporation [2,3] but it is necessary to heat the substrate up to a 680°C temperature, which leads to technological difficulties. R.F. reactive sputtering technology can overcome these problems [4,5]. We report here on the optimization of growth conditions for $CaS:Eu^{2+}$ thin films by this technique. We also report the results obtained when this active layer is integrated in E.L. devices.

2. SAMPLE PREPARATION

The $CaS:Eu^{2+}$ layers were deposited from a mixed powder target of CaS and 0.25 mol % of EuS on glass substrate. The radio-frequency power was varied between 1.9 and 2.2 W/cm^2 and the substrate temperature between 220 and 280°C. For the RF reactive sputtering, we used a mixed gas of Ar + x% H_2S (x = 0.5 ; 1 ; 2.5 ; 5). The deposition rate and the thickness of the films were controlled optically during the deposition. The thickness of the films is about 730 nm ; the deposition rate is about 20 Å/min for non reactive sputtering, and increases in presence of H_2S, up to 50 Å/min.

3. CHARACTERIZATION OF THE $CaS:Eu^{2+}$ THIN FILMS

3.1 X-ray diffraction

This X-ray analysis was done on $CaS:Eu^{2+}$ thin films which were all deposited at the same substrate temperature (280°C) and the same RF power (2.2 W/cm^2). The H_2S/Ar concentration was the parameter.

The X-ray diffraction patterns contain only the (200) and (400) lines.

Figures 1a and 1b display the dependence of the intensity I_{200} and the half-peak width $\Delta\theta_{200}$ of the (200) line on the H_2S/Ar concentration.

These results show that the presence of H_2S during the deposition improves the cristallinity of the $CaS:Eu^{2+}$ thin films, the optimal concentration being 2.5%. The films grown with this concentration have a good texture and are oriented with the (200) plane parallel to the substrate.

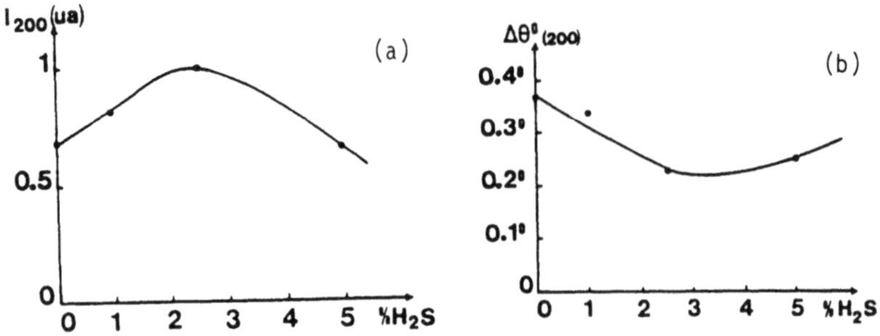

Fig. 1 : Evolution of the (200) line with the H_2S/Ar concentration.
a) Variation of the intensity I.
b) Variation of the half-peak width $\Delta\theta_{200}$.

3.2 X-ray photoelectron spectroscopy (XPS)

In order to characterize the quality of the $CaS:Eu^{2+}$ thin films, we have carried out XPS measurements. This technique concerns the 100 upper angstroms of the layer and is one of the appropriate methods for determining which elements are present in the film and their chemical state. For E.L. devices this technique has been used by H. Kobayashi et al. [3] for CaS, by Mito et al. [6] for ZnS:Tb,F and by Y. Kageyama et al. [7] for SrSe:Ce. Figure 2 shows the S 2p spectra of sulphur in CaS layers deposited with various concentration H_2S/Ar. The S $2p^{3/2}$ peak of Ca was used for the calibration ; this peak is at 348 eV for the equipment we used. When the $CaS:Eu^{2+}$ thin films are deposited in a pure argon atmosphere, two signals S 2p(a) and S 2p(b) are observed. The first peak S 2p(a) at 161 eV arises from a sulphur ion bounded to calcium to form CaS. The second peak S 2p(b) at 169 eV arises from another state of sulphur, we attribute it to a sulfate SO_4^{2-} [8]. When H_2S is added to the argon sputtering gas, we observe a decrease in the S 2p(b) peak and an increase in the S 2p(a) peak. For a concentration H_2S/Ar = 2,5%, the S 2p(b) peak disappears completely. Therefore, a reactive sputtering is necessary to prevent from a sulphur deficiency of the CaS layer, the necessary amount of H_2S being 2.5% of Ar.

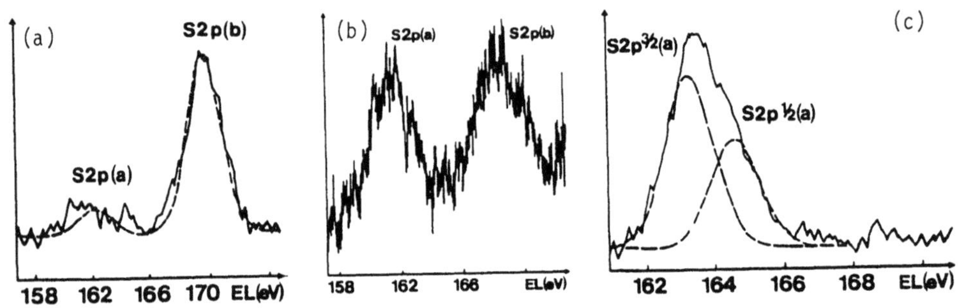

Fig. 2 : XPS of $CaS:Eu^{2+}$ layers : signal S 2p of S^{2-} ion versus H_2S concentration.
a = 0% ; b = 0.5% ; c = 2.5%

3.3 Cathodoluminescence spectra

The $CaS:Eu^{2+}$ (0.25 at.%) thin films grown by RF sputtering exhibit a red emission centered at 640 nm. The intensity of this emission depends on the deposition conditions. The most efficient thin films are deposited at a substrate temperature of 280°C, a RF power of 2.2 W/cm^2 and a H_2S/Ar concentration of 2.5%. The decay of the emission is very fast, with a decay time $\tau_e \leqslant 0.2$ μs, which is the resolution time of our apparatus (Fig. 3 a,b,c). Note that this cathodoluminescence emission does not present a slow decay component we could connect to some defects.

3.4 D.C. current measurements

In the literature, we found only one paper providing the resistivity of CaS thin films [9] which varies between 0.1 and 1 GΩ.cm. In order to determine the electrical properties of the CaS thin films, we have studied the current-voltage characteristics of ITO-CaS-Al capacitor. For all the samples, only the Al electrode exhibits an ohmic contact. Figure 4 represents the variations of the current density J versus the electrical field E. The J-E curve for ITO-ZnS-Al capacitor is shown for comparison.

In the ohmic region of the curve we find a resistivity of about 4.10^{13} Ωcm, a more realistic value according to the gap of CaS. The trap density is determined from the voltage at the slope discontinuity between the ohmic and the space charge regions [10] its value is about 4.10^{16} cm^{-3}. The J-E characteristics for CaS and ZnS are similar, except the fact that the traps in CaS are deeper : in the space charge region, the value of α being 2.5 instead of 2. On CaS capacitors, we have measured a value of 10 for the constant dielectric ϵ.

3.5 Conclusion

We have grown CaS:Eu^{2+} (0.25 mole %) thin films by R.F. reactive sputtering, the reactive gas being H_2S. We have studied the influence of the deposition

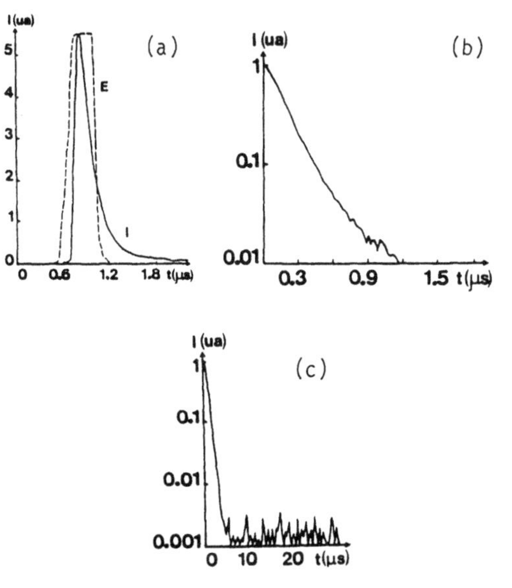

Fig. 3 : Decay of C.L. emission of CaS:Eu^{2+} films
E = electron flux
I = C.L. emission

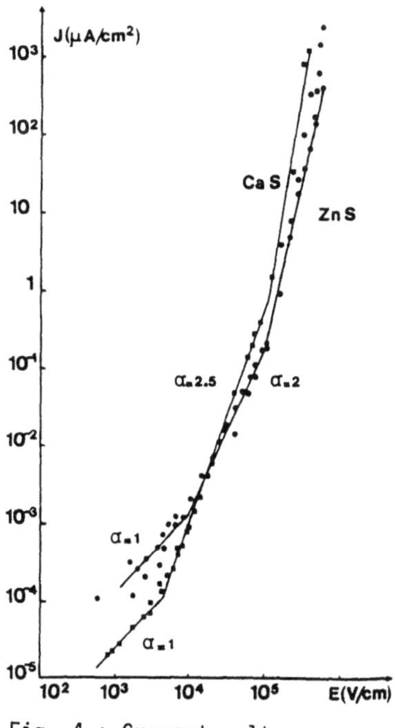

Fig. 4 : Current voltage characteristics.

conditions upon the properties of these layers by several techniques. We have established the necessity of a reactive sputtering and determined the optimal concentration H_2S/Ar : 2.5%. The thin films grown with this concentration have a good crystallinity and a good cathodoluminescence.

We have determined the resistivity and the trap density of the CaS thin films, these values are respectively $4.10^{13}\,\Omega$cm and 4.10^{16} cm^{-3}.

4. E.L. DEVICES

We have prepared E.L. devices with $CaS:Eu^{2+}$ active layer. These devices have the conventional structure with two insulators layers of Y_2O_3 or Ta_2O_5 deposited by R.F. sputtering. Just before the breakdown of the device by a high sinusoidal 5 kHz applied voltage, the brightness was only 2.10^{-2} cd/m^2 and the transferred charge across the device 6 nC/cm^2. For the critical voltage, we have calculated the electric field applied to each layer of the device. Taking into account the values of the thickness and the dielectric constant of Y_2O_3 and CaS layers, the values of the electric field are respectively 2 MV/cm and 2.6 MV/cm. Therefore $CaS:Eu^{2+}$ supports a very high electric field, but the transferred charge is too low to obtain a sizeable emission. We can attribute this result to the quality of the insulator-active layer interfaces.

This problem of a weak transferred charge has been also encountered by H. Ohnishi and al. [5] who solve it by codoping the active layer with copper.

Taking into account the value of the electric field supported by the active layer, in order to solve this problem of a low transferred charge, we have sandwiched the $CaS:Eu^{2+}$ active layer with two ZnS layers of 2000 Å thickness. This induces an increase of the transferred charge across the device and the brightness by a factor 200 ; these values are respectively 1.4 μC/cm^2 and 4 cd/m^2 at 5 kHz. This brightness is still too low compared to Ohnishi's result [5].

We have focused on the behaviour of this structure device, particularly the action of ZnS layers. By an analysis of the emission to a trapezoidal voltage and time resolved spectroscopy of EL emission, we can propose a model : ZnS layers take a prominant part as a source of energetic electrons and a mean to get good CaS interfaces ; the second point being the most important action. This result is in agreement with Gonzalez' [4] result on SrS:Ce devices encapsulated within two ZnS layers.

Partial support by the CNET is gratefully acknowledged.

REFERENCES

1. W.A. Barrow, R.E. Coovert and C.N. King, Digest 1984, SID Int. Symp., p.249
2. K. Tanaka, A. Mikami, T. Ogura, K. Taniguchi, M. Yoshida and S. Nakajima, Appl. Phys. Lett. 48, 1730 (1986)
3. H. Kobayashi, S. Tanaka, R. Konishi, H. Deguchi and Y. Mikami, Proceedings Int. E.L. Workshop, Kah-nee-ta, Oregon, 1986
4. C. Gonzalez, 7th Int. Display Research Conf., Eurodisplay's, p. B34 (1987)
5. H. Ohnishi and Y. Katayama, Proceedings Int. E.L. Workshop, Kah-nee-ta, Oregon, 1986
6. J. Mita, M. Koizumi, H. Kanno, T. Hayashi, Y. Sekido, I. Abiko and K. Niki Jap. J. of Appl. Phys. 26, L558 (1987)
7. Y. Kageyama, K. Kameyama and S. Oseto, Dig. 1986, S.I.D. Int. Symp. p.33
8. C.D. Wagner, W.M. Riggs, L.E. Davis, J.F. Moulder, G.E. Muilenberg, in Handbook of X-ray photoelectron spectroscopy, (Perkin Elmer Corp., Prairie Minnesota, 1979)
9. M. Ogawa, T. Shimouna, S. Nakada and T. Yoshioka, Jap. J. of Appl. Phys. 24, 168 (1985)
10. A. Rose, Phys. Rev. 97, 1538 (1955)

Improvement in Electro-Optical Characteristics of CaS:Eu Electroluminescent Devices

M. Ando[1], *Y.A. Ono*[1], *K. Onisawa*[1], *and H. Kawakami*[2]

[1]Hitachi Research Laboratory, Hitachi Ltd., 4026 Kuji-cho, Hitachi-shi, Ibaraki-ken 319-12, Japan

[2]Mobara Works, Hitachi Ltd., 3300 Hayano, Mobara-shi, Chiba-ken 297, Japan

Abstract

Electro-optical characteristics of rare-earth doped alkaline-earth-sulfide electroluminescent(EL) devices were studied. Measurement of luminance and transferred charge responses in CaS:Eu, CaS:Ce, SrS:Eu, and SrS:Ce EL devices revealed that rare-earth luminescent centers played a decisive role in controlling electro-optical characteristics. Especially, Eu was found to be the cause of the slow electro-optical response and a sharp decrease of luminance and transferred charge with a decrease of frequency in CaS:Eu. These characteristics were improved by co-doping a small amount of Ce to CaS:Eu. Especially, by doping 0.05mol% Ce to CaS:Eu(0.2mol%), the transferred charge response time decreased one-and-a-half orders of magnitudes and became almost the same as that of ZnS:Mn.

1. Introduction

Recently interest in multi-color thin film electroluminescent(EL) devices is growing rapidly and new phosphor layer materials, such as CaS and SrS, are employed to achieve higher luminance or better color purity than ZnS:Mn-based EL devices:[1]-[6] SrS:Ce for higher luminance even though the color is blue-green, and CaS:Eu for better red purity. However, electro-optical characteristics of these EL devices have been found to be quite different from those of commercially available ZnS EL devices.[7]-[14] Especially in CaS:Eu red EL devices, following characteristics have been found: (1) Luminance build-up with respect to time is very slow.[9] (2) Luminance vs. frequency (L-f) characteristics show a slow response of luminance at low frequencies, especially in low drive-voltage regions.[14] These characteristics are detrimental to matrix drive and have to be improved.

In the present paper, we clarify causes of these characteristics and propose a method of improvement. In order to determine whether the causes rest on the phosphor layer host material, CaS, or the luminescent center, Eu, four kinds of EL devices are fabricated: CaS:Eu, CaS:Ce, SrS:Eu, and SrS:Ce. We have studied the following electro-optical characteristics: Luminance vs. frequency (L-f) characteristics, transferred charge density vs. frequency (ΔQ-f) characteristics, and build-up characteristics of luminance and transferred charge with respect to time.

2. Experimental Results

The device structure is of the double insulating type; phosphor layers of CaS or SrS are fabricated by the electron beam deposition method at the substrate temperature of 500°C and Ta_2O_5 or Ta_2O_5/SiO_2 insulating layers were deposited by the rf-sputtering method. EL devices are driven with the voltage pulses of 200μs pulse width, of alternate polarity. Luminance L is measured by a photometer

Springer Proceedings in Physics, Vol. 38 **Electroluminescence**
Editors: S. Shionoya and H. Kobayashi

(Spectra Pritchard Photometer, Model 1980A), and response characteristics of transferred charge density ΔQ are measured by a newly devised transient charge measuring system composed of the Sawyer-Tower circuit[15] and a sampling circuit.

Figures 1 and 2 show luminance vs. frequency (L-f) characteristics and transferred charge density vs. frequency (ΔQ-f) characteristics, respectively. From these figures we observe the following: (1) In the case of Ce as a luminescent center, luminance L is proportional to frequency for all the voltages above the threshold voltage V_{th}, i.e. $L \propto f$, and the transferred charge density ΔQ takes an approximately constant value with respect to all the voltages above V_{th}. (2) In the case of Eu as a luminescent center, luminance becomes smaller than the values given by the $L \propto f$ relation at low frequencies and the transferred charge density decreases with the decrease of frequency. The latter behavior is more remarkable in CaS:Eu EL devices. In order to probe further the slow ΔQ response at low frequencies, time responses of L and ΔQ at 100Hz are measured. The results for CaS:Eu and CaS:Ce are shown in Fig.3. In the CaS:Eu case, L increases slowly and reaches a saturation level with a time constant 15s at 260V, just above the threshold voltage. With the increase in voltage, L response becomes faster and a time constant to reach a respective saturation level becomes smaller until it becomes approximately 1s at 280V. On the other hand, in CaS:Ce case, L response is found to be much faster than that of CaS:Eu. A detailed study has shown that the time constant to reach a respective saturation level is of the order of 100ms.

The temporal response behavior of ΔQ is almost identical to that of L in the respective EL devices. The similarity between L response characteristics and ΔQ response characteristics can be understood from the following argument. In the discussion of emission mechanism of thin film EL devices, it is argued that transferred electrons in the phosphor layer cause excitation of luminescent centers, which leads to eventual EL emission. Therefore, the luminance response should be governed by the transferred charge response. Based on the above argument, we discuss the transferred charge density characteristics in the following.

Figure 4 depicts voltage dependence of the transferred charge response time t_r, where t_r is defined by the 0%-90% rise time. In the figure, the abscissa is taken to be a voltage difference between an applied voltage and the threshold voltage

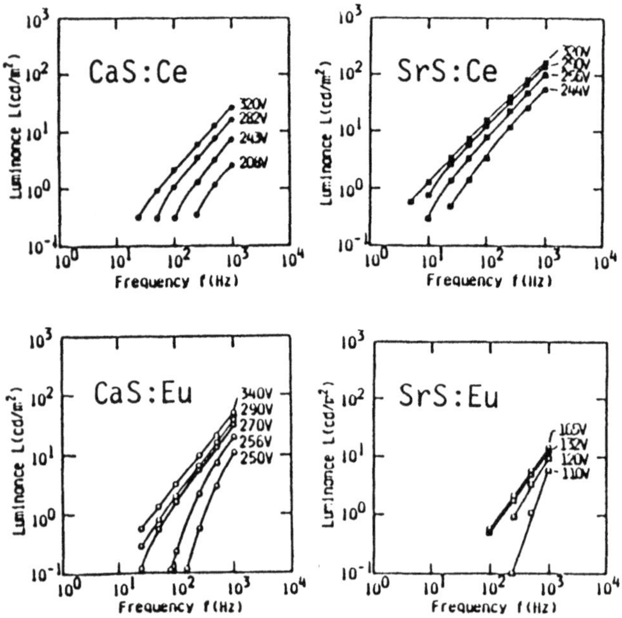

Fig.1 L-f characteristics

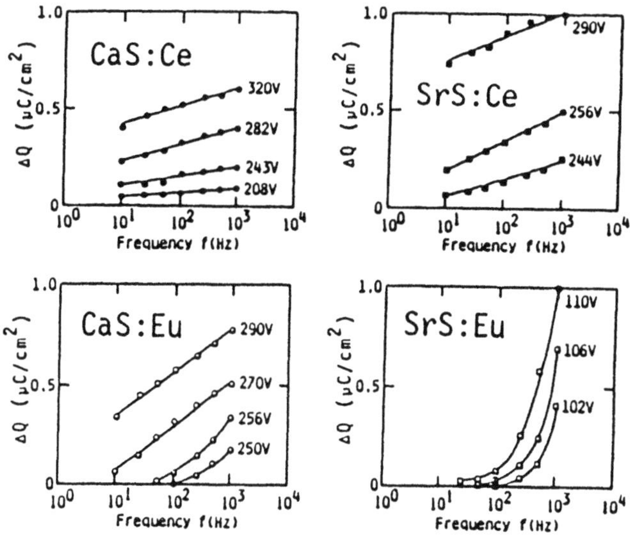

Fig.2 ΔQ-f characteristics

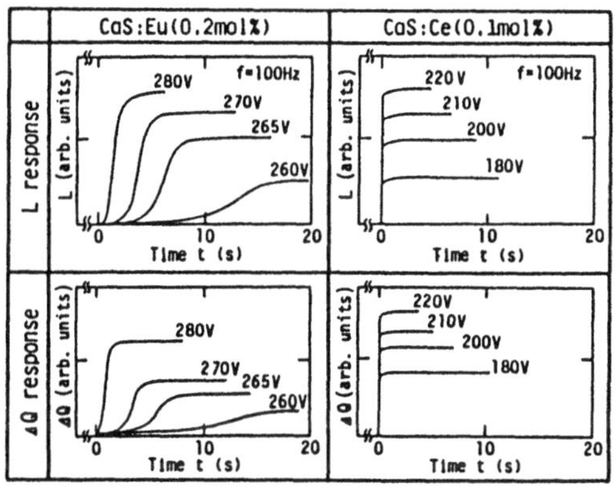

Fig.3 Time response characteristics of L and ΔQ of CaS:Eu and CaS:Ce

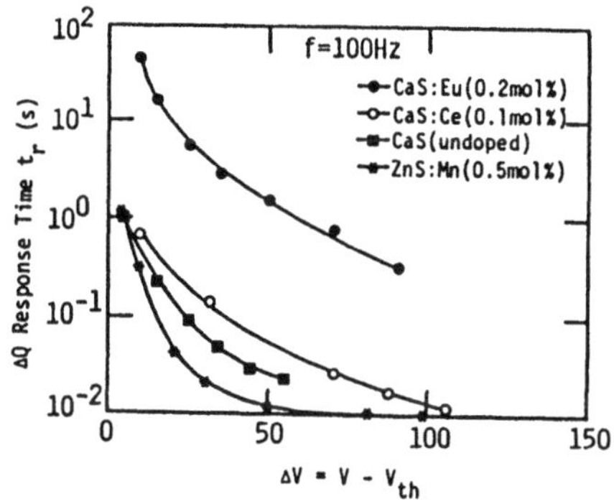

Fig.4 Voltage dependence of ΔQ response time t_r

which is denoted as ΔV, since only the voltage above the threshold voltages is relevant in creating transferred charges. In an undoped CaS, values for t_r are approximately equal to those of ZnS:Mn. Doping of Ce to CaS does not change t_r very much from that of undoped CaS. On the other hand, doping of Eu to CaS increases t_r substantially by one and a half orders of magnitudes. We note that the same kind of behaviors of t_r were observed in SrS, SrS:Ce and SrS:Eu EL devices; values for t_r in SrS and SrS:Ce were approximately equal to each other and values for t_r in SrS:Eu were one and a half orders of magnitudes larger than those of SrS and SrS:Ce.

From these results, we conclude that slow response characteristics of CaS:Eu EL devices depend on the luminescent center Eu, rather than on the phosphor layer host material CaS.

In order to improve response characteristics of CaS:Eu EL devices, a small amount of Ce was co-doped, which is written as CaS:Eu,Ce. Figures 5 and 6 show the results. In Fig.5, voltage dependence of the transferred charge response time

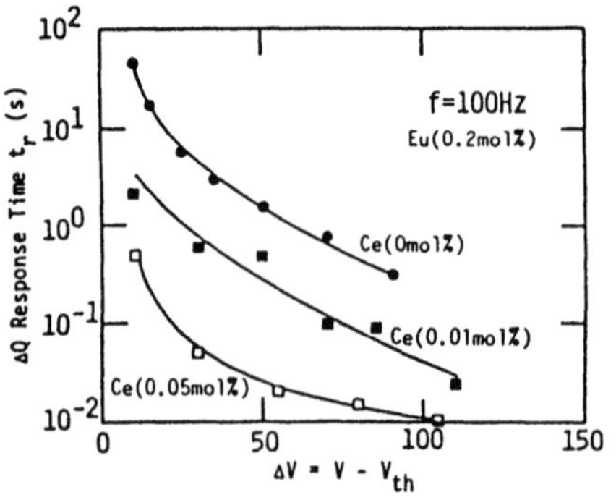

Fig.5 Effect of Ce co-doping on $t_r-\Delta V$ characteristics in CaS:Eu,Ce

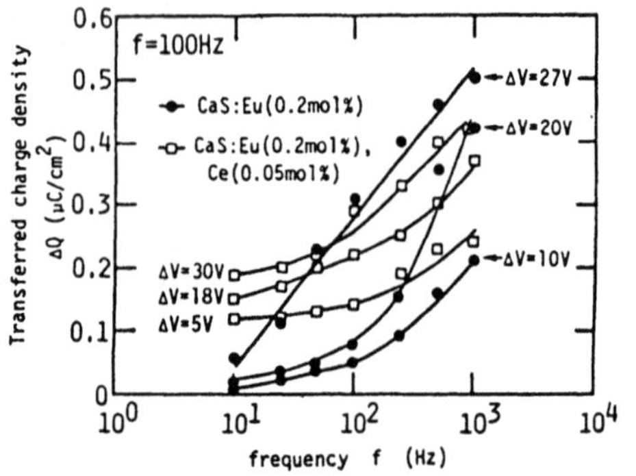

Fig.6 Effect of Ce co-doping on ΔQ-f characteristics

t_r is shown, where Ce concentration is varied from 0mol% to 0.05mol% while Eu concentration is fixed at 0.2mol%. From the figure, we observe that with the increase of Ce concentration t_r decreases and t_r of CaS:Eu(0.2mol%),Ce(0.05mol%) is one and a half orders of magnitudes smaller than that of CaS:Eu(0.2mol%). By comparing Fig.4 with Fig.3, we find that values for t_r of CaS:Eu(0.2mol%),Ce(0.05 mol%) are approximately equal to those of CaS:Ce or ZnS:Mn, indicating that response characteristics were improved to a practical level. Figure 6 shows ΔQ-f characteristics of CaS:Eu and CaS:Eu,Ce EL devices, with the voltage difference V as a parameter. From the graphs we observe that ΔQ-f characteristics are improved by co-doping Ce.

We note that EL emission color, however, did not change very much and remained deep red: The CIE color coordinates were x=0.690 and y=0.309 for CaS:Eu(0.2mol%) and x=0.674 and y=0.320 for CaS:Eu(0.2mol%), Ce(0.05mol%).

Since voltage dependence of transferred charge response time of CaS:Eu,Ce is approximately equal to that of CaS:Ce, it can be considered that Ce is excited first in CaS:Eu,Ce. The EL emission color, however, is red which is emitted from Eu luminescent centers. Therefore energy transfer process from Ce to Eu should exist.[12] From the above arguments, we conclude that co-doping a small amount of Ce to CaS:Eu improves response characteristics because Ce is excited first. This kind of effect of co-activators was also found in cathodoluminescence of phosphors, [16],[17] where this effect is called a sensitizing effect.

3. Discussions and Conclusions

Response time difference between CaS:Eu and CaS:Ce can be explained from the following arguments. From the electrical measurement, the threshold electric field E_{th} of CaS EL devices are as follows. CaS;E_{th}=1.6x10^{6}V/cm, CaS:Eu(0.2mol%); E_{th}=1.2x10^{6}V/cm, CaS:Ce(0.1mol%);E_{th}=1.0x10^{6}V/cm, CaS:Eu(0.2mol%),Ce(0.05mol%); E_{th}=1.0x10^{6}V/cm. These results indicate that the threshold electric field decreases from that of CaS by doping a luminescent center, Eu or Ce, and that the amount of decrease is larger for the case of Ce doping. From these, it may be concluded that Ce doped EL devices should be ionized more easily than Eu doped EL devices. Once an ionization of luminescent centers occurs, space charge within the bulk of phosphor layer should be formed and causes a bend in the band structure of phosphor layer.[9],[13] This band bending should cause an increase of carrier tunneling from the phosphor-layer/insulating-layer interface states. Therefore Ce' doping to CaS:Eu should cause a decrease of transferred charge response time.

In conclusion, we have shown that the electrical and optical response characteristics of rare-earth doped alkaline-earth-sulfide EL devices depend on rare-earth luminescent centers rather than on alkaline-earth-sulfide phosphor host materials. Furthermore, we have found that the slow response characteristics of CaS:Eu EL devices can be improved by co-doping a small amount of Ce.

Acknowledgements

We would like to thank Mr. K. Tamura and Dr. Y. Abe of Hitachi Research Laboratory for fabricating EL devices and for fruitful discussions. Thanks are also due to Dr. E. Kaneko, Dr. Y. Sugita, and Dr. M. Hanazono of Hitachi Research Laboratory and Dr. H. Yamamoto of Hitachi Central Research Laboratory for discussions and encouragement.

References

1. S. Tanaka et al.: Proc. SID 26/4, 255(1985).
2. Y. Tamura et al.: Jpn. J. Appl. Phys. 25, L105(1986).
3. K. Tanaka et al.: Appl. Phys. Lett. 48, 1730(1986).
4. S. Tanaka et al.: Proc. SID, 28, 357(1987).
5. S. Tanaka, J. Lumin.: 40&41, 20(1988).
6. W.A. Barrow et al.: 1988 SID Digest, 284(1988).
7. R.S. Crandall et al.: 1987 SID Digest, 245(1987).
8. R.S. Crandall: Appl. Phys. Lett., 50, 551(1987).
9. R.S. Crandall: Appl. Phys. Lett., 50, 641(1987).
10. R.S. Crandall and M. Ling: J. Appl. Phys., 62, 3074(1987).
11. V.P. Singh et al.: IEEE Trans. Electron Devices, ED-35, 38(1988).
12. S. Tanaka et al.: Appl. Phys. Lett., 51, 1661(1987).
13. S. Tanaka et al.: Proc. SID, 29, 77(1988).
14. Y.A. Ono et al.: J. Lumin., 40&41, 796(1988).
15. Y.A. Ono et al.: Jpn. J. Appl. Phys., 26, 1482(1987).
16. W. Lehmann and F.M. Ryan: J. Electrochem. Soc., 118, 477(1971)
17. H. Yamamoto and T. Kano: J. Electrochem. Soc, 126, 305(1979)

ZnS-like Behaviour of Efficient CaS:Eu Electroluminescent Devices

R. Mach[1], *H. Ohnishi*[2], *and G.O. Mueller*[1]

[1]Central Institute for Electron Physics, Academy of Sciences of GDR, Hausvogteiplatz 5, DDR-1086 Berlin, GDR
[2]Department of Electrical Engineering, Ehime University, Matsuyama, Ehime 790, Japan

1. Introduction

The mechanism of high field a.c. driven electroluminescence (EL) in MISIM structures (M-etal, I-nsulator, S-emiconductor) is rather well-understood in the case of S = ZnS:Mn [1,2,3,4,5]. Operation of these devices comprise as the main processes the tunnel emission of electrons from the interfaces between the I- and the S-films only at fields above the ballistic acceleration threshold, the ballistic accceleration of these electrons, their multiplication after reaching kinetic energies of about 7eV, and of course the direct impact excitation of Mn^{2+}. One important prerequisite of the first mentioned process is the existence of deep enough interface states of appropriate density (IF-DOS) [$eV^{-1}cm^{-2}$], into which the multiplied electrons are trapped after reaching the momentary anode, before the polarization field surges them back into the bulk after the end of the applied field pulse.
The action of the interface states could be demonstrated [5] by applying so-called irregular bipolar pulse bursts (i.e. ones which begin and end with pulses of the same polarity) to these structures, and even after intervals of some minutes between the bursts observing a missing response to the first pulse in brightness and transferred charge. This behaviour, which is strikingly different from that under regular bursts, could easily be explained by the fact that, in the last positive pulse of the preceding burst, transferred charge had been trapped in the interface of IS (naming the films in order from the substrate to the covering electrode) so deeply, that the starting positive pulse could find no electrons in the interface states of the SI interface. So no dissipative current, no transferred charge, no brightness can result. It is not until the second (negative) pulse - the IS now being cathode -, that tunnel emission occurs and response is observed.
Multiplication of the carriers, indicating a kinetic energy much larger than necessary to excite the dopant (Mn^{2+}), could be demonstrated [5] by observation of an infinitely steep portion of the current-voltage-characteristic of the S-film. This was done by measuring the dissipative current density D under the applied voltage U(t), and computing the voltage over the S-film

$$U_s(t) = F*d_s = a*U(t) - 1/C_i\int^t D(t')dt'$$

with d_s denoting the S-film thickness, $a = C_i/(C_i+C_s)$, C_i (C_s)-capacitances of the I(S)-films, and F the mean field strength in the S-film. The plots of D(F) are the normalized current-voltage-characteristics suitable for this judgement. The occurrence of hysteresis or, in other words, negative differential resistance, as a direct consequence of positive space charge accumulation, is another proof of multiplication.

Springer Proceedings in Physics, Vol. 38 Electroluminescence
Editors: S. Shionoya and H. Kobayashi

In the case of S = $CaS:Eu^{2+}$ SCHADE [6] was the first to compare its behaviour in MISIM structures with that of ZnS:Mn. He as many other authors investigating this and other earth alkaline sulphides with various dopants found drastic differences in the main features [7-13], the most easy to see of which are the light emission on the trailing edge of the exciting pulse, current reversal just there, and pseudo-memory in the B(V) characteristic.

2. Experiments and Results

Efficient $CaS:Eu^{2+}$ based EL devices were prepared by rf-sputtering as described in [14]. Their operating parameters were reported in [15]. They exhibited rather high efficiencies of light emission of about 0.3 lm/W, which however degraded rather quickly [15]. Checking these samples for their behaviour with just the methods mentioned in the Introduction, a rather good ZnS-like behaviour was observed. The pecularities reported in [6-12] could be observed too, but only at very low transferred charge levels - below about $0.2 \mu C/cm^2$. The brightness(B)-voltage(V) characteristics were hardly discernable from (low effiency) ZnS:Mn ones (Fig.1). Driving the samples at 'useful' operating conditions, i.e. in the range around $1 \mu C cm^{-2}$, they showed good multiplication, as evident in Fig. 2, 3 for both polarities. Evident too from the last figures is the comparatively low multiplication field of 1.8-2.0 $MVcm^{-1}$.

Applying regular and irregular pulse bursts respectively to these structures, precise ZnS-like behaviour was also obtained, as seen in Fig.4. In fact the collection efficiency of the interface states

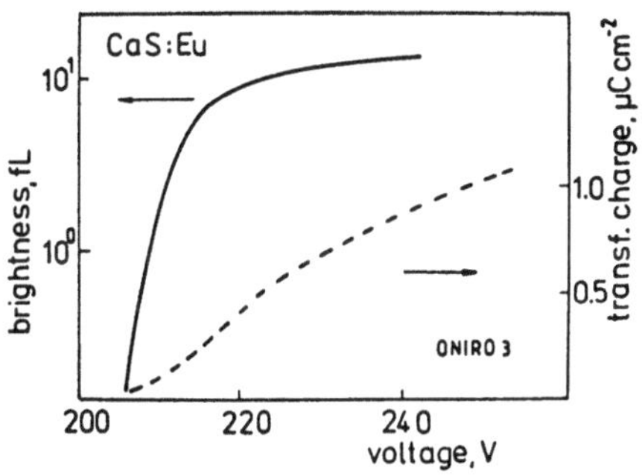

Fig.1 Brightness (B) and transferred charge (Q) on applied peak voltage (V), 200 Hz bipolar pulses

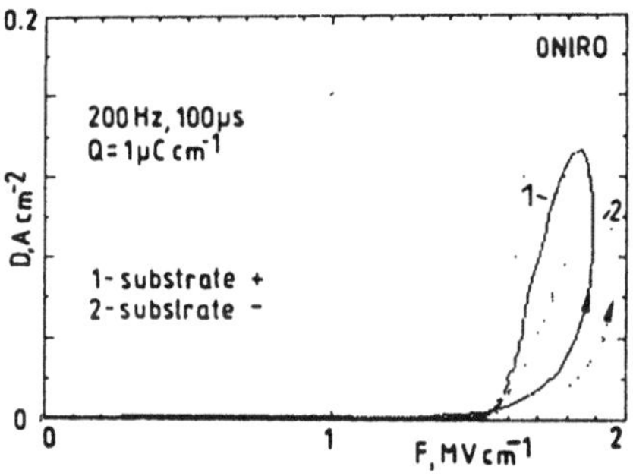

Fig.2 Current-voltage characteristic of the S-film; D - dissipative current density, and F - average electric field in the S-film resp.

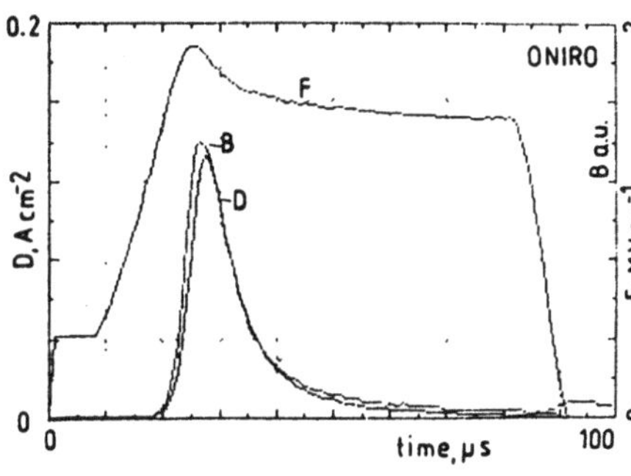

Fig.3 Time dependence of the mean field F, and current density D in S-film, and brightness B generated. The coincidence of B and D demonstrates 'all electrons excite'

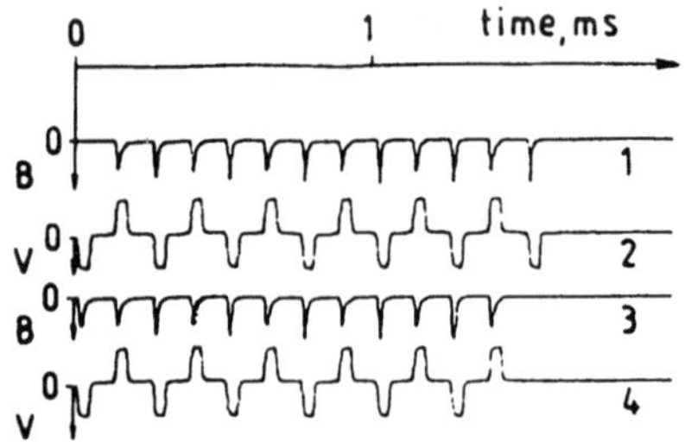

Fig.4 Periodic bursts of bipolar pulses V and the brightness (B) response to it, demonstrating that in the regular burst (3,4) the first pulse generates light, but not in the irregular burst (1,2)

might not be as high as in ZnS, giving rise to some charge floating back, which would account for the small reverse current at the trailing edge of the pulse, and the hump in the decay of the light pulse, which can be recognized in Fig.5, being more pronounced at lower drive.

3. Discussion

As the decay time of the Eu^{2+} emission - presumably almost its radiative decay time - is smaller than 10^{-6}s, the generation function of excited carriers within this time resolution equals the light pulse. So, as obvious from Fig.2, all the transferred electrons excite with equal probability, and without a time lag. Really, as is evident from Fig.1, the field does not vary very much during the current flow time. This of course does not prove impact ionization, but as during all the current transfer the field is high enough to ionize electrons from the valence band, no reason can be seen for the assumption of no impact excitation for dopants, the cross section of which should be rather large. So very tentatively following the procedure outlined in [14,15] we estimate this cross section,

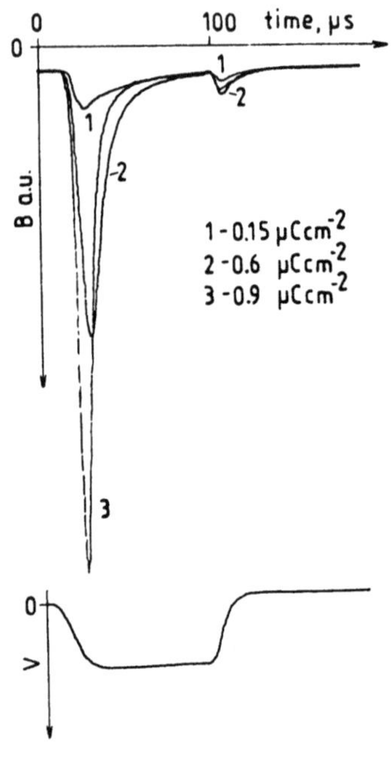

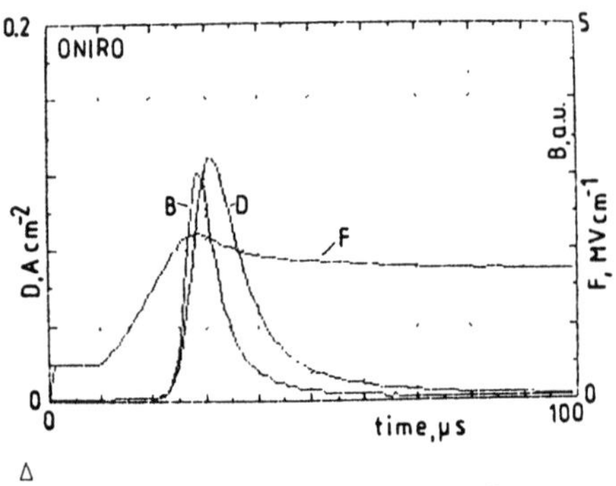

Δ

Fig.6 Brightness B and current density D upon time for the same sample as in Fig.3 after aging for 11h at 1kHz: only the 'early' electrons excite, in spite of an increase of the mean field F

◁

Fig.5 Brightness B and voltage pulse V upon time at various transferred charges

assuming that all the excitation proceeds via Coulomb impact. The Eu concentration, measured by Rutherford backscattering, in the sample with an efficiency of 0.3 lm/W at a transferred charge of $0.4 uC/cm^2$ is $3.5*10^{-19} cm^{-3}$. Assuming an outcoupling efficiency of the produced photons of 10% an inner efficiency of 3 lm/W has to be used for the estimate. The luminescence efficiency of the Eu^{2+} transition, in the sense of decay time/radiatitive decay time, is assumed to be 0.8. With these rather probable values we get an impact cross section no smaller than $10^{-15} cm^2$.
Comparing this value with the one of Mn^{2+} in ZnS - $4*10^{-16} cm^2$ - it appears rather small. Considering the fact that the lifetimes, and so the quantum mechanical transition probabilities W, differ by at least three orders of magnitude,and that the electron loss function is proportional to Im(1/DF), with DF being the dielectric function, which is monotonic with W, we would expect at least an order of magnitude difference between the cross sections of Eu^{2+} and Mn^{2+}, host effects very probably being minimal. Any decrease in the assumed luminescence efficiency of course would increase the value given above. Concentration quenching however is not likely at the concentration used, and this would be the only loss process we could imagine, to reduce the emissive yield. Going one step further, and assuming the estimated cross section to be valid, the sample used had been underdoped. The optimum doping would be $5*10^{20} cm^{-3}$ and an (outer) efficiency of 5 lm/W would result (no concentration quenching assumed).

What has to be overcome in these efficient CaS:Eu EL devices is the rapid degradation mentioned in [14]. It clearly relates to a drastic change in the nice coincidence of D(t) and B(t) shown in Fig.2, proving that all electrons excite. Fig.5 shows on the contrary, that this is no longer the case in the degraded (0.1 lm/W) condition. Simultaneously however, the mean field F(t) in the S-film has increased, pointing to a possibly heavy deviation from this average in some parts as a cause for the degradation. A change in the rather sensitive intrinsic defect structure of CaS under operation with its presence of very energetic carriers could account for this.

4. Acknowledgement
The authors gratefully acknowledge the careful RBS-determination of Eu concentrations by Dr. B. Selle.

5. References
/1/ T.Inoguchi, S.Mito,Springer Topics in Appl. Phys. 17, 197 (1977)
/2/ R. Mach, G.O. Mueller,phys. stat. sol. (a) 69, 11..66, (1982)
/3/ P.M. Alt, D.B. Dove, W.E. Howard, J. Appl. Phys. 53, 1982, 5186
/4/ R. Mach, G.O. Mueller, W. Gericke, J. v. Kalben, phys. stat. sol. (a) 75, 187 (1983)
/5/ G.O. Mueller,SID Seminar Lecture Notes,New Orleans,USA,1987
/6/ H. Schade,Proc. 3rd. EL Workshop, Kah Nee Ta, Oregon, 1986
/7/ R.S. Crandall,Proc. 3rd. EL Workshop, Kah Nee Ta, Oregon, 1986
/8/ B. Tsujiama, Y. Tamura, J. Ohwaki, H. Kozawaguchi, SID 86 Digest, p. 37
/9/ C. Gonzalez, Eurodisplay 1987, p. 21
/10/ S. Tamaka, H. Yoshiyama, Y. Mikami, J. Nishiura, S. Oshio, H. Kobayashi, Japan Display 86 Digest, p. 242
/11/ V. Singh, D.C. Morton, SID 88 Digest, p. 27
/12/ R. Crandall, M. Ling, J. Appl. Phys. 62 (7), 1987, 3074
/13/ S. Tomaka, H. Yoshiyama, Appl. Phys. Lett 50 (3), 1987, 119
/14/ H. Ohnishi, Y. Katayama, Proc. 3rd EL Workshop, Kah Nee Ta 1986
/15/ H. Ohnishi, R. Iwase, Y. Yamasak, SID 88 Digest, p. 289
/16/ G.O. Mueller, R. Mach,Proc.of the 17th ICPS, 647 (1985)

Bright SrS TFEL Devices Prepared by Multi-Source Deposition

S. Tanda, A. Miyakoshi, and T. Nire

Research Division, Komatsu Ltd., 1200 Manda, Hiratsuka, Kanagawa 254, Japan

Bright SrS electroluminescent devices with 1100 cd/m^2 at 5 kHz have been obtained using a new method of multi source deposition (MSD). Four types of SrS EL devices were studied.

1. Introduction

Recently, alkaline-earth sulfides such as SrS and CaS thin films with rare-earth dopants have been extensively studied to apply to color thin film electroluminescent (TFEL) devices. As preparation techniques for these thin films, a post deposition anneal method /1/, electron beam deposition /2/ and CVD method /3/ have been used. It is important to obtain alkaline-earth thin films with high crystallinity for taking high luminance color TFEL devices and making clear the mechanism of EL phenomenon in these films with dopants. We showed that the MSD method can prepare ZnS film with high crystallinity. The devices had high luminance and low threshold voltage /4/.

In this paper, we report on crystal properties of SrS thin films prepared by the MSD method and EL characteristics of SrS:Ce, SrS:CeCl$_3$,SrS:Ce,KCl and SrS:Ce,K$_2$S. In particular, we have obtained bright blue-light in a SrS:Ce,KCl TFEL device.

2. Experiments and Results

2-1 Crystal Properties and Orientation of SrS Thin Film

The host phosphor layer SrS was deposited by the MSD method, in which Sr and S were evaporated at the same time from separate sources. Stoichiometric SrS can be obtained by controlling Sr and S source temperatures independently. S atomic concentration is 0.50 ± 0.01 as examined by X-ray micro analysis. Figure 1 shows X-ray diffraction patterns of the SrS thin films. Each of them indicates a NaCl-type crystal structure. The preferred orientation of the SrS film varied from the (111) axis to the (200) axis and finally to the (220) axis, irrespective of the substrate temperature, as the S/Sr ratio in the source beams is increased. The width of diffraction peak is very narrow (0.18° FWHM) when the film is oriented to the (200) axis.

2-2 Blue TFEL Devices

We have fabricated blue TFEL devices of the typical double insulating layer structure shown in Fig.2, using the (200)-oriented SrS films deposited at the substrate temperature of 450℃. To investigate the effect of Cl or K against Ce on the

Springer Proceedings in Physics, Vol. 38 **Electroluminescence**
Editors: S. Shionoya and H. Kobayashi

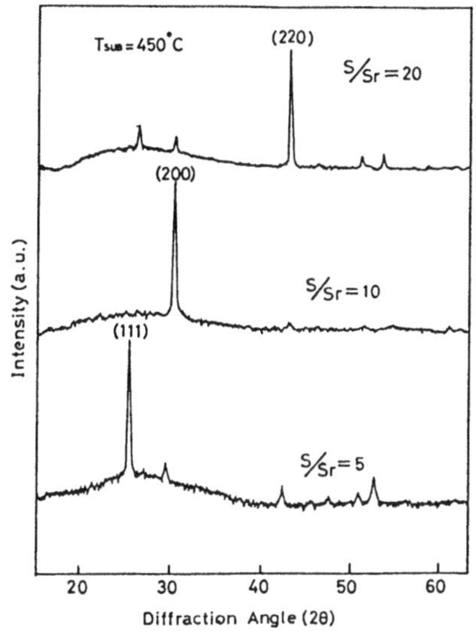

Fig.1
X-ray diffraction patterns using CuKα line.
S/Sr is ratio of source beams intensity.

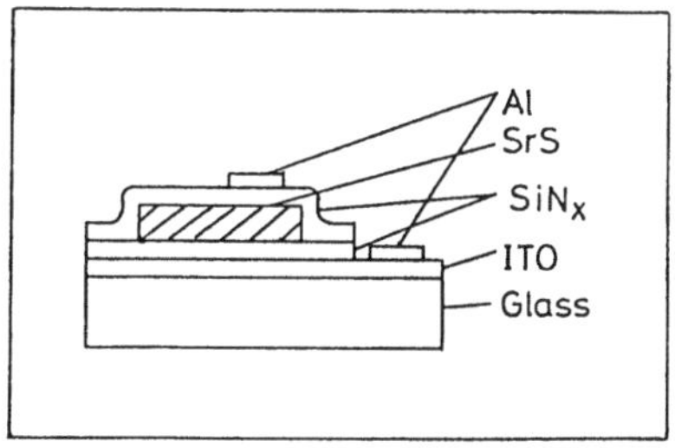

Fig.2 Structure of TFEL devices

luminance, we studied dopants of Ce, $CeCl_3$, Ce+KCl and Ce+K_2S. For Ce+KCl and Ce+K_2S, the MSD apparatus was operated with four crucibles (Sr, S, Ce, KCl or K_2S) Chen-Krupka circuit was used to measure transported charge density (ΔQ).

The EL spectra, luminance(L)-Voltage(V) curves, and ΔQ-V curves are shown in Fig.3 through 6.

The SrS:Ce TFEL device shows that ΔQ is almost linear to V and correspondingly low luminance.

The SrS:$CeCl_3$ TFEL device shows a sudden increase in ΔQ at the voltage corresponding to the threshold voltage in the L-V curve. This suggests that Cl contribtes the luminescence, probably to carrier magnification. In order to find an optimum concentration of Cl, KCl concentration was varied by keeping the Ce concentration at a constant value. A preliminary result shows an optimum ratio at

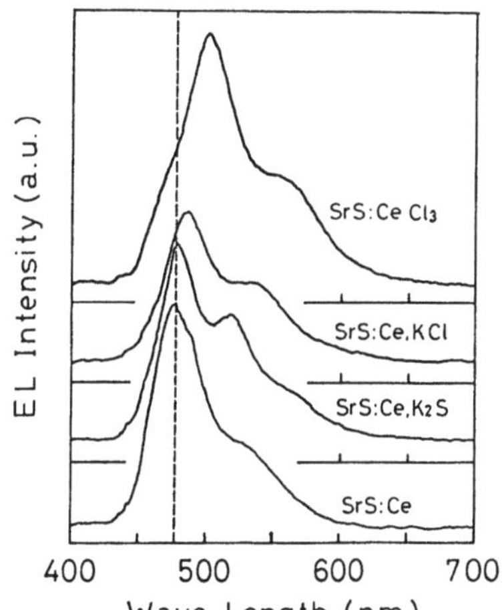

Fig.3 EL spectra of SrS-based TFEL devices

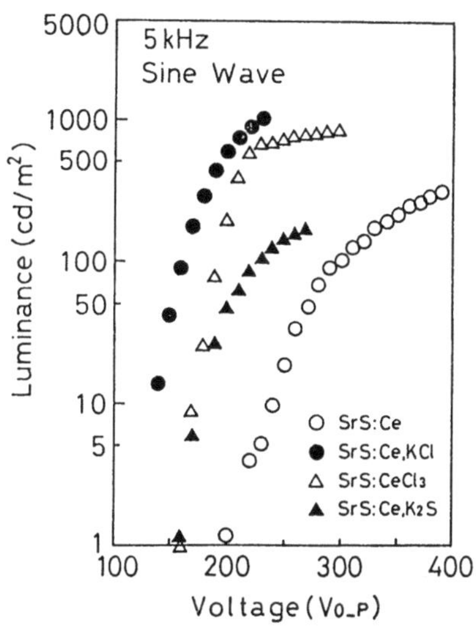

Fig.4 L-V curve at 5 kHz sine wave

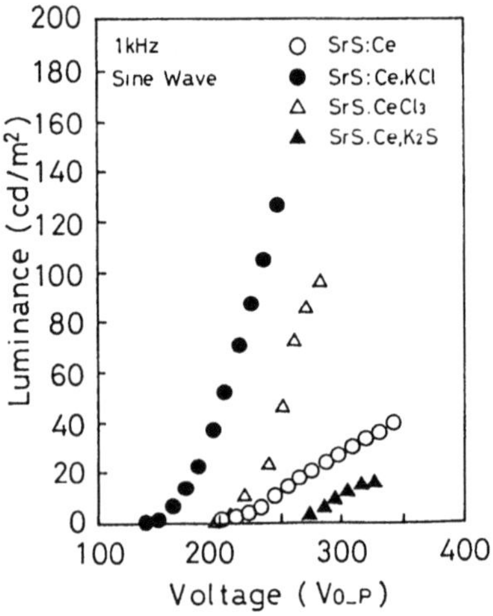

Fig.5 L-V curve at 1 kHz sine wave.

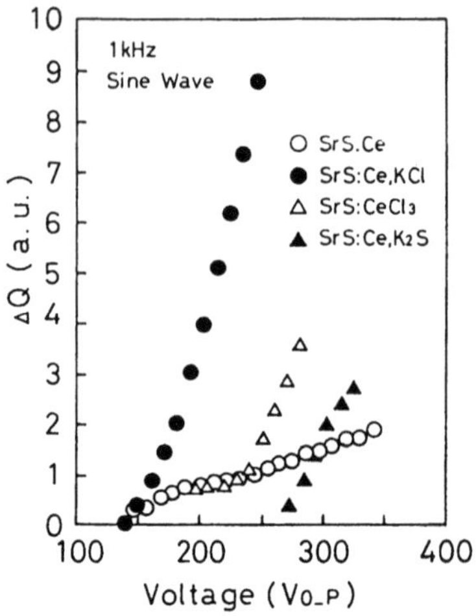

Fig.6 ΔQ-V curve at 1 kHz sine wave.

about Ce:Cl=1:1. At this ratio we find the following (1) ΔQ and L increase rapidly against the applied voltage, (2) the threshold voltage in the L-V curve is the lowest among the devices studied, (3) the maximum luminance of 1100^2cd/m was obtained at 5 kHz sine wave drive.

The SrS:Ce,K_2S TFEL device shows the lowest L among the studied devices. All SrS:Ce,K_2S TFEL devices were unstable and we observed a reduction of luminance with time. Potassium had no effect in improving luminance.

3. Summary

We have shown that SrS film with very high crystallinity can be obtained by the MSD method. The mechanism of crystal growth in the SrS thin film is a affected not only by surface energy but also by the S/Sr supply ratio.

The SrS:Ce,KCl had high luminance of 1100 cd/m^2 . This indicates that MSD method is a very powerful method for fabricating future full-color TFEL displays.

We have also shown that Cl atoms in SrS:Ce,KCl are playing an important role of carrier magnification. This may lead to the full understanding of the luminescing mechanism in TFEL devices.

REFERENCES

(1) W. A. Barrow, R. E. Coovert and C. N. King, Digest 1984 SID Int. Symp. p.249
(2) S. Tanaka, V. Shanker, M. Shiiki, H. Deguchi and H. Kobayashi, Digest 1985 SID Int. Symp. p.218.
(3) J. Kane, W. E. Harty, M. Ling and P. N. Yacom, Conf. Record 1985 Int. Display Research Conf. (IEEE New York), p.163.
(4) T. Nire, T. Watanabe, S. Tanda and S. Sano, Digest 1987 SID Int. Symp. p.243.

Sulphur Defects and Deep Levels in SrS:Ce Thin Films

Y. Tamura and H. Kozawaguchi

NTT Opto-Electronics Laboratories,
Tokai-Mura, Naka-Gun, Ibaraki-Ken, 319-11, Japan

The integrated charge-applied voltage (Q-V) characteristics, voltage induced light waveforms, photo-excitation efficiency and photo-induced capacitance change in SrS:Ce thin films have been studied. Low clamp electric field strength is not an extrinsic characteristic caused by sulphur defect but an intrinsic characteristic caused by Ce doping. Sulphur defects reduce emission intensity generated by recombination of electrons and ionized Ce. Cerium generates deep levels in SrS and acts as a carrier source.

1. Introduction

Cerium-doped strontium sulphide (SrS:Ce) is the most promising host material for blue-emitting thin-film electroluminescent (TFEL) devices. To achieve a higher luminance level, electroluminescence mechanisms in TFEL devices have been intensively studied and some characteristic behaviors have been found. For example, these devices show a very low clamping electric field strength (E_{cl}=0.8~1.0 MV/cm) despite their large band gap energy (4.3 eV) and they emit three times during a half period of applied alternative polarity voltage/1/. Native defects and deep activator levels in SrS are suggested to cause these behaviors /2/. However the effect of these levels on E_{cl} and emission characteristics have not yet been clarified. This report discusses the effect of sulphur defects and deep levels on electroluminescent characteristics.

2. Sample fabrication

TFEL devices were fabricated with the structure shown in Fig. 1. A SrS:Ce thin film was sandwiched between SiN thin films deposited by ECR-PCVD. A Ta_2O_5 film, deposited by RF-sputtering, was used as a top insulator. Al was used as a back electrode and ITO was used as a transparent electrode. SrS:Ce thin films were deposited by electron beam evaporation using a SrS pellet mixed with 0.1 mol% $CeCl_3$ or Ce_2S_3. Substrate temperature was kept at 500 °C during deposition. SrS:Ce thin films were deposited on silicon wafers for photo-excitation efficiency measurement.

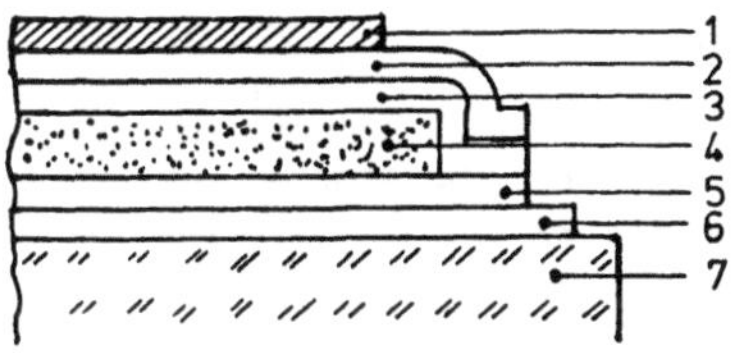

Fig. 1 Device structure

1. Al electrode 2. Ta_2O_5 (300nm)
3. SiN (100nm) 4. SrS:Ce (1μm)
5. SiN (100nm) 6. ITO
7. Glass substrate

3. Results and discussion

3-1. Electrical behavior

The most suitable parameter for describing the electrical behavior of TFEL devices is the integrated charge(Q) in response to the drive voltage(V)/3/. This Q-V characteristic is given by using a Sawyer-Tower circuit.

The Q-V figures for operation well above the threshold voltage are shown in Fig. 2. For devices with many sulphur defects, the Q-V characteristic becomes ellipsoidal. This is also observed in the lower voltage region, indicating that the SrS:Ce active layers start to conduct current in the low electric field region. Therefore, as suggested by Crandall/2/, TFEL devices start to emit light in the low electric field region. However net effect of above effect, on luminance-voltage characteristics, is only decreasing slope, dL/dV, in luminance-voltage characteristics.

Rapid generation of carriers above the threshold voltage is also observed in SrS:Ce TFEL devices by decreasing sulphur defects. This characteristic is well known as field clamping in ZnS based TFEL devices/3/. This phenomena appears in Q-V loop as clear clamping, as shown in Fig. 2(b). However, the E_{cl}, derived from the Q-V loop, is still smaller than the value expected by its large band gap energy. This suggests the possibility of another factor lowering the E_{cl}.

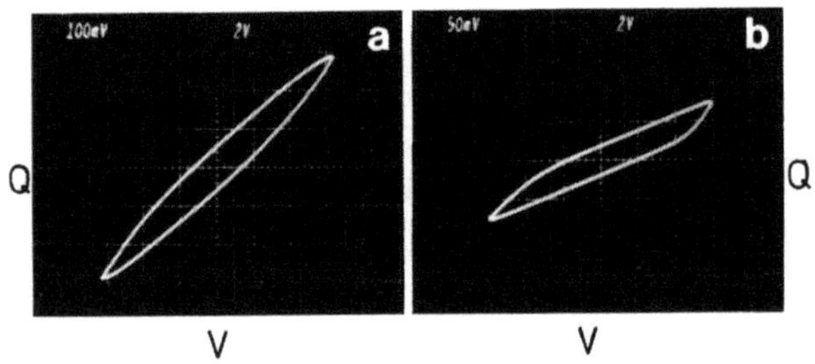

(a) S/Sr=0.88 (b) S/Sr>0.98

Fig. 2 Q-V figures

(a) The Q-V figure for a device with many sulphur defects. (b) The Q-V figure for a stoichiometric device. S/Sr is the XPS signal ratio of S_{2p} and $Sr3p_{3/2}$ peaks.

Fig. 3 shows E_{cl} dependency on Ce concentration for stoichiometric devices. The E_{cl} of samples without Ce is 1.4~1.8 MV/cm. On the other hand, the E_{cl} is drastically decreased to 0.7~0.9 MV/cm by Ce doping. This implies Ce doping is the main factor lowering E_{cl}.

The most plausible mechanism lowering E_{cl} by Ce doping is carrier multiplication by two-step ionization of impurities. E_{cl} can be decreased by this kind of carrier generation/4/. To confirm this effect, photo-capacitance characteristics were studied. Fig. 4 shows the devices' photo-capacitance signals. Monochromated light was irradiated to the sample from the devices' glass side and capacitance was measured under 1kHz 1V sinusoidal voltage excitation. Signal strength depends on Ce concentration, and non doped samples showed no signal. The signals are unaffected by Cl co-activator and they show a maximum peak at about 2.8 eV. This value coincides closely with that measured by Keller/5/ for Ce doped SrS powder phosphor. This shows that SrS:Ce has deep levels, peaks at 2.8 eV, and it is caused by Ce doping.

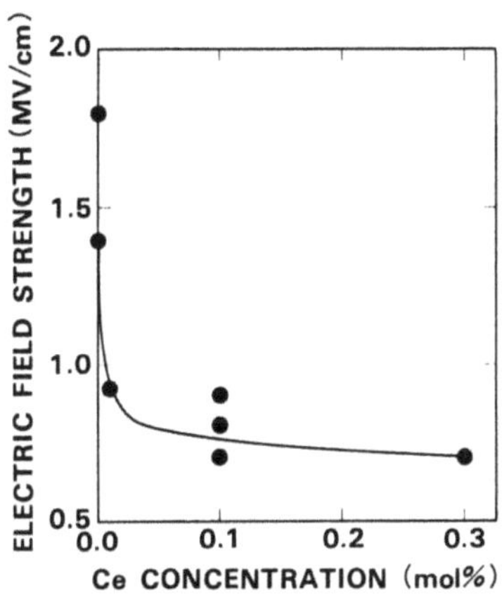

Fig. 3 E_{cl} dependency on Ce concentration

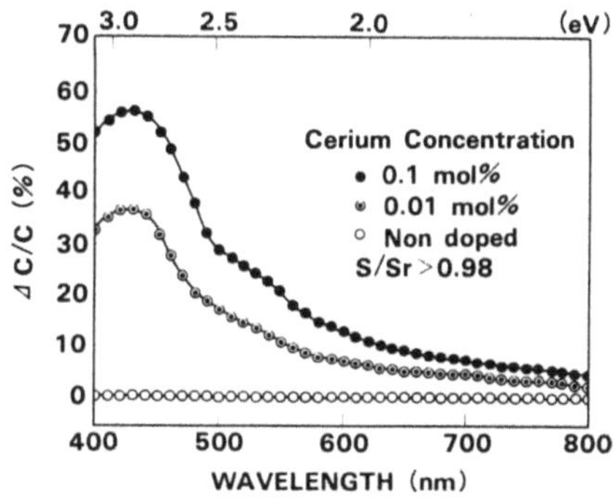

Fig. 4 Photo-capacitance signals

Signal intensity depends on Ce concentration and their maximum peak at about 2.8 eV. Non-doped sample shows no signal.

These results suggest that low E_{cl} in SrS:Ce EL devices is due not simply to sulphur defects but also in large measure to Ce doping.

3-2. Emission characteristics

Fig. 5 shows voltage induced emission wave forms for the devices under alternative polarity, 1KHz, trigonal voltage excitation. Three emission peaks, observed during a half period of exciting voltage, are called First Peak (F.P), Second Peak (S.P) and Third Peak (T.P), in time order/1/. T. P and a part of F.P are suggested to be caused by recombination of ionized cerium with electrons/6/. This figure clearly shows that with increasing sulphur defects F.P and T.P drastically decrease.

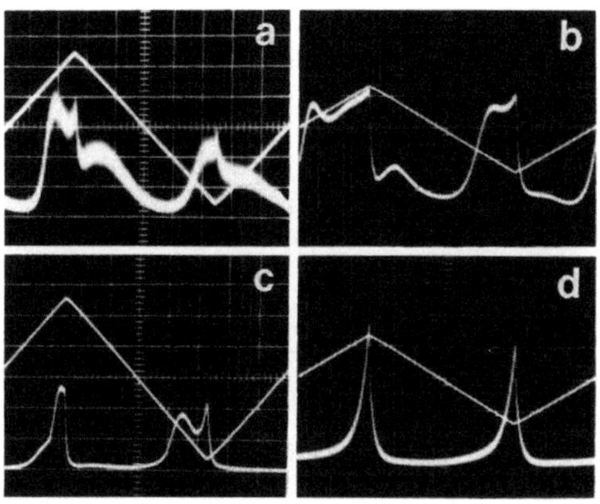

Fig. 5 Voltage induced emission wave forms for the devices with different amount of sulphur defect. Increasing the sulphur defect((a) to (d)) emission intensity of F.P and T.P are drasticaly decreased.

(a) S/Sr=0.98 (b) S/Sr=0.92
(c) S/Sr=0.91 (d) S/Sr=0.88

Fig. 6 shows excitation efficiency for Ce^{3+} 480 nm emission. The filled circles show the excitation efficiency via band to band excitation of SrS. The open circles show the excitation efficiency of direct excitation of Ce^{3+}. The efficiency of band to band excitation increases with decreasing sulphur defects.

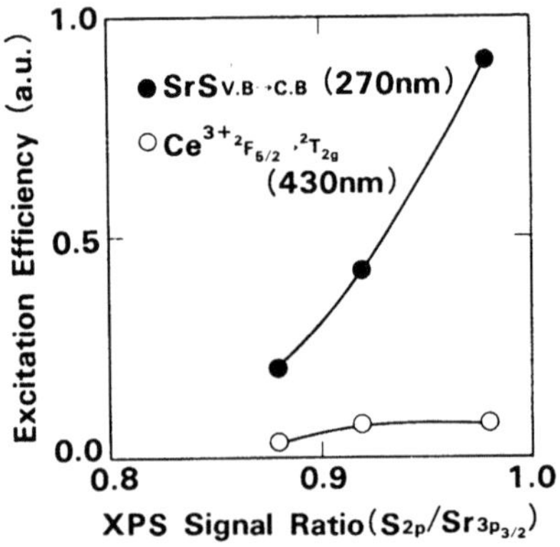

Fig. 6 Photo-excitation efficiency for Ce^{3+} 480 nm emission band.

Filled circles show the excitation efficiency via valence band to conduction band excitation of SrS. Open circles show the excitation efficiency via Ce^{3+} grand state($^2F_{5/2}$) to excited state($^2T_{2g}$).

These results show that sulphur defects reduce the recombination of electrons and ionized cerium.

4. Conclusion

Sulphur defects reduce the intensity of emission generated by the recombination of electrons and ionized cerium. As a result of this effect, F.P and T.P, which account for more than 50 % of total emission, are decreased. Then luminance level is decreased by sulphur deffects.

Low clamping electric field strength is not an extrinsic characteristic caused by sulphur defect but an intrinsic characteristic caused by Ce doping. Cerium generates deep levels in SrS, and reduces the clamping electric field strength of active layer.

5. References

1. B. Tsujiyama, Y. Tamura, J. Ohwaki and H. Kozawaguchi: 1986 SID Digest Papers p. 37.
2. R. S. Crandall, M. Ling, J. Kane, and P. N. Yocom: 1987 SID Digest Papers p.245
3. D. H. Smith: J. Lumin. 23 (1981) 209
4. J. W. Allen: J. Lumin. 23 (1981) 127
5. S. P. Keller and G. D. Pettit: Phys. Rev. 15 (1958) 1533
6. S. Tanaka, H. Yoshiyama, Y. Mikami, J. Nishiura, S. Ohshio, H. Kobayashi: 1986 Japan Display p.242

$Sr_{1-x}Zn_xS$:Ce, F Phosphor for Thin Film Electroluminescent Devices

K. Takahashi, K. Utsumi, Y. Ohnuki, and A. Kondo

Tosoh Corporation, 2743-1 Hayakawa, Ayase-shi, Knagawa-ken 252, Japan

1. Introduction

SrS is one of the most noticed thin film electroluminescent (TFEL) materials. At first, SrS:Ce was found to be an efficient blue phosphor/1,2 /. Although it emits bluishgreen light, its efficiency is 100 times greater than ZnS:Tm blue phosphor. Recently, Tanaka et al./3/ reported efficient white phosphors which are SrS:Ce,K,Eu and SrS:Pr,K. SrS:Ce,K,Eu especially has a potential of a full color display, because it has a wide range of electroluminescent spectrum from blue to red.

Stoichiometry, crystallinity and charge compensation are main factors of an efficient TFEL device. Typical deposition at substrate temperature in the range of 500 to 700 ℃ with sulfur atmosphere is required to obtain the proper stoichiometric and crystallized SrS thin film, because SrS has a refractory and easily hydrolyzing nature /1,2/. Refering to ZnS, it has a nature of sublimation and is chemically stable. Thus the substrate temperature of ZnS deposition is as low as 150 to 300℃.

In this paper, a new TFEL phosphor $Sr_{1-x}Zn_xS$:Ce,F is reported, which is a solid solution of ZnS in SrS. ZnS was considered to have low solubility in alkaline-earth sulfides generally. Brightwell et al./4/ reported that the solubility of ZnS in MgS is up to 5 mol% and Lehmann/5/ reported that the solubility limit of ZnS in CaS is only 0.35 mol%. We have found much higher solubility of ZnS in SrS (more than 30 mol%) relative to other alkaline-earth sulfides . It has been also observed that the $Sr_{1-x}Zn_xS$:Ce,F thin film has a good stoichiometry even when deposited at relatively low substrate temperature and shows efficient electroluminescence after proper post deposition annealing.

2. Device Preparation

The schematic cross section of a $Sr_{1-x}Zn_xS$:Ce,F TFEL device is shown in Fig.1. A $Sr_{1-x}Zn_xS$:Ce,F active layer is deposited by an electron beam evaporation method using $Sr_{1-x}Zn_xS$ pellets doped with CeF_3. The substrate temperature is maintained at 200℃ during deposition.

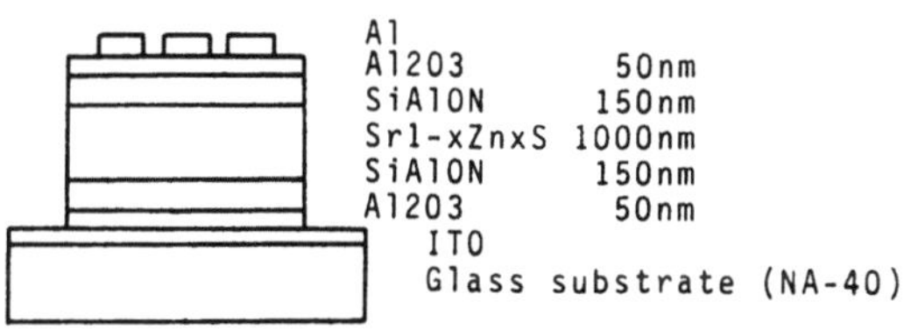

Fig.1 Schematic cross section of the Sr1-x Znx S:Ce,F TFEL device

Springer Proceedings in Physics, Vol. 38 **Electroluminescence**
Editors: S. Shionoya and H. Kobayashi

The sample is subsequently annealed at 600-800 ℃ for 10 minutes. The upper and lower insulating layers consisting of Al_2O_3 and SiAlON, were prepared by sputtering method.

3. Experimental Results and Discussion

3.1 Stoichiometry of an Active Layer

Figure 2 shows the S/(Sr+Zn) atomic ratio of active layers measured by X-ray fluorescent method . As Zn concentration decreases, the S/(Sr+Zn) atomic ratio is far from stoichiometry. In case that Zn concentration is less than 10 mol%, the film is sometimes black before annealing. Stoichiometric thin film is obtained when Zn concentration is more than 15 mol%.

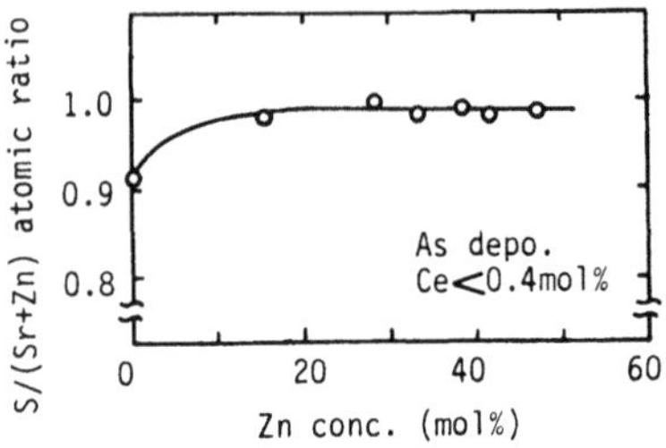

Fig.2 Stoichiometry of active layers

3.2 X-ray Diffraction

SrS has the NaCl structure and ZnS has the zinc brend structure normally and the wurzite structure as a high temperature phase. Figure 3 shows the X-ray diffraction patterns of SrS:Ce,F, $Sr_{0.69}Zn_{0.31}S$:Ce,F and $Sr_{0.5}Zn_{0.5}S$:Ce,F thin films. These films are deposited on the first insulators and annealed at 700 ℃ for 10 minutes. The diffraction line corresponding to (200) plane of the NaCl structure is observed at any value of X between 0 and 0.5. The (100), (002) and (110) lines of the wurzite type ZnS structure appear in the $Sr_{0.5}Zn_{0.5}S$:Ce,F film.

The spacings of (200) and (111) planes of the NaCl structure decrease with the increase of Zn concentration. (Fig.4)

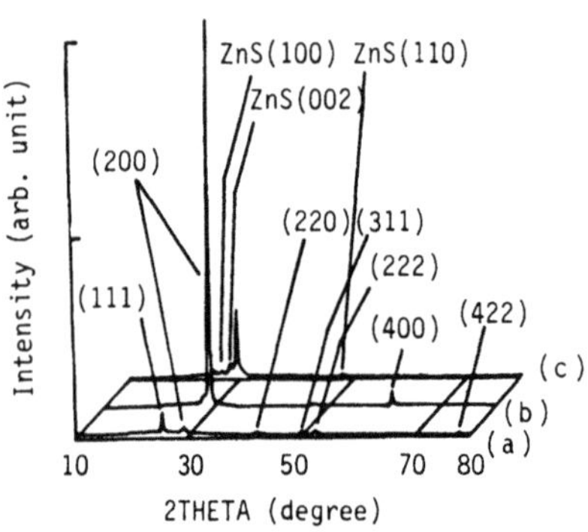

Fig.3 X-ray diffraction patterns of Sr1-xZnxS:Ce,F thin films
(a)x=0, (b)x=0.31, (c)x=0.5

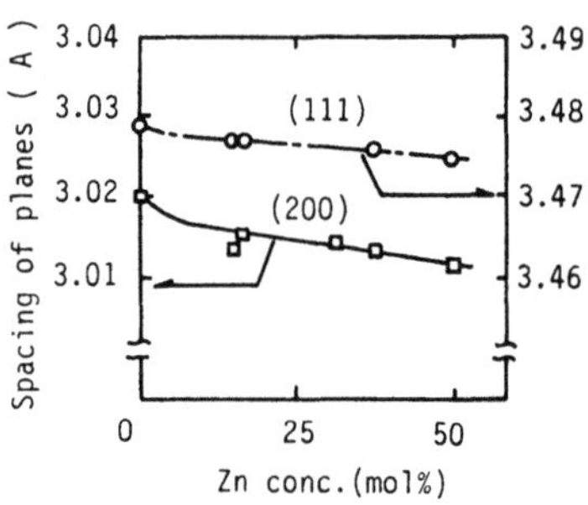

Fig.4 Dependence of a spacing of planes on Zn concentration

Figure 5 shows relationship between X-ray integral intensity and Zn concentration. The intensity of (002) line of ZnS steeply increases as Zn concentration exeeds about 33 mol%. Other peaks originate in ZnS have a similar tendency. Concerning the peaks assigned to the NaCl structure, the intensity of (200) plane is maximum at around 30 mol% of Zn concentration, while the intensity of (111) and other lines are minimum.

These results indicate that ZnS is soluble in SrS up to about 33 mol% and the thin film of a solid solution $Sr_{1-x}Zn_xS$ is strongly oriented to the [100] direction.

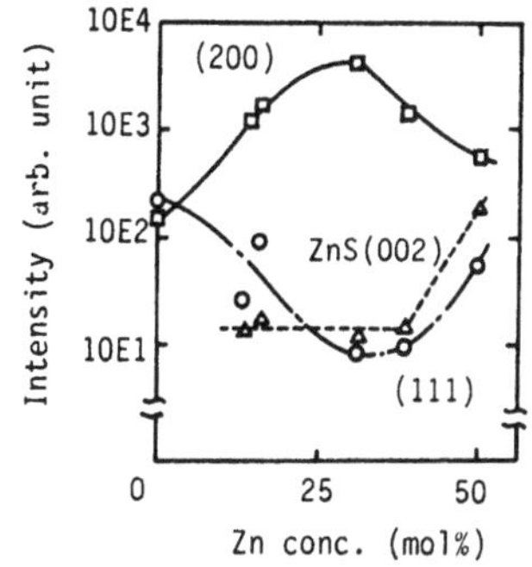

Fig.5 Dependence of X-ray integral intensity on Zn concentration (annealed at 700 ℃ for 10 min.)

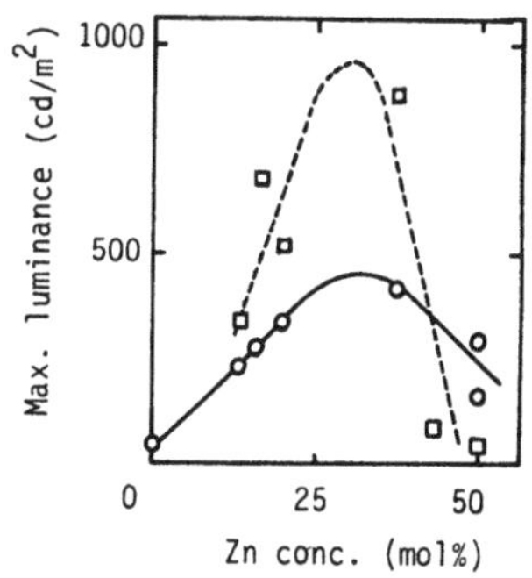

Fig.6 Dependence of maximum luminance at 5kHz drive on Zn concentration ○annealed at 700 ℃, □annealed at 800 ℃

3.3 Electroluminescent Characteristics of $Sr_{1-x}Zn_xS$:Ce,F Devices

A series of EL devices were prepared, in which the active layer had approximately the same Ce concentration (~0.2 mol%) while the Zn concentration was varied from 0 to 50 mol%. Figure 6 shows the dependence of maximum luminance driven at 5 KHz on Zn concentration. The tendency of luminance level having a maximum at around 33 mol% of Zn concentration almost coincides with the strength of [100] orientation as shown in Fig.5. Figure 7 shows typical EL spectra of these devices. The spectrum shifts toward longer wave length as Zn concentration increases. Thus the color of luminescence varies from blue-green to green depending on Zn concentration.

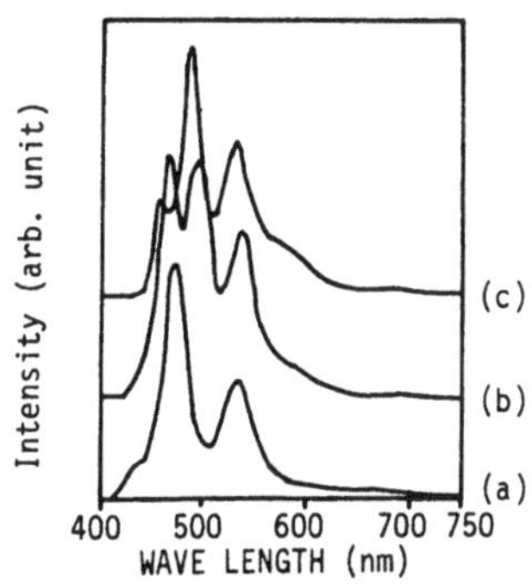

Fig.7 EL spectra of Sr1-xZnxS:Ce,F EL device (a)x=0,(b)x=0.31 (c)x=0.5

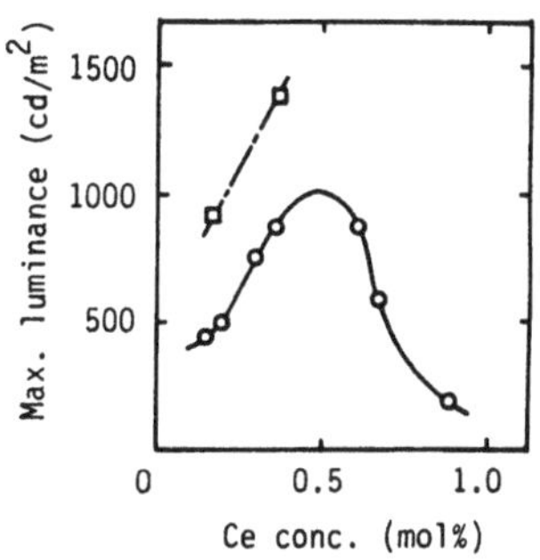

Fig.8 Dependence of maximum luminance at 5kHz drive on Ce concentration ○annealed at 700 ℃, □annealed at 800 ℃

Zn concentration in an active layer is almost the same as the $Sr_{1-x}Zn_xS:CeF_3$ pellet when the substrate temperature is maintained at 200 ℃. Figure 8 shows the relationship between the maximum luminance of a $Sr_{1-x}Zn_xS:Ce,F$ EL device deposited using a $Sr_{0.67}Zn_{0.33}S:CeF_3$ pellet and Ce concentration. The luminance level shows maximum between 0.3 and 0.6 mol% of Ce concentration. At 0.36 mol% Ce concentration, the device exhibits a luminance level of 860 cd/㎡ when annealed at 700 ℃ for 10 minutes and 1400 cd/㎡ when annealed at 800 ℃ for 10 minutes (5KHz, sinusoidal wave).

4. Conclusion

A new $Sr_{1-x}Zn_xS:Ce,F$ TFEL device has been investigated. X-ray diffraction analysis indicates that ZnS is soluble in SrS up to about 33 mol%. A $Sr_{1-x}Zn_xS$ thin film with good stoichiometry is obtained when X is more than 0.15. At solubility limit of ZnS, the $Sr_{1-x}Zn_xS:Ce,F$ thin film has a strong [100] orientation and exhibits the maximum luminance level. The $Sr_{1-x}Zn_xS:Ce,F$ electroluminescent device shows blue-green to green luminescence depending on Zn concentration.

References

1. W.A.Barrow, R.E.Coovert and C.N.King: SID 84 Digest, p249
2. S.Tanaka, H.Deguchi, Y.Mikami, M.Shiiki and H.Kobayashi : Proceedings of the Society for Information Display, 28, p21 (1987)
3. S.Tanaka, H.Yoshiyama, J.Nishiura, S.Ohshio, H.Kawakami and H.Kobayashi : SID 88 Digest p293.
4. J.W.Brightwell, B.Ray and S. White: Journal of Materials Science Letters 3, p951(1984)
5. W.Lehmann: Journal of Luminance 5, p87 (1972)

SrSe:Ce Thin Film Electroluminescent Device

S. Oseto, Y. Kageyama, M. Takahashi, H. Deguchi, K. Kameyama, and I. Fujimura

R&D Center, Ricoh Co., Ltd., 16-1 Shinei-cho, Kohoku-ku, Yokohama 223, Japan

1 Introduction

The emission spectra corresponding to 5d-4f transition of Ce^{3+} ion vary with the host material [1]. Ce^{3+} ion is considered to be one of the useful emitting centers for multi-color electroluminescence (EL). Recently, SrS:Ce and CaS:Ce thin film EL (TFEL) devices have attracted much attention because of their brightness [2,3]. As compared with the emission spectra of CaS:Ce devices, these of SrS:Ce devices shift to shorter wavelength by about 30 nm, and consequently exhibit blue EL. This fluctuation on the emission spectra is explained as follows; The 5d excited state of Ce^{3+} ion is very sensitive to the crystal field because the 5d state is not shielded by the $5s^2p^6$ closed shells [4,5]. However, the EL colors of SrS:Ce devices are not sufficient for the primary blue. We have been carrying out the attempts to replace SrS by SrSe [6]. In this paper, the characteristics of the SrSe:Ce TFEL devices are reported with the emphasis on EL spectra.

2 Sample Preparation

The TFEL devices were prepared with SrSe:Ce and SrS:Ce as the emitting layer. The device structure is shown in Fig.1. Since some oxide insulating layers are thought to react with SrS and SrSe layers easily, AlN insulating layers were formed on both sides of the emitting layers.

The SrSe:Ce layers were deposited by coevaporation of Sr, Se and $CeCl_3$. To investigate the effect of oxygen contamination, two kinds of SrSe:Ce layers (a) and (b) were prepared. Measured by SIMS, the layer (b) contained about 20-times as much oxygen as the layer (a). The SrS:Ce layers were deposited by electron beam evaporation of SrS: $CeCl_3$. The AlN and SiO_2 insulating layers and the ZnO:Al transparent conducting layers were deposited by RF magnetron sputtering.

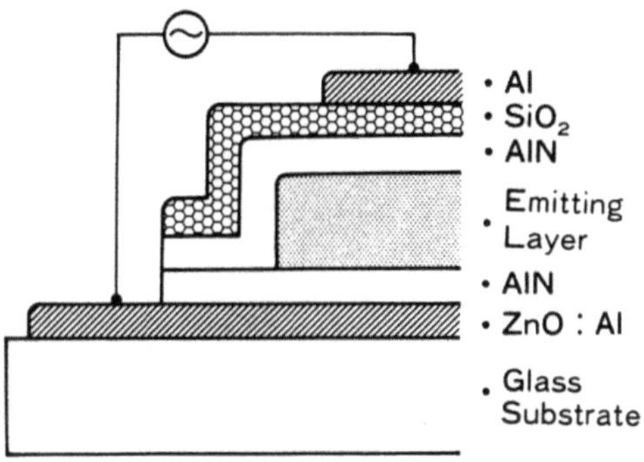

Fig.1. Structure of the TFEL devices

Electroluminescence

3 EL Characteristics of the SrSe:Ce TFEL devices

As shown in Fig.2, typical EL spectra of the SrS:Ce and SrSe:Ce devices consist of two emission bands (1) and (2) corresponding to the transitions from the $5d(^2T_{2g})$ excited state to the $4f(^2F_J)$ ground state (J=5/2,7/2) of Ce ion. The SrSe:Ce spectra (a) and (b) were obtained from the SrSe:Ce layers (a) and (b), respectively. As compared with the SrS:Ce spectrum, the SrSe:Ce spectrum (a) is characterized as follows; For the emission bands (1) and (2), each peak wavelength shifts to shorter wavelength by about 10 nm, and the relative intensity of the emission band (1) to band (2) increases. Because of these two differences in the spectra, the EL colors of the SrS:Ce and SrSe:Ce devices are classified into bluish green and greenish blue, respectively. On the other hand, comparing the two SrSe:Ce spectra (a) and (b), one can see both of two peak wavelengths are unchanged, but the relative intensity of the emission band (1) to band (2) decreases with oxygen concentration of the SrSe:Ce layer.

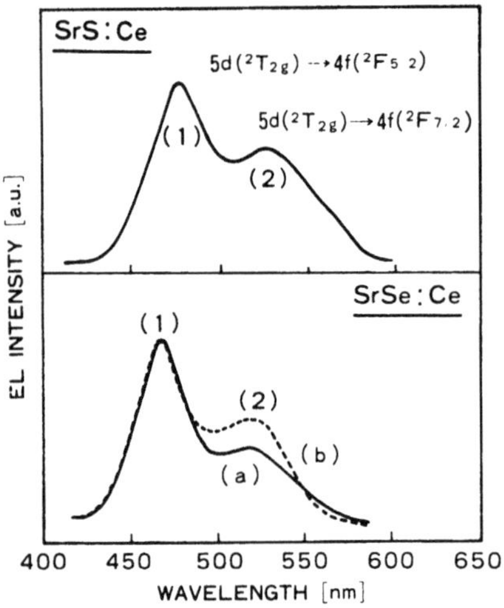

Fig.2. EL spectra of SrS:Ce and SrSe:Ce. Spectrum (b) was obtained from the SrSe:Ce layer (b) contained about 20-times as much oxygen as the layer (a)

Table 1 shows the comparison of the EL characteristics between the SrSe:Ce devices (a) and (b). The oxygen concentration in the SrSe:Ce layers affect both the luminance and the color. Even though the characteristics of the SrSe:Ce device (a) are not sufficient yet for practical use, these are expected to be improved by decreasing oxygen contamination.

Table 1. EL characteristics of SrSe:Ce TFEL devices

	CIE Chromaticity	Luminance at 1kHz
SrSe:Ce (a)	x=0.15 y=0.25 (greenish blue)	40 cd/m^2
SrSe:Ce (b)	x=0.18 y=0.28 (greenish blue)	5 cd/m^2

4 Shift of Emission Spectra of Ce^{3+} ion

Figure 3 shows the emission and excitation spectra of the CaS:Ce, SrS:Ce and SrSe:Ce thin films measured at 300 K. The SrS:Ce and SrSe:Ce films were prepared by the same method as the TFEL devices. The CaS:Ce film was deposited by electron beam evaporation of CaS:$CeCl_3$. Each emission spectrum is very similar to the EL spectrum, and consists of two emission bands (1) and (2). On the other hand, each excitation spectrum takes the form of only one band corresponding to the absorption transition from the $4f(^2F_{5/2})$ ground state to the $5d(^2T_{2g})$ excited state of Ce^{3+} ion. Therefore, the Stokes shift for transition between the $4f(^2F_{5/2})$ ground state and the $5d(^2T_{2g})$ excited state is obtained as the difference in peak wavelength between the excitation band and the emission band (1). The values of the Stokes shifts represented by ΔCaS, ΔSrS and ΔSrSe are 48 nm, 45 nm and 24 nm. In general, Stokes shift is closely related to the magnitude of the radiationless energy released by phonons, in other words, the degree of the electron-phonon interaction during the excitation and emission processes [7]. Because of the observed trend in magnitude of the shift(ΔCaS≃ΔSrS >ΔSrSe), it is considered that the electron-phonon interaction around Ce^{3+} ion becomes weak by means of replacing CaS or SrS by SrSe. It is assumed that the wavelength shift of the EL spectra of SrSe:Ce to SrS:Ce is mainly caused by the difference in the above-mentioned electron-phonon interaction, unlike the case of CaS:Ce.

Recently, the excitation and emission spectra of the CaS:Ce, SrS:Ce and SrSe:Ce powder phosphors have been measured at low temperature [8, 9]. The phonon sidebands with zero-phonon lines were observed in these spectra. Each zero-phonon line of the transition from the $5d(^2T_{2g})$ excited state to the $4f(^2F_{5/2})$ ground state appears at 475 nm for CaS, 449 nm for SrS and 447 nm for SrSe. In general, a zero-phonon line

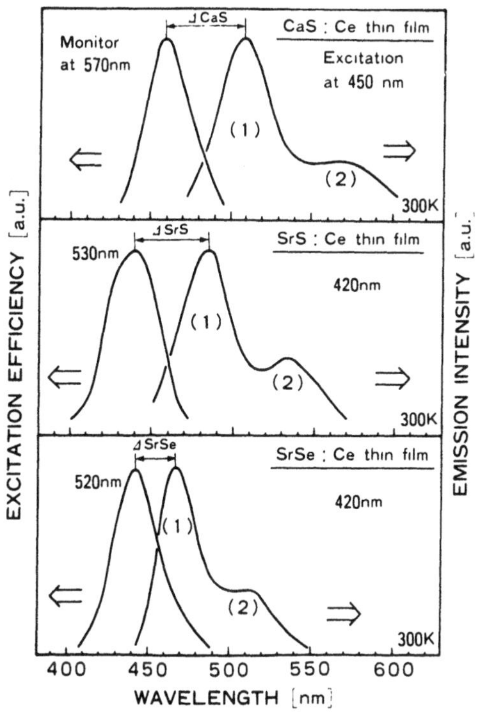

Fig.3. Emission and excitation spectra of CaS:Ce and SrSe:Ce thin films measured at 300 K. The emission spectra were obtained under the excitation at 450 nm for CaS:Ce, at 420 nm for SrS:Ce and SrSe:Ce, respectively. The excitation spectra were monitored by emission at 570 nm for CaS:Ce, at 530 nm for SrS:Ce and at 520 nm for SrSe:Ce. ΔCaS, ΔSrS and ΔSrSe show the values of Stokes shift

appearing on the emission or excitation spectra corresponds to the energy gap between the lowest vibrational level of the ground state and the lowest vibrational level of the excited state [7]. Therefore, it seems that these reported energy gaps between the 5d($^2T_{2g}$) state and the 4f($^2F_{5/2}$) state of Ce^{3+} ion does not make much difference by means of replacing SrS by SrSe, but becomes small by means of replacing SrS by CaS. These results appear to support our present results on the wavelength shift of the EL spectra of thin films.

5 Conclusion

The SrSe:Ce TFEL devices are superior to SrS:Ce devices in blue chromaticity. This results from two differences in the EL spectra, the shift to shorter wavelength by 10 nm and the increase in the relative intensity of the emission band of shorter wavelength. Together with other photoluminescence results of CaS:Ce, SrS:Ce and SrSe:Ce powder phosphors, it is considered that the wavelength shift of EL spectra is mainly caused by decrease in the radiationless energy released by phonons, unlike the case of the CaS:Ce devices. The influence of the electron-phonon interaction around Ce^{3+} ion should be considered in order to improve the blue chromaticity of TFEL phosphors.

The oxygen concentration in the SrSe:Ce layers appears to affect both the color and the luminance. Even though the highest luminance, 40 cd/m^2 at 1kHz, is not sufficient yet for practical use, it is expected to be improved by decreasing oxygen contamination.

Acknowledgments

The authors would like to thank Professor S.Tanaka of Tottori University for useful suggestions during this work. The authors would also like to thank Professor H.Saito and Professor M.Oishi of Okayama University of Science for comments and advice concerning the measurement of photoluminescence.

References

1. T.Hoshina: J. Phys. Soc. Jpn., 48, 158 (1980)
2. W.A.Barrow, R.E.Coovert and C.N.King: 84 SID Int. Symp. Digest, 249 (1984)
3. S.Tanaka, V.Shanker, M.Shiiki, H.Deguchi and H.Kobayashi: 85 SID Int. Symp. Digest, 218 (1985)
4. M.Ogawa, T.Shimouma, S.Nakada and T.Yoshioka: Jpn. J. Appl. Phys., 24, 2, 168 (1985)
5. S.Tanaka: J. Lumin., 40/41, 20 (1988)
6. Y.Kageyama, K.Kameyama and S.Oseto: 86 SID Int. Symp. Digest, 33 (1986)
7. For example, B.Di Bartolo: Optical Interactions in Solids, Wiley, New York, 422 (1968)
8. S.Yokono, T.Abe and T.Hoshina: J. Phys. Soc. Jpn., 46, 351 (1979)
9. N.Yamashita, Y.Michitsuji and S.Asano: J. Electrochem. Soc., 2933 (1987)

Electroluminescence of Rare-Earth Activated SrS Thin-Films

S. Okamoto, E. Nakazawa, and Y. Tsuchiya

NHK Science & Technical Research Laboratories,
1-10-11, Kinuta, Setagaya-ku, Tokyo 157, Japan

1. Introduction

Recently, SrS:Ce phosphor was actively studied as an emitting layer in thin-film electroluminescence (TFEL) devices, and the host, SrS, seems to be a good material for this use. This paper reports emission spectra and other EL properties of TFEL devices with SrS emitting layers activated with all rare-earth group ions (Ln) except for La, Pm, and Lu. The difference of the emission spectra between the SrS- and ZnS-based devices is discussed in relation to excitation mechanisms.

2. Fabrication of Devices

The TFEL devices, prepared on an ITO (indium-tin-oxide)-coated glass substrate, consist of a SrS emitting layer EB-deposited at a substrate temperature of 450 - 500°C with coevaporation of sulfur, two ZnS buffer layers deposited at a substrate temperature of 130°C by electron-beam (EB) evaporation, two Ta_2O_5 insulating layers deposited by rf-magnetron sputtering. As the deposition sources of the emitting layer, high-grade SrS powder (Mitsui Mining and Smelting Co.) was mixed with 0.1 - 1 mol% of rare-earth fluorides and formed into pellets.

3. Spectra and Waveforms of EL Emissions

In the observed EL emission spectra shown in Fig.1 most of the emission lines can be assigned to the f-f transitions of the trivalent dopant rare-earth ions, indicating that the dopant ions are active in the trivalent state. The broad emission bands in SrS:Ce and SrS:Eu, however, are due to the d-f transition of Ce^{3+} and Eu^{2+} ions. In the cases of Gd and Yb dopants, although ultraviolet and infra-red emissions were expected respectively from Gd^{3+} and Eu^{2+} ions, no EL emission has been observed.

It is noticed that the emission spectra of these SrS-based TFEL devices, as compared with those of ZnS-based devices with the same rare-earth activators/1/2/, are rich with short-wavelength lines, which in some cases result in different emission colors in the two hosts. For instance, emission lines extending from UV to blue in the spectrum of SrS:Tb in Fig.1, which originate from the $^5D_3$ emitting level, are suppressed in ZnS:Tb devices, and the emission spectrum of the latter is dominated by the transitions originating from the lower $^5D_4$ level. And also, the orange emission of Nd^{3+} in ZnS-based devices is changed to orange-white in SrS:Nd devices.

Waveforms of EL emission responding to 100μs-duration pulses repeating alternately at 1kHz were measured with a transient recorder. In the exciting-term, the waveforms showed fast-rises and fast-decays in the cases of Ce, Nd and Eu, a fast-rise and a slow-decay in the case of Tm, and slow-rises and slow-decay in the cases of Pr, Sm, Tb, Dy, and Er. The build-up rates of the slow-rises were dependent on applied voltages. In the long term after the termination of the excitation pulse most emissions showed exponential decays, with decay time constants (see Table 1) corresponding to life times of emitting levels of dopant ions, while in the initial short-term

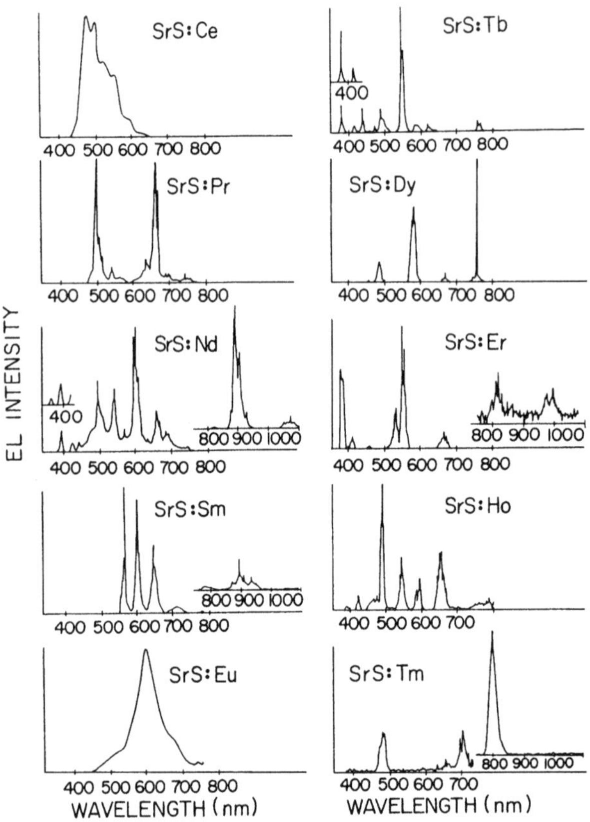

Fig. 1. Emission spectra of SrS:Ln,F TFEL devices (Ln= rare-earth ions)

after the pulse they did not always follow exponential decay, even showing a flash at the trailing edge of the excitation pulse in the cases of Ce, Nd, and Eu.

4. Discussion

The transitions assigned to the main emission lines in Fig.1 are listed in Table 1. These results indicate that in the cases of Pr, Sm, Dy, and Tm the most active emitting level is the same in SrS- and ZnS-based devices/1/, /2/, and the color differences between the two devices are caused by the difference of the branching ratios from the emitting level. In the case of Nd, Tb, Er and Ho, however, highly-located 4f-levels are more active in SrS-devices than in ZnS-devices. This shift of emitting levels to the high energy side in SrS-devices may be caused by the difference of crystal field effects, activator concentration or excitation mechanism between the two hosts. It seems unreasonable to expect the crystal field effect in SrS to shift the emitting levels only toward the high energy side. The second conceivable reason will be neglected, because our activator concentration is not much smaller than the case of the ZnS-based devices. In regard to the third reason, the excitation mechanism of ZnS-based devices is usually thought as direct impact excitation by hot electrons. Krupka/3/ observed in ZnS-devices that the intensity of the emission from an upper emitting level relative to the one from the lower level was increased in accordance with applied voltage, and concluded that the impact of hotter electrons accelerated by the higher field induces emissions from the higher emitting level. Although we performed the same experiment on SrS:Tb, SrS:Er, and SrS:Nd, such effect as indicates the direct impact excitation mechanism was not observed. Very recently Tanaka et. al./4/ proposed another excitation mechanism for SrS:Ce,K device-

Table 1. Optical properties of SrS TFEL devices

rare-earth	color	chromaticity coordinates x	y	luminance (cd/m^2) (5kHz)	wave-length (nm)	transitions	decay time (μs)
Ce	blue-green	.20	.38	650	480	$5d\text{-}4f(^2F)$	4
Pr	white	.33	.44	210	490	$^3P_0\text{-}^3H_4$	60
					660	$^3P_0\text{-}^3F_2$	
Nd	white	.40	.42	240	490	$^4G_{9/2}\text{-}^4I_{9/2}$	
					530	$^2G_{7/2}\text{-}^4I_{9/2}$	
					600	$^4G_{5/2}\text{-}^4I_{9/2}$	70
					900	$^4F_{3/2}\text{-}^4I_{9/2}$	
Sm	orange	.57	.42	235	580	$^4G_{5/2}\text{-}^6H_{5/2}$	
					600	$^4G_{5/2}\text{-}^4H_{7/2}$	580
					650	$^4G_{5/2}\text{-}^4H_{9/2}$	
Eu	orange	.54	.42	45	610	$4f^65d\text{-}4f^7$	16
Tb	green	.29	.54	30	380	$^5D_3\text{-}^7F_6$	720
					550	$^5D_4\text{-}^7F_5$	940
Dy	yellow	.45	.44	200	580	$^4F_{9/2}\text{-}^6H_{13/2}$	740
					760	$^4F_{9/2}\text{-}^6H_{9/2}$	
Ho	white	.30	.39	110	490	$^5F_3\text{-}^5I_8$	520
					550	$^5S_2\text{-}^5I_8$	
					660	$^5F_3\text{-}^5I_7$	
Er	green	.30	.66	210	380	$^4G_{11/2}\text{-}^4I_{15/2}$	70
					550	$^4S_{11/2}\text{-}^4I_{15/2}$	1000
Tm	blue	.13	.15	15	480	$^1G_4\text{-}^3H_6$	470
					800	$^1G_4\text{-}^3H_5$	

s, based on the concept of the impact ionization of luminescent centers. The recombination of the ionized centers with conduction electrons will be able to supply sufficient energies for 4f-electrons to be excited up to high energy levels. Therefore this mechanism may work in our devices.

The appearance of the flash at the trailing edge of excitation pulses in the SrS:Ce devices has been explained also with the ionization model/4/. As described in the previous section, we observed the same flashes in the waveforms of Ce, Nd, and Eu emissions. When the relaxation times of activators are longer than the pulse duration, as is the cases of Pr, Sm, Tb, Dy, Ho, Er, and Tm, the activators still remain in the excited states at the pulse end, and, therefore, may not produce flashes. The waveform of Tb activators after the pulse end showed, instead of a flash, an abrupt decrease to two thirds of the intensity in the end of the pulse. We consider that the decrease is caused by Auger effects of the released electrons to the excited Tb^{3+} ions.

5. Summary

All rare-earth ions except for La, Pm, and Lu have been tested for the use as activators to SrS hosts in emitting layers of TFEL devices. Color-coordinates, luminance-levels, emission spectra, and waveforms of the EL-emissions of the devices have been measured. The color and spectral region of the emission are for many cases shifted to shorter wavelengths compared with those of the same ions in ZnS-based TFEL devices. These features and the waveforms measured suggest different excitation mechanisms between SrS- and ZnS-based TFEL devices.

6. References

1. E. W. Chase, R. T. Hepplewhite, D.C. Krupka, and D.Kahng: J. Appl. Phys. 40(1969), 2512.
2. T. Yamamoto, R. Fukao, and Y. Hamakawa: Abstract 126th meeting, The 125 Research Committee, Japan Society for the Promotion of Science (1986), p.25 (in Japanese).
3. D. C. Krupka: J. Appl. Phys. 43(1972), 476.
4. S. Tanaka, H. Yoshiyama, Y. Mikami, J. Nishiura, S. Ohshio and H. Kobayashi: IEICE Tech. Report 86 (1987) No. 368, 9 (in Japanese).

Multi-Color Electroluminescent Devices Utilizing SrS:Pr, Ce Phosphor Layers and Color Filters

Y. Abe, K. Onisawa, K. Tamura, T. Nakayama, M. Hanazono, and Y.A. Ono

Hitachi Research Laboratory, Hitachi, Ltd.,
4026 Kuji-cho, Hitachi-shi, Ibaraki-ken, 319-12 Japan

Abstract

A new phosphor material, SrS doped with both Pr and Ce, with bright white emission was proposed for an application to multi-color thin film electroluminescent (TFEL) devices. Spectral changes by varying a concentration ratio of the activators and color separation with filters for SrS:Pr,Ce TFEL cells were studied. It was shown that three primary colors with suitable color coordinates and with balanced luminance levels could be obtained by utilizing SrS:Pr,Ce phosphor layers and color filters.

1. Introduction

Recently much attention is directed to developing multi- or full-color thin film electroluminescent (TFEL) displays.[1]-[6] Three types of techniques for realizing multi-color TFEL displays were proposed [7] and studied: (I) stacking of TFEL cells for three primary colors, (II) patterning of three primary color phosphors, and (III) color filtering of broad-band or white light emitting phosphors. In case (III), the device fabrication process is much simpler than that in the other cases. In order to achieve the type (III) devices, development of bright white light emitting phosphors is required. Several investigations to obtain these phosphors have been reported. [8]-[12]

In the present paper we propose a new phosphor material, SrS doped with both Pr and Ce, denoted by SrS:Pr,Ce, with bright white light emission. Spectral changes by varying the concentration ratio, C_{Ce}/C_{Pr+Ce}, and the results of color separation with filters are reported.

2. Experimental

TFEL cells used in the present study consist of a SrS:Pr,Ce phosphor layer of about 1 μm thickness sandwiched by two insulating layers made of Ta_2O_5 with about 0.5 μm thickness. To prevent phosphor-layer oxidation and peeling-off from the insulating layers, ZnS buffer layers of about 0.1 μm thickness were inserted between the phosphor layer and the insulating layers. The SrS:Pr,Ce phosphor layers were grown by the electron-beam (EB) evaporation method under a sulfur pressure of $2\text{-}3\times10^{-2}$ Pa on glass substrates kept at 500°C. A deposition rate of the phosphor layer controlled by the electron beam current was as high as 6nm/s, because higher deposition rate, which corresponds to higher surface temperature of evaporation sources, was found to provide better quality of the phosphor layers and higher luminance.[13] Pressed SrS pellets mixed with both $CeCl_3$ and $PrCl_3$ were used as evaporation sources. The total concentration of activators, C_{Pr+Ce}, was 0.1-0.3 mol%. The Ta_2O_5 insulating layers were grown by the RF sputtering method.

3. Results and discussion

EL spectra of SrS:Pr,Ce TFEL devices are summarized in Fig. 1. Figure 1(a) shows an EL emission spectrum of SrS:Pr with two main peaks: One peak around 490nm (blue-green) is due to $^3P_0$ -$^3H_4$ transition and another peak around 660nm (red) is due to $^3P_0$ -$^3F_2$ transition of $(4f)^2$ electron configuration in Pr^{3+} ions. These two peaks satisfy approximately the complementary color relationship, resulting in a white emission. However, the emission peak around 490nm is too narrow to obtain bright blue and green emission components by using color filters. On the other hand, as shown in Fig. 1(e), an EL emission spectrum of SrS:Ce has a broad-band blue-green emision. This should give larger blue and green emission components with color filters. Figures 1(b)-(d) show spectra of the phosphor layers doped with both Pr^{3+} and Ce^{3+}. From these it is found that by increasing the concentration ratio, C_{Ce}/C_{Pr+Ce}, the peak around 490nm broadens and reaches the shape of Ce^{3+}, and the intensity of the peak around 660nm decreases.

In order to determine an optimum concentration ratio for the multi-color TFEL cells with color filters, the luminance ratios of each primary color component to the total luminance, L_R/L_T for red, L_G/L_T for green, L_B/L_T for blue, are estimated and compared with color balance of the color CRT, where $L_R:L_G:L_B$ is approximately equal to 2:7:1 to display white. Here L_R, L_G, L_B, and L_T are calculated from the EL emission spectrum $I(\lambda)$.

$$L_{R,G,B} = \int_{390}^{730} I(\lambda)F_{R,G,B}(\lambda)V(\lambda)d\lambda \qquad (1)$$

$$L_T = Km\int_{390}^{730} I(\lambda)V(\lambda)d\lambda \qquad (2)$$

where $V(\lambda)$ is the CIE photopic relative luminous efficiency function, Km is the maximum luminous efficacy (=683 lm/W), and $F_{R,G,B}(\lambda)$ are the transmission characteristic functions of red, green and blue filters, respectively. We note that $I(\lambda)F_{R,G,B}(\lambda)$ are equal to the measured spectral power distribution, $P(\lambda)$, by using color filters. The CIE chromaticity coordinates x and y are computed from tristimulus values, X, Y, Z as follows:

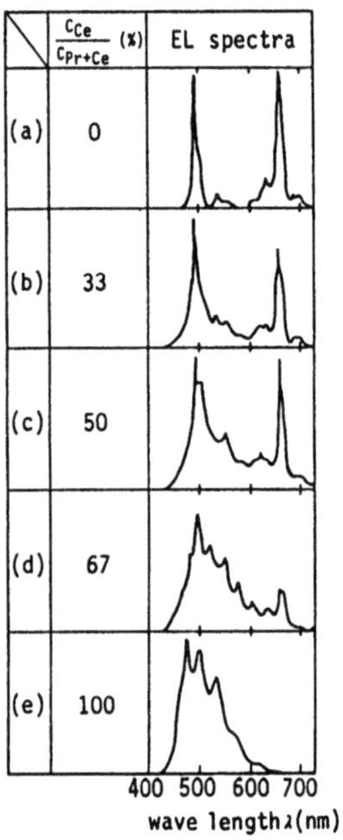

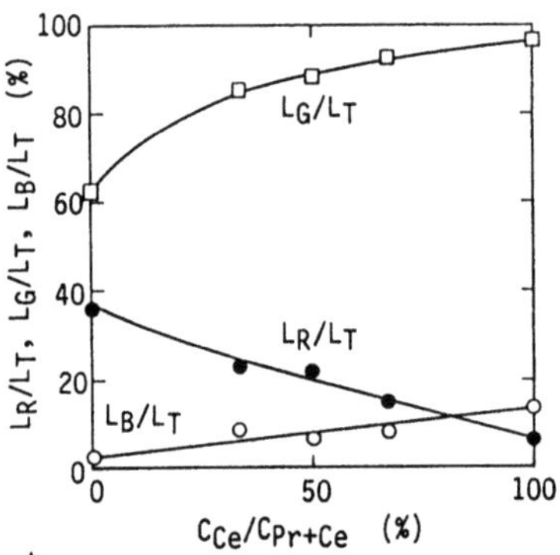

△
Fig. 2 Dependence of luminance ratio for three primary colors on concentration ratio, C_{Ce}/C_{Pr+Ce}

◁
Fig. 1 EL spectra of SrS:Pr,Ce TFEL devices for several concentration ratio, C_{Ce}/C_{Pr+Ce}

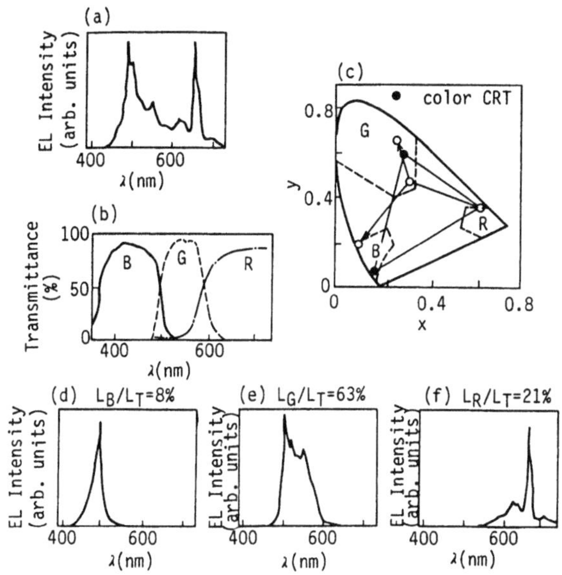

Fig. 3. Results of color separation with filters for SrS:Pr (0.1mol%),Ce(0.1mol%) (a) Original spectrum (b) transmission properties of used filters, (c) CIE chromaticity diagram, (d) spectrum after blue filtering, (e) spectrum after green filtering, and (f) spectrum after red filtering

$$x = X/(X+Y+Z), \quad y = Y/(X+Y+Z) \tag{3}$$

$$X = K_m\int_{390}^{730}P(\lambda)\bar{x}(\lambda)d\lambda, \; Y = K_m\int_{390}^{730}P(\lambda)\bar{y}(\lambda)d\lambda, \; Z = K_m\int_{390}^{730}P(\lambda)\bar{z}(\lambda)d\lambda \tag{4}$$

where $\bar{x}(\lambda)$, $\bar{y}(\lambda)$, $\bar{z}(\lambda)$ are the color-matching functions. We note that $\bar{y}(\lambda)$ is by definition identical to $V(\lambda)$.

Model calculation is performed by assuming ideal color filters with a 100% transmission between λ_1 and λ_2. By varying λ_1 and λ_2 for $F_{R,G,B}(\lambda)$ so that the chromaticity coordinates are fixed at x=0.65, y=0.35 for red, x=0.28, y=0.59 for green, and x=0.10, y=0.15 for blue, L_R/L_T, L_G/L_T, and L_B/L_T are evaluated. The results are shown in Fig. 2 as a function of the concentration ratio, C_{Ce}/C_{Pr+Ce}. From this graph it is found that the green and the blue components increase and the red component decreases with increasing C_{Ce}/C_{Pr+Ce} and that color balance is most favorable at $C_{Ce}/C_{Pr+Ce} \fallingdotseq 50\%$.

Figure 3 summarizes the results of color separation with practical filters for SrS:Pr(0.1mol%), Ce(0.1mol%). Figure 3(a) shows the original spectrum. The CIE chromaticity coordinates for red, green, and blue as well as for the original white are plotted in Fig. 3(c), when the color filters with transmission properties given in Fig. 3(b) are used. The coordinates for three primary colors of the color CRT are also plotted in Fig. 3(c), demonstrating that the three coordinates of the multi-color TFEL cell are fairly close to those of the color CRT. The spectra after color filtering and values for L_B/L_T, L_G/L_T, and L_R/L_T, are shown in Fig. 3(d)-(f). It is found that the color balance of the TFEL cell similar to that of the color CRT can be obtained with these filters.

Figure 4 shows the luminance vs. drive voltage characteristic of TFEL cell with SrS:Pr(0.1mol%),Ce(0.1mol%) under 1kHz sinusoidal wave drive condition. It is shown that L_{60} is 280cd/m^2 and Lmax is 950cd/m^2 at 350V.

4. Summary

White light emitting SrS:Pr,Ce phosphor materials for the multi-color TFEL devices with filters were investigated. It was found that three primary colors

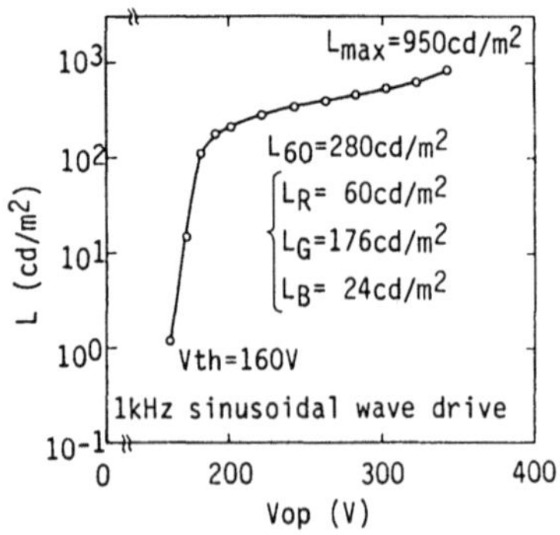

Fig. 4. Luminance vs. voltage characteristic of SrS:Pr(0.1mol%), Ce(0.1mol%)

with suitable color coordinates and with balanced luminance levels could be obtained by utilizing color filters and SrS:Pr,Ce phosphor layers with the activator concentration of C_{Pr}=0.1mol%, C_{Ce}=0.1mol%.

Acknowledgments

The authors thank Dr. Y. Sugita and Mr. M. Ando of Hitachi Research Laboratory, Hitachi, Ltd. and Dr. H. Matsumura, Dr. H. Yamamoto, and Mr. O. Kanehisa of Central Research Laboratory, Hitachi, Ltd. for valuable comments and encouragement.

References

1. S. Tanaka et al.: Proc. SID, 26, 255(1985).
2. Y. Tamura et al.: Jpn. J. Appl. Phys. 25, L105(1986).
3. K. Tanaka et al.: Appl. Phys. Lett. 48, 1730(1986).
4. S. Tanaka et al.: Proc. SID, 28, 357(1987).
5. S. Tanaka: J. Lumin. 40&41, 20(1988).
6. W. A. Barrow et al.: 1988 SID Digest, 284(1988).
7. L. E. Tannas, Jr.: 1986 SID Seminar Lecture Notes, Vol.1, S-3.2(1986).
8. S. Tanaka et al.: Jpn. J. Appl. Phys., 25, L225(1986).
9. S. Tanaka et al.: Appl. Phys. Lett., 51, 1661(1987).
10. Y. A. Ono et al.: J. Lumin. 40&41, 796(1988).
11. K. Onisawa et al.: Extended Abstract of Electrochemi. Soc., 1987 Fall Meeting, 1701(Honolulu, 1987), and J. Electrochemi. Soc., 135, 2631(1988).
12. G. Zhong et al.: SID 88 Digest, 287(1988).
13. K. Onisawa et al.: Submitted to J. Electrochemi. Soc.

Part V

Processing Technology

Electroluminescent Materials Grown by Atomic Layer Epitaxy

M. Leskelä

Laboratory of Inorganic and Analytical Chemistry,
Helsinki University of Technology, SF-02150 Espoo, Finland
Permanent address: Department of Chemistry, University of Turku,
SF-20500 Turku, Finland

1. INTRODUCTION

In the very near future the first book on Atomic Layer Epitaxy (ALE) will be published containing a separate chapter on chemical aspects of the ALE process /1/. Therefore the present paper is only a brief summary on electroluminescent (EL). materials grown by the ALE method.

2. ATOMIC LAYER EPITAXY

The history of the ALE began with the growth of polycrystalline ZnS films from elemental Zn and S for use as matrix material in electroluminescent devices /2/. The ALE-method can be described as an epitaxial deposition method where the reactants are alternatively pulsed onto the substrate. During the pulse the reactor chamber contains an excess of reactant and the substrate temperature and gas flow have been adjusted so that one monolayer of reactant can be chemisorbed on the substrate. The excess of the reactant in the gas phase as well as loosely-bound, physisorbed molecules are purged out from the reactor by an inert gas pulse. The second reactant is then dosed and it chemisorbs and reacts with the first reactant forming a solid molecular film and gaseous side-product which is purged out by the following inert gas pulse. Repeating this reaction cycle results in a controlled layer by layer thin film growth.

In principle the ALE method relies upon the difference between chemisorption and physisorption. In chemisorption only one atom or molecule can bind to the surface. The growth is insensitive for pressure and substrate material but the substrate temperature plays an important role in the growth process. The temperature must be sufficient for chemisorption but not too high to lead to the desorption of the monolayer.

ALE growth has been demonstrated in several reactor types, viz. vacuum evaporation apparatus, gas flow type reactor and ultra high vacuum MBE-type reactor /3/. Large area EL panels are prepared in a flow type reactor /4/.

3. EL DEVICE

Besides the luminescent layer, an EL device contains several other layers which are very important for the function of the device. For example, the commercial ACTFEL device produced by Finlux consists of ion barrier (Al_2O_3), transparent electrode ($In_xSn_yO_z$), dielectric ($Al_xTi_yO_z$), luminescent (ZnS:Mn) and passivation (Al_2O_3) layers /5/. All these layers, except the electrodes, are made by ALE. In deposition of luminescent materials based on rare earth doped alkaline earth sulfides the ALE-method has been employed as well /6/.

Springer Proceedings in Physics, Vol. 38 Electroluminescence
Editors: S. Shionoya and H. Kobayashi

4. SOURCE MATERIALS

The ALE process accepts source materials of all types: elements, inorganic compounds, metal complexes and organometallics. The source material has to fulfil the following requirements: (i) volatilize at reasonable temperature, (ii) not to decompose but chemisorb on the substrate surface, (iii) react with the counter reactant and form a solid film of desired composition and a gaseous product.

The use of elements as source materials is mainly limited to II (Zn,Cd,Hg) - VI (S,Se,Te) compounds because of the small number of elements which are volatile under available conditions /1/. Other elements which have vapor pressure high enough in conditions of flow type ALE reactor are the alkaline metals, phosphorus and arsenic. Metals having low vapor pressure can be used in MBE-type reactor if a special procedure is adopted /3/.

The number of binary inorganic compounds which fulfil the requirements presented above is also sparse. Metal halides, in particular chlorides, are the most commonly used source materials in vapor phase depositions. In ALE process, $ZnCl_2$ ($MnCl_2$) combined with H_2S has been the most widely studied system for the preparation of ZnS:Mn. Other metal chlorides, $AlCl_3$, $TaCl_5$, $SnCl_4$, $TiCl_4$ combined with H_2O have been used in the preparation of corresponding oxides /1,3,5/.

Large electropositive metals, such as the alkaline earth and rare earth metals, form no stable inorganic compounds which are volatile under conditions usually used in the film growth. The use of metal complexes and organometallics offers a possibility to overcome this problem. β-diketonato complexes of alkaline earth and rare earth metals have proven to be applicable in both growth of alkaline earth sulfide matrices for EL devices /6,7/ and in doping of ZnS and MS (M = Ca,Sr,Ba) films by rare earth ions /8,9/. From other metal complexes studied, zinc acetate in growth of ZnS and ZnO can be mentioned /10/.

The advantages of the metal complexes and organometallics include their high vapor pressures and the possibility to prepare both sulfides and oxides from the same metallic source. Their disadvantage is the slow growth rate which is due to the large size of the molecule chemisorbed at the surface. Furthermore, there exist no thermodynamic data on these complexes in the literature. In addition the commercial availability of these complexes is often limited.

5. PREPARATION of SULFIDES

Zinc chloride (manganese chloride) and hydrogen sulfide are the source materials used in the production of commercial EL panels having yellow emitting $ZnS:Mn^{2+}$ as active layer. The reaction is complete enough as the traces of Cl^- can not be detected by Auger spectrometry (Fig. 1). Well oriented (00.2) hexagonal 2H phase

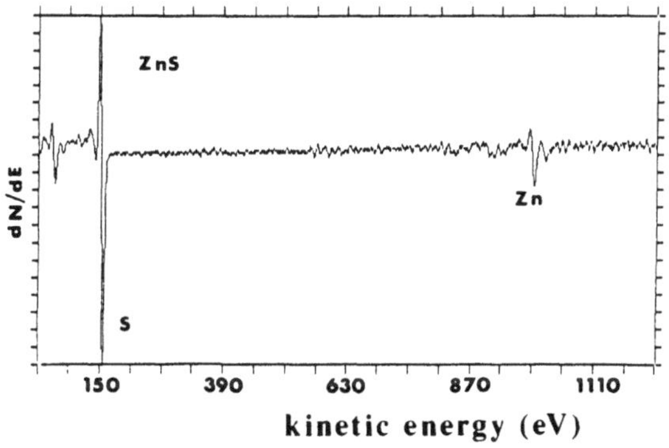

Figure. 1. An Auger survey spectrum of ZnS prepared from $ZnCl_2$ and H_2S

of ZnS can be grown from $ZnCl_2$ at 500 °C (Figs. 2,3). Luminance of the ALE grown $ZnS:Mn^{2+}$ films exceeds the level required for device applications and the efficiency obtained is sufficient (Fig. 4) /11/.

Alternative zinc sources for the growth of ZnS thin films include zinc acetate, β-diketonates and dimethylzinc (DMZ). When compared to the reaction $ZnCl_2 + H_2S$, the ALE experiments made with zinc acetate showed that in acetate system the growth occurs at lower temperature, the growth rate is higher and the structure of the film is different (Figs. 2,3) /10/.

The first MOALE depositions of $ZnS:Mn^{2+}$ have been made from DMZ and H_2S at ambient pressure /12/. Films of good quality have been obtained at substrate temperatures of 250-500 °C.

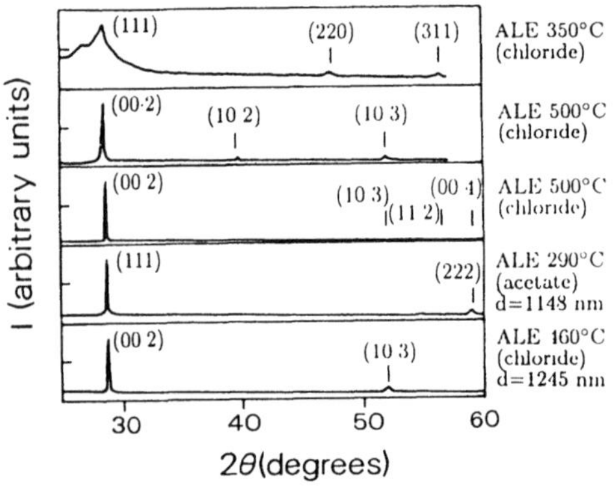

Figure. 2. A comparison of diffraction patterns of five ZnS films grown by ALE at various temperatures /17/

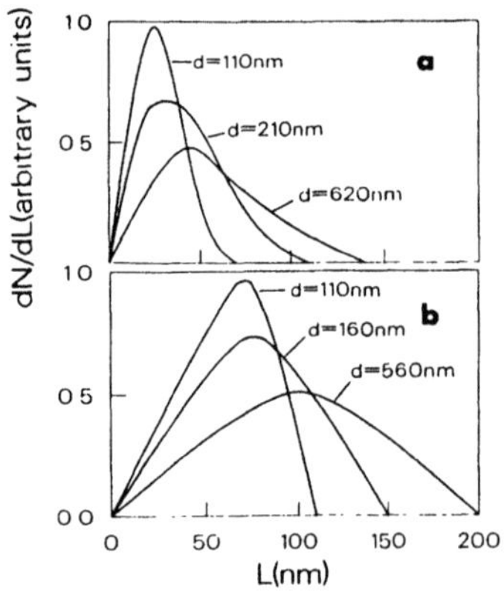

Figure. 3. A comparison of crystallite site distribution functions dN/dL in two ZnS films grown by ALE from (a) zinc acetate and (b) zinc chloride /17/

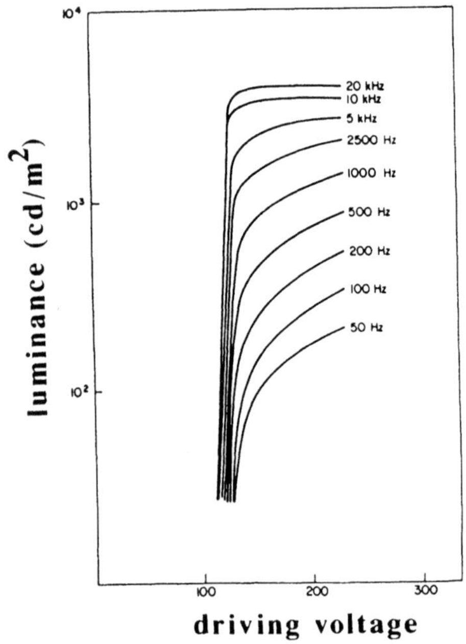

Figure. 4. Luminance vs. driving voltage at different frequencies in a ALE grown $ZnS:Mn^{2+}$ thin film /11/

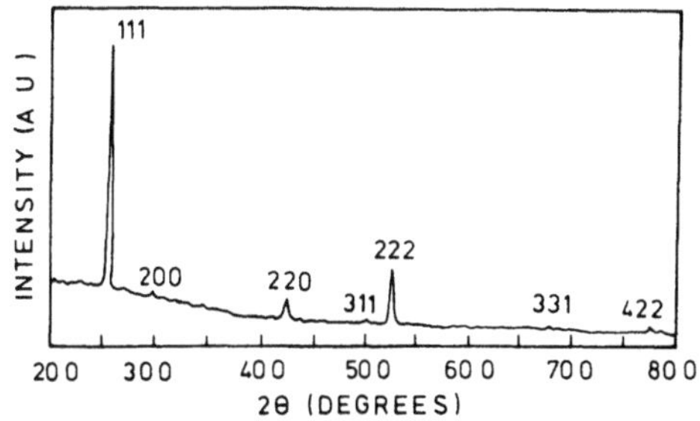

Figure. 5. A X-ray diffraction pattern for SrS thin film grown on a glass substrate /7/

The worldwide interest in development of multicolor ACTFEL device has led to the research of alkaline earth sulfides doped with rare earth ions /13/. 2,2,6,6-tetramethyl-3,5-heptanedione (thd) complexes of alkaline earth metals react under ALE conditions with H_2S and epitaxially grown MS (M=Ca,Sr,Ba) films can be obtained in a wide temperature range (Fig. 5) /7/.

6. DOPING

Luminescent films require a dopant or activator ion which gives the emission colour. The element used as dopant depends on the colour desired. In yellow emitting devices the activator is Mn^{2+}. The full colour device needs blue, green and red phosphors and the rare earth doped ZnS and MS films are possible candidates. The best blue, green and red EL phosphors are so far $SrS:Ce^{3+}$, $ZnS:TbF_x$ and $CaS:Eu^{2+}$, respectively (Fig. 6) /13,14/. Other rare earth ions have been tested, too /13/.

In the ALE process doping can be arranged in two ways: the host ion and dopant ion can be pulsed simultaneously or they can be given as separate pulses. In epitaxial gas phase growth processes the controlled doping can be relative easily achieved.

In ALE growth of EL films the most convenient way to choose the dopant source material is to use similar compound type as for host ion. For zinc chloride $MnCl_2$ is a good choice whereas $Mn(thd)_2$ chelate is usable with Zn acetate and $Zn(thd)_2$. If DMZ is employed possible Mn sources are $Mn_2(CO)_{10}$ and $Mn(CO)_3(C_5H_5)_2$. For rare earth ions the thd chelates are good source materials in particular with $M(thd)_2$ host source materials.

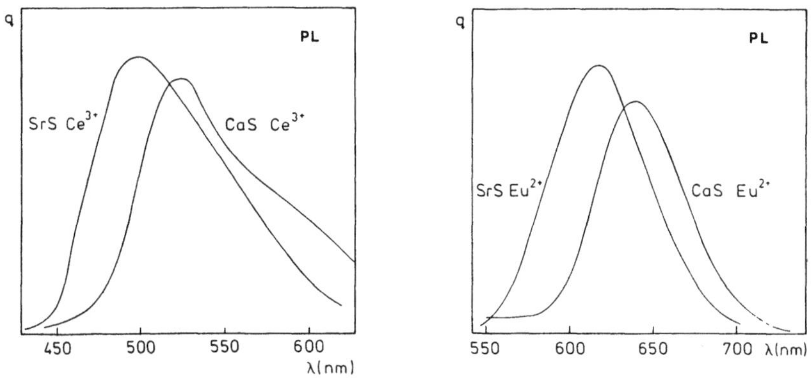

Figure. 6. Room temperature photoluminescence spectra of $MS:Ce^{3+}$ and $MS:Eu^{2+}$ (M = Ca,Sr) films /18/

7. PREPARATION of OXIDES

The interest in oxide thin films originates from their use in ACTFEL devices as ion barrier, dielectric and passivation layers. Al_2O_3 films may be used for all these purposes. However, in commercial devices when used as dielectric layer Al_2O_3 contains some titanium. The mixing of Ti to aluminium oxide enhances the dielectric strength.

Al_2O_3 can be grown by ALE from $AlCl_3$ and H_2O source materials (Fig. 7). $AlCl_3$ has a high vapour pressure and the source temperature can be significantly lower than that of reactor. At 450-500 °C growth rate is not very fast, mainly because of the size of the $AlCl_3$ molecule. Small amounts of chlorine tend to remain in the film but has no clear effect on the properties of Al_2O_3. The oxide formed is amorphous.

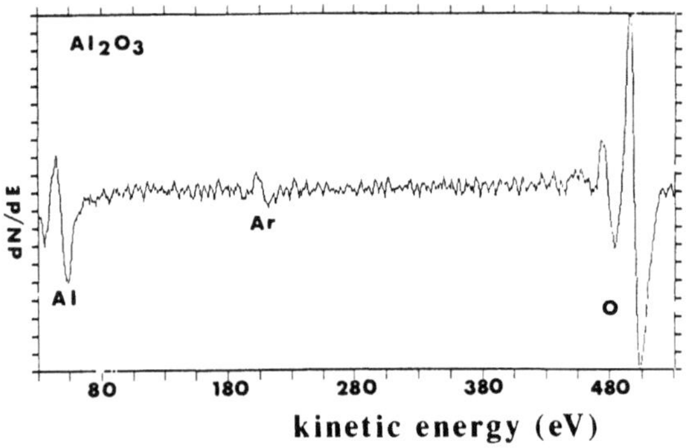

Figure. 7. An Auger survey spectrum of Al_2O_3 grown from $AlCl_3$ and H_2O

Other oxides which have been grown by ALE from chlorides and water are tantalum oxide for the use as dielectric layer and indium and tin oxides for the use as transparent electrodes /1/.

8. OUTLOOK

The use of metal complexes with large organic ligands as source materials has so far been modest. Despite the difficulties, the research on the use of metal complexes in thin film growth will undoubtedly expand in the future.

In the ALE process organometallic source materials have mainly been used to grow the III-V semiconductors /15/. Organometallics, despite the difficulties in their handling, may offer some advantages for ALE process and their use will probably increase, too.

Organic compounds can also be used in ALE. Potential organic chalcogen sources may be tetrahydrofuran (C_4H_4O), thiophene (C_4H_4S), selenophene (C_4H_4Se) and dialkyltellurides already employed in MOCVD /16/.

Acknowledgement

Prof. L. Niinistö (Helsinki University of Technology) is gratefully acknowledged for valuable discussions.

References

1. M. Leskelä, L. Niinistö: In Atomic Layer Epitaxy, ed. by T. Suntola, Blackwell and Sons (Blackwell, Glasgow 1988) in press
2. T. Suntola, J. Antson: U.S. 4 058 430 (1977)

3. C.H.L. Goodman and M.V. Pessa, J. Appl. Phys. 60 R65 (1986)
4. T. Suntola, A. Pakkala, S. Lindfors: U.S. 4 389 973 (1983)
5. T. Suntola, J. Hyvärinen: Ann. Rev. Mater. Sci. 15 177 (1985)
6. M. Leskelä, M. Mäkelä, L. Niinistö, E. Nykänen, M. Tammenmaa: Chemtronics 3 113 (1988)
7. M. Tammenmaa, H. Antson, M. Asplund, L. Hiltunen, M. Leskelä, L. Niinistö, E. Ristolainen: J. Cryst. Growth 84 151 (1987)
8. M. Tammenmaa, M. Leskelä, L. Niinistö, T. Koskinen: J. Less-Common Met. 126 209 (1986)
9. M. Asplund, J. Hölsä, M. Leskelä, L. Niinistö: Inorg. Chim. Acta 139 261 (1987)
10. M. Tammenmaa, T. Koskinen, L. Hiltunen, M. Leskelä, L. Niinistö: Thin Solid Films 124 125 (1985)
11. T. Sutela: Display Technol. Appl. 5 73 (1984)
12. A. Hunter, A.H. Kitai: J. Cryst. Growth 91 111 (1988)
13. M. Leskelä, M. Tammenmaa: Mater. Chem. Phys. 16 349 (1987)
14. S. Tanaka, V. Shanker, M. Shiiki, H. Deguchi, H. Kobayashi: Proceedings of SID 26/4 255 (1987)
15. S.M. Bedair: In Atomic Layer Epitaxy, ed. by T. Suntola, Blackwell and Sons (Blackwell, Glasgow 1988) in press
16. J.B. Mullin, S.J.V. Irvine, D.J. Ashen,:J. Cryst. Growth 55 92 (1981)
17. M. Oikkonen: Thesis, Helsinki University of Technology, Espoo 1988, 46 pp.
18. M. Leskelä, M. Mäkelä, L. Niinistö, E. Nykänen, P. Soininen: in Compound Semiconductor Growth Processing and Devices for the 1990's. Gainesville FL 1987, p. A7

The Role of Chemical Vapour Deposition in the Fabrication of High Field Electroluminescent Displays

A. Saunders and A. Vecht

Department of Materials Science and Physics, Thames Polytechnic, Wellington Street, London SE18 6PF, UK

Introduction

High field thin film electroluminescent (TFEL) displays are undoubtedly one of the most promising emissive display technologies for the future. As is well known, the TFEL device may be AC or DC driven. The AC driven display (ACTFEL) has adopted the basic five layer MISIM configuration /1/, M = Metal/conductor, I = insulator, S = semiconductor (ZnS.Mn) and has achieved considerable success in recent years. Its operation now has a well developed theoretical basis /2-13/ and continuing, albeit sporadic, work is under way to correlate device characteristics with materials properties and therefore the deposition technology that produced them, /14, 15/. DC driven devices (DCTFEL), not having the benefit of a double insulating layer to protect them against high transient current 'spikes' most frequently suffer catastrophic breakdown and consequently have a rather short operational life. Recent work /16-20/ has reiterated the importance of current controlled negative resistance in the simple DCTFEL structure and with this may come sufficient understanding for real progress to be made. In recent years, thin film coating technology for TFEL has been heavily dependent on the techniques of thermal evaporation and sputtering, but these present many problems; poor stoichiometry control, high macroscopic defect densities and lack of conformal coating being but a few. Indeed, the low yield apparently inherent in these processes has, in recent years, forced a number of production operations to fall by the wayside. Now, however, the application of chemical vapour deposition (CVD) to TFEL device fabrication is receiving more attention, and it is the purpose of this review to highlight this aspect, and perhaps point the way in which CVD can make a significant contribution. We shall consider the current and potential role of CVD in terms of the general MISIM structure, i.e. the conducting thin film/substrate, the insulating layers and the ZnS.Mn light emitting layer.

The Chemical Vapour Deposition of Transparent Conducting Thin Films

Currently, sputtered tin doped indium oxide (ITO) is the predominant transparent conductor employed in TFEL device fabrication. The specifications for this initial coating are rather stringent. It should have good

a) adhesion to glass,
b) optical transparency ($>$ 80% in visible range),
c) electrical conductivity (sheet resistance $<$20 ohm/sq),
d) thermal stability,
e) barrier properties to underlying ions in glass,
f) compositional and microstructural continuity, and,
g) surface morphology.

The most promising candidates are

i) In_2O_3.Sn (ITO)
ii) SnO_{2-x} doped with Sb or F, and
iii) ZnO doped with Al or In

Springer Proceedings in Physics, Vol. 38 Electroluminescence
Editors: S. Shionoya and H. Kobayashi

The preparation of transparent conducting thin films has been the subject of a number of excellent reviews /21, 22/. The spray pyrolysis process has been reviewed in detail by Mooney and Radding /23/ and further good general accounts are to be found in /24-26/. Essentially, in this process a solution of an inorganic or organometallic precursor is sprayed directly onto a heated substrate (Fig. 1) such that under the best processing conditions, the solution droplets vapourize and the solute undergoes thermal decomposition near the substrate surface to give the desired thin film. In a traditional CVD process, the active species are transported as a vapour/gas from a volatile liquid or solid source, or directly from a gaseous source. In this case, thin film deposition can occur at atmospheric or reduced pressure conditions, the latter generally ranging from 0.1 torr - 100 torr. The deposition systems employed are many and varied, Fig. 2 showing a selection of such systems that may be employed for an all gaseous reactant system. Similar systems may be employed for less volatile source materials, but heated transport lines are usually required.

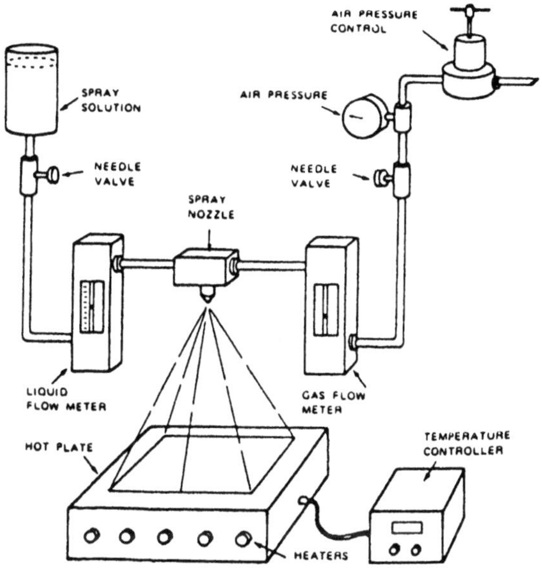

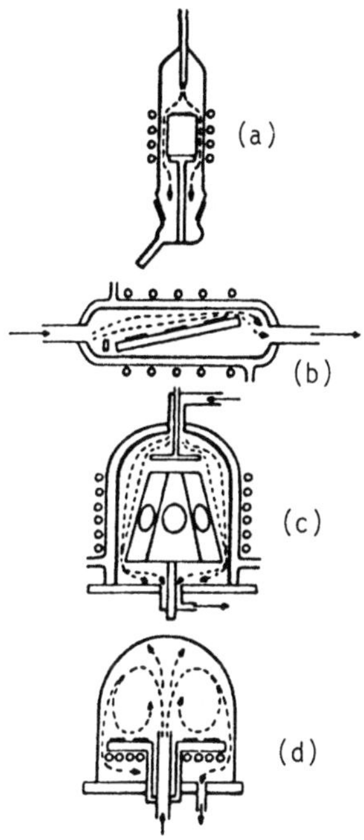

△ Fig. 1. Spray pyrolysis system /23/

▷ Fig. 2. A schematic of the (a) pedestal; (b) horizontal; (c) barrel; and (d) pancake-type RF-heated silicon deposition system /31/

Typically, ITO thin films are deposited by CVD employing $In(thd)_3$ with dibutyl tin diacetate (DBTD) as the tin source /27/.

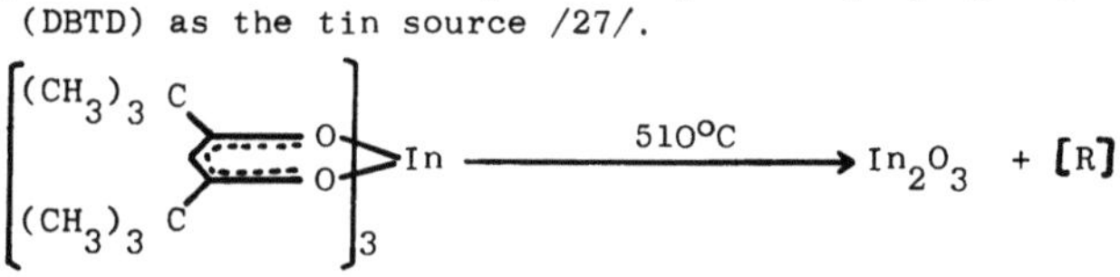

$In(thd)_3$, (volatile solid).

$$(C_4H_9)_2\ Sn\ (OOCCH_3)_2 \longrightarrow SnO_2 + [R]$$

DBTD volatile liquid

The best films produced by this method have a sheet resistance of 3 ohms/sq and 80-90% transmission in the visible range with a composition of 4.6 $^m/_o$ SnO_2 in In_2O_3. These reactants may be employed in a spray pyrolysis system, but it is currently more common and more economical to use an ethanolic solution of the chlorides, $SnCl_4$ and $InCl_3$ to obtain similar transmission and conductivity properties. Thin films of SnO_x may be prepared with equal ease, either by conventional CVD or spray pyrolysis. Traditional CVD generally employs anhydrous $SnCl_4$ /29/ or any of a host of volatile organotin compounds, e.g. $(CH_3)_4Sn$, /29/ with a conventional bubbler system to generate the vapour and a water vapour, air or oxygen supply to oxidise the tin compound to SnO_2 in the hot zone in the range 400-600°C. Doping with fluorine or antimony is preferred for the highest conductivity. For example, trichlorofluoroethane, CF_3CCl_3 and $SbCl_5$ have been successfully employed /30/. Spray pyrolysis is generally carried out via ethanolic solutions of $SnCl_2.2H_2O$ or $SnCl_4.5H_2O$ with an oxygen bearing carrier gas and NH_4F or CF_3COOH as the source of fluorine and $SbCl_5$ (with hydrochloric acid added to assist its solubility) for Sb doping /28/. SnO_x thin films that satisfy the requirements of TFEL may be readily prepared in this manner. Finally, we have the choice of zinc oxide, (ZnO) thin films, usually with aluminium or indium added to give high conductivity. There has been a recent upsurge in interest in this material due to its excellent and versatile acoustic, optical and electrical properties /31/. It is now being considered as an important component in heterojunction solar cells /32, 33/ and as a transparent conductive window layer /34/. In addition, it has the important advantage of being a low cost and relatively non-toxic material. Two principal methods have been applied to the deposition of ZnO thin films:

a) the oxidation of dialkyl zinc compounds (R_2Zn), and
b) spray pyrolysis of a zinc chelate or zinc alkanoate solution.

Wright, /35-38/ has recently reviewed the growth of wide band gap II-VI compounds by MOCVD, including ZnO. Essentially, a gaseous dialkyl zinc compound, or adduct is reacted with oxygen or an oxygen bearing compound /39/ in a conventional cold-wall MOCVD reactor, Fig. 3 being typical of such a system. Growth can occur at atmospheric or reduced pressure, the latter being preferred in order to minimize pre-reaction problems. Resistivities of $\sim 10^{-3}$ ohm cm have been reported /40/. There are no reports relating to the growth of highly conducting doped ZnO thin films via the alkyl route. Spray pyrolysis as applied to ZnO has been widely reported /41/, growth occurring via the decomposition of zinc acetate, zinc chloride or a zinc chelate (e.g. zinc acetylacetonate). Bube et al /42/ report the fabrication of highly transparent films with resistivities as low as $\sim 10^{-3}$ ohm cm, following annealing at 400°C in hydrogen. Rahman et al /43/ deposited ZnO from aqueous zinc acetate and obtained similar resistivities following vacuum annealing. Work in the authors' laboratory concerning the thermal decomposition of indium doped zinc acetylacetonate has yielded high transparency thin films (> 85% transmission) with unoptimised sheet resistances of $\sim$30 ohms/sq. No post-deposition heat-treatment was employed with these films.

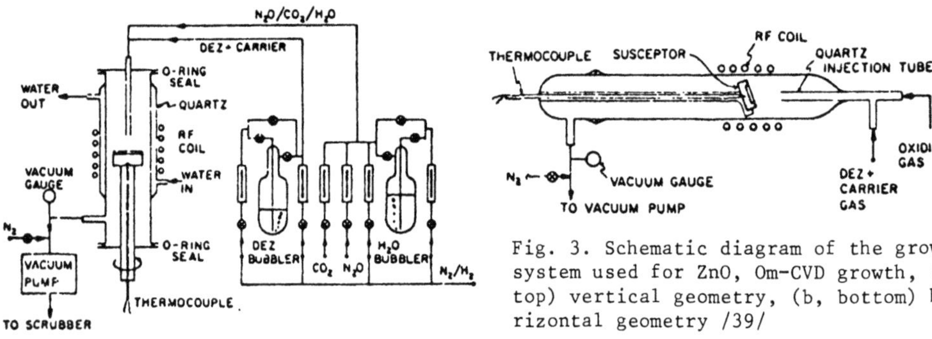

Fig. 3. Schematic diagram of the growth system used for ZnO, Om-CVD growth, (a, top) vertical geometry, (b, bottom) horizontal geometry /39/

The Chemical Vapour Deposition of Dielectrics for TFEL Displays

The selection of suitable dielectric materials for TFEL displays has been extensively reviewed, /4, 44-48/. Thin film dielectrics most commonly employed are the refractory binary and ternary oxides, silicon nitride and various composites such as silicon nitride-aluminium oxide, (SiAlON).

To-date, the input of CVD technology to this aspect of TFEL displays has been minimal, although CVD techniques for dielectric materials are quite well established. From a CVD viewpoint, the deposition of many binary and ternary oxide thin films is a relatively simple matter and it only remains to determine if the deposition conditions are likely to be detrimental to the device characteristics. Most dielectric oxide thin films employed in ACTFEL devices can be prepared by the (oxidative) vapour phase pyrolysis of acetylacetonates (I), alkoxides (II) or alkanoates (III) at temperatures in the range 200-500°C.

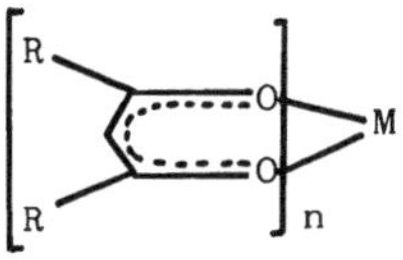

(I)

(II)

(III)

Ryabova /49/ has reviewed the use of these types of vapour source materials for the deposition of metal oxides. Shapir /50-51/ has described the deposition of ZrO_2 thin films from zirconium trifluoroacetylacetonate over the temperature range 350-550°C with a N_2/O_2 carrier gas. Growth rates of 250Å min^{-1} are obtained. In addition to a study of the deposition parameters, a detailed account is presented of the electrical characterisation for application of ZrO_2 in MOS capacitors. A relative dielectric constant of 16-30 is obtained and breakdown field of 3MV/cm. Balog and Schreiber et al /52/ extended this work to the deposition of HfO_2 thin films. Aboaf /53/ and Woodward et al /54/ have described the application of aluminium isopropoxide for the deposition of Al_2O_3 at temperatures up to 420°C. Dielectric constants of 8-10 were obtained for the range $1MH_z$ to $1H_z$. Y_2O_3 thin films have been deposited by the pyrolysis of yttrium acetylacetonate derivatives volatilised under reduced pressure /55/, with deposition temperatures in the range 400-500°C.

The deposition of Si_3N_4 thin films has been reviewed by Morosanu /56/. In general, deposition temperatures are too high (500-1300°C) for display purposes but it should be noted that plasma enhanced CVD /57/ and the more recently developed laser induced CVD /58/ and photo-CVD /59/ are capable of depositing Si_3N_4 thin films in the range 200-500°C.

There are a few reports of ternary oxide thin films by CVD. Recently, however, there has been increased interest in the alkoxides and mixed alkoxides /60/ for application in CVD. Nasu /61/ has demonstrated the use of the mixed alkoxide $MgAl_2$ $(o\text{-}i\text{-}Pr)_8$ (MAI) for spinel doping ZrO_2 thin films to promote the cubic phase. Thin films were grown over the temperature range 300-700°C by atmospheric pressure CVD, an N_2/O_2 carrier gas being employed for the liquid Zirconium alkoxide and MAI sources. In the author's laboratory, a spray pyrolysis process has been employed to grow thin films of binary and ternary oxides. In this process, a fine mist is produced from a solution bearing the required compound (e.g. acetylacetonates, alkanoates or alkoxides) and the mist is transported to the reactor via a pre-heat zone to fully vapourise the mist just prior to its arrival at the substrate. Typical depositions are:

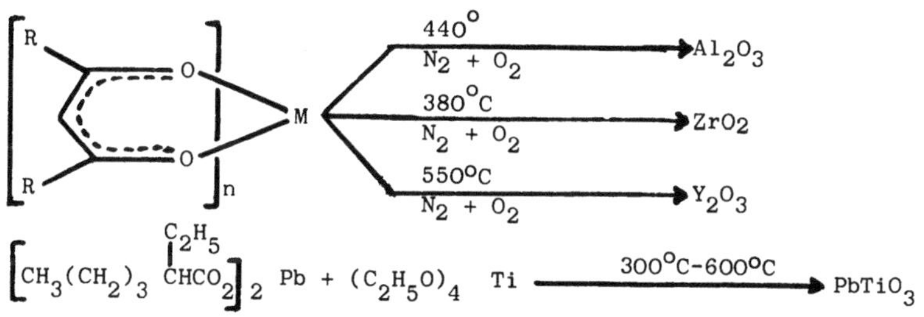

$PbTiO_3$ deposited at 300^oC is amorphous, but crystalline (perovskite) at deposition temperatures of 550^oC.

The Chemical Vapour Deposition of ZnS(Mn) Thin Films

The vapour phase reaction between dimethyl zinc (DMZ) and H_2S to deposit thin films of ZnS was first reported by Manasevit /62/ in 1971. Wright /63/ further developed the technique to demonstrate the growth of opto-electronic quality ZnS thin films using an atmospheric pressure cold wall, inductively heated reactor. Deposition temperatures as low as 350^oC could be employed, and this clearly opened the way for the deposition of ZnS onto glass based substrates. Wright /64/ first demonstrated that manganese could be incorporated into the growing ZnS film by simultaneously cracking the vapour from tricarbonylmethyl cyclopentadienyl manganese (TCM). This was found to crack most efficiently at a temperature of 450^oC but could be used at a ZnS growth temperature of about 400^oC. This led to the first reported demonstration of thin film electroluminescence from a ZnS.Mn film prepared by metal-organic chemical vapour deposition /65/. Following this work, progress in ZnS MOCVD has been somewhat limited in its application to TFEL device fabrication. Much of the work has centred on circumventing the pre-reaction problem encountered with DMZ and H_2S and this has involved development of alternative precursors such as dialkyl zinc adducts /66/, CS_2 /67/, alkylsulphides /68/ and sulphur bearing heterocyclic molecules /35, 69/. Developments in precursors has been the subject of some recent reviews by Wright /36-38/. MOCVD via the dialkyl zinc route has been reported for the fabrication of ACTFEL device structures /70-75/, the general view being that MOCVD yields higher brightness devices than those of other methods.

There have been relatively few reports of MOCVD for ZnS.Mn via zinc chelates. Vecht /76/ first reported the use of zinc dialkyldithiocarbamates chelates for the deposition of ZnS by vapour phase pyrolysis. Takahashi et al /77/ indicated the use of dithiophosphinate zinc chelates $(R_2PS_2)_2$ Zn, but only reported the deposition parameter for CdS from the corresponding cadmium chelate under reduced pressure. Svechnikov /78/ has described the growth of ZnS thin films, primarily by spray pyrolysis of a zinc chelate dissolved in pyridine. Strangely, they make no attempt to identify the chelated compounds employed in their work. In a further report /79/ Svechnikov et al give a very brief account of ACTFEL device characteristics using an (unknown) chelate derived ZnS.Mn coating. They present voltage-luminance characteristics and compare them very favourably with ACTFEL data presented by Cattell et al /65/. In the authors' laboratory, extensive investigations have been carried out to determine ZnS.Mn growth conditions from zinc chelates of the 1,1-dithiolato type (IV and V).

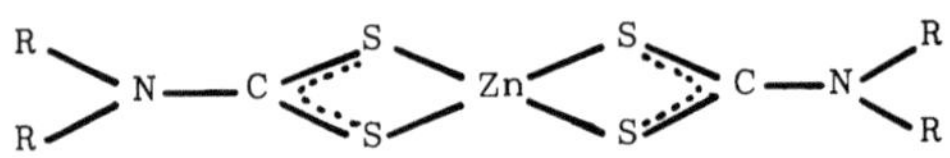

Zinc dialkyl dithiocarbamate (IV)

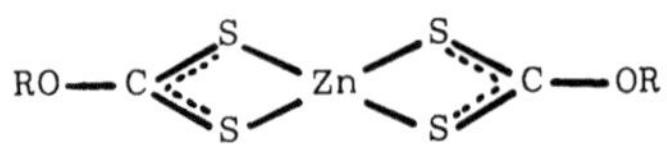

Zinc xanthate (V)

The zinc dialkyldithiocarbamates have been successfully used to deposit ZnS.Mn thin films (Mn derived from TCM or manganese diethyl dithiocarbamate) from atmospheric /80/ or reduced pressure /81/. Under atmospheric pressure conditions (APCVD), with an N_2 carrier gas, the source material is heated to temperatures in the range $190-250^oC$ and the vapour carried downstream to the heated substrate ($380-450^oC$), where the notional reaction

$$(C_2H_5)_2NCS_2\,_2\,Zn \xrightarrow[N_2]{400^oC} ZnS + \left[(C_2H_5)_2NCS\right]_2\,S$$

occurs. Growth rates of 0.2-3.0 μmh^{-1} are obtained in the temperature range $380-450^oC$. Low pressure CVD (LPCVD) is carried out in the pressure range 0.1-20 torr, with a nitrogen bleed gas. Source and substrate conditions are similar to those employed in APCVD and with similar growth rates. Currently, LPCVD ZnS.Mn can be deposited over an area of 100 cm^2. X-ray diffraction studies indicate a strong preferred orientation on glass, with a single sharp peak assigned to the (00.2) hexagonal modification, the hexagonal structure being indicated via optical band gap measurements. More recently, complete ACTFEL stacks (Fig. 4) have been fabricated by LPCVD and spray pyrolysis, as part of a collaborative programme with the US Army, Fort Monmouth. Preliminary results of this work will be presented at this workshop.

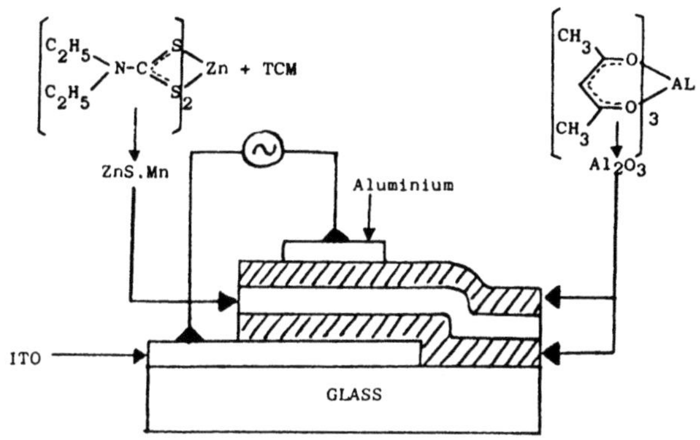

Fig. 4. CVD ACTFEL stack

Discussion

We have seen that CVD can be employed in almost every aspect of TFEL device fabrication. The current application of ITO (sputtered) is a relatively expensive process and could easily be replaced by a low cost spray pyrolysis system, with a continuous, large area deposition facility. Improved spray technology is available and should be considered as a low cost alternative to sputtering. Perhaps more interesting is the development of ZnO (In,Al) thin films. For similar conductivity, these can have superior optical properties and the spray pyrolysis/CVD precursors are readily available at low cost, and have the added advantage of being of low toxicity. Coating large, uniform areas should not be a problem. The deposition of dielectric materials has been given added impetus by recent developments in high temperature superconductors. The requirements of substrate barrier layers such as ZrO_2, $SrTiO_3$ have led people to reconsider spray/CVD sources and processes and has been further strengthened by the requirement for volatile rare earth and alkaline earth compounds in the fabrication of thin film superconductors. Common binary and ternary oxide thin films can be readily prepared by spray pyrolysis techniques, many at temperatures below 500°C. Perhaps, initially, one of the great advantages of this process is its flexibility. The production of thin films that require extensive, costly and time-consuming development work, e.g. $BaTa_2O_6$, $SrTiO_3$ etc., could be produced within 1-2 days on a research basis, when suitable source materials are available.

Good progress has been made in the development of spray/CVD processes for the deposition of ZnS.Mn. The improved light emitting properties of CVD thin films (via dialkyl zinc route or ALE) would appear to be advantageous, although Mach & Muller /82/ have attributed this to increased surface roughness. The application of chelates (especially dialkyldithiocarbamates) in spray pyrolysis or more conventional CVD systems, (atmospheric or reduced pressure) does put forward the possibility for the volume production of large area, low cost devices, but further work is required to improve device efficiency. In particular, the correct application of a spray deposition system with low thermal stability source materials (e.g. zinc xanthate) offers the potential for low temperature deposition ($\sim$200°C), and the benefits derived therefrom.

Overall, the picture emerging is most encouraging and the application of spray/CVD processes should gain more widespread acceptance as its full potential in TFEL technology is realised. If one considers the enormous contribution CVD technology has made in the silicon industry, it should dispel any reluctance on the part of production management to enter into this field for the production of high field TFEL displays.

References

1. T. Inoguchi, S. Mito: Topics in Appl. Phys. Vol. 17 (Springer 1977) 197.
2. F. Williams: Journal of Luminescence 23 (1981) pp.1-16.
3. D.H. Smith: Journal of Luminescence 23 (1981), 209-235.
4. R. Mach & G. O. Muller: Phys. Stat. Sol. (a) 69 (1982), 11-66.
5. R. Mach, G.O. Muller, W. Gericke & J. von Kalben: Phys. Stat. Sol. (2) 75 (1983) 187-197.
6. R. Mach, J. von Kalben, W. Gericke, G.O. Muller, G.V. Reinsperger: Phys. Stat. Sol. (a) 75 (1983) 489-494.
7. G.O. Muller: Phys. Stat. Sol. (a) 81 (1984) 597-608.
8. R. Mach & G.O. Muller: Phys. Stat. Sol. (a) 81 (1984) 609-623.
9. R. Mach, G.O. Muller & G. Schultz: Phys. Stat. Sol. (a) 81 (1984) 723-732.
10. R. Mach, G.O. Muller, G. Schultz, J. Von Kalben, W. Gericke: Phys. Stat. Sol. (a) 81 (1984) 733-738.
11. J.W. Allen: Journal of Luminescence 31 & 32 (1984) 665-670.
12. R. Mach in Polycrystalline Semiconductors, Springer Series in Solid State Sciences, Vol. 57, Ed. G. Harbeke (Springer Berlin 1985) p.186.
13. G.O. Muller: Proceedings of SID 1987.
14. D. Theis, H. Oppolzer, G. Ebbinghaus & S. Schild: Journal of Crystal Growth 63 (1983) 47-57.
15.T D. Theis: Phys. Stat. Sol. (a) 81 (1984) 647-655.
16. J. Kirton, G.W. Blackmore, J.M. Blackmore, A.F. Cattell, N.G. Chew, A.G. Cullis, K.F. Dexter, P. Lloyd, M.J. Slater, P.J. Wright, J.C. Inkson, D. Kirk: Proceedings of SID 1985, pp. 222-235.
17. A.P.C. Jones, D.Y. Watts, A.W. Brinkman, G.J. Russell, J. Woods: Journal of Crystal Growth 86 (1988) 880-884.
18. A.P.C. Jones, R.E. Jones, A.W. Brinkman, D.Y. Watts, J. Woods, D.J. Robbins: Journal of Crystal Growth 86 (1988) 895-899.
19. J.M. Blackmore, A.F. Cattell, K.F. Dexter, J. Kirton, P.Lloyd: J. Appl. Phys. 61 (2) 15 January 1987, pp. 714-721.
20. A.F. Cattell, J.C. Inkson, J. Kirton: J. Appl. Phys. 61 (a) 15 January 1987, pp. 722-733.
21. J.L. Vossen in Physics of Thin Films - Advances in Research & Development, Vol. 9, Ed. Georg Hass, Academic Press, 1977.
22. Z.M. Jarzebski: Phys. Stat. Sol. (a) 71 (1982) 13-41.
23. J.B. Mooney & S.B. Radding: Ann. Rev. Mater. Sci. 1982, 12, 81-101.
24. R.R. Chamberlin & J.S. Skarman: J. Electrochem Soc. Vol. 113, No. 1, 86-89.
25. J.W. Gilliland & M.S. Hall: Electrochem. Technol. 5 (1967) 303-306.
26. J.C. Viguie & J. Spitz: J. Electrochem. Soc. Vol. 122, No. 4, 585-588.
27. D.K. Ranadive, F.J.J. Smith, R.P. Khosla: Proceedings of ECS, CVD 1977, 448-460.
28. J.C. Manifacier, J.P. Fillard, J.M. Bine: Thin Solid Films, 77 (1981) 67-80.
29. R.N. Ghoshtagore: Proceedings of ECS, CVD 1987, 433-447.
30. D. Belanger, J.P. Dodelet, B.A. Lombos, J.I. Dickson: J. Electrochem. Soc. Vol. 132, No. 6, 1398-1405.
31. G.W. Cullen, C.C. Wang: Heteroepitaxial Semiconductors for Electronic Devices, Ch. 4, Springer-Verlag 1978.
32. A.M. Mancini, P. Pierini, A. Valentini, L. Vasanelli, A. Quironi: Thin Solid Films, 124 (1985) 85-92.
33. L. Vasanelli, A. Valentini, A. Losacco: Solar Energy Materials 16 (1987) 91-102.
34. K. Zweibel & R. Mitchell: 18th IEEE PV Specialists Conf. (1985).
35. P.J. Wright, R.J.M. Griffiths, B. Cockayne: Journal of Crystal Growth, 66 (1984) 26-34.
36. B. Cockayne & P.J. Wright: Journal of Crystal Growth, 68 (1984) 223-230.
37. P.J. Wright & B. Cockayne: Chemtronics, Vol. 2, June 1987, 49-53.
38. A.C. Jones, P.J. Wright, B. Cockayne: Chemtronics, Vol. 3, March 1988, 35-37.
39. Chi Kwan Lau, Shiban K. Tiku, K.M. Lakin: J. Electrochem Soc. Vol. 127, No. 8, 1843-1847.
40. P. Souletie, S. Bethke, B.W. Bessels, H. Pan: J. Crystal Growth, 86 (1988), 248-251.
41. S. Major, A. Banerjee, K.L. Chopra: Thin Solid Films 168 (1983) 333-340.
42. J. Aranovich, A. Ortiz, R. Bube: J. Vac. Sci. Tech. 16 (4), 1979, 994-1003.

43. M.N. Islam, M.O. Hakim, H. Rahman: J. Materials Sci. 22 (1987) 1379-1384.
44. Richard D. Ketchpel: Proceedings of SID 1982.
45. Y. Fujita, J. Kuwata, M. Nishikawa, T. Tohda, T. Matsuoka, A. Abe, T. Nitta: Proceedings of Japan Display 1983.
46. M.G. Clark: Proceedings of IX Int. Vac. Congress and V. Int. Conf on Solid Surfaces, Madrid 1983.
47. S. Tiku & G. Smith: IEEE Trans. Electron Devices, Vol.Ed-31, No.1 (1984) 105-108.
48. A. Vecht: SID Seminar Lecture Notes, Vol. 1, S.1.1. (1984).
49. L.A. Ryabova & Y.S. Savitskaya: Thin Solid Films, 2 (1968) 141-148.
50. L. Ben-Dor, A. Elshtein, S. Halabi, I. Pinsky, J. Shappir: J. Elect. Materials, Vol. 13, No. 2, 1984, 263-272.
51. J. Shappir, A. Anis, I. Pinsky: IEEE Transactions on Electron Devices, Vol. Ed-33, No. 4 (1986) 442-449.
52. M. Balog, M. Schieber, M. Michman, S. Patai: J. Electrochem. Soc. Vol. 126, No. 7 (1979) 1203-1207.
53. J.A. Aboaf: J. Electrochem Soc. Vol. 114, No. 9 (1967), 948-952.
54. D.C. Cameron, L.D. Irving, G.R. Jones, J. Woodward: Thin Solid Films, 91 (1982) 339-347.
55. J.P. Dismukes, J. Kane, B. Binggeli, H.P. Schweizer: Proceedings of CVD, 4th Int. Conf. (275-286).
56. C.E. Morosanu: Thin Solid Films, 65, (1980) 172-207.
57. R. Gereth & W. Scherber: J.E.C.S. Vol. 119, No. 9 (1972), 1248-1254.
58. E. Pan, J. Flint, D. Adler, J. Haggerty: J. Appl. Phys. 61, 9 (1987) 4535-4539.
59. N Hayafuji, K. Nagahama, H. Ito, T. Murotani, K. Fujikawa: Extended Abstracts 16th (1984) Int. Conf. on Solid State Devices and Materials, Kobe, 663-666.
60. D.C. Bradley, R.C. Mehrotra, D.P. Gaur: Metal Alkoxides, Academic Press (1978).
61. U. Takahashi, T. Kawae, M. Nasu: J. Crystal Growth 74 (1986), 409-415.
62. H. Manasevit & W. Simpson: J.E.C.S. Vol. 118, No. 4 (1971) 644-647.
63. P.J. Wright &. B. Cockayne: J. Crystal Growth, 59 (1982), 148-154.
64. P.J. Wright, B. Cockayne, A.F. Cattell, P.J. Dean, A.D. Pitt, G.M. Blackmore: J. Crystal Growth, 59 (1982) 155-160.
65. A.F. Cattell, B. Cockayne, K. Dexter, J. Kirton, P.J. Wright: IEEE Transactions on Electron Devices, Vol. Ed-30, No. 5 (1983) 471-475.
66. B. Cockayne, P.J. Wright, A.J. Armstrong, A.C. Jones, E.D. Orrell: J. Crystal Growth 91 (1988) 57-62.
67. S. Takata, T. Minami, T. Miyata, H. Nanto: J. Cryst. Growth 86 (1988) 257-262.
68. S. Fujita, M. Isemura, T. Sakamoto, N. Yoshimura: J.Crys. Growth 86 (1988) 263-7.
69. B. Cockayne, P.J. Wright, M.S. Skolnick, A.D. Pitt, J.D. Williams, T.L. Ng: J. Crystal Growth, 72 (1985) 17-22.
70. A. Yoshiwaya, S. Yamaga, K. Tanaka: Japanese J. Appl. Phys. 23 6 (1984) L388-390.
71. K. Hirabayashi & D. Kogure: Japanese J. Appl. Phys. 24, 11 (1985) 1484-1487.
72. K. Hirabayashi & H. Kozawaguchi: Japanese J. Appl.Phys. 25, 5 (1986) L379-L381.
73. K. Hirabayashi & H. Kozawaguchi: Japanese J. Appl.Phys. 25, 4 (1986) 711-713.
74. K. Hirabayashi, H. Kozawaguchi, B. Tsujiyama: Proc. Japan Display 1986.
75. B. Tsujiyama, K. Hirabayashi, H. Kozawaguchi: Proc. EL Workshop, Oregon, 1986.
76. A. Vecht, G.B. Patent 20499636 (1980).
77. Y. Katahashi, R. Yuki, M. Sugiura, S. Motojima, K. Sugiyama: J. Cryst. Growth 50, (1980) 491-497.
78. L. Zharovsky, L. Zavyalova, G. Svechnikov: Thin Solid Films, 128 (1985) 241-249.
79. S.V. Svechnikov, N.A. Vlasenko, L.V. Zavyalova, A.K. Savin: Sov. Phys. Tech. Phys: 30, 12 (1985) 1431-1432.
80. A. Saunders & A. Vecht: Extended Abstracts, ECS Meeting, Cincinatti (1984).
81. A. Saunders, A. Vecht, G. Tyrrell: Proceedings of 7th Int. Conf. on Ternary & Multinary Compounds, Colorado (1986) MRS. 213-219.
82. R. Mach & G.D. Muller: J. Crystal Growth, 86 (1988) 866-872.

Multi-Source Deposition Method for ZnS and SrS Thin-Film Electroluminescent Devices

T. Nire, T. Watanabe, N. Tsurumaki, A. Miyakoshi, and S. Tanda

Research Division, Komatsu Ltd., Manda 1200, Hiratsuka, Kanagawa 254, Japan

A new multi-source deposition (MSD) method for thin-film electroluminescent (TFEL) devices has been developed. Bright yellow-orange, green and blue EL emissions are obtained in ZnS:Mn, ZnS:Tb,F and SrS:Ce,KCl TFEL devices, respectively, prepared by this method.

The crystallinity of these thin films has been investigated. The crystal structure of the ZnS film is wurtzite-type with strong preferred orientation to the (001) axis. The SrS film has NaCl-type structure and the preferred orientation varies as the S/Sr ratio in the source beams. The dependence of characteristics of the ZnS TFEL device on Mn, Tb, and F concentration and the SrS TFEL device on the dopant materials are also reported.

1. Introduction

Various fabrication methods for thin-film electroluminescent (TFEL) devices have been developed and are even now under investigation. Yellow-orange light emitting ZnS:Mn TFEL panels prepared by electron beam (EB) /1/ and atomic layer epitaxy (ALE) method /2/ are already commercially available. Furthermore, bright green EL emission has been obtained in ZnS:Tb,F devices prepared by sputtering method /3/. However, luminance levels of the other colors are insufficient for practical use. In order to improve luminance for three primary colors in TFEL devices, the following are needed:

(1) The mechanism of EL phenomenon is clarified.
(2) Host material thin films with high crystallinity are fabricated.
(3) Dopant materials and their concentrations are optimized.
(4) Fabrication techniques to accomplish the above are developed.

Recently, we have developed a new multi-source deposition (MSD) method, which is applied to TFEL devices /4/. In this paper, the crystallinity of ZnS and SrS thin films prepared by this method and the dependence of characteristics of their TFEL devices on the dopants are reported.

2. Experimental Methods

The TFEL device structure consisted of the ZnS or the SrS active layer sandwiched between two insulating layers. Ta_2O_5, $BaTa_2O_6$ and Si_3N_4 prepared by sputtering method were used as the insulating layers for ZnS:Mn, ZnS:Tb and SrS:Ce devices, respectively. The ZnS and the SrS thin films were fabricated by the MSD method.

Springer Proceedings in Physics, Vol. 38 Electroluminescence
Editors: S. Shionoya and H. Kobayashi

2-1 MSD Technique

Figure 1 shows a schematic picture of MSD equipment. Evaporation sources for each constituent elements of an active layer are put in a vacuum chamber, which can be pumped down to 10^{-5} to 10^{-7} Torr. Compound thin films are obtained on this equipment by appropriately controlling the temperatures of these sources independently. In this work, the active layers of the TFEL devices were fabricated by using three sources or four sources as follows:

(1) Three sources (Zn, S, Mn) for yellow-orange light emitting ZnS:Mn.
(2) Three sources (Zn, S, Tb), (Zn, S, TbF_3) and four sources (Zn, S, Tb, ZnF_2) for green light emitting ZnS:Tb.
(3) Three sources (Sr, S, Ce), (Sr, S, $CeCl_3$) and four sources (Sr, S, Ce, KCl),

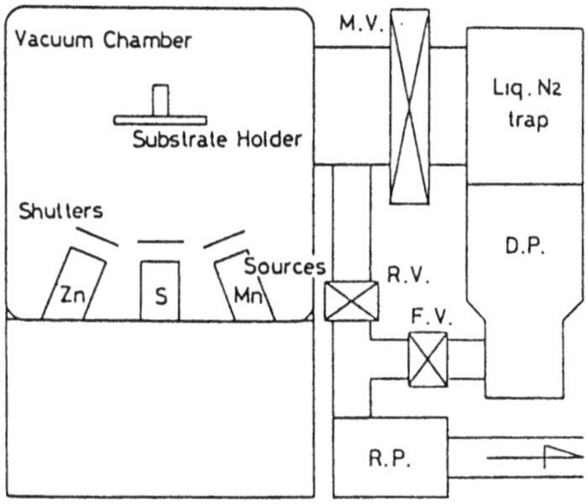

Fig 1. MSD equipment

2-2 Measurements

Crystal properties of thin films were investigated by X-ray diffraction, transmission electron diffraction and transmission electron microscopy. Stoichiometry and composition in the films were measured by X-ray micro analysis. In the present study, light emitting area of the TFEL devices was 2mm in diameter. The luminance (L) - voltage(V) characteristics and the EL spectra were measured under the condition of driving the devices with sinusoidal excitation at room temperature.

3. Results and Discussion

3-1 Crystallinity of ZnS Thin Films Prepared by the MSD Method

X-ray diffraction pattern of the ZnS thin film prepared by the MSD method on glass at a substrate temperature around 200 ℃ is shown in Fig.2. The diffraction pattern consists of a peak with high intensity and its overtone, and the full width of half maximum (FWHM) of the peak is less than 0.18° . The diffraction angle of the peak corresponds to either (111) of zinc-blende cubic structure or (002) of wurtzite hexagonal structure. Figure 3 shows a transmission electron diffraction pattern of a cross section of the ZnS thin film. This pattern indicates that the film has hexagonal crystal structure. Furthermore, the following have been observed by means of transmission electron microscopy:
(1) columnar polycrystals of the ZnS extend from the bottom to the top surface, (2) the grain size of these columns is about 100 to 200 nm, (3) crystal faces in the column are arranged periodically /4/. Therefore it is concluded that the ZnS

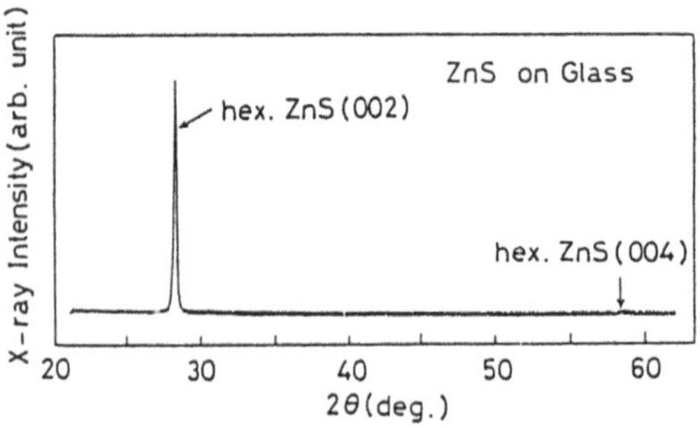

Fig.2 X-ray diffraction pattern of the ZnS thin film prepared by the MSD method using CuKα line.

Fig.3 Transmission electron diffraction of a cross section of the ZnS thin film prepared by the MSD method.

films prepared by the MSD method have high crystallinity and strong prefered orientation to the (001) axis of wurtzite structure.

3-2 Crystallinity of SrS Thin Films Prepared by the MSD Method

Stoichiometric SrS thin films can be obtained by controlling Sr and S source temperatures independently. Figure 4 shows X-ray diffraction patterns of the SrS thin films grown at a substrate temperature of 450℃. Each of them indicates a NaCl-type crystal structure. The preferred orientation of the film varies from the (111) axis to the (200) axis and finally to the (220) axis,as the S/Sr ratio in the source beams is increased /5/. The FWHM of the highest diffraction peak is less than 0.18° when the film has strong orientation to the (200) axis.

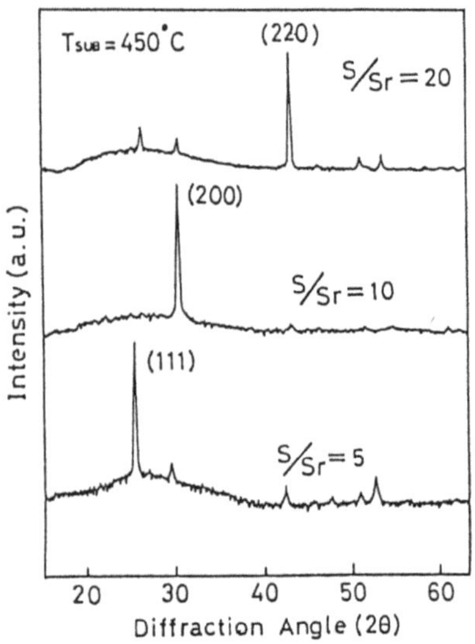

Fig.4 X-ray diffraction patterns of the SrS thin films prepared by the MSD method using CuKα line.

3-3 Dependence of Characteristics of the ZnS:Mn TFEL Devices on Mn Concentration

In this experiment, the devices with the ZnS:Mn active layer of the thickness 600 nm sandwiched between Ta_2O_5 insulating layers of the thickness 600 nm were used. L-V characteristics of the devices are shown in Fig.5. It suggests that the maximum luminance is obtained at Mn concentration around 0.5 atomic %. Figure 6 shows EL spectra of the devices. The EL intensity around 580nm decreases and that around 720 nm increases relatively with increasing Mn concentration between 0.42 and 13 atomic %. On the other hand, the dependence of X-ray diffraction on Mn composition in $Zn_{(1-x)}Mn_xS$ ($0 \leq x \leq 1$) films prepared by the MSD method has been studied /6/. It shows that the crystal structure, orientation and crystal quality do not change in the region of Mn concentration less than 15 atomic %. Therefore, the variation in EL spectra is thought to be due mainly to Mn-Mn interaction.

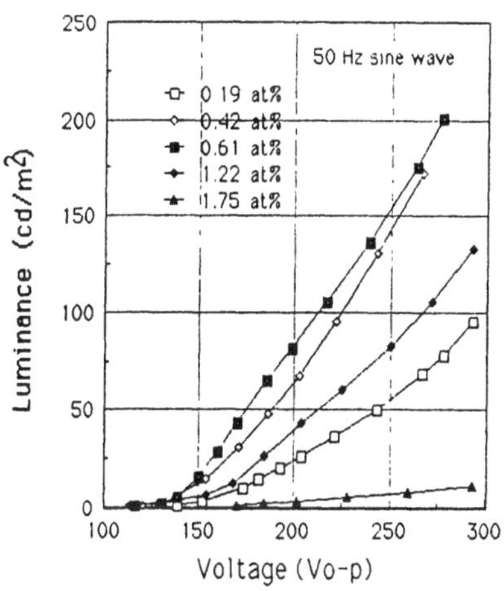

Fig.5 L-V characteristics of the ZnS:Mn TFEL devices

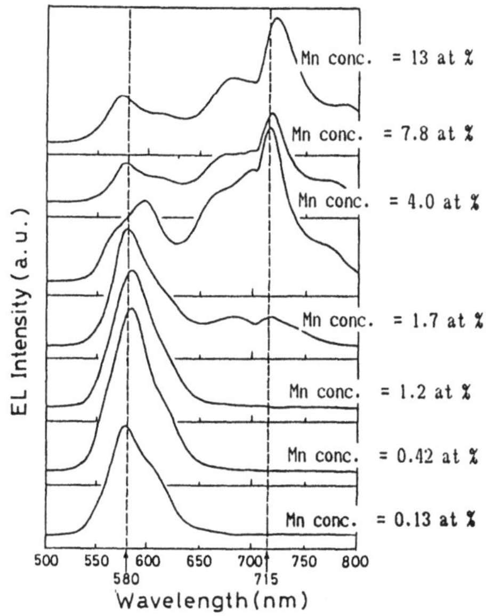

Fig.6 EL spectra of the ZnS:Mn TFEL devices

3-4 Dependence of Characteristics of the ZnS:Tb,F TFEL Devices on Tb and F Concentration

The device structure consisted of the ZnS:Tb,F active layer of the thickness around 550 nm sandwiched between $BaTa_2O_6$ insulating layers of the thickness 400 nm. Figure 7 shows L-V characteristics of the ZnS:Tb devices prepared by using three sources (Zn, S, Tb). Relative high luminance for a device without coactivators such as F was obtained. Figure 8 shows L-V characteristics of the ZnS:Tb,F devices prepared by using four sources (Zn, S, Tb, ZnF_2). These results show that the highest luminance can be obtained in the ZnS:Tb,F devices of Tb concentration around 1 atomic % and of the ratio of F/Tb around 1.

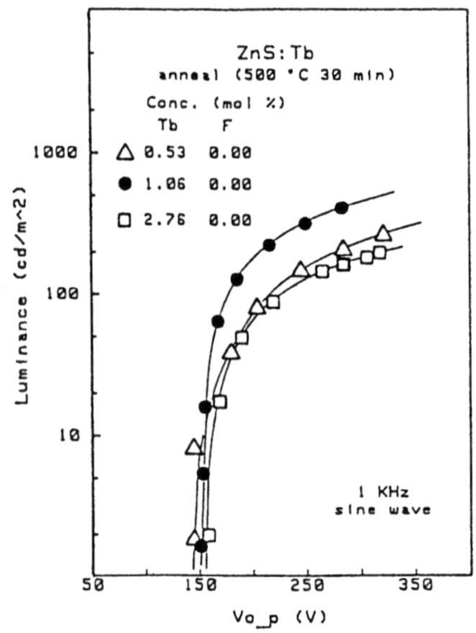

Fig.7 L-V characteristics of the ZnS:Tb devices

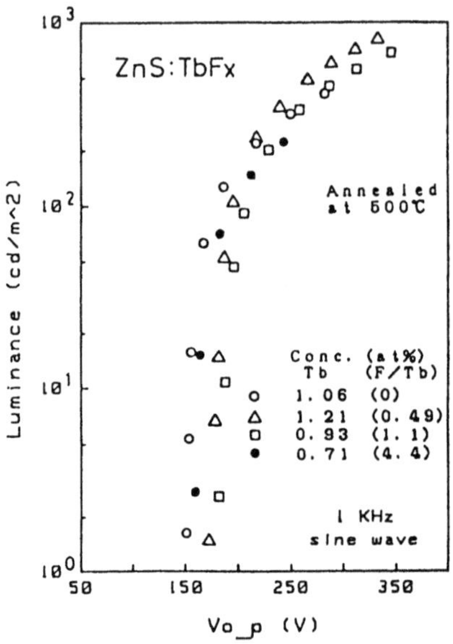

Fig.8 L-V characteristics of the ZnS:Tb,F devices

3-5 Dependences of Characteristics of the SrS TFEL Devices on Dopant Materials

In this experiment, the devices with the SrS active layer of the thickness around 500 nm sandwiched between Si_3N_4 insulating layers of the thickness around 250 nm. EL characteristics of the SrS devices prepared by using three sources (Sr, S, Ce), (Sr, S, $CeCl_3$) and four sources (Sr, S, Ce, KCl), (Sr, S, Ce, K_2S) have been studied /5/. L-V characteristics and EL spectra of these devices are shown in Fig.9 and Fig.10, respectively. These results indicate that chlorine contributes

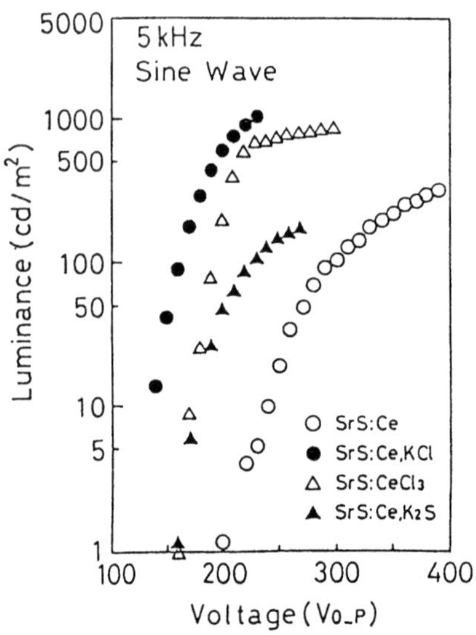

Fig.9 L-V characteristics of the SrS devices

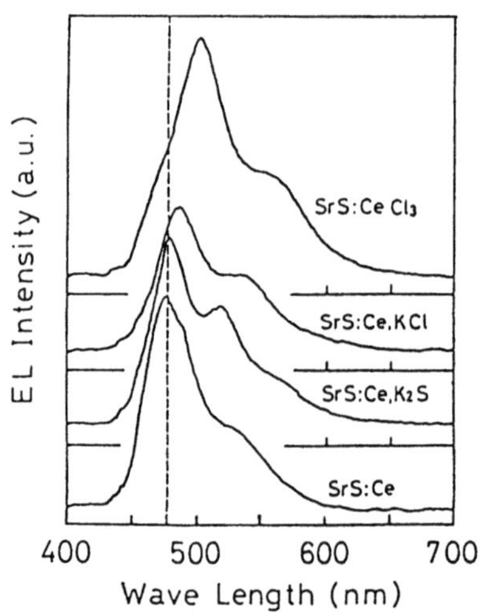

Fig.10 EL spectra of the SrS devices

to luminescence but potassium has no effect in improving luminance. The highest luminance of 1100 cd/m^2 was obtained in the SrS:Ce,KCl device at the ratio of Cl/Ce around 1 with 5 kHz sinusoidal drive.

4. Summary

We have developed the multi-source deposition (MSD) method as a new fabrication technique for TFEL devices. Crystallinity of the ZnS and the SrS thin films and EL characteristics of the devices fabricated by this method were investigated. The ZnS thin film was wurtzite type crystal structure with high crystallinity and strong preferred orientation along the (001) axis. The SrS thin film was NaCl-type crystal structure.The preferred orientation varied as the ratio of S/Sr in the source beams.The film had the highest crystallinity at the orientation to the (200) axis. High luminance level was obtained in the yellow-orange light emitting ZnS:Mn device,the green light emitting ZnS:Tb,F device and the blue light emitting SrS:Ce,KCl device. In particular, the blue device at the ratio of Cl/Ce around 1 showed the luminance level of 1100 cd/m^2 with 5 kHz sinusoidal drive.

ACKNOWLEDGMENT

Thanks are due to Mr. K. Fuchiwaki, Mr. C. Ueno and Mr. Y. Niwatsukino for help with the experiments,and Miss N. Naoki for assistance in preparing the manuscript.

REFERENCES

(1) T. Inoguchi et al., 1974 SID Int. Symp. Dig. Tech. Papers, P.84.
(2) T. Suntola et al., 1980 SID Int. Symp. Dig. Tech. Papers, P.108.
(3) H. Ohnishi et al., Conf. Record 1985 Int. Display Research Conf., P.159.
(4) T. Nire et al., 1987 SID Int. Symp. Dig. Tech. Papers, P.242.
(5) S. Tanda et al., to be published.
(6) T. Nire at al., to be published.

Efficient ZnS:Mn Electroluminescent Films Grown by Metal Organic Chemical Vapor Deposition

M. Shiiki, M. Migita, O. Kanehisa, and H. Yamamoto

Central Research Laboratory, Hitachi Ltd., P.O. Box 2, Kokubunji, Tokyo 185, Japan

Electroluminescent ZnS:Mn films grown by reduced pressure MOCVD are characterized in comparison with films deposited by the EB method. An important feature of the crystallinity is that the mean grain size as observed by X-ray diffraction line widths and by TEM images is larger in the MOCVD films than in the EB films. The intensities of PL and EL reach their maximum at Mn concentrations from 0.8 to 1.0 wt%, which is considerably higher than that for EB deposition (0.5 wt%). The MOCVD technique can provide about twice the EL efficiency as the EB method. This can be ascribed to the large grain size and the uniform Mn distribution.

1. Introduction

Metal organic chemical vapor deposition (MOCVD) is a promising film growth technique for thin film electroluminescent (EL) devices, because it can provide high-quality crystals doped with an activator. This technique was applied to the formation of ZnS:Mn films with TCM (tricarbonylmethylcyclopentadienyl manganese) as a Mn source /1, 2/. However, the TCM has poor thermal decomposition characteristicseven at a temperature (550℃) much higher than the optimum growth temperature of ZnS (280-350℃). By using CPM (di-π-cyclopentadienyl manganese) and BCPM (bis-methylcyclopentadienyl manganese) as new Mn sources, we have obtained highly efficient ZnS:Mn EL devices with a double insulating structure /3/. The saturated luminance has reached 4300 cd/m^2 with CPM and 3600 cd/m^2 with BCPM under 1-kHz sinusoidal wave excitation. The maximum efficiency is 4.8 lm/W for CPM. The efficiency of MOCVD films is higher than that of the films by electron beam (EB) deposition at the same conduction charge density.

In this work, the active layers of ZnS:Mn for EL devices have been characterized to elucidate the factors determining such high efficiencies.

2. Experimental

The ZnS films doped with Mn were grown on non-alkaline glasses or on stacked layers of Ta_2O_5/SiO_2 by the reduced pressure MOCVD method. Diethyl zinc (DEZ) and H_2S were used as source reactants of Zn and S. The ratio of the molar flow rate of S to that of Zn was fixed at 6. BCPM was used as a Mn source in this paper. The flow rate of the carrier gas, H_2, was varied from 5 to 40 cc/min. The Mn concentration in the films was analyzed by atomic absorption spectroscopy or by fluorescence X-ray analysis. The growth conditions are summarized in Table I.

As references, ZnS:Mn films were prepared by EB deposition at the substrate temperature of 190℃ and then annealed at 550℃ for 1 hour. The annealing process was omitted for the MOCVD films, because it does not increase the brightness significantly, as it does for the EB films.

Photoluminescence (PL) spectra were obtained under the excitation by a N_2 laser, and EL intensities were measured at 80 V above the threshold under 1-kHz sinusoidal wave excitation.

Springer Proceedings in Physics, Vol. 38 Electroluminescence
Editors: S. Shionoya and H. Kobayashi

Table I
Growth conditions of ZnS:Mn films by reduced pressure MOCVD

Flow Rate of H_2S/H_2(cc/min)	800
Flow Rate of DEZ(cc/min)	40
[S]/[Zn] Ratio	6
Growth Rate(μm/hr)	1.2
Flow Rate of BCPM/H_2(cc/min)	5-40
Substrate Temp.(℃)	320-350
Deposition Pressure(Torr)	1
Back Pressure(Torr)	1×10^{-6}

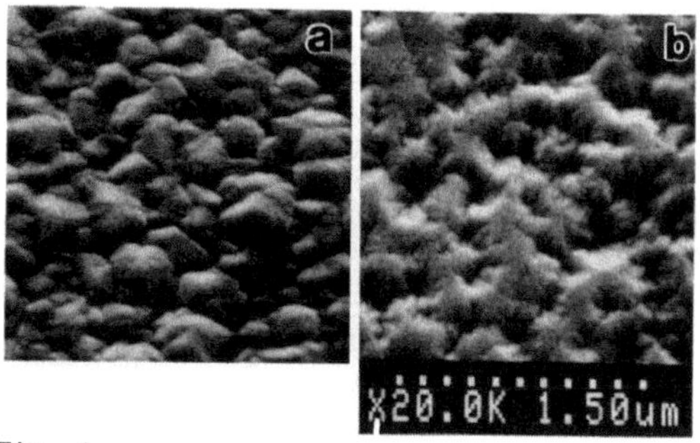

Fig. 1.
The SEM images of the surface:
(a) MOCVD film; (b) EB film.

3. Results and Discussion

3.1 Effect of the Surface Structure

Observation by a Normarski interference microscope has shown that the MOCVD films have rough surface morphology. Such a structured surface may help the emission escape from the ZnS layer by reduction of the optical trap effect.

Figures 1(a) and (b) show the surface images of the ZnS:Mn films by MOCVD and EB deposition, respectively. The scanning electron microscope (SEM) image of the MOCVD film shows that the ZnS:Mn film consists of many smooth surfaced grains 150 to 400nm in size. The EB film, however, has many pointed projections on the surface. To clarify the effect of these differences in surface morphology on EL brightness, the light output from the front side of the EL cells was compared with the total light output measured in an integrating sphere. The intensity ratios of the total output to the output from the front side are 5.3 with MOCVD film and 5.1 with EB film.

3.2 Crystallinity of ZnS:Mn Films

The EL efficiency is affected by the crystallinity of ZnS, particularly in the boundary region of the first insulator, where injected electrons are accelerated most efficiently by an enhanced electric field. The relationship between the film thickness and the average size of the crystallites (i.e. the smallest domains in a particle having the same orientation contributing to the X-ray diffraction line widths) was investigated in a previous paper /3/. It was shown that the MOCVD films have considerably larger crystallites than the EB films even at the bottom of the ZnS layer. In 500-nm-thick films, the crystallite size is 50-60 nm with MOCVD films and 25-30 nm with EB films.

Observation of the cross section by transmission electron microscope (TEM) was made on films grown on (100)Si wafers. As shown in Fig. 2, the grain size is 200-300 nm in diameter near the substrate in the MOCVD films, but is much smaller at a thickness less than 150 nm in the EB films. In the EB films, a columnar structure is observed at the thickness of more than 150nm.

This difference is thought to be the reason of the high efficiency of the MOCVD films. This is because large grains and/or crystallites mean less boundaries, which work as the quenching centers of carriers. At the present stage, however, it is difficult to qualitatively determine this size effect due to grain size and/or crystallite size.

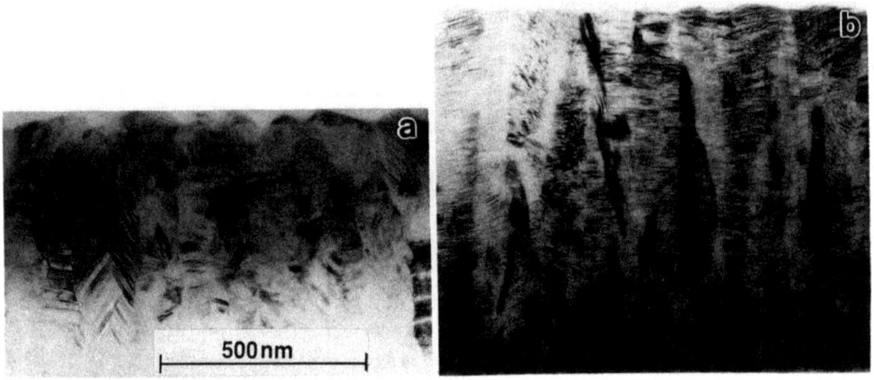

Fig. 2.
TEM images of the ZnS:Mn cross section on the (100)Si wafer:
(a) the MOCVD film (600nm-thick); (b) the EB film (830nm-thick).

3.3 Distribution of Mn^{2+} Ions

Another possible factor affecting the efficiency is the distribution of Mn ions in ZnS, because spots with a high Mn concentration can work as energy sinks. It was found that Mn^{2+} is more uniformly distributed in the MOCVD films than in the EB films as described below.

(1) Auger electron spectroscopy has shown that the depth profiles of Mn, Zn, and S are uniform in the MOCVD films, but not in the EB films.
(2) The ratio of Auger line intensities S/Zn is 0.95 for EB films when normalized to the value of the MOCVD films /3/, or the MOCVD films have less Zn vacancies.
(3) SEM images of EB films have revealed fine particles deposited on the film surface. These particles were found to be MnS by X-ray microprobe analysis. Such particles are not found in MOCVD films.
(4) At Mn concentrations higher than 1 wt%, PL spectra show a subpeak at 650nm, which was previously ascribed to Mn-Mn pairs /4, 5/. The critical Mn concentration where this subpeak appears is around 1.4-1.8wt% for MOCVD films, which is between the value reported for EB films (1.0-1.2 wt%) and the value for single crystals grown by the iodine transport method (2.5 wt%).

The PL spectra of MOCVD films are shown in Fig. 3 for different Mn concentrations. The integrated intensity of the Mn emission band was measured as a function of the Mn concentration for MOCVD films. First, PL intensities were measured on ZnS:Mn films deposited on the first insulator layers, and then EL intensities were measured on the same films after the second insulator layers and the electrodes were deposited. It was found that both PL and EL have the maximum intensity at around 1.0 wt%. This optimum Mn concentration is about twice as high as that of commonly obtained EB films and the MOCVD films reported by Hirabayashi /2/.

The high optimum Mn concentration indicates a uniform Mn distribution, which is effective for restraining the concentration quenching in ZnS:Mn films. High efficiency in the MOCVD films is due to the decrease in density of Mn-Mn pairs , which can function as energy sinks.

4. Summary

The ZnS:Mn electroluminescent films grown by reduced pressure MOCVD have been characterized in comparison with EB films. There is little difference in the

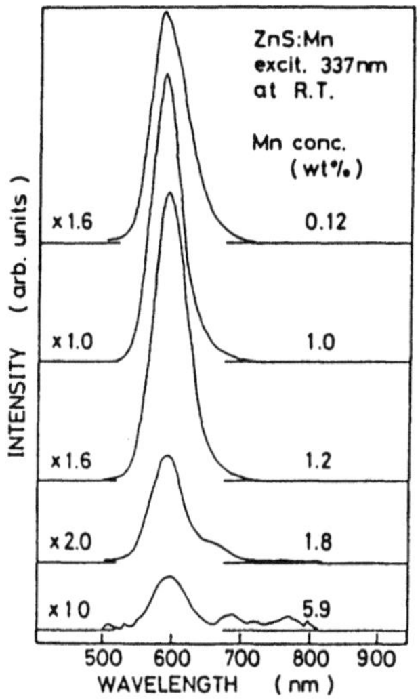

Fig. 3.
PL spectra of MOCVD films with different Mn concentrations.

intensity ratios of the total output to the output from the front side between the MOCVD film and the EB film. The structured surface is, therefore, not related to the improved efficiency of ZnS:Mn films grown by the MOCVD method.

The characteristics of the MOCVD films are large grain size, uniform Mn distribution, and the relatively S-rich composition, which can decrease the energy sinks such as boundaries, Mn pairs, and defects. Such characteristics are presumably related to the growth process by surface reactions and independent supply of each source material of the MOCVD method. As a result, the MOCVD method provides EL efficiency about twice that of EB deposition.

Acknowledgement

The authors would like to thank Mr. T. Yoshida and Mr. K. Takemura, Tsukuba Research Laboratory, Hitachi Chemical Co., Ltd., for the analysis of the Mn content and the measurement of SEM images. The authors are also greatly indebted to Mr. K. Onisawa, Hitachi Research Laboratory, Hitachi, Ltd., for fabrication of the EL devices.

References

1. P. J. Wright, B. Cockayne, A. F. Catell, P. J. Dean, A. D. Pitt and G. W. Blackmore, J. Crystal Growth 59 (1982) 155.
2. H. Hirabayashi and O. Kogure, Jpn. J. Appl. Phys. 24 (1984) 1484.
3. M. Migita, O. Kanehisa, M. Shiiki and H. Yamamoto, Abstract of 4th Intern. Conf. MOVPE (1988) 144.
4. D. D. Thong and O. Goede, Phys. Stat. Sol. (b) 120 (1983) K145.
5. D. D. Thong, W. Heimbrodt, D. Hommel and O. Goede, Phys. Stat. Sol. (a) 81 (1984) 695.

Chemical Vapor Deposition of Thin Films for ACEL

D.C. Morton, M.R. Miller, A. Vecht, A. Saunders*, G. Tyrell*, E. Hryckowian, R.J. Zeto, L. Calderon, and R.T. Lareau*

Organization address: US Army, LABCOM, Fort Monmouth, NJ 07703, USA
*Aron Vecht Associates, 95 Corringham Rd., London NW11 7DL, UK

1. INTRODUCTION

The work described here represents a cooperative experimental effort between Aron Vecht Associates and the US Army Electronics Technology and Devices Laboratory to evaluate the applicability of chemical vapor deposition (CVD) for the fabrication of thin film electroluminescent (TFEL) display devices. Device fabrication and process development have been carried out at Aron Vecht Associates and ETDL has performed some final process steps, device characterization and analysis of the films. Several iterations of fabrication and evaluation have yielded devices comparable to those being produced by conventional methods. Further process development and some inherent advantages of the process could make this an attractive manufacturing approach.

This paper presents a deposition method, which should, in principle, allow more direct control over the molecular species deposited and the energy environment under which deposition takes place. The results are somewhat preliminary, but devices have been made which compare favorably with those from more conventional methods. Potential advantages stem from the fact that in deposition of materials such as ZnS, both the Zn and the S are carried to the substrate in the same organic molecule. Stoichiometry, then is dependent only on the reaction conditions for this molecule and does not vary with small changes in gas mixture at the substrate, flow rate and other difficult to control process parameters. The process does not require a high vacuum and therefore is suitable for low cost manufacturing processes and deposition over large areas. TFEL stacks using MOCVD deposited ZnS:Mn have been reported before however the dielectric layers were deposited using conventional methods (ref. 1). With these techniques the complete dielectric-phosphor-dielectric stack can be fabricated using the same types of processes.

2. SAMPLE FABRICATION

In the work reported here, thin films have been deposited by the following methods:

2.1 Aerosol Spray Pyrolysis

Thin film dielectric layers of Al_2O_3 and ZrO_2 have been deposited by the vapor phase pyrolysis of a fine mist of acetylacetonate chelates of aluminum and zirconium dissolved in suitable solvent mixtures. The mist, generated by a reduced space saturation technique, is transported via a N_2:O_2 carrier gas to a heated narrow rectangular section tube, such that it passes over a horizontal glass/ITO substrate. Vapor concentrations in the range 10^{-4} ~ 10^{-5} [mole/litre] and carrier gas flow rates ~4 [liters/minute] are typical.

Springer Proceedings in Physics, Vol. 38 **Electroluminescence**
Editors: S. Shionoya and H. Kobayashi

Deposition temperatures of ~ 380[^{o}C] and 440[^{o}C] are employed for ZrO_2 and Al_2O_3 respectively. Under these conditions, growth rates of >500 [nm/hour] are obtained with good uniformity (<10% variation) across a 27[cm^2] deposition area.

2.2 Low Pressure CVD

Thin films of ZnS:Mn are deposited onto dielectric coated substrates by the vapor phase pyrolysis of zinc diethyldithiocarbamate {$Zn[(C_2H_5)_2NCS_2]$} [$Zn(dtc)_2$] according to the idealized equation 1 :

$$Zn[(C_2H_5)_2NCS_2]_2 \xrightarrow{400^oC} ZnS + \{[(C_2H_5)_2NCS]_2S\} \qquad \text{Eq.1}$$

Manganese is introduced via the pyrolysis of methyl - cyclopentadienyl manganese tricarbonyl, [TCM], [$CH_3:C_5H_4.Mn.(CO)_3$], this being employed as a 5% solution in decalin. A two zone horizontal reactor is employed, this being evacuated to a base pressure of $5x10^{-2}$ [torr] with the substrate platform at 400[^{o}C]. The zinc diethyldithiocarbamate vapor source is employed in the temperature range 160 ~ 200[^{o}C], with an N_2 carrier gas introduced via a calibrated needle valve. The manganese bearing vapor is introduced via the exposure of the manganese solution to the reduced pressure via a series of needle valves. Total system pressures up to 25 [torr] are employed with vapor concentration control being obtained via the $Zn(dtc)_2$ source temperature and the N_2 (main inlet): N_2/TCM inlet pressures. Typically, growth rates of 150[nm/hour] are obtained. Optical band gap measurements indicate the hexagonal modification of ZnS,(Eg_{hex} ~ 3.56 [eV] and measured as 3.60 [eV]. Note Eg_{cubic} ~ 3.80 [eV]) with good crystallinity and a high degree of preferred orientation indicated by the presence of a single sharp peak in the XRD spectrum corresponding to the **(002)** reflection. Completion of the MISIM stack, i.,e. deposition of the top insulating layer is carried out by aerosol spray pyrolysis, but in this case, the exposed ZnS:Mn layer is heated in an inert atmosphere to minimize surface oxidation. After the deposition of an initial thin protective coating, oxygen is introduced into the spray system to increase the growth rate of the top oxide coating.

2.3 OXYNITRIDE DEPOSITION

Silicon oxynitride was deposited by PECVD (Plasma Enhanced Chemical Vapor Deposition) using a LFE, Pnd-301 plasma system (silane and nitrogen). Deposition temperature was 300[^{o}C] and average deposition rate of 500 [A/minute] was used. Standard photoresist and etching technique were used for pattern definition.

2.4 ALUMINUM ELECTRODES

Top Al electrodes were defined by patterning photoresist stripes across the sample and then 1000 A of aluminum was deposited over the entire sample. Unwanted Al was then lifted off.

3. TESTING

After the samples are completed electro-optical evaluation is accomplished using an automated test and measurement system (Ref. 2). All timing, voltage parameters and output measurements are under computer control. Data which is normally taken on all samples is: luminance vs. voltage, efficiency vs.

voltage, spectrum, charge transferred vs. voltage, drive and output waveforms. From the luminance and efficiency vs. voltage data the applicability of the device to a practical display can be discerned. These tests provide the standards on which TFEL displays are compared and judged. The spectrum is studied to determine if the electroluminescent emission is correct for the phosphor material. In this case the width and position of the peak can provide information on the crystallinity of the sample and the doping level of the Mn. The charge vs. voltage curves can provide the dielectric constant of the insulating layers of the dielectric-phosphor-dielectric stack. The waveforms show the response of the device and the emission decay curve can provide an estimate of the doping level of the Mn. (ref. 3).

3.1 ELECTRO-OPTICAL TEST RESULTS

Figure 1 shows the luminance voltage characteristics for a sample having ZrO dielectric layers. Using SION and Al2O3 for the dielectric layers showed similar results. Although these unoptimized samples obtain reasonable luminance and efficiency values they are below those obtained using conventional methods. The drive conditions for this data was: frequency of excitation 60 [Hz], pulse width 30 [us], delay between positive and negative pulses 8.3×10^{-3} [sec].

Figure 2 shows typical voltage and current waveforms for the sample of figure 1. The current waveform is the real component of the total current.

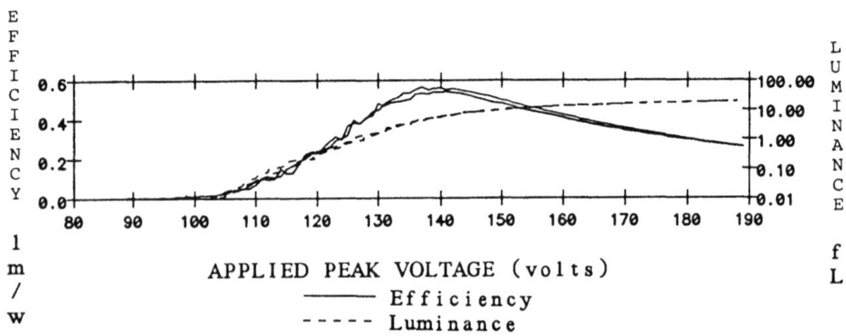

Fig. 1. Luminance & efficiency vs voltage (zro die.)

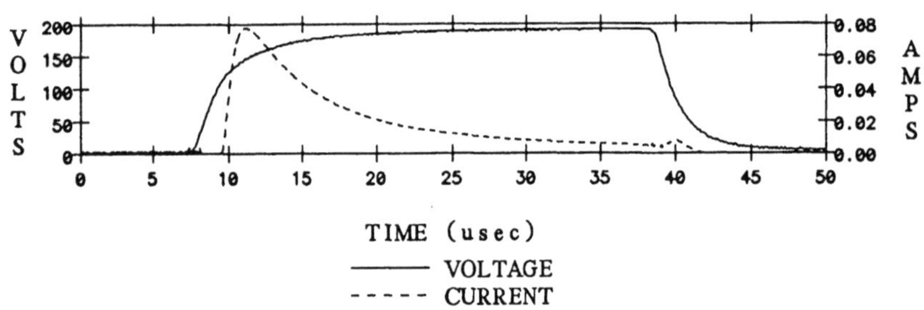

Fig. 2. Voltage & current waveforms

5. ANALYSIS

Several samples were analyzed by secondary ion mass spectrometry (SIMS) to obtain highly sensitive elemental depth profiles of the CVD layers. SIMS analyses were obtained using a quadruple-based ion microanalyzer (PHI 6300) at ETDL. The cesium primary ion beam was rastered over a 500 [um] x 500 [um] area, with secondary ions collected from the inner 9% of the sputtered crater. The impact energy of the primary ions was 9 keV, with an angle of incidence of 60 degrees to the surface normal; electron neutralization was used to compensate for sample charging. A typical SIMS depth profile for the CVD structures is presented in Figure 3. The CVD thin films were grown on a glass substrate in the following order: InSnO (ITO) transparent electrode/Al_2O_3 lower dielectric/ZnS:Mn phosphor/$Si_xO_yN_z$ upper dielectric. This structure was subjected to a 450[^{o}C], 30 second rapid thermal anneal (RTA) in an argon atmosphere. Very little diffusion is observed from this profile, indicating that the films are fairly stable at short anneal times. Both oxygen and carbon in the CVD films are present at qualitatively high levels, as would be expected by the CVD processes. The rise in the In signal at the ZnS:Mn/Al_2O_3 interface is due to a molecular ion interference, rather than true In outdiffusion. It does appear, however, that sulfur is enriched at the leading and trailing interfaces of the ZnS:Mn layer. The enrichment in sulfur at the trailing interface could explain the change in the Mn ion signal which increases due to an apparent matrix change (SIMS matrix effect,Ref.4). The depths indicated on Fig.3 are estimates from growth conditions during the CVD thin film fabrication. Thus, the 'sputter time' scale on this depth profile is not linear, due to changes in sputtering rates with varying matrices.

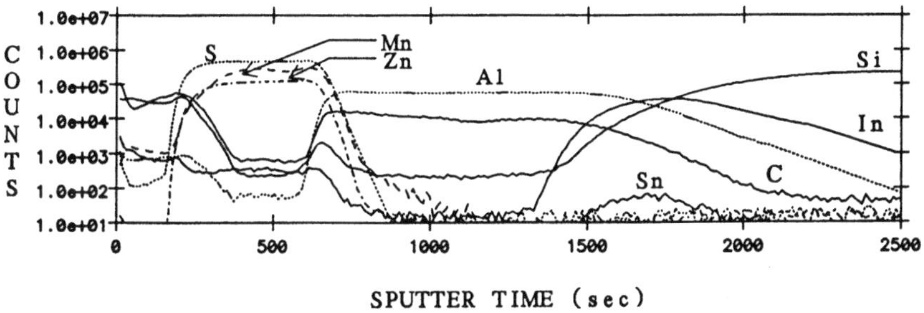

Fig. 3. SIMS analysis of a TFEL device

6 CONCLUSIONS

Complete TFEL stacks up to rear electrode have been fabricated by the MOCVD process. Luminance values of 20 [fL] have been measured at 60 [Hz] in these devices with efficiencies of 0.5 [Lumens/Watt]. Improvements in these values are expected with further process development, but these results are very encouraging considering the limited amount of time and resources that have been applied so far. Questions concerning the effects caused by the incorporation of carbon in the films and the possibility of eliminating it will be the subject of further study. Future devices will be completed at one location rather than after days and thousands of miles' travel which should eliminate sources of contamination which may have an effect on some of the above results.

Ref. 1 A.F. Cattell, B. Cockayne, K. Dexter, J. Kirton, P.J. Wright, IEEE Trans. on Electron Dev. Vol. ED-30, No. 5,(1983).

Ref. 2 D.C. Morton, M.R. Miller and M.J. Jusinski, Proceedings of U.S. Army Test Technology Symposium. (1988).

Ref. 3 H. Sasakura, H. Kobayashi, S. Tanaka, J. Mita, J. Appl. Phys. 52(11) (1981).

Ref. 4 G. Blaise and M. Bernheim, Surf. Sci. 47, 324 (1975)

AC Thin-Film ZnS:Mn Electroluminescent Device Prepared by Intense Pulsed Ion Beam Evaporation

Y. Shimotori[1], *M. Yokoyama*[2], *K. Masugata*[1], *and K. Yatsui*[1]

[1]Laboratory of Beam Technology, Nagaoka University of Technology, Nagaoka, Niigata 940-21, Japan

[2]Nippon Seiki Co., Ltd., Nagaoka, Niigata 940, Japan

Abstract

Electroluminescent ZnS:Mn thin films have been deposited by Zn, S and Mn ions from high-density, high-temperature plasma produced by an intense (> GW/cm^2), pulsed (~ tens of ns), ion beam onto the ZnS:Mn target. X-ray diffraction analysis shows the ZnS films have a polycrystalline, hexagonal structure. Deposition rate is typically ~ 0.3 μm per beam shot. Using a simple model, the instantaneous deposition rate is estimated to be ~ 2 cm/s, which exceeds five orders of magnitude higher than those of conventional deposition method. The electroluminescent device exhibits the maximum brightness of 195 cd/m^2 under a 10 kHz sinusoidal excitation at 102 V (rms).

1. Introduction

Recently, electroluminescent (EL) ZnS thin films have been widely investigated as the flat panel display devices [1-4]. By now, many successful techniques have been used to prepare these films such as an electron-beam evaporation [1], sputtering deposition [2], atomic-layer epitaxy (ALE) [3], and metalorganic chemical vapor deposition (MOCVD) [4].

The technique of intense pulsed ion beam evaporation (IBE) [5-8] is a novel method for the preparation of thin films. Since the stopping-range of the high-energy (~ MeV) ions (e.g., protons) in a solid target is quite short (< tens of μm), only the surface of the target is effectively heated by the irradiation of an intense pulsed ion beam [9-10], which produces a high-density, high-temperature plasma.

The ions secondarily generated from such a target plasma are considered to be very useful for the vacuum deposition, since the ions with energy in the range of 1 ~ 100 eV are known to be effective for the production of high-quality films [11]. Furthermore, high-speed deposition might be expected. These ions are collected at a substrate suitably placed in a vacuum chamber.

The features of the IBE are summarized as follows:
1) Deposition rate is very high.
2) Most of the atoms to be deposited are ionized.
3) These ions have kinetic energy of more than few eV.

In this paper, the experimental results of the IBE for the preparation of ZnS thin films, and the characteristics of the ZnS:Mn EL device are described.

2. Experimental Procedure

Figure 1 shows the cross-sectional view of the IBE system. The experiment is carried out by using the pulse-power generator, "ETIGO-I" (output parameter: 1.2 MV, 240 kA, 0.3 TW, 50 ns, 14.4 kJ). The intense pulsed ion beam (maximum beam

Springer Proceedings in Physics, Vol. 38 Electroluminescence
Editors: S. Shionoya and H. Kobayashi

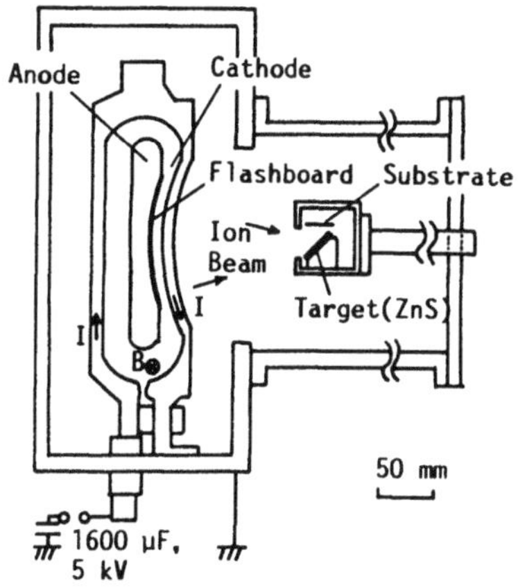

Fig. 1
Cross-sectional view of the IBE system

energy ~ 940 keV, ion-current density ~ 7 kA/cm², and pulse width ~ 30 ns (Full Width at Half Maximum; FWHM)) is extracted from a magnetically-insulated diode (MID). A polyethylene sheet is attached to the anode as a flashboard.

The cathode works as a one-turn theta-pinch coil, producing a magnetic field (~ 1 T) to prevent electron current in anode-cathode gap (10 mm). The anode and cathode are spherically shaped to achieve a geometric focusing at the target. The vacuum chamber is pumped to ~ 10^{-4} Torr.

The target box consisting of the target and substrate is located at z = 130 ~ 140 mm downstream from the anode. The diameter of the entrance is 9 ~ 20 mm. The ZnS target is set to be 45 degree with respect to the beam axis. The distance from the entrance to the center of the target is 20 mm. The substrate (glass) is located at y = 16.5 ~ 18 mm above the central line of the beam.

3. Preparation of ZnS Thin Films

Figure 2 shows the thickness (d) of the ZnS films prepared on glass plotted against the number of shot (N). In this experiment, the entrance (9 mm in diameter) of the box is located at z = 140 mm. The distance between the target and substrate is 18 mm. As can be seen from Fig. 2, we see the thickness increases almost linearly with increasing shot. Moreover, the deposition rate of ~ 0.3 μm/shot is obtained.

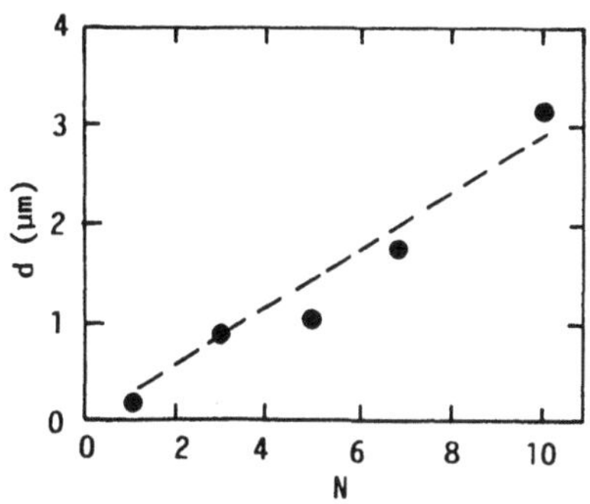

Fig. 2
Film thickness (d) plotted against the number of shot (N)

Figure 3 shows an X-ray diffraction pattern for the film obtained by one shot. In Fig. 3, a sharp peak appears at 2θ = 28.5°, which corresponds to the diffraction angle for the cubic (111) or hexagonal (002) ZnS. The another peak at 2θ =

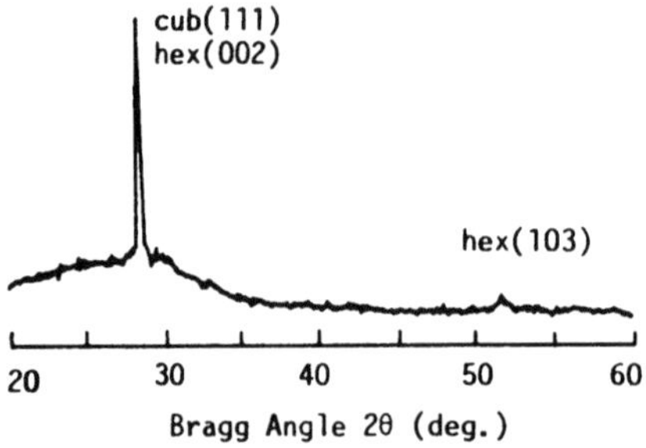

Fig. 3
X-ray diffraction pattern of the films (using Cu-K_α)

51.8° corresponds to the hexagonal (103). Furthermore, the X-ray diffraction analysis of ZnS powder obtained from the films shows the structure to be hexagonal. From these results, we see the films prepared have a polycrystalline, hexagonal structure with the strong orientation in the (002) direction.

4. ZnS:Mn EL Device

The structure of the ac thin-film EL device which we have studied is shown in Fig. 4. This is the double insulating layer structured EL device. It has been sequentially deposited with $BaTiO_3$, ZnS:Mn, HfO_2, Ta_2O_5, HfO_2 and Al onto an indium-tin-oxide (ITO) coated glass. The first insulating layer of $BaTiO_3$ (~ 580 nm) is deposited by an rf-sputtering. The second insulating layer of $HfO_2/Ta_2O_5/HfO_2$ (~ 200 nm in total thickness) is fabricated by an electron-beam evaporation. The Al back-electrode (~ 200 nm) is also deposited by an electron-beam evaporation. The ZnS:Mn luminescent layer (~ 200 nm) is deposited by the IBE method. The Mn concentration in ZnS:Mn target is 0.85 wt%. The substrate is kept at room temperature.

Figure 5 shows the brightness plotted against the applied voltage (sinusoidal, rms) with operation frequency as a parameter. The color of the emission is observed to be yellow-orange, which is caused by the presence of Mn as a luminescent center. The brightness increases with increasing voltage in the range of 40 ~ 80 V. Furthermore, it increases with increasing frequency in 100 Hz ~ 10 kHz. The maximum brightness is 195 cd/m² under a 10 kHz sinusoidal excitation at 102 V.

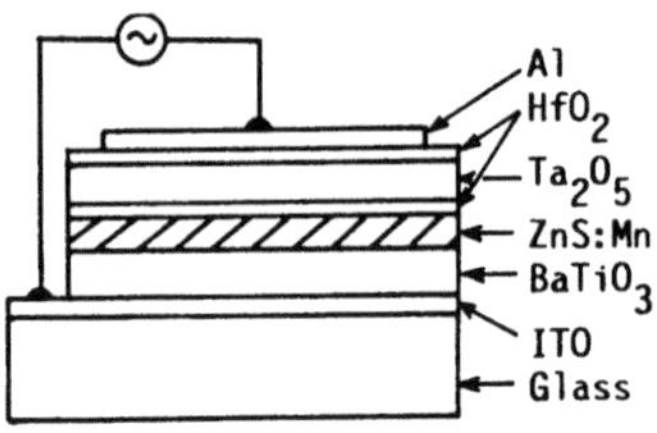

Fig. 4
Structure of EL device

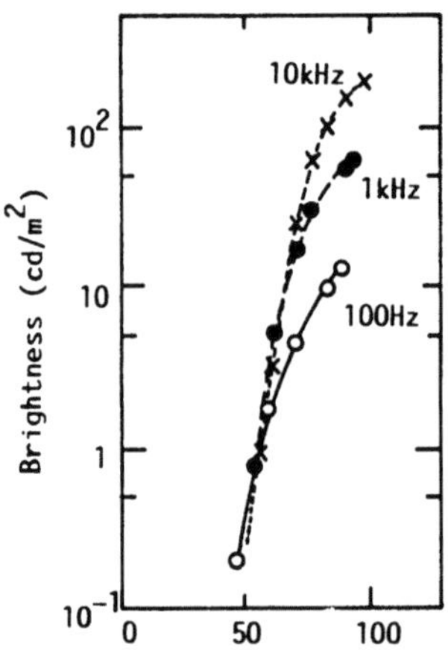

Fig. 5
Brightness vs. applied voltage

5. Conclusions

We have successfully developed the IBE system for the preparation of polycrystalline, hexagonal-ZnS thin-films. Film thickness of ~ 0.3 μm is obtained per one shot. The instantaneous deposition rate is evaluated to be ~ 2 cm/s, which exceeds five orders of magnitude higher than those of conventional evaporation techniques. Furthermore, we have prepared ZnS:Mn EL device, which has the maximum brightness of 195 cd/m² under 10 kHz sinusoidal excitation at 102 V (rms).

Acknowledgement

The authors would like to thank Professor C. Y. Chang of National Cheng Kung University, Republic of China and Messrs S. Ohta and T. Tadokoro of Nippon Seiki Co., Ltd.for their valuable discussions and assistance.

References

1. H. Sasakura, H. Kobayashi, S. Tanaka, J. Mita, T. Tanaka, and H. Nakayama: J. Appl. Phys., 52, 6901 (1981).
2. J. Mita, M. Koizumi, H. Kanno, T. Hayashi, Y. Sekido, I. Abiko, and K. Nihei: Jpn. J. Appl. Phys., 26, L558 (1987).
3. C. H. L. Goodman, and M. V. Pessa: J. Appl. Phys., 60, R65 (1986).
4. K. Hirabayashi, and O. Kogure: Jpn. J. Appl. Phys.: 24, 1484 (1985).
5. M. Yokoyama, Y. Shimotori, K. Masugata, and K. Yatsui: Proc. of World Conf. on Advanced Materials for Innovations in Energy, Transportation and Communications (Chem. Soc. Jpn., Tokyo, 1987), IA03 (1987).
6. Y. Shimotori, M. Yokoyama, H. Isobe, S. Harada, K. Masugata, and K. Yatsui: J. Appl. Phys., 63, 968 (1988).
7. Y. Shimotori, M. Yokoyama, S. Harada, K. Masugata, and K. Yatsui: Oyo Buturi (Applied Physics in Japanese), 57, 1078 (1988).
8. Y. Shimotori, M. Yokoyama, S. Harada, K. Masugata, and K. Yatsui: Bull. Nagaoka. Univ. Tech., No. 10, 57 (1988).
9. K. Yatsui, A. Tokuchi, H. Tanaka, H. Ishizuka, A. Kawai, E. Sai, K. Masugata, M. Ito , and M. Matsui: Laser & Part. Beams, 3, 119 (1985).
10. K. Yatsui, Y. Shimotori, M. Ikeda, A. Takahashi, T. Tanabe, K. Aga, A. Kanai, Y. Araki, K. Masugata, S. Kawata, and M. Murayama: Laser & Part. Beams, 5, 415 (1987).
11. T. Takagi: J. Vac. Sci. Technol., A2, 382 (1984).

Part VI

Thin Film Electroluminescent Panels

Review of Flat Panel Displays: Electroluminescent Displays, Liquid Crystal Displays, Plasma Displays, etc.

H. Uchiike

Hiroshima University, Shitami, Saijo-cho, Higashi-Hiroshima 724, Japan

1. Introduction

Applications of flat panel displays for laptop computers and TV receivers have been in the spotlight of attention over the years because of their inherent characteristics of low volume and light weight. With the advance of display technology, we are able to list many kinds of flat panel displays. In the group of emissive flat panel displays, there are electroluminescent displays, plasma displays, vacuum flourescent displays, light emitting diodes, and flat cathode ray tubes. In that of non emissive ones, there are liquid crystal displays and electrochromic displays, Until several years ago, all these display devices coexisted, as they each had their own specific application . Since the middle of 1980, sales of laptop computers have been increasing very rapidly and the demand for quality in flat panel displays has been very great with regard to viewing angles, high speed addressability, contrast ratio, luminance, and power consumption.

Flat panel displays have been researched and developed against the background described above. This review will describe the trend of research and development in the field of flat panel displays with regard to two major applications of the displays, for computer terminals and for full-color televisions.

2. Flat Panel Displays for Computer Terminals

In the early 1980s, Grid Systems designed and manufactured laptop personal computers using electroluminescent displays and later they added plasma displays and liquid crystal displays to their products. Twisted-nematic liquid crystal displays were very attractive for character display because of low cost and low power comsumpution. In 1987, Toshiba 3100 was commercially successful by supplying DC plasma displays with 400 x 640 pixels. This fact indicates that emissive flat panel displays have an advantage over nonemissive ones.

2.1. Liquid Crystal Displays

In the mid- 1980's, twisted-nematic liquid crystal displays could not hold their pre-eminent position in laptop computers, since they did not satisfy expectations of rapid scrolling, graphic displays, and high speed addressability with a value of 100 msec by using a mouse in the computer terminal systems. In order to compete with plasma displays and electroluminescent displays in display quality,

Springer Proceedings in Physics, Vol. 38 Electroluminescence
Editors: S. Shionoya and H. Kobayashi

the supertwisted birefringence mode of liquid crystal display was researched and developed by Sheffer.[1] However, it didn't satisfy the demands of computer terminals because of the existence of blue or yellow color interference.

Seiko Epson developed double layer supertwisted-nematic liquid crystal displays, which resulted in a black and white mode with a high contrast ratio.[2] They, however, still suffer from insufficient performance. One such limitation is a narrow viewing angle which is not suitable for supervisor use , that is, two simultaneous users on the computer terminals.

Very recently in place of the second super twisted-nematic layer for compensation, polycarbonate film has been developed to reduce costs and increase transparency.[3] The quality of compensation in color interference by polycarbonate film has not changed essentially.

2.2. Plasma Displays

Plasma displays can be divided into DC and AC types, depending on electrode material, structure, and related operation mechanisms. Since 1986, DC plasma display manufacturers have been comercially successful. Now laptop computers with plasma displays leave a favorable and high-grade impression upon us, while those with liquid crystal displays tend to make a poor one. Panasonic manufactures 30,000 units per month at peak output. Oki and Dixy also manufacture DC plasma displays.

DC plasma displays use the priming effect to minimize the statistical and formation delay of the discharge. Panasonic and Oki use self priming discharge mechanisms for their DC plasma displays to which short width pulse voltage is applied. Dixy uses the pilot priming mechanisms by which the trigger discharge is generated between the cathode and the trigger electrode.

AC plasma displays have the inherent characteristic of memory function due to the electrode structure, where metal electrodes are insulated from gas by a dielectric layer. AC plasma displays have an excellent display quality, an especially high contrast ratio, a very wide viewing angle of 360 degrees, and a capability of large area display in memory operation. The largest AC plasma display is manufactured by Photonics Technology with a 1.5 meter diagonal and 2,048 x 2,048 pixels.

The life of AC plasma displays is 350,000 hours; that of DC ones is about 10,000 hours. Despite the superiority of AC plasma displays over DC ones concerning length of life, DC sales are much higher. This is due to the high price of IC drivers used in AC plasma displays, since the complimentary type of transistors must be used in the final step of the AC driving circuits.

2.3. Electroluminescent Displays

Electroluminescence is the emission of light from a polycrystalline phosphor due to the application of an electric field. Development of electroluminescence started in the late 1940s. In the 50s, the initial intended use was for luminaires. In the 60s, interest in the use of electroluminescence was focused on information display. There are four kinds of electroluminescent displays, DC powder, thin-film, AC powder and thin-film electroluminescent displays. DC powder electroluminescent displays were developed for the Apollo vehicle and LM vehicles for NASA's lunar program.

All products of electroluminescent displays are thin-film AC type except for the DC powder one reported by Cherry.[4] Thin-film AC electroluminescent displays close to the present type were developed by Soxman of Servomechnisms.[5] Sharp improved the fabrication technology; they used electron beam evaporation of the ZnS:Mn and applied better dielectrics.[6] In 1978, they demonstrated a 240 x 320 thin-film electroluminescent display monochromatic TV for the first time at the consumer Electronics Show in Chicago. Sharp's disclosures have offered a solution to the problems of longevity and matrix addressability which had proved very difficult. Sharp's success with thin-film AC electroluminescent displays has prompted research and development into its practical applications.

The phosphors of zinc sulfide activated with less than one mole percent of manganese has thus far given the best performance in terms of luminous efficiency, longevity, and luminance. The color is a pleasant yellowish orange. This type of thin-film AC electroluminescent displays is its only practical use at this time.

The performance of electroluminescent displays is very process-dependent. There are six technologies used in fabricating thin-film AC or DC electroluminescent displays. They are Vapor Deposition, Sputtering, Atomic Layer Epitaxy(ALE)[7], Chemical Vapor Deposition(CVD), Metal Organic Chemical Vapor Deposition(MOCVD)[8], and Multi Source Deposition(MSD).[9] Discussions on the advantages and disadvantage of these fabrication methods have been interesting because of the recent research and development in the field of color electroluminescent displays. Two of the six fabrication technologies, Vapor Deposition and Atomic Layer Epitaxy are used in the practical fabrication process at this time. The other fabrication technologies are still only at the research and development stage. Sputtering, MOCVD, and MSD attract attention due to their capacity to improve the performance of color thin-film electroluminescent displays.

3. Full-Color TV Display

It is one of the ultimate dreams of mankind to create a TV display by using the flat panel display, that is, "TV-on-the-wall".

3.1. Full-Color Liquid Crystal Displays

In order to overcome the disadvantages of double layer supertwisted-nematic liquid crystal displays, multi-color ones have been researched and developed very recently.[10] Color filters have been added to the structure of double layer supertwisted-nematic liquid crystal displays. However, almost all disadvantages, such as narrow viewing angles and slow addressing speed remain, despite excellent static operation performance. As a result, multi-color double layer supertwisted-nematic liquid crystal displays cannot produce a full-color TV display with a high level gray scale.

Full-color active matrix addressing liquid crystal TVs have advanced very rapidly due to improvements in process technology. 3, 4, and finally 5 inch diagonal ones were marketed in 1987. Very recently Sharp demonstrated a 14inch diagonal full-color active matrix addressing liquid crystal display.[11] What kind of semiconductor device is suitable for the production of full-color liquid crystal TV with a large display area? There are a lot of problems and several fabrication processes involved in producing thin-film transistors, such as amorphous silicon, polysilicon, and MIM technologies. With MIM it is hard to keep a large display area uniform, while MIM is superior to the thin-film transistor technologies in terms of cost reduction. Of the fabrication process of thin-film transistors, that of polysilicon is the most promising, since the external driving circuits and the thin-film transistors for active matrix addressing are fabricated simultaneously.

Liquid crystal specialists forecast that 20 to 25 inch diagonal full-color active matrix liquid crystal TV displays will be produced within five years. It is indeed very pertinent to research and develop a full-color active matrix liquid crystal TV display. However, there are a lot of very difficult problems for increasing low yield, developing transparent electrode material with low resistivity, new alignment material, and new liquid material to obtain high speed addressability.

3.2. Full-Color Plasma Displays

AC plasma displays have the advantage of large scale display. This is confirmed by the success of the Photonics Technology in that they manufacture a 1.5 m diagonal AC plasma display, which itself is monochromatic.

Color in plasma displays is achieved by placing phosphors in plasma panels and exciting those phosphors with the ultraviolet rays of gas discharge. A major problem with color plasma displays is designing them to achieve high brightness and luminous efficiency. This can be achieved with fluorescent lamps, but does not have favorable scaling laws when applied to the very small pixels in plasma displays. In addition, almost all plasma displays use light from the negative glow, which is known to be considerably less efficient than light from the positive column in a typical discharge region.

The DC plasma display which has been researched and developed by Dr.Mikoshiba at Hitachi Central Labs. uses light from the positive column to excite the phosphors.[12] This panel has the highest brightness and luminous efficiency with values of 200 fL and 1.6 lm/W at white balance, respectively.

The most impressive of all of the color displays demonstrated was developed at NHK Technical Labs. Their 20 inch diagonal DC plasma display was demonstrated at the recent NHK show and their results were presented at the 1988 SID Symposium.[13] NHK's DC plasma displays have memory function due to pulse memory mechanisms, which results in the ability to fabricate plasma displays with a large display area. The recent NHK's DC plasma panels have a planar type structure in order to fabricate them by using thick film technology as shown in Fig.1. The brightness of the NHK display is sufficient for practical purposes. The gray scale of NHK's plasma panel is 256 levels. This value is sufficient for the High-Definition TV.

In order to attain a gray scale of 256 levels, the plasma displays must be operated at a high brightness level. This results in a short life due to the degradation of metal cathode material caused by the sputtering. This is the primary reason why the full-color DC plasma displays have not been of practical use, even though NHK and Hitachi Labs have been engaged in researching and developing them for more than 15 years.

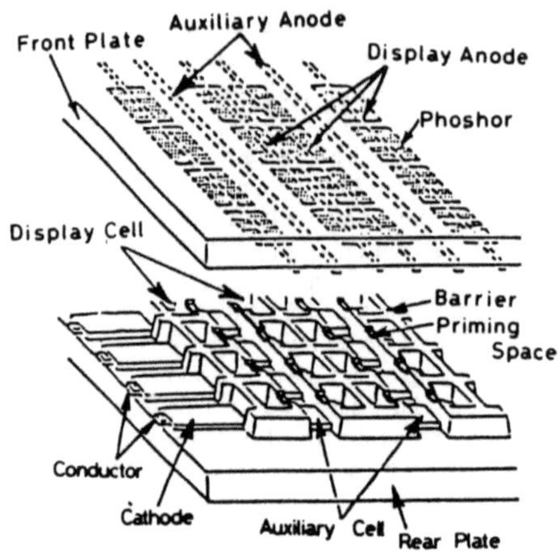

Fig.1 Electrode structure of dc plasma display developed at NHK Tech Labs. All metal electrodes and barriers are fabricated by using thick-film technology.

Figure 2 shows the electrode structure of the color surface-discharge AC plasma displays which are researched and developed at Hiroshima University.[14] This electrode configuration is used to reduce the ionic bombardment of the phosphor that results in degradation of the phosphor and short display lifetime. The highest results are 100 fL and 1 lm/W for the luminance and luminous efficiency respectively, with a capability of 256 gray scale.[14] As a result of applying the distinctive electrode configuration, the surface-discharge AC plasma displays are expected to be operative in the full-color TV display. Now more than five companies including Thomson CSF are researching or interested in surface-discharge AC plasma displays for full-color TV image display.

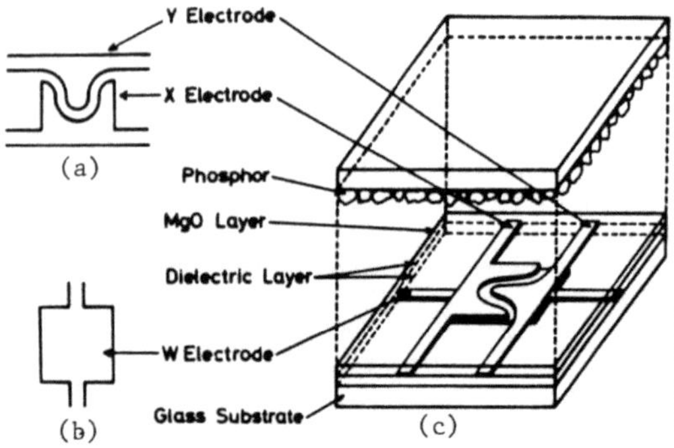

Fig.2 Electrode structure of surface-discharge AC plasma display researched and developed at Hiroshima University

3.3. Color Electroluminescent Displays

Very active research and development of red, green, and blue color phosphors for the electroluminescent displays are being undertaken in full-color electroluminescnet display.

Thin-film electroluminescent multicolor displays are a technology that is now emerging from the laboratory. Since 1980, we can see that significant progress has been made in thin-film electroluminescence color phosphors.

Progress in improving the luminance of the green-emitting ZnS:Tb thin-film electroluminescent displays has attracted our attention during the last few years. Investigations by Prof Ohnishi into the sputter deposition process for this material have increased the green luminance with a value of 26 fL at 60 Hz to nearly the brightness level of the ZnS:Mn phosphor.[15] This material has the problems of charge compensation and lattice size mismatch. These problems, however, have been overcome by changing dopant form Tb to Tb-F complex and by using the sputter deposition process. From these efforts, ZnS:TbF phosphor is expected to be of practical use in monochrome as well as color thin-film electroluminescent displays.

For red color, ZnS:Sm and CaS:Eu phosphor have attracted attention. At the present time, the CaS:Eu system, with a luminance with a value of 6.5 fL at 60 Hz[16], has achieved a higher luminance level than the ZnS:Sm phosphor, 3.5 fL at 60 Hz.[17] The ZnS:Sm phosphor, however, has advantages in terms of ease of processing, both in terms of film deposition and film etching.

For blue color, the hundred fold improvement in the luminance of SrS:Ce phosphor over ZnS:Tm has stimulated much of the recent investigations into both SrS and CaS as thin-film electroluminescent displays. Prof Kobayashi's group has been engaged in this field and has achieved significant results in terms of theorectical discussion on lattice matching, charge compensation and experiments to obtain high luminance.[18] They achieved luminance of 10 fL at 60 Hz for SrS:Ce[19] by using a high processing temperature of 500 to 700 C and excess sulfur or ZnS buffer layers to

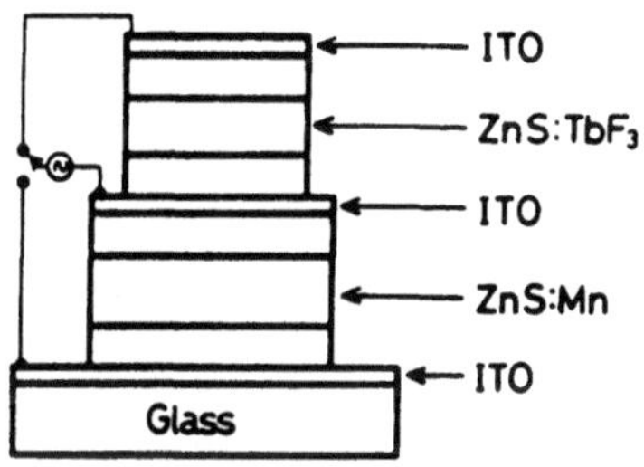

Fig.3(a) Layered structure of dual color AC thin-film electroluminescent display

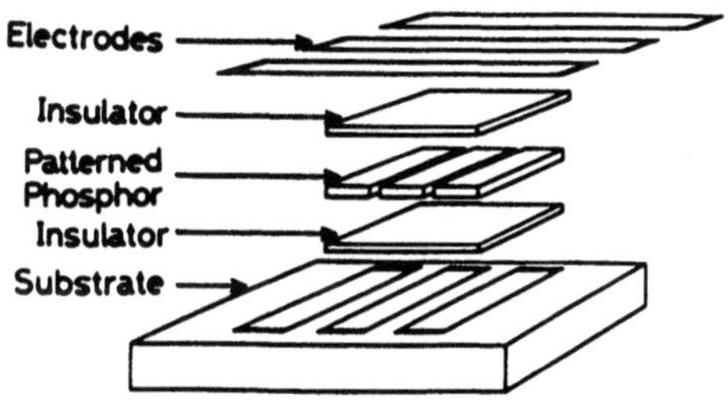

Fig.3.(b) Red-Green-Blue patterned phosphor structure

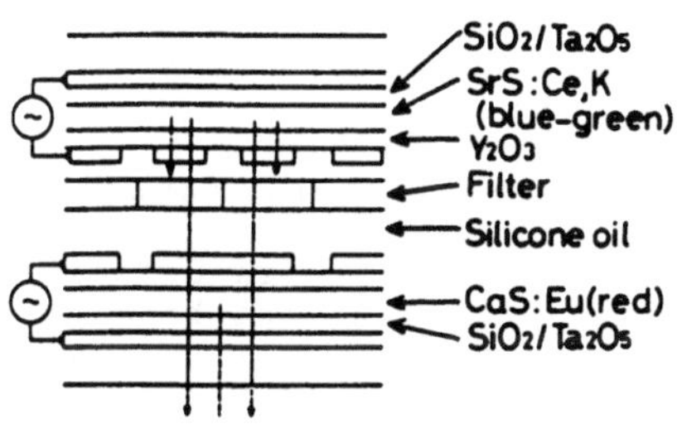

Fig.3(c) Hybrid structure of full-color electroluminescent display

achieve the proper stoichiometry. However, there is a problem with this host material as the blue phosphor, is blue-green rather than pure blue in color chromaticity. For a video display this blue-green restricts the color gamut that is achievable. One approach to this problem has been to filter out the green component[19]. However, this is at the expense of reducing the blue luminance by a factor of 10 and increasing cost. Even so this would still be the brightest blue phosphor available but the luminance level would only be about one-half of the minimum expected value.

Concerning the multicolor device structures of thin-film AC electroluminescent displays, there is a layered structure, patterned structure, and a hybrid structure. A layered structure is essentially a stack of two electroluminescent displays as shown in Fig.3(a). A major advantage of the layered type structures is the ability to fabricate a multicolor display without limiting the resistance of the transparent electrodes. For a patterned structures the traditional approach to fabricated multicolor displays is shown in Fig.3(b). The advantage of the patterned phosphor structure is the ease of use of standard electrodes;its disadvantage is a lower than average brightness due to the small pixel fill factor.[20] A hybrid structure is a combination of the above two structures. This structure has an advantage in the ease of fabricating full-color by using the present technologies of thin-film AC electroluminescent displays as shown in Fig.3(c).[19] When phosphors of red, green, and blue are improved to the point of having sufficient luminance and luminous efficiency, full-color thin-film AC electroluminescent displays will be produced using the device technologies described above.

4. Conclusion

The biggest threat to full-color electroluminescent displays as well as to full-color plasma displays is the distraction of interest due to the promise of full color liquid crystal displays that use thin-film transistors. While the liquid crystal results have been dramatic, there is considerable risk from liquid crystal technology when one considers large liquid crystal displays with many pixels. Plasma displays have few problems producing a full-color TV display with a large display area. Forutunately the recent market success of DC plasma displays in the personal computer market will stimulate more research funding for color.

In contrast to the success of plasma displays, there is little to attract attention to electroluminescent displays. We must search for an innovative use for electroluminescnet displays. As shown in Fig.1, electroluminescent displays have an essential advantage of high resolution. Even color phosphors of the electroluminescent displays have not been achieved with such exact color chromaticity as that of the cathode ray tubes. We must not let this distinctive property of the high resolution of electroluminescent displays be lost. Applying this distinctive high resolution, the electroluminescent displays have a very attractive application in their individual use. If we used full-color electroluminescent displays with very high resolution, mounted on eyeglasses like an eyecamera, we would appear to be in a natural land and the scene would appear natural. If such a high definition of eyedisplay were manufactured, display size might not be a problem because the viewing field becomes wider than that of the human eye. We will not need large scale TATAMI size display for personal use. Laptop computers will only be keyboards and eye-displays in the near future. It is anticipated that the electroluminescent display will fulfil this dream use in the very near future!

In addition to these excellent properties of eye-displays, the eyedisplay system has a three-dimensional capacity. In the 21st century it is expected that highly integrated computer network systems will be personalized and so displays must be also personalized. So, electroluminescent displays in computer terminal use and TV display will replace liquid crystal displays and plasma displays, except for those viewed by a group of people at one time.

5. References

1. T.J.Scheffer et al., SID 85 Digest, 1985, pp.120-123
2. K.Katoh et al., Japan J.Appl.Phys. vol.26, 1987, pp.784-786
3. S.Matsumoto et al.,The 12th Int'l Liquid Crystal Conf.,Aug. AP03,1988
   K.Nagae et al., ITEJ Tech.Report, vol.12,ID88-75, 1988,PP.29-34
   I.Fukuda et al.,1988 IDRC,Oct.,1988, pp.159-160
   S.Matsumoto et al., ibid, pp.182-183
   H.Odai et al., ibid,pp.195-198
4. D.Glaser et al., SID Digest, 1987, pp.292-294
5. E.J.Soxman, Army/Navy Instrumentation Program Phosphor Research, Final Report. Servomechanisms, Inc.
6. T.Inoguchi et al., SID 74 Digest, 1974, pp.84-85
   S.Mito et al., SID 74 Digest, 1974, pp.86-87
7. T.Suntola et al., SID 80 Digest, 1980, pp.109
8. A.F.Cattel et al., IDRC 25, Oct., 1982
9. T.Nire et al., SID 87 Digest, 1987, pp.242-244
10. H.Koh et al., SID 88 Digest, 1988, pp.53-56
11. T.Nagayasu et al., 1988 IDRC, Oct., 1988, pp.56-58
12. S.Mikoshiba et al., SID 84 Digest, 1984, pp.91-94
13. H.Murakami et al., SID 88 Digest, 1988, pp.142-145
14. H.Uchiike et al., SID 88 Digest, 1988, pp.146-149
15. H.Ohnishi et al., SID 87 Digest, 1987, pp.238-241
16. S.Tanaka et al., SID 86 Digest, 1986, pp.29-32
    M.Yoshida et al., SID 86 Digest, 1986, pp.41-43
17. T.Tohda et al., Appl.Phys.Lett. vol.48,1986, pp.95-96
    K.Hirabayashi et al., Jpn.J.Appl.Phys. vol.26,1987,pp.1472-1476
18. S.Tanaka et al.,SID 85 Digest, 1985, pp.218-221
19. S.Tanaka et al.,SID 87 Digest, 1987, pp.234-237
20. N.yamauchi et al., SID 87 Digest, 1987, pp.230-233

Design Rules for Thin Film Electroluminescent Display Panels

L.L. Hope

GTE Products Corporation, 60 Boston Street, Salem, MA 01970, USA

1 Introduction

The formal procedures for optimization of nonlinear systems in the presence of constraints are applied in this paper to the design of the thin film electroluminescent (TFEL) display panel. A general method is presented and specific examples worked through.

The ideal TFEL system would be bright, perfectly uniform, and have unlimited life and reliability. Many possibilities exist for realization of TFEL systems. The task of the designer is to arrive at the best design for a particular purpose. He has specifications for system performance, lifetime, and cost. Thin film materials with particular physical properties and semiconductor drive elements with particular specifications are available. A method is required to resolve the conflicting requirements and choose a design.

Rules are developed below for determination of the optimum TFEL structure in a design situation in which material properties (including a figure of merit for the phosphor) and drive electronics capabilities are given. Specification of operating luminance for on pixels, maximum driver voltages and currents, and tolerable nonuniformity due to column fade do not define a unique set of film thicknesses, but provide an set of constraints. Within this constraint envelope the design rules allow selection of the structure that provides best reliability in terms of resistance to electric breakdown.

Equivalent circuit models for TFEL systems have been discussed extensively in the literature.[1-7] Equivalent circuits are developed by lumping of electrical parameters, i.e. by averaging spatial variations of electrical and photometric variables in single layers, and representing the layers by circuit elements. For simple devices the appropriate element, e.g. a capacitor for a dielectric, is obvious upon examination. For more complex devices the circuit can be derived. The nonlinear electrical behavior of the zinc sulfide phosphor is modeled by including switching elements such as zener diodes in the circuit. The resulting models have been used in analysis of EL panels and drive circuitry.[8-9]

TFEL structures more complex than the simple dielectric/semiconductor/dielectric stack have been introduced. Multiple dielectric systems are in development for enhanced performance and improved film adhesion and device stability.[10-11] Additional layers may be used for control of ion migration during device life.[12-13] Some designs for color devices have multiple EL cells in intimate contact.[14-15] Dark layers may be added for contrast enhancement.[16] For complex structures the optimum stacks and driving conditions may be difficult to formulate from intuition and rules of thumb. Explicit modeling for these situations is a necessity since operational behavior is sometimes counterintuitive. The methods used here are readily extended.

Performance and reliability factors make conflicting demands on the designer. Thin layers of transparent conductor allow good step coverage and breakdown protection for subsequent layers but thick layers avoid column fade. Thin dielectric layers enhance optoelectric performance but reduce breakdown strength. Thick phosphors are brighter but increase drive voltage, which is limited by driver capability. Novel device designs, e.g. transparent conductor planarization either by recessing the lines into the glass or filling in gaps between lines, can in principle change the trade-off balances, but the conventional TFEL structure[17] without electrode planarization and with the basic dielectric/phosphor semiconductor/dielectric working layers has been used for

Springer Proceedings in Physics, Vol. 38 **Electroluminescence**
Editors: S. Shionoya and H. Kobayashi

many years in device development and commercialization. For this reason the structure examined in detail in the paper is of this type.

2 Design Parameters

The procedures are used to optimize film properties to meet design goals given parameters available for semiconductors and materials. Table 1 below lists the parameters used as input to the optimization.

Table 1. TFEL design input parameters

Symbol	Parameter	Unit
n_c	Number of Columns	
n_r	Number of Rows	
p_c	Column Pitch	m
p_r	Row Pitch	m
w_c	Column Width	m
w_r	Row Width	m
V_{rmax}	Row Driver Maximum Voltage	V
V_{cmax}	Col Driver Maximum Voltage	V
i_{max}	Row Driver Maximum Current	A
ρ	ITO Sheet Resistance	Ω/□
f_r	Refresh Rate	Hz
κ_p	ZnS Relative Dielectric Const	
E_T	ZnS Threshold Field	V/m
β	Phosphor Figure of Merit	fL-m/C
κ_d	Insulator Relative Diel Const	
B_{op}	Design Operating Brightness	fL
B_{min}	Minimum Brightness	fL

The optimization is carried out using the phosphor thickness ℓ_p and the total thickness of all dielectric layers ℓ_d as variables. With these known, the threshold and modulation voltages and pixel capacitances are determined. Table 2 lists the parameters obtained as output.

Table 2. TFEL design output parameters

Symbol	Parameter	Unit
ℓ_p	Phosphor Thickness	m
ℓ_d	Dielectric Thickness	m
V_{thr}	Threshold Voltage	V
V_{mod}	Modulation Voltage	V
A	Pixel Area	m^2
C_d	Pixel Dielectric Capacitance	F
C_p	Pixel Phosphor Capacitance	F
C_t	Pixel Total Off Capacitance	F

The pixel area A is $w_c * w_r$. The phosphor figure of merit β is an intrinsic material parameter characterizing the ability of the phosphor to produce light from transported charge. The brightness is given by:

$$B = \beta \frac{\Delta Q}{A} \ell_p, \qquad \Delta Q = C_d (V - V_{thr}). \tag{1}$$

Normal operating brightness is obtained with V_{mod} as the modulation voltage:

$$B_{op} = \beta \frac{C_d V_{mod}}{A} \ell_p. \tag{2}$$

The capacitances are given by:

$$C_d = \kappa_d \epsilon_o A / \ell_d, \qquad C_p = \kappa_p \epsilon_o A / \ell_p, \tag{3}$$

$$C_t = \frac{C_d C_p}{C_d + C_p}. \tag{4}$$

3 Threshold Analysis

At threshold, the voltage across the pixel has reached the value providing a field E_T in the phosphor. It is easily shown that the relation between the pixel capacitance in the off state and the threshold voltage is given by

$$C_t = \frac{\kappa_p \epsilon_o A E_T}{V_{thr}}. \tag{5}$$

4 Pulse Length

The maximum time available to charge the row is the reciprocal of the refresh rate times the row count. If a split column design is used to allow independent drive of two groups of rows, the row count for a group is used. Some of the time is used for switching of semiconductors and phosphor luminescence processes. As a general rule the phosphor is fast but drive semiconductor rise time must be considered.

5 Constraints and Optimization

Relations between feasible values for the thicknesses of the phosphor and dielectric films will be derived from the limitations on column voltage, row voltage, row current, and column fade. These are constraints (dominance relations) rather than equations and represent the limitations usually encountered. Additional constraints can be added if further limiting conditions are noted, e.g. on column currents. The constraints determine the range of feasible values of the phosphor and dielectric thicknesses ℓ_p and ℓ_d. In the event that no feasible thicknesses are obtained, it is inferred that the design is too ambitious for the materials and drivers available.

Panels built with any feasible choices of phosphor and dielectric thickness can be operated at the required brightness with appropriate row and column voltages. Optimization proceeds by using the largest feasible dielectric thickness to afford maximum coverage of pinholes and asperities, thus allowing optimum protection against electric breakdown.

6 Column Driver Voltage Constraint

The first constraint is derived from the requirement that the modulation voltage be less than the column driver maximum V_{cmax}. This condition may be written as

$$\frac{B_{op} A}{\beta \ell_p C_d} \le V_{cmax}, \tag{6}$$

and rewritten in terms of the thicknesses ℓ_p and ℓ_d as

$$\ell_d \leq \frac{\kappa_d \epsilon_o \beta}{B_{op}} V_{cmax} \ell_p . \tag{7}$$

The constraint may be understood by noting that above threshold the phosphor capacitance is switched out, leaving only the capacitance due to the dielectrics alone to provide impedance to the voltage source.

For a given phosphor thickness, if the dielectric capacitance is too low the charge at full modulation voltage is insufficient to reach the specified brightness. Therefore the feasible dielectric thickness has an upper limit. For a thicker phosphor higher brightness is obtained for a given charge, and lower dielectric capacitance is allowable, providing a higher upper limit.

Figure 1 shows the constraint graphically. The arrow points into the feasible region. Considering only column driver voltage, any point to the right of the constraint line is feasible.

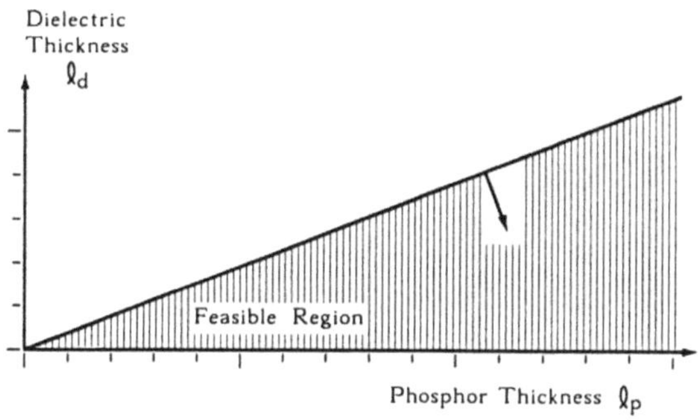

Figure 1. Column Driver Voltage Constraint

7 Row Driver Voltage Constraint

The row driver has to operate at the sum of the threshold and modulation voltages, leading to the constraint:

$$V_{thr} + V_{mod} \leq V_{rmax} . \tag{8}$$

Combining (2), (3), (4) and (5) a quadratic constraint on ℓ_p and ℓ_d is obtained:

$$\ell_d \leq \ell_p \frac{(V_{rmax}/E_T) - \ell_p}{[B_{op}/(\beta \kappa_d \epsilon_o E_T)] + (\kappa_p/\kappa_d) \ell_p} . \tag{9}$$

Figure 2 shows this constraint. The arrow indicates that the feasible region lies between the curve and the ℓ_d axis. For reasons similar to those for the column driver an upper limit on ℓ_d is again obtained for a given ℓ_p. A limit on ℓ_p itself is also obtained where the curve crosses the axis on the right. For values of ℓ_p larger than this limit, device capacitance is so small that charge required to meet the brightness specification cannot be reached at maximum voltage even with no dielectric present.

8 Row Driver Current Constraint

The worst case for row driver current is that for which all pixels are selected on, since the row capacitance is largest for that condition. The row functions as a transmission line of length L, driven at $x = 0$ with voltage V_o applied at time $t = 0$. Voltage along the line is the solution to

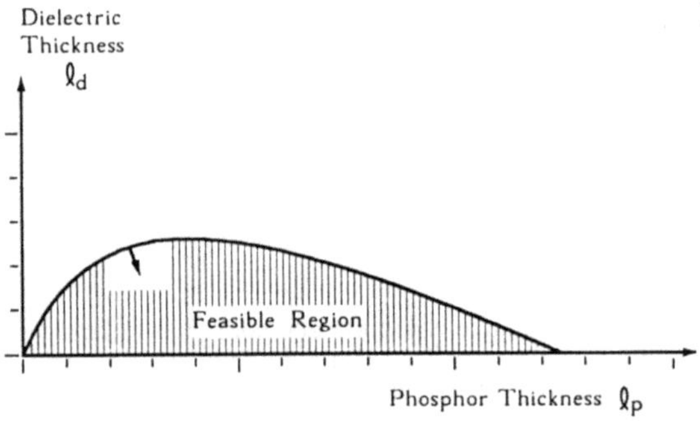

Figure 2. Row driver voltage constraint

$$\left[\frac{\partial^2}{\partial x^2} - \Gamma \frac{\partial}{\partial t}\right] V(x,t) = 0\,, \qquad \Gamma = \begin{Bmatrix} \Gamma_1 = R_p\, C_d/p_c^2 & \text{for } V(x,t) > V_T \\ \Gamma_2 = R_p\, C_t/p_c^2 & \text{for } V(x,t) < V_T \end{Bmatrix}, \tag{10}$$

where R_p is the resistance of the pixel (length p_r) along the row. Equation 10 represents a form of nonlinear diffusion in which the diffusivity increases precipitously from Γ_1 to Γ_2 at a threshold concentration. The point x along the row that is just at threshold moves from $x = 0$ at $t = 0$ to $x = L$ at a time t_L. For most displays t_L is short compared with the pulse lengths, but for large systems this need not be the case. Figure 3 depicts the value of the dimensionless parameter $\beta_L = \pi^2 t_L/\Gamma_1 L^2$ as a function of the voltage ratio V_T/V_o.

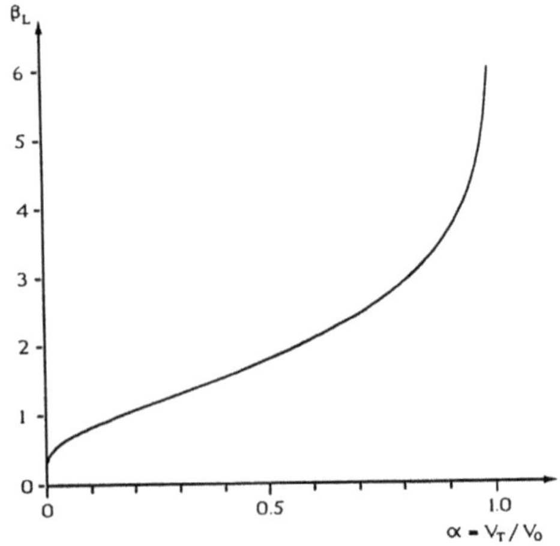

Figure 3. Dimensionless parameter β_L vs threshold to drive voltage ratio

An estimate of the peak current is provided by

$$i = n_c\, C_d \frac{V_{thr} + V_{mod}}{\tau_o} \leq i_{max}\,, \tag{11}$$

where τ_o is a characteristic time for charge inrush. It may be estimated from the transmission line model above. This leads to another quadratic constraint:

$$\ell_d \geq \frac{(\kappa_d/\kappa_p)\ \ell_p^2}{\lambda\, \ell_p - [B_{op}/(\beta\, \kappa_p\, \epsilon_o\, E_T)]}\,, \tag{12a}$$

$$\lambda = \frac{\tau_o\, i_{max}}{n_c\, \kappa_p\, \epsilon_o\, A\, E_T} - 1. \tag{12b}$$

This constraint does not provide upper bounds on ℓ_p and ℓ_d since large thickness values will reduce line capacitance and decrease row current.

9 Column Fade Constraint

A final constraint is provided by the requirement that minimum brightness exceed B_{min}. Even with perfect uniformity of material properties and film thicknesses, transmission line losses can reduce brightness at column ends to unacceptable levels.

A full treatment requires transmission line analysis.[1, 5] An estimate of the constraint curve is provided by

$$C_t \le \frac{\tau_c}{n_r R}, \tag{13}$$

where τ_c is a characteristic time for the end of the column to reach sufficient voltage to attain brightness B_{min}. The column resistance R is given by

$$R = \rho\, n_r\, p_r / w_c\,. \tag{14}$$

This translates to

$$\ell_d \ge \kappa_d\, \epsilon_o\, A\, n_r\, R/\tau_c - (\kappa_d/\kappa_p)\, \ell_p\,. \tag{15}$$

The constraint insures that the column capacitance is low enough to allow the column end to reach B_{min} in the available time.

10 Practical Results

The combined effect of the constraints is best illustrated with an example. Assume that a TFEL panel is to be built with 512 rows and 640 columns using line pitch of 0.25 mm and line width of 0.125 mm for all electrodes. The device is to run at 30 fL with a maximum of 5 fL loss from column fade.

The phosphor has $E_T = 185\times10^6$ V/m, $\kappa_p = 12$, and $\beta = 5\times10^9$ fL-m/C. The dielectric is silicon oxynitride with $\kappa_d = 6.5$. ITO with sheet resistance of 10 Ω/□ is used.

Semiconductor drivers are available with maximum column voltage 40 V, maximum row voltage 230 V, and maximum row current 50 mA. The system is to be refreshed at 60 Hz.

Figure 4 shows all of the constraints for this example. The optimum is determined by the intersection of the constraint curves for row and column voltage. ITO sheet resistance is low

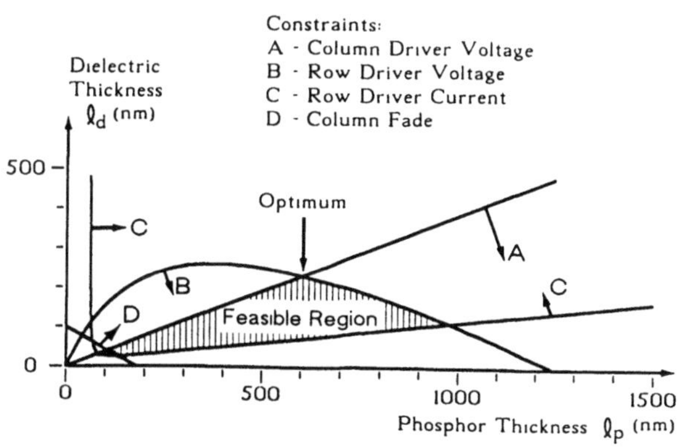

Figure 4. Constraints on film thicknesses

enough to keep column fade from limiting the thicknesses. Row current capability is also good enough to prevent limitation.

It can be shown that for the case illustrated by the example, where row current and column fade do not play a role, that the best feasible thicknesses are those that correspond to operation at $V_{mod} = V_{cmax}$ and $V_{thr} + V_{mod} = V_{rmax}$. The highest voltages compatible with driver reliability should be used. If the best available row current or ITO sheet resistance is not good enough, the designer may be forced use lower voltages and lose breakdown protection.

11 Conclusion

This work provides the device designer means to determine feasible operating performance parameters. Performance issues specific to given device types, i.e. issues that fall within the designer's charter, are isolated from those of importance to those conducting research on improved materials. Extension to more complex devices, e.g. with multiple colors on the same substrate, is carried out by developing or modifying constraints. Optimization variables other than ℓ_d and ℓ_p may be introduced, with the analysis carried out in a space of higher dimension than two. Performance may be optimized for devices not yet built.

12 References

1. L.L. Hope, Proceedings of the Sixth International Display Research Conference, Tokyo, 1986, pp. 260-263.

2. J. Kuwata, Y. Fujita, T. Tohda, T. Matsuoka, M. Nishikawa, T. Tsukada, and A. Abe. Proceedings of the Sixth International Display Research Conference, Kobe, 1983, pp. 128-131.

3. W.E. Howard, IEEE Trans Electron Devices ED-24, No. 7, July 1977, pp. 903-908.

4. M.R. Miller, E. Schlam, R.P. Tuttle and R.M. True. Society for Information Display Digest, 1982, pp. 120-121.

5. L.L. Hope, J.L. Plumb, and D.H. Baird. Proceedings of the Third International Display Research Conference, Kobe, 1983, pp. 582-585.

6. P.M. Alt, Proceedings of the Society for Information Display, 25/2, 1984, pp. 123-146.

7. W.G. Runyan and G.L. Vick, SPIE Advances in Display Technology VI (1986), 624, pp. 66-72.

8. M.R. Miller and R.P. Tuttle, SID International Symposium Digest, XII, 1981, pp. 26-27.

9. G.W. Draper, MSEE Thesis, Southern Methodist University, Dallas, TX, 1986.

10. Y. Fujita, J. Kuwata, M. Nishikawa, T. Tohda, T. Matsuoka, A. Abe, and T. Nitta. Proceedings of the Third International Display Research Conference, Kobe, 1983, pp. 76-79.

11. Y. Sekido, M. Koizumi, J. Mita, T. Hayashi, M. Kazama, and K. Nihei, SID International Symposium Digest, XVII, 1986, pp. 171-176.

12. J. Watanabe, M. Wakitani, S. Sato, and S. Miura, SID International Symposium Digest, XVIII, 1987, pp. 288-291.

13. M. Nishikawa, T. Matsuoka, T. Tohda, Y. Fujita, J. Kuwata, and A. Abe, SID International Symposium Digest, XIX, 1988, pp. 19-22.

14. Y. Oishi, T. Kato, and Y. Hamakawa, Proceedings of the Third International Display Research Conference, Kobe, 1983, pp. 57-60.

15. W.A. Barrow, R.T.Tuenge, and M.J. Ziuchkovski, SID International Symposium Digest, XVII, 1986, pp. 25-28.

16. T. Matsuoka, M. Nashikawa, T. Tohda, and A. Abe, IEEE Transactions on Electron Devices, ED-33, No. 9, September 1986, pp. 1290-1293.

17. T. Inoguchi and S. Mito, in Electroluminescence, ed. by J.I. Pankove, (Springer, Berlin, Heidelberg, 1977) pp. 197-210.

Power Consumption of Thin-Film Electroluminescent Matrix Display

J. Kuwata, E. Ozaki, Y. Fujita, T. Tohda, T. Matsuoka, M. Nishikawa, T. Tsukada, and A. Abe

Central Research Laboratories, Matsushita Electric Ind. Co., Ltd.,
3-15, Yagumo-Nakamachi, Moriguchi, Osaka 570, Japan

1. Introduction

In order to reduce power consumption (driving power) of AC thin-film electroluminescent (TFEL) matrix displays, various driving schemes have been studied by several authors |1-5|. Such a demand to reduce the power is increasing for high information-content TFEL panels with pixels more than 1000 x 1000 which are employed as the terminal displays for integral personal computers and engineering workstations. We have already suggested in the previous papers |6,7| that low driving power and low voltage of the ACTFEL matrix displays can be realized by the thin-film structure consisting of high capacitive dielectric layers. Here dielectric layers are composed of the perovskite type oxides with high dielectric constant of more than 100 and the tungsten-bronze type $BaTa_2O_6$ with the dielectric constant 22.

It is also proposed by us that an optimum TFEL structure with small power consumption could be constructed by the following procedure. The allowed panel charging periods, which are determined by the electrode resistance of the TFEL panel, the electric resistance of the driving transistor, and the electric capacitance of electroluminescent pixel, require the capacitance of the dielectric layers to be smaller than a certain value. On the other hand, the power consumption of the panel is determined approximately by only two design parameters, the capacitance of the dielectric layers and the thickness of the phosphor layer. If the luminance and the luminous efficiency are assumed to be kept constant, smaller power consumption of the panel could be attained by employing the dielectric layers with capacitance as high as possible. An optimum thin-film structures with both the desired emission characteristics and the power consumption as low as possible must satisfy the two requirements described above. Prior to the design described above, characteristics of these TFEL panels must be measured in order to clarify the relation between the power consumption and luminous efficiency. After the determination of the capacitance of the dielectric layers, the thickness of the phosphor layer is determined for the desired luminance |8,9|.

In this paper, we describe the comparison of the powers between the calculation based on the suggested theory and the prototype TFEL panel with 1024 x 768 pixels having a pitch of 0.2 mm. We further discuss the relationship between the power consumption and display fill-factor for the TFEL characteristics and driver IC characteristics.

2. Theoretical Calculation for an ACTFEL Matrix Display with 1024 x 768 Pixels

Figure 1 shows the thin-film structure of a typical ACTFEL matrix display. It consists of a phosphor layer sandwiched between two dielectric layers. The relationship between the luminance L and the charge transferred through the phosphor layer ΔQ is represented by the following equation,

$$L = Ls\ (\ 1 - \exp(-\Delta Q/\Delta Qs)), \tag{1}$$

Springer Proceedings in Physics, Vol. 38 **Electroluminescence**
Editors: S. Shionoya and H. Kobayashi

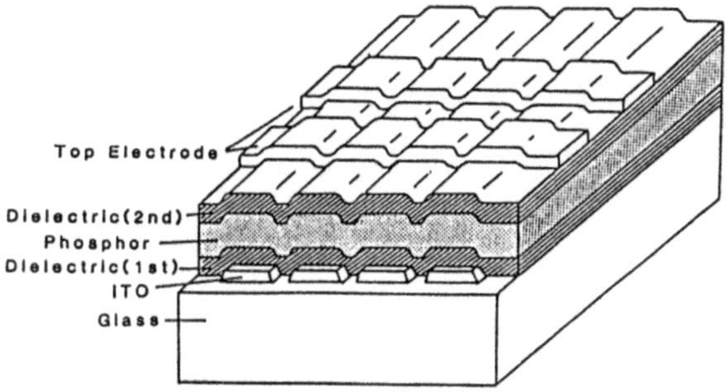

Fig.1 Structure of an ACTFEL panel

where Ls and ΔQs are constants determined by the experimental result as a function of the driving frequency F in each of TFEL structures. Then the luminous efficiency η can be represented by the following equation,

$$\eta = \frac{\pi\ Ls}{4\ Eth\ dz\ F\ \Delta Qs} \exp(-\Delta Q/\Delta Qs) \ , \qquad (2)$$

where Eth is the electric field strength at emission threshold of the phosphor layer, dz, the thickness of the phosphor layer, and F, the driving frequency.

In our previous report |2|, the power consumption (driving power) P was fully approximated by the following equation:

$$\begin{aligned} P = F\ A\ [\ & K_1\ \Delta Q^2 (Ci^{-1} + dz/\varepsilon_z)^{-1} Ci^{-2} \\ & + K_2\ (Ci^{-1} + dz/\varepsilon_z)\ \varepsilon_z^{\ 2}\ Eth^2 \\ & + K_3\ (Ci^{-1} + dz/\varepsilon_z)\ \Delta Q\ \varepsilon_z\ Eth \\ & + K_4\ \Delta Q\ Ci^{-1}\ \varepsilon_z\ Eth\ ]\ , \end{aligned} \qquad (3)$$

where A is a luminous area; $Ci = (d_1/\varepsilon_1 + d_2/\varepsilon_2)^{-1}$; ε_z, ε_1, and ε_2, the dielectric constants of the phosphor, the 1st and 2nd dielectric layers, respectively; d_1, d_2, the thickness of the 1st and 2nd dielectric layers, respectively. The coefficients K_i (i=1,2,3,4) are determined from the driving schemes represented in Table 1 |3-5|.

Table 1 Coefficients of relational equations of driving power for three driving methods.

Driving Scheme	K_1	K_2	K_3	K_4	
1	2N(1-m/M)+1	2	4m/M	2-6m/M	Ref.3
2	N	1+m/M	2m/M	-2m/M	Ref.4
3	4m/M(1-m/M)(N-1)	2	4m/M	-2(1+m/M)	Ref.5

m: number of 'ON' pixels, M: number of columns, N: number of rows

From the viewpoint of low power consumption and short charging period, we have found out previously that the 3rd type driving scheme was the best of the three |2|. Figure 2 shows the dz dependence of P(m/M=0.5) with Ci as parameter. The value of P has the minimum as a function of dz in each Ci. Increasing the Ci value, the minimum P values decrease, although the value of Eth, total capacitance of EL pixel per unit area Ct, $\Delta V(=\Delta Q/Ci)$ and threshold voltage Vth depends on the value of dz as shown in Fig.3.

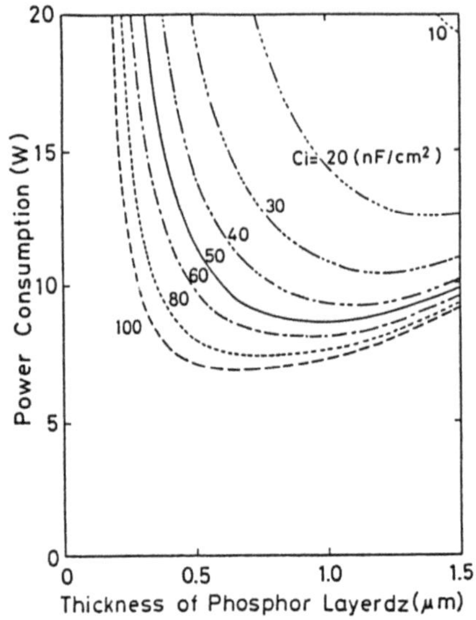

Fig.2 Power consumption vs. dz as a function Ci.

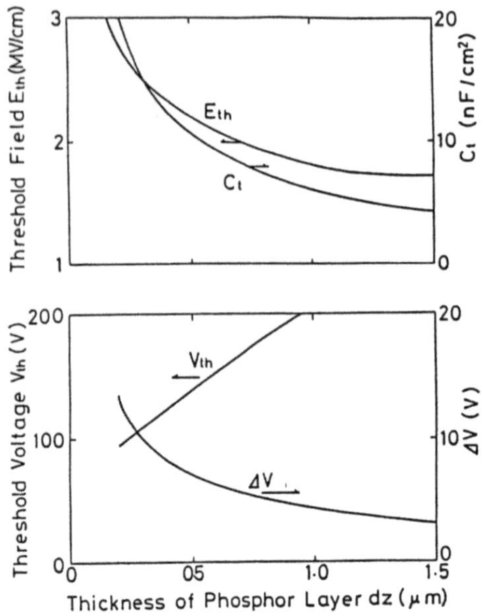

Fig.3 Eth, Ct, Vth, and ΔV as a function of dz.

3. ACTFEL Panel Charging Period

The uniformity of luminance within luminous area in the 1024 x 768 pixel ACTFEL panel depends on the spatial deviation of the electric properties from the specified ones for each thin film, and also spatial difference of the charging period T(x). T(x) is the charging period till the luminous area of the ACTFEL matrix display is charged up to x % of the saturated value of ΔQ. The allowable pulse width to drive this ACTFEL matrix panel when operated at 60 Hz must be less than 12 μs, which should be determined from the driver IC characteristics. The T(x) could be approximately simulated by using a π-type transmission line model coupled by the electric capacitance of each EL pixel and the resistances of the electrodes and the driver ICs. The characteristics of the charging period T(95%) as a function of Ci are shown in Fig.4, where the following conditions are used: Al sheet resistance is 0.2 Ω/square, ITO sheet resistance is 2 Ω/square, row driver resistance is 150 Ω, the maximum row current is 100 mA, column driver resistance is 200

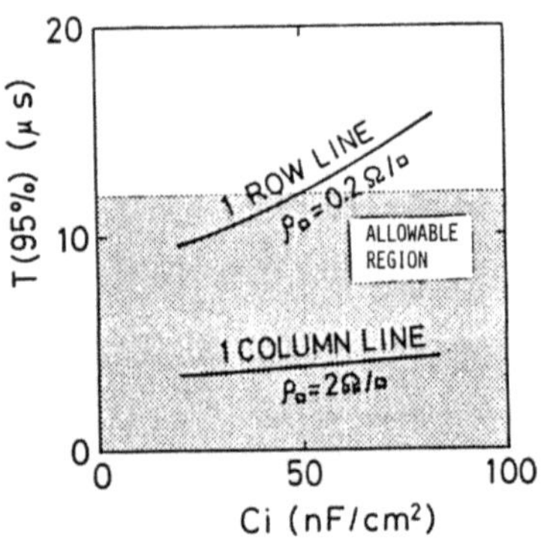

Fig.4 Charging period T(95%) vs. Ci.

Ω, and the maximum column current is 15 mA. Comparing the charging period T(x) calculated in Fig.4 with the allowable value of 12 μs, the allowable maximum value of Ci is determined to be less than 50 nF/cm^2.

4. Prototype Results

Figure 5 shows the power consumption of the developed 1024 x 768 ACTFEL panel consisting of ITO(indium tin oxide)-$Sr(Zr,Ti)O_3$-CaS-ZnS:Mn-CaS-$BaTa_2O_6$-Al. Good agreement is obtained between the calculated (bold solid line shown in Fig.5) and the measured (shadow region in Fig.5) values. However, a difference was observed between the calculated and the measured values at the point of all 'OFF' state. Optimum driving circuits and the mounting arrangement are now under development. We have also obtained that the P depends strongly on the uniformity of Ci and dz values within the luminous area as shown in Fig.6. It should be noted that the uniformity of dz value is more significant than that of Ci value.

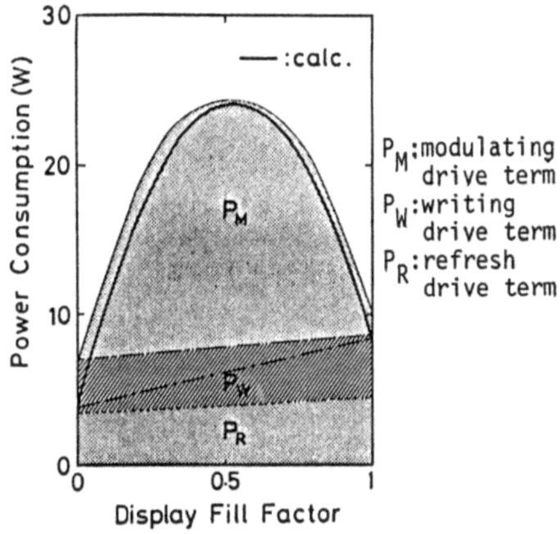

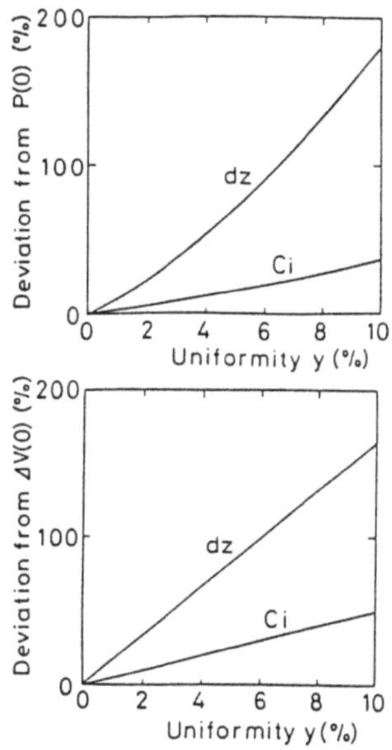

△
Fig.5 Power consumption vs. display fill-factor.

▷
Fig.6 Deviation of P and ΔV vs. uniformity of Ci and dz.

5. Conclusion

Operation of 1024 x 768 ACTFEL panel at low power consumption of 24 W has been accomplished by optimizing the TFEL device structure and drive conditions: Capacitance of dielectric layers is as high as 50 nF/cm^2, phosphor layer thickness is 700 nm, threshold voltage is 155 V and modulating voltage is 30 V.

References

1. M.L.Higgins: Digest 1985 S.I.D. Int. Symp., p.226 (1985)
2. J.Kuwata, Y.Fujita, T.Tohda, T.Matsuoka, M.Nishikawa, T.Tsukada, and A.Abe: Proc. of the 6th Int. Display Research Conf. (Kobe, Japan, 1986) p.128
3. Y.Kanatani, H.Kishishita, and M.Kawaguchi: NIKKEI Electronics, 209, 118 (1979)
4. T.Ohba, S.Harada, Y.Fujioka, Y.Kanatani, and H.Uede: Television Society Technical Report, IPD 95-2, p.7 (1985) (in Japanese)
5. K.Takahara, T.Kawada, H.Yamaguchi, and S.Andoh: Proc. of the 3rd Int. Display Research Conf., (Kobe, Japan, 1983) p.578

6. Y.Fujita, J.Kuwata, M.Nishikawa, T.Tohda, T.Matsuoka, A.Abe, and T.Nitta: Proc. of the S.I.D., 25/3 , 177 (1984)
7. J.Kuwata, Y.Fujita, T.Matsuoka, T.Tohda, M.Nishikawa, and A.Abe: Jpn. J. Appl. Phys., 24 Suppl. 24-2, 413 (1985)
8. L.L.Hope and M.I.Abdalla: Digest 1987 S.I.D. Int. Symp., p.303 (1987)
9. Y.A.Ono, H.Kawakami, M.Fuyama, K.Onisawa: Jpn. J. Appl. Phys. 26, 1482 (1987)

TFEL Matrix Display Design Rules Based on a 3 Part Electrical Model

M.R. Miller and T.G. Kelley

US Army, LABCOM, Fort Monmouth, NJ 07703, USA

I. Introduction

Driving a matrix display device is a very complicated situation to analyze. The thousands of cross coupled components, voltage pulses applied to hundreds of rows and columns and resistive, capacitive and non-linear currents defy straightforward circuit analysis. This paper will show that simplifying assumptions can be chosen which separate the problem into analyzable sub-parts with negligible loss of precision. Taken together, the parts of the model are sufficient to allow accurate calculation and design of panel drive parameters such as maximum refresh rate, minimum pulse width, power consumption, brightness and contrast.

II. Equivalent Circuits

A. The DC case, the 4 capacitor model

If we assume that all pixel capacitances C_e are equal and row and column electrode resistances can be neglected as can the non-linear current in the on pixels, then we can calculate the voltage distribution among the pixels at the time when the drive pulse has been on long enough that the panel is fully charged. A panel of N rows and M columns becomes a purely capacitive array of $M \times N \times C_e$. In a typical drive method, one row is driven along with the columns corresponding to the desired "on" pixels in that row. All of the pixels fall into one of four categories of on and off rows and columns. Let m be the number of "on" columns, then the pixels can be lumped into their categories as in the diagram below.

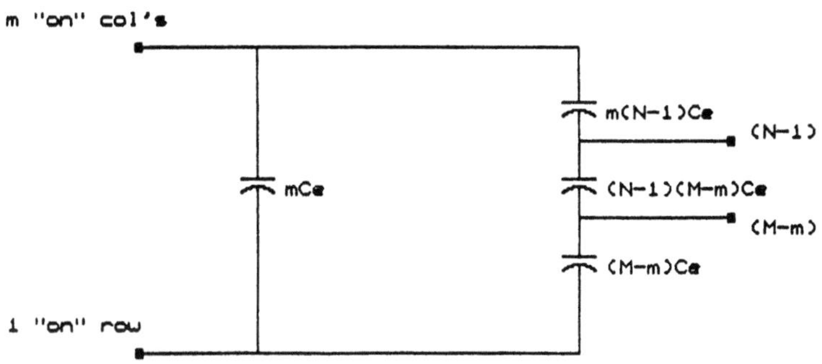

Figure 1. Four Capacitor Model

B. Row and Column electrodes, distributed R-C

Having established the steady state voltages on the pixels after a long pulse, the next logical question is how long that pulse must be to reach the steady state condition. For this calculation, it is only necessary to consider one row or column at a time. The row or column is approximated by an R-C ladder network where each R is 1/N of the total resistance of an electrode and each C is C_e. The ladder network has been analyzed many times in terms of its impedance and transfer function for sinusoidal inputs but this does not directly apply to pulse waveforms. We have used the computer programs, "Sceptre" and "Spice" to provide the resulting waveforms and can thus show that the time constants of this circuit are much smaller than previously estimated. Calculations for the last node in the ladder indicate that for any number of nodes above 25, the network closely approximates the totally distributed case. The last node in the network reaches 85% of

III. Voltage Division

A. Avoiding crosstalk

The ability to matrix address a display depends primarily on the threshold slope of the display medium and the division of the voltages in the model of figure 1. The division of the drive voltage among the capacitances of the "off" pixels depends on the number of "on" columns m, and on the condition of the "off" electrodes. In a typical drive approach, the "off" rows are floating and the "off" columns are grounded, the (M-m) "off" pixels in the "on" row will have the row drive voltage applied to them. At the same time, the "off" pixels in all of the remaining rows will divide the "on" column modulation voltage V_m among them as:

$$V(\text{"on" col's - "off" rows}) = ((M-m)/M)V_m$$

and

$$V(\text{"off" col's - "off" rows}) = (m/M)V_m$$

B. Energy and power

To calculate power consumption in a display we can start with the energy produced by the voltages on the four capacitors of the model of figure 1 times the row addressing rate. Defining a dissipation factor D as total energy in, divided by energy stored, D would have a maximum of 2 for a step function. The minimum case is for a constant charging current I = CV/T where T is the charging time. With recovery of the energy stored on the capacitor, a minimum of 4RC/T can be approached.

IV. Electrode R-C analysis

When a step voltage function is applied through a lumped resistor, R, to a lumped capacitor, C, a time period equal to 3RC is needed for the voltage on the capacitor to reach 95% of the applied voltage. When the resistance and capacitance are distributed as in the ladder model, the charging time is reduced for the same total values of resistance and capacitance. Graphs of computer simulated waveforms as a function of time at various points along the ladder indicate full response in much less time than 3RC, even at the far end of the ladder. This means that pulses on the order of RC provide sufficient time. Because of the low resistance of the aluminum row electrode the pulse waveform is not significantly degraded in passing through the ladder network. The column electrode represents a very different situation.

The analysis of the pulse waveform propagated down a single column depends on both the 4-capacitor model and the ladder network model. The m "on" columns cross only 1 "on" row which can be thought of as an ac ground with a dc bias and we will ignore the dc in the ac waveform analysis. These m columns have another path to ground, through the N-1 -"off" rows to the M-m grounded "off" columns. Since the row drivers, in the "off" state, have a high impedance, the voltage of these rows will float. The "on" column to "off" row capacitance is therefore in series with the "off" row to "off" column capacitance in this path. The effective capacitance of the m "on" columns depends on the value of m and is the sum of the capacitance to the "on" row plus the series combination to the "off" columns which equals:

$$mC_e + m(M-m)(N-1)C_e/M$$

The effective capacitance for one "on" column is 1/m times this value, or:

$$C_{eff} = C_e + C_e(N-1)(M-m)/M$$

which can be very large when m is small but decreases toward C_e as m approaches M. Using this C_eff in the ladder network produces a worst case (largest capacitance) when m = 1, that is, there is only one "on" pixel in the driven row. Assuming a somewhat limiting case where N = 512, M = 640, C_e = 7 PicoF and the total resistance of the column, R = 10,000 Ohms, the voltage waveforms for m = 1 are shown in figure 2. A considerable improvement in the waveforms, shown in Figure 3. results when m = M/2, or one half of the pixels in the driven row are turned on. Observed luminance differences between the column electrodes driven at the near end and those driven at the far end near the edges of a display can be partially explained as resulting from the delay seen in Figure 2., putting the column pulse partially outside of the row pulse, rather than being totally caused by a degradation of column voltage. This can be corrected to some extent by using a row pulse delayed by an amount equal to the maximum delay of the column pulse.

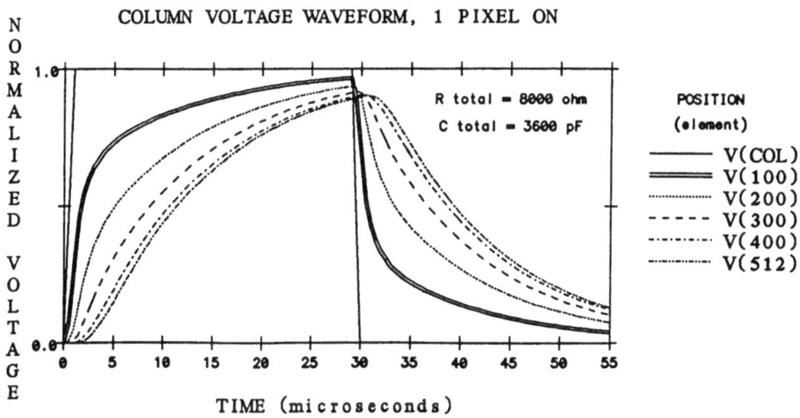

Figure 2. Column Voltage Waveforms - 1 Pixel on

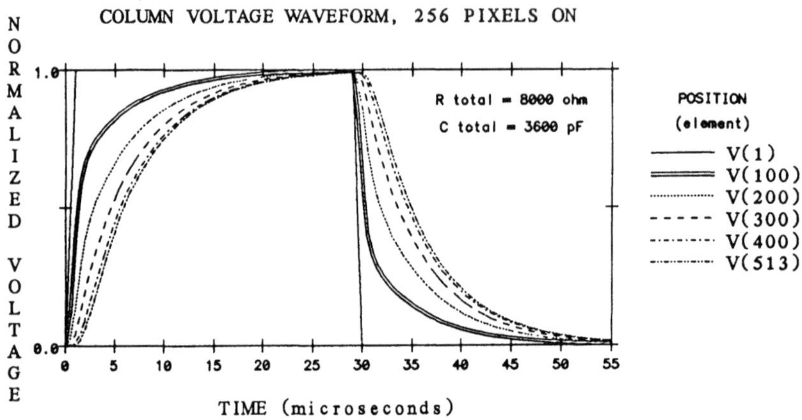

Figure 3. Column Voltage Waveforms - 256 pixels on

V. Lighting the pixel

A. Effect of Real Current on the RC Model

The RC model, above, ignored effects caused by the real current in the "on" pixels. An "effective capacitance" for an "on" pixel is Q/V in the on condition. In a typical example, Q/V for a device at luminance saturation is measured at 60% higher than the normal capacitance for the device. In the RC model for the column, this effective capacitance will affect only one pixel in the column, since only one row is driven during any column pulse. For the column model, then, the real pixel current is truly negligible in calculating the charging waveforms. The effective capacitance for the row will depend on the number of "on" pixels in the row, m, according to:

$$C_{row} = (M-m)c_e + mQ/V$$

B. The pulse train

A pixel does not achieve full brightness with the first pulse of "on" voltage but takes several pulses to build up. At a 60 Hz frame rate this takes only a fraction of a second. Similarly, because "off" pixels are still being addressed with row pulses below threshold, light output does not immediately fall to zero when an "on" pixel is turned off. It, too, takes several pulse times for the excitation to decay. This effect does not alter the steady state or R-C models and so will not be addressed here. It was covered previously in Reference 2.

VI. Putting the models together

The overall electrical model comes down to analyzing how the R-C effects of the electrodes can be related to the four capacitor model of the total panel since the real current in the pixels, being a small fraction of the total can be ignored. The row resistance is typically very low and, for analysis of the row voltages and currents, the column resistances are in parallel and therefore, appear very low also. This effective low resistance makes the R-C delays in the row circuit very small in comparison to the columns. The limitations on the panel refresh speed and other parameters are therefore defined by the columns. The columns can be analyzed independently of the row pulse because the one row being driven has little effect compared to the currents to the other N-1 rows that each column crosses. Since the "off" rows are open circuited, they will float to a voltage that is an average of all the column voltages, $V_m m/M$. The major part of the current in a panel is that which flows into the "on" columns and out of the "off" columns through the capacitance equal to:

$$C=C_e m(M-m)(N-1)/M.$$

The value of m determines the final voltage across all of the capacitors in any column except for the one capacitor in the "on" row which can be neglected. The model becomes the R-C network with an input pulse having a voltage of $V_m(1-m/M)$. The pulse waveforms along the network follow the waveforms derived from the Sceptre and Spice models.

REFERENCES

1) Miller and Tuttle, "High Efficiency Drive Method for EL Matrix Displays", pg 85, Proceedings of the SID, Vol. 23/2, 1982

2) Miller, Morton and Schlam, "Time Response Characteristics of TFEL Displays", Digest of the International Display Research Conference, Oct 1985.

Degradation Processes in Thin Film Electroluminescent Devices

R. Mach and G.O. Mueller

Central Institute for Electron Physics, Academy of Sciences of GDR, Hausvogteiplatz 5, DDR-1086 Berlin, GDR

1. Introduction

In his famous paper INOGUCHI /1/ as early as 1974 considered lifetimes of thin film electroluminescent devices (ELD), well aware of the fact that Destriau's EL powder samples had dis-credited EL by their low maintenance. What he claimed in /1,2/ - 10.000 h operation at surprisingly high frequencies and even at elevated temperatures - was to raise interest and stimulate new hopes and activities worldwide. Nevertheless it was a long way to production of information displays, not only because of the driver problems. The fact that even today new insulator materials for the 'simple' ZnS:Mn based ELD are invented (patented) and published, claiming better lifetimes and/or higher storable charge, proves that an international standardization has not yet been reached /3/. Several papers on degradation /1, 2, 4, 5/ have been published descibing the main symptom - an almost rigid shift of the brightness(B)-voltage(V)-characteristic along the V-axis. No explanations however have been offered.

While ZnS is a relatively simple host material for Mn, special effects might be expected, if it is doped with badly soluble lanthanide ions Ln^{3+}. To our knowledge no account of it was given in the literature. Reactivity and interdiffusion is liable to cause problems in 'other color' displays with earth alkaline sulphides as S-film. In one case however a very good result of almost 1000 h of 5kHz operation has been achieved by incorporating thin Si-nitride films adjacent to $CaS:Eu^{2+}$ /6/.

2. Aims and Methods

The main aim of this paper, following a first one /7/ along these lines, is to go beyond the symptoms, and analyze according to today's knowledge of the operational principles, degradation in its processes and causes. This has become possible by using the well-established description of the B(V) characteristic of the ELD

$$B(V) = eta(B,Q)*2f*Q(V) = eta(B,Q)*2f*C_i*(\alpha V - E_t d_s) \quad (1)$$
$$= eta(B,Q)*2f*C_i\alpha(V - V_{th}) \quad (2)$$

comprising the 3 parameters α, E_t, and eta; eta standing for the light yield or efficiency (depending upon B or Q, see Figs.7,11), $\alpha = C_i/(C_i + C_s)$, denoting the voltage fraction, acting below threshold on the active or S-film, and E_t being the threshold field, at which electrons start to be tunnel-emitted from states at the interface between S- and I-films. By measuring B(V) **and** Q(V), the transferred charge per pulse, and fitting a straight line to it, the three parameters can be determined **non-destructively** during the aging process

Springer Proceedings in Physics, Vol. 38 **Electroluminescence**
Editors: S. Shionoya and H. Kobayashi

of every sample /8/. And this has been done, decomposing the degradation into 'α-aging', 'E_t-aging', and 'aging of eta', the light yield. Taking into account the easily measurable total capacitance of a simple test sample, α can be decomposed into C_i and C_s, and further conclusions about degradation causes can be drawn.

A very important further tool in the analysis has been the highly time-resolved measurement of the dissipative current (together with the applied voltage), and the on-line computation of the mean electric field strength in the S-film /7, 8/, yielding the (normalized) current-voltage characteristic of the S-film (as shown e.g. in Figs.6 and 8). The latter readily allows for the classification of the interface into two classes - controlling, in the sense of limitting, the current ('ifc' for interface controlled current) or not ('cc' for charge controlled current). Again the emphasis is on the non-destructiveness of the diagnostics for the changes in **any** of the interfaces /8/, while E_t measures some average over both.

So using these tools, and in special case seeking support from others, it has been possible to gain some insight in various degradation mechanisms of ELDs, which of course allow for the lifetimes stated by the commercial producers for their products, but might be further suppressed, knowing their origin, and/or avoided in other materials combinations.

In /7/ the hypothesis has been put forward, that the degradation is proportional to the total amount of charge transferred through the device in each direction, i.e. proportional to

$$Q_{tot} = f * Q * T$$

with T denotes the time of operation at frequency f with a charge Q transferred in each pulse. This relation, which is likely to apply as long as no appreciable heating occurs, allows for accelerated degradation tests, and so most of the runs have been made under constant charge operation using sinusoidal voltages and a Q-controller /9/.

3. Results

The most general result is, as mentioned above, the shift of B(V) along the V-axis. In a matrix display any shift, depending upon operation time, results in a spotty appearance of the display as soon as the addressing times of pixels have become enough different. As evidenced by the formulae (1), (2) such shifts might have their origin in changes of α or/and of E_t.

3.1 Aging of Thin Film Dielectrics

One of the most unexpected results of degradation studies on ELDs was the proof of electrochemical aging of the used insulator films. As just mentioned, one possible reason for shifts of the B(V) characteristic is a change in 'α'. As 'α' is determined by the I- and S-film capacitances C_i and C_s, the question whether both or which one changes has to be answered. By virtue of equation (1) they can be determined independent of each other, and really we never found any appreciable change of C_s. So we are left with the task to explain rather dramatic changes of C_i, as shown for instance in Fig.1. Within the 6000 50Hz-equivalent-hours both C_i and E_t alter, but in the initial part the decrase of C_i via a decrease of 'α' brings about more than a compensation of the decrease of E_t in its influence on the B(V) shift, at later times the increase of E_t overbalances and V_{th} rises. Dielectrics like this one have been termed 'soft' /7/. In

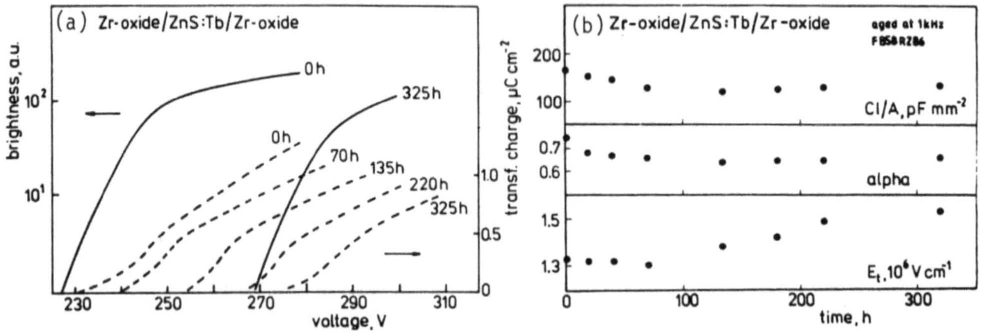

Fig.1 A typical example of a 'soft'insulator sample (a) as seen in B(V) after aging times given; (b) decomposed into the aging dependences of α and E_t, with α changing by C_i-changes only

general they are somewhat lossy, resulting in loss angles around 0.05 at 1kHz, if measured in a complete MISIM stack. This conductance decreases with increasing frequency and with decreasing temperature. Even more pronounced is the temperature dependence of C_i at slightly elevated temperatures (Fig.2), with the total capacity C of the MISIM measured at 1kHz and C_i evaluated using $C_s(T)$ = const, which has been independently verified.

In the sample, to which Fig.1 relates, the 'soft' dielectric is ZrO_2, but Al_2O_3 may be as soft /7/. The technologies checked - EBE, dc and rf sputtering and CVD - might be different with regard to the ease with which they can be controlled not to produce nonstochiometry, but all of them tend to soft films, if not carefully monitored. An extreme case is given in Fig.3. Ta-oxide, rf-sputtered is the material of both I-films in this sample. Even after the first measuring run, dark spots - presumably bubbles of O_2 - have developed. The oxygen excess was directly seen in RBS: the Ta-oxide corresponded to $Ta_2O_{5.9}$! /10/. Further indications and even the direct visual evidence for an electrochemical **reaction** of O-excess from Al-oxide with an Al electrode have been given in /7/.

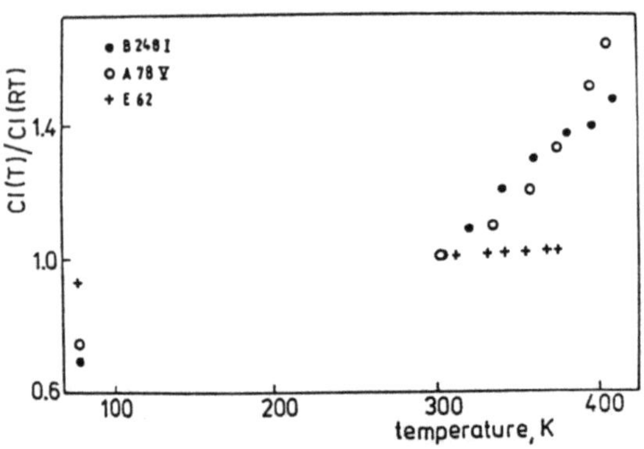

Fig.2 Temperature dependence of C_i, in ratio to the room temperature value, for two 'soft and one 'hard' insulator resp.

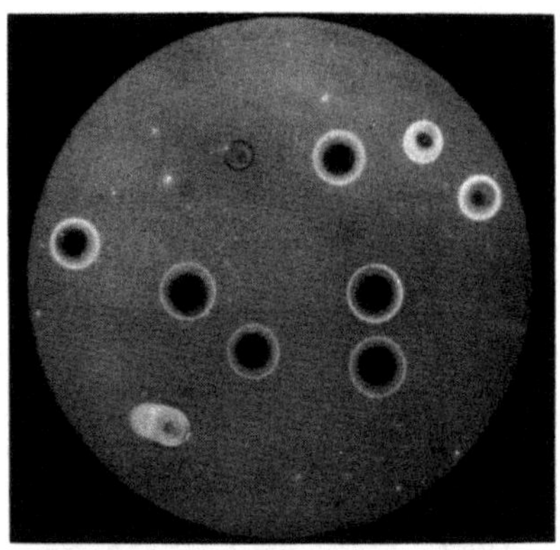

Fig.3 Microphotograph of a sample with soft dielectric films - $Ta_2O_{5.9}$ - after some minutes of operation

Summing up this part and drawing some conclusions, one can state that the low temperature, especially the so called physical deposition techniques (PVD), are prone to produce nonstochiometric, often oxygen-excess oxide films. Oxygen atoms out of these films are able to migrate /7/ in the very high fields of the order of 10^6V/cm even in the short pulse or period times (as the polarization fields are in the same order) to the electrodes and/or interfaces, react, and/or form O_2 bubbles, possibly with the help of available electrons. The net oxygen concentration will tend to equilibrium, thereby changing the total dielectric constant towards smaller values, reducing C_i, and consequently 'α'. All the details and qualitative features will strongly depend upon the material, the interface conditions etc. 'Differential aging'/11, 12/ and even the deterioration of ITO in contact with Ta-oxide /13/ can be understood on the same footing.

Starting from this model for the 'α'-aging, it was tempting to look for the aging with the sample in liquid nitrogen (LNT). Again an extremely soft dielectrics containing sample was used, and really the shift after 100 h/1kHz at LNT is smaller than the one after 6 h/1kHz at room temperature.
Even the E_t changes are slowed down at LNT (Fig.5). However it is far too early to estimate thermal activation energies.

As mentioned already the same starting material can form hard or soft I-films. Figure 4 shows a hard ZrO_2. Almost no changes of C_i and/or 'α' have been found in this case. E_t drifts however to higher values, pushing B(V). Careful control of stochiometry is the key to hard reliable I-films.

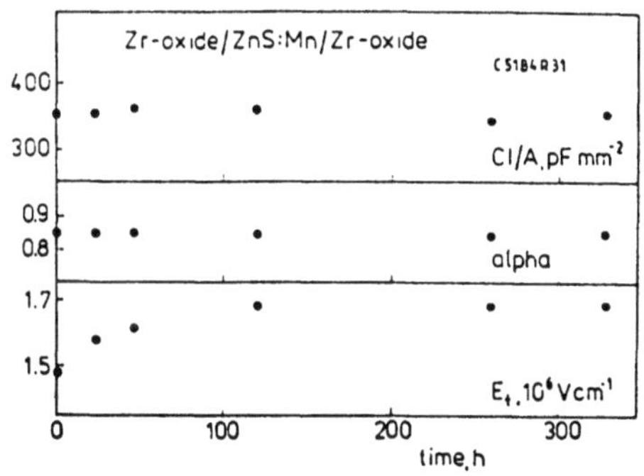

Fig.4 Aging of this sample with a hard dielectric is only affecting E_t

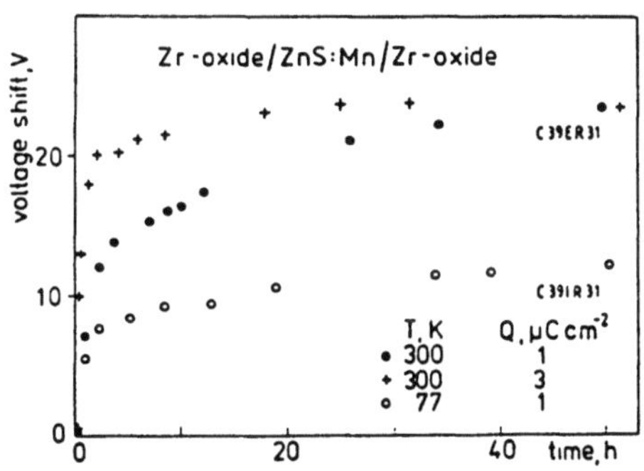

Fig.5 Changes of constant-Q-voltages depend strongly on temperature T and on Q itself

3.2 Aging of Insulator-Semiconductor Interfaces

Changes in the tunnel threshold E_t have their origin in changes of the density of interface states (IF DOS), as outlined in /8/. The most general case of an increase of E_t with operation time can be thought of as an increase of the IF DOS, causing a decrease of the filling margin (quasi Fermi level) for a certain transferred charge. As mentioned in /7/ the two I/S interfaces, even if made from the same materials, are neither equivalent in their genesis nor necessarily equal in their behaviour. While the substrate near interface is formed by condensation of the semiconductor material on a glassy substrate, the substrate-far one grows from insulator material on rather well crystallized semiconductor grains. Nevertheless rather

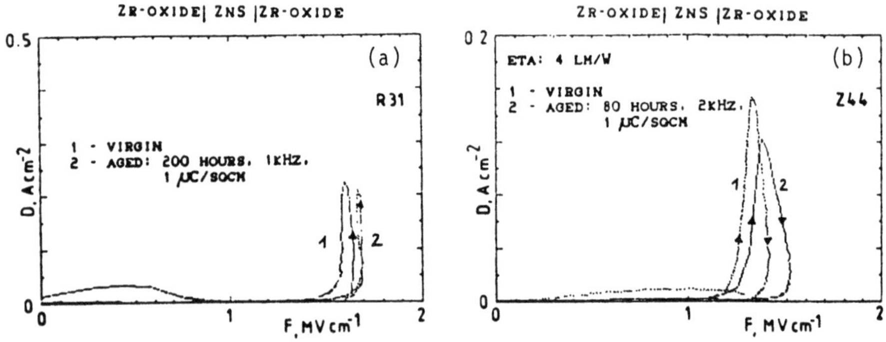

Fig.6 D(F) characteristics showing conservation of (a) **cc**, and (b) **ifc** behaviour and of symmetry. The curves for different polarities coincide within drawing accuracy

symmetric samples exist, and as a very crude rule one can say that asymmetries tend to disappear. No pronounced differences in aging have been observed between cc and ifc interfaces. Figure 6a gives an example where symmetry and cc character are conserved. Figure 6b on the other hand shows conservation of ifc. Quite similar behaviours have been found for S = ZnS:Tb and CaS:Eu resp./14/.

Summing up on interface aging: There is always some initial shift of E_t and B(V), in general to higher voltages, even if hard insulators are used. Forced aging or 'burn-in' might be used to stabilize the interface as quickly as possible, before the pixels go to their individual history. Figure 5 demonstrates that 'stability' is reached sooner, the higher the temperature and the transferred charge are in the initial hours. While the 77K/1μCcm^{-2} aging only results in a 'metastable' state - the shift increases drastically during further room temperature 200 Hz operation - the 300K/3μCcm^{-2} final value on the other hand is a stationary one. The quantitative details again depend strongly on materials and technologies used.

A rather strange but illustrative example of interface degradation was given in /7/. Up to now this kind of aging appears to be limited to samples incorporating CVD/ALE prepared ZnS:Mn. Starting from an asymmetric behaviour with very different E_t values and different

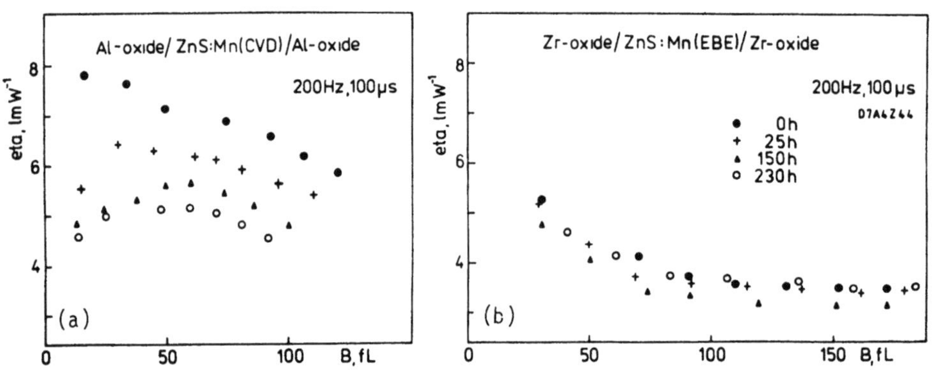

Fig.7 Degradation of EL efficiency eta(B) for (a) CVD grown ZnS:Mn films and (b) EBE deposited ZnS:Mn films in ELD samples

modes, the initially high efficiency after a rather short 50 Hz equivalent operation time had degraded appreciably (Fig.7a). The reason becomes obvious from Fig.8: E_t of the substrate-near interface has decreased to such low values, that the lossfree acceleration of the electrons is no longer guaranteed, and the brightness in this polarity falls off. A possible explanation is the following: because of the CVD process some chlorine is inevitably present in these S-films. The built-in asymmetry mentioned, imposes during operation a static elctric field which tends to drift the Cl^- ions to the sub-strate-near interface, deteriorating it further. This might explain the findings in /15/ too.
During the same aging period there is a seldomly found feature developed in this sample: As Fig.9 demonstrates the high initial steepness of B(V) deteriorates, because of the same premature supply of electrons, which gives rise to an increase of efficiency with voltage, completely unusual in well-made ELDs.

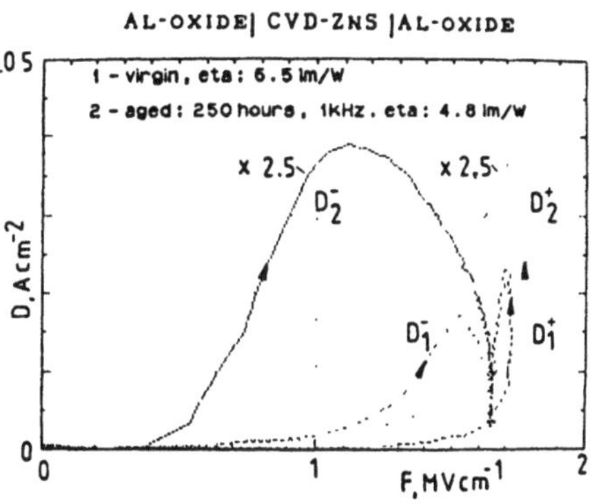

Fig.8 The dramatic degradation of the substrate-near-interface caused the loss of efficiency shown in Fig.7a

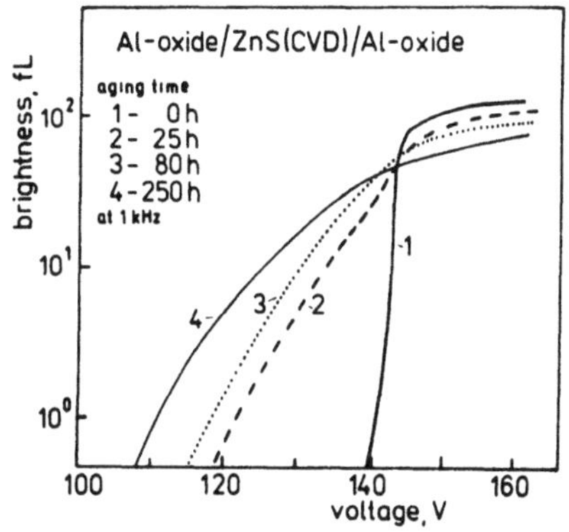

Fig.9 B(V) of the sample of Fig.8 showing a decrease in steepness because of an only gradual rise of efficiency with voltage above E_t

3.3 Degradation of Efficiency

ZnS:Mn : The above described very special mechanism is the only one observed in ZnS:Mn based ELDs, which reduces the efficiency; it closely resembles the one described in /16/ for the drift of Na^+ ions to the (opposite) interface. In the sense of the factorization of efficiency /17/, eta = $eta_{out}*eta_{lum}*eta_{exc}$, it is not the lumines-cent yield which degrades, but eta_{exc}, the fraction of the transfer-red charge, which is able to excite.
It is easy to monitor degradation of $eta_{lum} = t/t_{rad}$, with t being the decay time of emission and t_{rad} the radiative decay time, which is just the inverse of the quantum mechanical transition probability. Figure 7b shows typical behaviour of the efficiency in EBE prepared ZnS:Mn samples. It shows the normal decrease of eta, due to eta_{lum}, with increasing brightness or in physical terms with the concentra-tion of excited dopants /18/, but it does not degrade.

ZnS:Ln^{3+} : There has been some curiosity about the degradation of efficiency in rare earth doped ZnS, but no reports on experimental findings. The curiosity stems from the fact that the solubility of Ln^{3+} is very low in ZnS. Even samples with 0.1 mol% are overdoped and far from thermodynanical equilibrium (TDE), which is a phase mixture comprising Ln-saturated ZnS and other Ln-richer compounds. As any non-TDE system is thought to approach TDE, if energy is deposited, special unwanted effects are suspected. There are no estimates of the time scale on which the approach to TDE should occur.

Really in some cases rather stable operation has been observed in ZnS:Tb,F. Only a slight decrease of eta has been determined over more than 6000 50Hz-h on some samples. More drastic changes also were found, but again not eta_{lum} degraded (sometimes it even increased). So, eta_{exc} must be to blame . Checking for E_t /14/, it increased too, so no parallel to the CVD samples can be drawn. If all carriers are still able to excite, either the concentration or the cross section of the dopants must have decreased, as eta_{exc} = cross-section * concentration /17/. The most likely explanation is that the concentration of the atomically dispersed Tb^{3+} has been diminished by an approach to TDE in the S-film.
Rather characteristic for ZnS:Tb, at least for films prepared by EBE of ZnS and TbF_3, and ones rf-sputtered from doped targets, is a shift of E_t (virgin) with the Tb concentration to higher values. Figure 10 relates to a series of samples with the SIMS and/or RBS determined Tb concentrations given. Micro-inhomogeneities might be responsible.

CaS:Eu : The earth alkaline sulphides CaS and SrS doped with Eu^{2+} and/or Ce^{3+} have attracted much attention as the most likely realisation of the EL full color display /19, 20, 21/. Plenty of work has been published about non-ZnS like behaviour but vigorously contradicted in /22/ for samples using rf-sputtered CaS films. A very special, not readily understood aging effect mentioned in /20/, is described by Fig.11. As it is our firm belief that these and only these ZnS-like earth alkaline sulphide samples will survive the lab-stage, we have included early comments on this special aging feature into /22/.

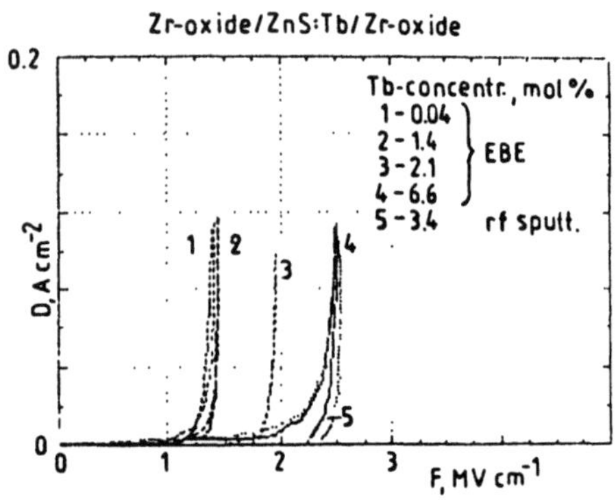

Fig.10 The virgin E_t monotonuously increase with the Tb content, as evidenced by the D(F) curves

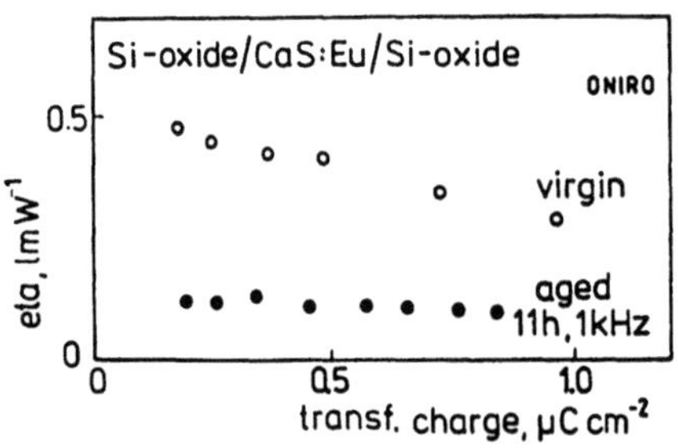

Fig.11 Rapid degradation of efficiency eta in a CaS:Eu sample, accompanied by a change of its Q dependence

4. Conclusions

Some of the prominent degradation mechanisms have been identified. As usual an understanding can contribute to an elimination of the reasons and/or the invention of counter-measures.

Summing up shortly:

- the degradation of 'α' caused by electrochemical processes in the dielectrics can be avoided by close control of their stochiometry
- the continuous increase of E_t is not understood in the sense of possible modelling, but can be reduced to tolerable values by burn-in
- rare earth doping of ZnS leads to rather special problems connected with the low solubility of these elements
- the tendency of earth alkaline sulphides to form native defects and to react at the interfaces with the dielectrics has to be taken into account in technology and materials selection
- side effects, as e.g. reactions of the dielectrics with ITO, especially at thin edges of the latter, where heating occurs, have to be eliminated by proper design.

A large scale industrial production of ZnS:Mn based ELDs of high complexity proves that degradation problems are no inhibitive issue.

6. Acknowledgement

The confidence provided by numerous colleagues from all over the world donating samples, the behaviour of which was not very different from that of our own, is gratefully acknowledged . And of course the 'makers' of the used samples, Dr. U. Reinsperger, Dr. R. Reetz, Dr. E. Schnuerer, Dr. H. G. Eberle, E. Halden, R. Herrmann, K. Erb, G. Gers, and B. Reinsperger have done well, if the results are appreciated by the reader. To E. Lenz and H. Schneider we are indebted for assistance in the measurements, to G. Schulz for design and construction of most of the equipment, and to Dr. B. Selle for RBS analyses.

7. References

/1/ T. Inoguchi, M. Takeda, Y. Kakihara, Y. Nakata, and M. Yoshida, SID 1974 Digest, p.84
/2/ T. Inoguchi, S. Mito
Topics in Appl. Physics, Springer, Heidelberg 1977, p. 222
/3/ R.A. Boudreau, J.E. Conelly, B. Dale, SID 88 Digest, p. 12
/4/ K. Okamoto, Y. Nasu, Y. Hamakawa,
IEEE Trans. Electron Devices ED 28, 1981, 698
/5/ P.M. Alt, D.B. Dove, W.E. Howard,
J. Appl. Phys. 53 (7), 1982, 5168
/6/ M. Yoshida, A. Mikami, T. Ogura, K. Tanaka, SID 86 Digest, p.41
/7/ G.O. Mueller, R. Mach, R. Reetz, G.U. Reinsperger,
SID 88 Digest, p. 23
/8/ G.O. Mueller, R. Mach, B. Selle, G. Schulz,
phys. stat. sol.(a), submitted
/9/ G. Schulz, to be published
/10/ B. Selle, e.a., to be published
/11/ C. King, 3th EL Workshop 1986, Kah-Nee-Ta, Oregon
/12/ R.T. Flegal, C.N. King, SID 86 Digest, p. 177
/13/ Y. Shimizu, T. Matsudaira, SID Res. Conf.1985,p.101
/14/ R. Mach, e.a.,to be published
/15/ J. Karila, J. Hyvaerinen, R. Toernqvist,
Eurodisplay 1987, p. 236
/16/ J. Watanabe, W. Wakitani, S. Sato, S. Miura,
SID'88 Digest, p.228
/17/ R. Mach, G.O. Mueller, phys. stat. sol.(a) 81, 609 (1984)
/18/ G.O. Mueller, J. Neugebauer, R. Mach, G.U. Reinsperger,
J. Cryst. Growth 86, 890 (1988)

/19/ S. Tanaka, H. Yoshiyama, J. Nishiura, S. Ohshio, H. Kawakani, H. Kobayashi, SID 88 Digest, p. 293
/20/ H. Ohnishi, R. Iwase, Y. Yamasaki, SID 88 Digest, p. 289
/21/ C. Gonzalez, Eurodisplay London 1987, p. 23
/22/ R. Mach, H. Ohnishi, G.O. Mueller, 4th EL Workshop, Tottori 1988

ZnS:Mn Electroluminescent Devices with High Performance Using $SiO_2/Ta_2O_5/SiO_2$ Insulating Layer

M. Yoshida, T. Yamashita, K. Taniguchi, K. Tanaka, T. Ogura, A. Mikami, H. Nakaya, S. Yamaue, and S. Nakajima

Central Research Laboratories, Sharp Corporation, Tenri, Nara 632, Japan

1. Introduction

In the development of thin film electroluminescence (TFEL) with double insulating layer configuration (insulating layer / EL active layer / insulating layer), a lot of effort has been made for the EL active layer.

But, much more attention has to be paid to the insulating layer, because the insulating layer takes an important role in realizing the TFEL with high stabilityand high reliability. It is not exaggerated to say that the suitable insulating layer for TFEL bring about good results into practical use as information display panel [1].

There are many reports on the trial of insulating materials for TFEL. However, few insulating materials are practically used for TFEL except SiO_2, Si_3N_4, Al_2O_3 and TiO_2 [2, 3].

Ta_2O_5 is thought to be a suitable material for the insulating layer of three layered EL device because of its high dielectric constant; ~25 and high break down electric field; ~ 3×10^6 V/cm. Therefore, TFEL device with Ta_2O_5 insulating layer was tried in some laboratories [4,5,6].

Up to now, however, many problems still remain unsolved in appling this material into an actual information display. First, Ta_2O_5 insulating material shows poor adhesion to lower side transparent electrode of ITO. Second, TFEL device with Ta_2O_5 insulating layer breaks down in propagating mode, but not in self-healing mode. Third, the reaction between Ta_2O_5 and ZnS:Mn active layer takes place during their annealing process.

Considering above mentioned problems, we developed TFEL device with high performance using SiO_2 / Ta_2O_5 / SiO_2 as an insulating layer, and the results obtained are described in this paper.

2. Sample Preparation

Lower insulating layer of our newly developed TFEL device has a structure of SiO_2 / Ta_2O_5 / SiO_2 as shown in Fig.1. The three thin films were deposited sequentially on transparent electrode of ITO by the same reactive sputtering method. Si metal and Ta metal were used as the sputtering target for SiO_2 and Ta_2O_5, respectively. After formation of the lower insulating layer, ZnS:Mn active layer was deposited by electron beam evaporation method using ZnS pellet doped with Mn, and then, annealed in vacuum atmosphere in order to improve the crystalinity and to diffuse the Mn center. After that, Si_3N_4 and Al_2O_3 were successively deposited as the upper insulating layer by the reactive sputtering

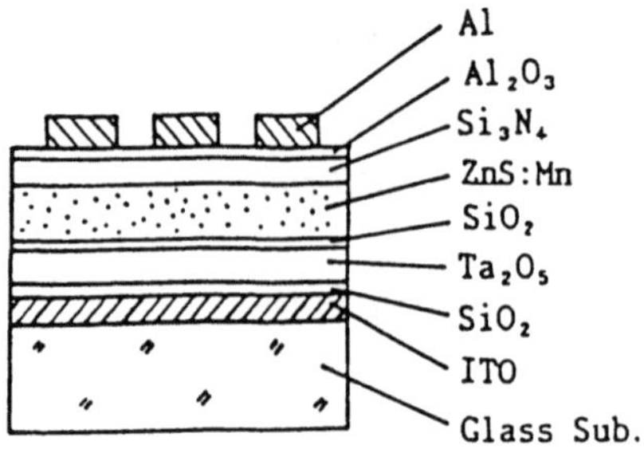

Fig.1 Structure of newly developed EL device

method Finally, the upper electrode of Al was deposited by conventional thermal evaporation method.

3. Development of the Lower Insulating layer Using Ta_2O_5 Material

TFEL using only Ta_2O_5 layer as the lower insulating layer had three problems as mentioned before. Each problem was solved as follows.

3.1 Poor Adhesion of Ta_2O_5 Layer to ITO

Many bubbles were observed in the Ta_2O_5 layer when the Ta_2O_5 layer was directly deposited on the ITO. Same phenomenon was also observed when Si_3N_4 was directly deposited on the ITO

But, such a phenomenon could not be observed in the case of SiO_2. In the system of Si_3N_4-SiO_2 as shown in Fig.2, compressive internal stress of SiON film is decreasing as constituent element of SiON is changing from Si_3N_4 to SiO_2. In this figure, notations of O and N indicate oxygen and nitrogen atomic concentrations measurered by Auger spectroscopy analysis, respectively.

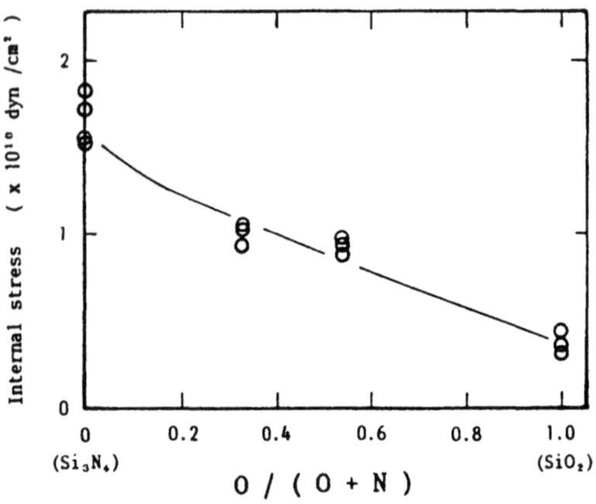

Fig.2 Compressive internal stress in the system of Si_3N_4-SiO_2

The observed bubbles can be concluded to arise from the internal stress of insulating layer. Therefore, SiO_2 film with low internal stress was inserted between Ta_2O_5 film and ITO film. As a result, no bubble could be observed in Ta_2O_5 film on SiO_2/ITO.

Considering the driving voltage, the thickness of SiO_2 with low dielectric constant of ~ 4 is optimized to be around 400 Å.

3.2 Reaction Between Ta_2O_5 and ZnS:Mn

A TFEL device excluded SiO_2 film adjacent to EL active layer in the structure of Fig.1 shows inferior L-V characteristics. As shown in in Fig.3, brightness decreases and steepness of L-V characteristics become lower as compared with those of TFEL with SiO_2 insulating film.

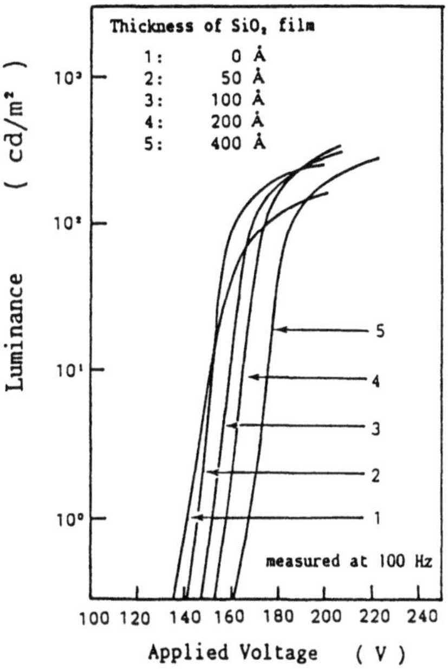

Fig.3 Thickness dependence of SiO_2 thin film adjacent to EL active layer on EL characteristics

These phenomena are due to the reaction between Ta_2O_5 insulating layer and ZnS:Mn active layer. As reported in reference[7], such a kind of phenomenon was also observed in the EL device with CaS active layer. In order to prevent the reaction between ZnS:Mn active layer and Ta_2O_5 layer, SiO_2 layer was inserted between these layers and the thickness dependence of SiO_2 layer on the EL characteristics was examined. Increasing the thickness of SiO_2, the luminance of the EL device increases until 200 Å as shown in Fig.3, and the dielectric break down mode changes from propagating mode to self-healing mode above 200 Å. Therefore, considering the shift of driving voltage to the higher voltage side, the thickness of SiO_2 layer is optimized at around 200 Å.

4. EL Characteristics

EL device with the structure as shown in Fig.1 was stabilized by EL operation at driving frequency of 1 kHz under room temperature for 20 hours from the initial stage. After the stabilization process, stable aging characteristics were observed as shown in Fig.4.

The luminance at a 60 V higher than threshold voltage remains in the same level, and the shift of threshold voltage to higher voltage side is settled within 5 volts for 200 hours under the accelerated aging condition of 1 kHz driving frequency at ambient temperature of 75 ℃. This means that the life of EL device is more than 2 x 10^4 hours under the actual driving frequency of 60 Hz at room temperature.

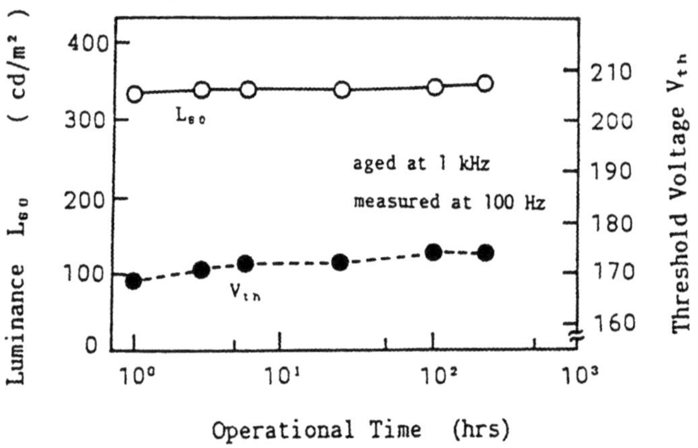

Fig.4 Aging characteristics of newly developed EL device

References

1. M. Takeda et al.: Digest of '80 SID International Symposium , 66 (1980)
2. H. Uede et al.: Digest of '81 SID International Symposium , 28 (1981)
3. H. Antson et al.: Fresenius Z Anal Chem, 322, 175 (1985)
4. H. Kozawaguchi et al.: Jap. J. Appl. Phys. 21, 1028 (1982)
5. H. Kobayashi et al.: Jap. J. Appl. 12 759 (1973)
6. Y. Shimizu et al.: Digest of '85 Inter. Dis. Res. Conf., 101 (1985)
7. M. Yoshida et al.: Digest of '86 SID International Symposium, 41 (1986)

Thin Film Photoconductor-Electroluminescent Memory Display Devices

P. Thioulouse

Centre National d'Etudes des Télécommunications, Laboratoire de Bagneux, 196 avenue Henri Ravera, F-92220 Bagneux, France

I - Introduction

The potential of memory electroluminescent (EL) devices for the display of highly complex information has long been recognized. Extensive work has been devoted to ac thin-film EL devices with inherent memory. However, this technique turned out to be non viable industrially, mainly because of stability and reproducibility problems |1|.

A novel memory EL structure was introduced recently | 2-5 |: this is a monolithic device integrating a thin film EL structure and a photoconducting (PC) thin film. The PC layer and the EL structure are connected in series ; the optical excitation of the PC layer by the EL emissive film produces a memory effect of the so-called photoconductor-electroluminescent (PCEL) type. The PCEL memory has the significant advantage over the inherent memory of enabling independent control and optimization of the emission characteristics, related to the EL structure, on the one hand and of the hysteresis characteristics, related to the PC layer, on the other hand.

We first estimate the performance of a memory EL matrix panel and we predict a drastic improvement compared to a conventional EL panel, provided that an efficient EL memory technique is available. We then describe the structure and fabrication technique of our PCEL devices. The performance of an experimental matrix panel is reported. We present a detailed study of the steady-state electrical behaviour of the PC layer, which plays a dominant role in the memory characteristics of the PCEL device. Finally, we describe the dynamic response of the PC film.

II - Potential performance of a memory EL matrix panel

The principle of operation of a memory EL matrix panel is basically different from that of a non-memory display. A pixel with memory can exhibit at least two different operating states : an emitting ON state and a non-emitting OFF state under a steady-state voltage excitation of a given amplitude chosen inside the hysteresis loop of the luminance vs voltage curve. Under such conditions, this pixel can be switched ON (resp. OFF) by temporarily increasing (resp. decreasing) the applied voltage. Consequently, the operation of a memory EL panel will essentially consist of two phases, similarly to ac plasma panels. First, an ac voltage is applied continuously to the whole display in order to hold all the pixels within the memory loop. This is the sustain mode, which is used to hold the displayed image between changes. In the case of emissive displays, this mode has the additional and essential role of providing the energy to the pixels. Second, the writing mode is used to modify the displayed information : voltage variations will be applied to selected pixels by the conventional one-line-at-a-time method.

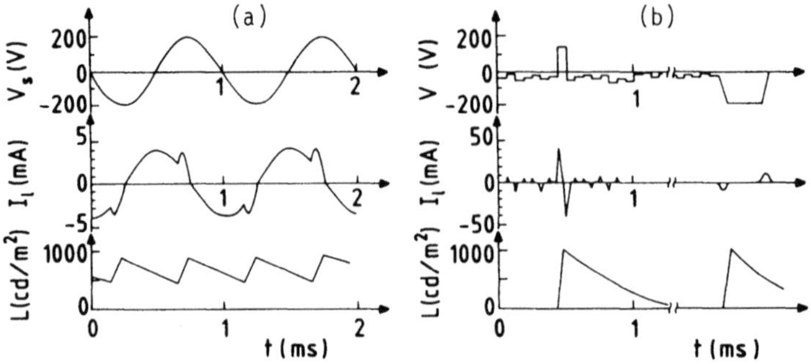

Fig. 1 : Schematic waveforms of the potential of a row electrode, of the corresponding row current and of the pixel light emission for : (a) memory EL matrix panel (sustain mode) and (b) a non-memory EL matrix panel.

We estimated the performance of a memory EL panel, for comparison with a conventional EL display. We chose the ideal case of memory pixels with the same electrical (current vs time) behaviour as for non-memory and with an EL efficiency reduced by a factor of 2 (1 lm/W). Fig. 1 shows the schematic waveforms of the voltage applied to the pixels, the row current and the pixel light emission for a memory EL panel and for one without memory in the case of a still image displayed. It is evident that, by choosing a sustain voltage of the memory panel of a relatively high frequency (1 kHz) and with a 100 % duty cycle, the pixel luminance would be significantly increased whereas the current surge would be drastically reduced.

Table I summarizes the main characteristics computed for the ideal memory panel in the sustain mode compared to a conventional EL display. We studied two types of memory panel : one with the conventional 50 % pixel filling factor (high luminance option), the other with a filling factor reduced to 7 % to obtain a standard average luminance of 45 cd/m^2 |6|. The implementation of a novel multi-phase sustain mode, which would reduce the column peak current by a factor of 4 at least |7|, and of an energy recovery system were taken into account in the computation |8|. We predict, for a given average luminance of 45 cd/m^2 a reduction of almost one order of magnitude for the power consumption and almost two orders of magnitude for the peak current requirements on the drivers, compared to the non-memory display. Besides, in the case of conventional computer

Table I

Format : 256×512 pixels - $2dm^2$

	Non-Memory EL panel	Memory EL panel	
		High L option	Regular L option
Average luminance (cd/m^2)	45	325 (high contrast)	45 (high contrast)
Consumption (typical) (W)	11	9	1.3
Row current (peak) (mA)	40	4	0.6
Column current (peak) (mA)	6	0.5	0.07
Switched voltage (V)	60	30	30

applications with moderately high information refresh rates, the writing mode should not degrade the above performance significantly |8|. Finally, as will be shown later, memory EL pixels can be switched ON and OFF by voltage modulation amplitudes as low as 30 V, so that low cost CMOS drivers could be used.

III - Device structure and fabrication

The fabrication of the ac thin film EL structure is conventional. The ZnS:Mn is electron-beam evaporated. The dielectric layers are reactively magnetron-sputtered Ta_2O_5 unless otherwise stated.

The first generation PCEL structure |2,4| was obtained by depositing a photoconducting thin film directly onto the EL film stack before the top Al electrode deposition (fig. 2 (a)). We used no intermediate electrode between the EL structure and the PC layer.

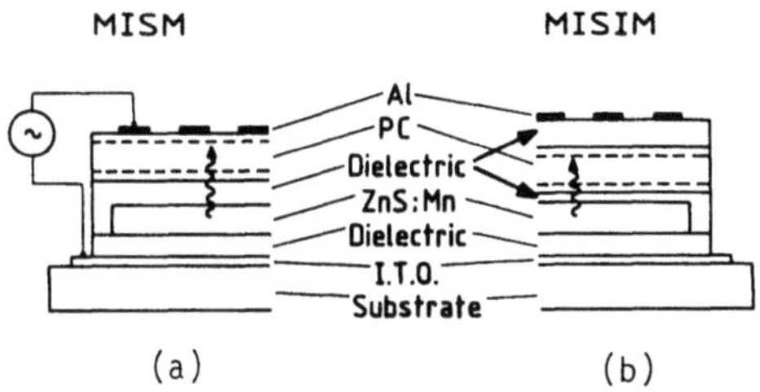

Fig. 2 : Cross section of the PCEL structure of : (a) the first generation and (b) the new type.

Here, the PC layer is a 0.9 to 2µm thick n^+ n n^+ structure, deposited in a capacitance glow-discharge reactor. n and n^+ layers are of an a-$Si_{1-x}C_x$:H alloy |9|, obtained from a gas mixture of SiH_4-CH_4, with a gas flow rate ratio $[CH_4]$ / $[CH_4]$ + $[SiH_4]$ ranging from 45 to 62.5 %. The n^+ layers are obtained by adding phosphine (PH_3) to the gas mixture ; they provide quasi-ohmic contacts for electron injection in the n layer. The thickness of each n^+ layer is 60 nm.

We present here an improved PCEL structure, obtained by inserting the PC layer between the top dielectric layer and the emissive film, but keeping a very thin dielectric interlayer (fig. 2 (b)) |3|. This interlayer, about 20 nm thick, is thick enough to protect the ZnS film from reaction with the glow discharge plasma during the PC layer deposition and to keep a good quality interface with the ZnS film for proper EL injection. But it is thin enough to avoid optical reflection of the EL light emission at the dielectric-ZnS diopter related to refractive index differences. Thereby, we insure that almost 90 % of the EL emission is theoretically transferred to the PC layer, the refractive index of which is higher than that of ZnS. Moreover, light piping in the ZnS film is suppressed and optical cross-talks in a matrix display are prevented. Finally, the virtuous couple dielectric layer-Al electrode is restored and the breakdown self-healing process is enhanced. The thin dielectric interlayer is magnetron-sputtered SiO_xN_y.

IV - Device performance

The high resolution capability of the new PCEL structure was demonstrated by fabricating a 60x60 pixel matrix panel with a 0.3 mm pitch and with a pixel size of 0.15x0.15 mm^2 (fig. 3 (a)). The luminance vs voltage (L-V) characteristics of this display, fig. 3 (b), show a truly bistable behaviour with vertical transitions from the OFF to the ON state and back. Main features are a memory margin of more than 20 V, an ON pixel luminance of about 600 cd/m^2 and an intrinsic contrast ratio greater than 100. Additionally, the PC layer acts as an

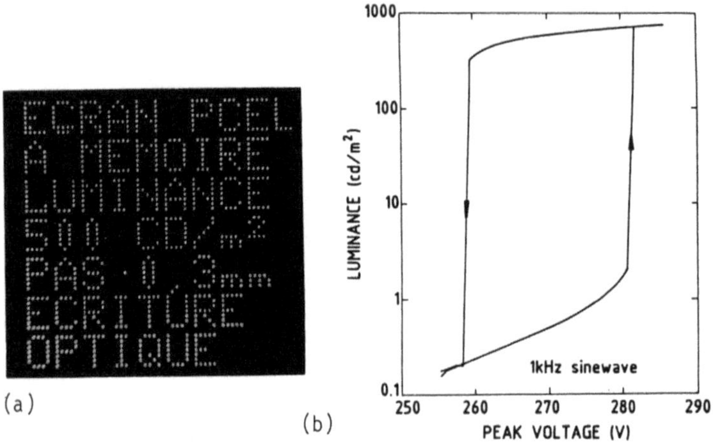

(a) (b)

Fig. 3 : Matrix PCEL panel of 60x60 pixels with a 0.3 mm pitch : (a) image obtained by optical writing through a mask ; (b) luminance versus voltage characteristics of a pixel.

efficient dark backing layer and drastically improves the contrast ratio under ambient illumination. Moreover, such a bistable electro-optical behaviour was demonstrated for a pixel size as small as $50x50 \mu m^2$.

V - Electrical behaviour of the PC film (steady-state)

The PC layer was studied in three different structures. First, the PC film is sandwiched between SnO_2 and Al electrodes in a Metal-Semiconductor-Metal (MSM) structure (fig. 4 (a)) ; second, a dielectric layer is inserted below the PC layer to form a Metal-Insulator-Semi-conductor-Metal (MISM structure : fig. 4 (b)) ; third, a second dielectric layer is added on top of the PC layer to simulate the new PCEL device (MISIM structure : fig. 4 (c)).

We first describe the behaviour of the PC layer in the dark (OFF state of the PCEL device). The PCEL operation is first simulated by connecting in series the PC sample (fig. 4 (a)) and a capacitor representing the EL element. The resulting device is called C-PC. The measured curves of the current I and the voltage V_{PC} across the PC sample vs time (fig. 5) show that the PC layer abruptly switches from a non-conducting state to a conducting state when V_{PC} exceeds a threshold V_{SC}. Above the threshold, V_{PC} is clamped to V_{SC} and the conducted current I_C is controlled by the series capacitor C. The current vs voltage (I-V) characteristics of the PC sample are ruled by the mechanism of space charge limited conduction (SCLC). The SCLC has been precisely described under constant voltage operation |10|, and the characteristic voltage is found to be approximately proportional to the space charge (SC) density in the n layer. When the C-PC device is operated under steady-state conditions (fig. 5 : ac voltage with a constant amplitude), the SC density results from the balance between injected electrons trapping in the gap of the n layer during the conducting phases II and thermal detrapping during the non-conducting phases I.

The time-dependence of the current and the voltage across the PC film V_{PC} in the MISIM structure under similar triangular voltage excitation (fig. 6) suggests a new mechanism, in addition to the previously described SCLC process. After an initial conducting phase, the current drops although V_{PC} increases. The study of asymmetrical MISM structures shows that the conduction is then primarily controlled by the cathodic interface. More precisely, the n^+ thin film, sandwiched between the n PC layer and an insulating film must not only act as a

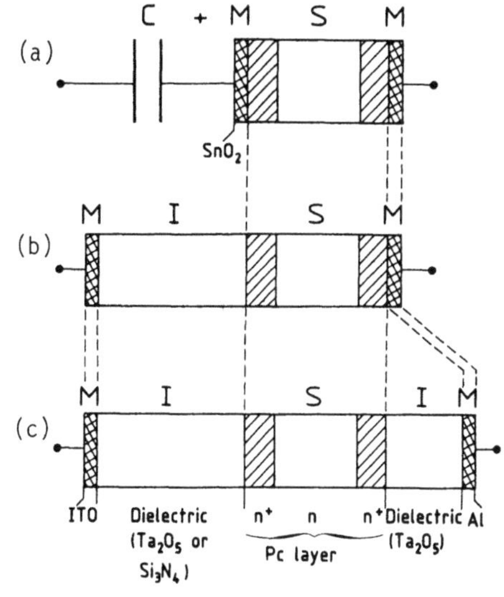

Fig. 4 : Cross sections of the various PC structure used for the electrical study : (a) PC film with metallic electrodes (MSM structure)in series with an external capacitor ; (b) PC film with a dielectric layer on its lower side (MISM structure) ; (c) PC film with a dielectric layer on each side (MISIM structure)

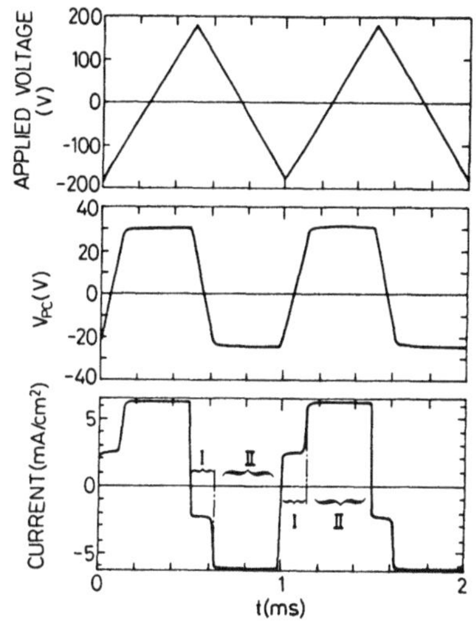

Fig. 5 : Time-dependence of the applied voltage (1 kHz, triangular), the voltage across the PC film V_{PC} and the current for the simulating circuit : PC film (MSM structure) + series capacitor.

good ohmic contact with the n layer, it must also provide all the electrons injected and those trapped to form the space charge. Practically, after a quasi-ohmic injection of a moderate amount of charge Q_{OI}, we observe a starving of free or shallow trapped electrons from the n^+ layer. We call it the Blocked Injection (BI process). After a significant voltage increase V_{BI}, a high level injection is restored by a field assisted emission of more deeply trapped electrons (fig. 6). The decrease of V_{SC} observed, correlating with the increase of V_{BI} when increasing the applied voltage amplitude, indicates that part of these BI electrons comes from the space charge in the n layer itself.

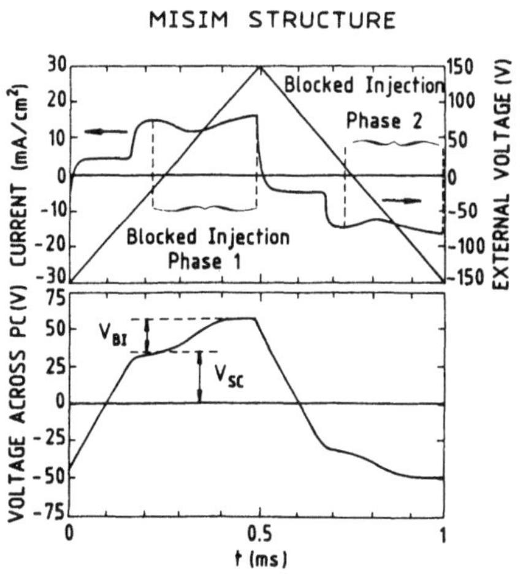

Fig. 6 : Time-dependence of the applied voltage (1 kHz, triangular), the current and the voltage across the PC film in the MISIM structure.

We studied the influence of the carbon content in the n and n^+ layers - more specifically the gas flow rate G = $[CH_4]$ /($[CH_4]$+$[SiH_4]$) - on the SCLC threshold voltage V_{SC} and on the BI voltage V_{BI}. First, we observe a steady increase of V_{SC} with G for the n layer (fig. 7 (a)). This is accounted for by three combining effects : carbon incorporation increases the density of states at the Fermi level at the equilibrium E_{Fo} and deepens the Fermi level with respect to the conduction band |11, 12 |, it also broadens the band-tails |12|. Second, fig. 7 (b) points out a threshold effect in the Blocked Injection voltage vs the carbon content of the n^+ layers. This confirms the predominant role of the n^+ layer in the BI process. The observed abrupt increase of V_{BI} above G = 55 %, correlated with a drastic quenching of the charge Q_{OI} available for a quasi-ohmic injection, probably arises from trapping of the electrons originated from the P donors in the additional trap states generated by the excess carbon incorporation.

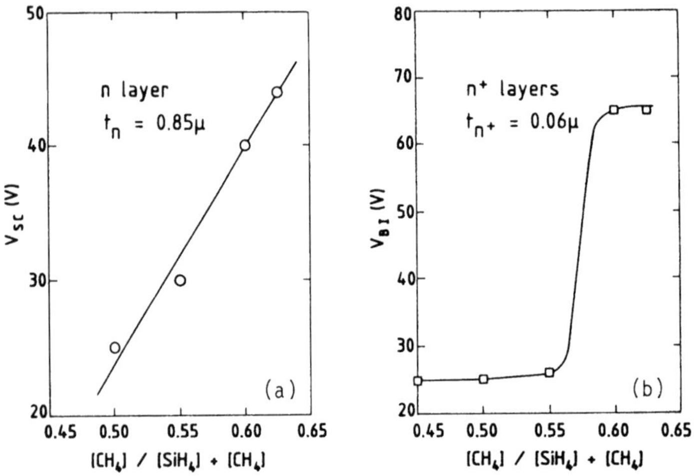

Fig. 7 : Influence of the methane concentration in the gas during deposition of (a) the n layer upon the space charge limited injection voltage V_{SC} ; (b) the n^+ layers upon the blocked injection voltage V_{BI}.

The electrical behaviour of the PC layer when exposed to light was also studied. The SCLC is modified by light by the process of photoexcitation of trapped electrons, resulting in a decrease of the net negative space charge density and of the related threshold voltage V_{SC} (fig. 8). The BI process is sensitive to relatively high light intensities only. The photoexcitation of trapped electrons combines with the field assisted detrapping, resulting in a decrease of V_{BI} for a given overall injection level. Fig. 9 shows that the carbon content G has a strong influence on the sensitivity to light of the voltage across the PC layer V_{PC}.

Present optimum fabrication conditions for a PCEL device are a G value of 55 % for the n^+ layers for a maximum planar resistivity (about 10^5 Ω.cm) with a minimum BI effect and a G value of 60 % for the n layer to minimize the influence of ambient light on the L-V hysteresis.

VI - Dynamic measurements

The electrical response of the PC layer in the MISIM structure to voltage transients or to a light excitation pulse was studied in order to predict the behaviour of PCEL displays under writing conditions. This also provides additional data on the underlying mechanisms.

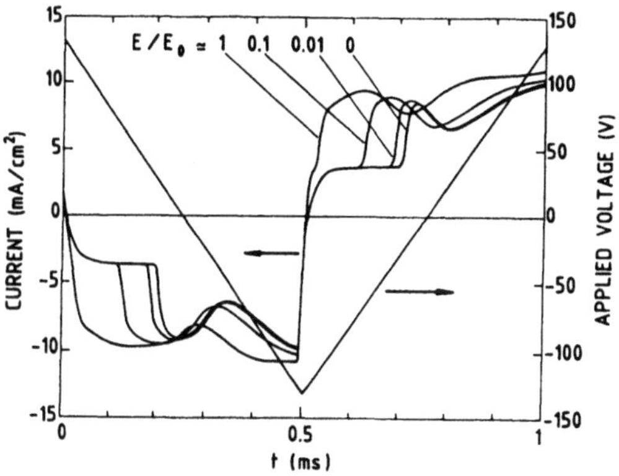

Fig. 8 : Effect of an optical excitation of intensity E (λ = 577 nm) on the time - dependence of the current in the MISIM structure (1 kHz triangular excitation voltage). The illumination level of reference E_o is about 30,000 lux.

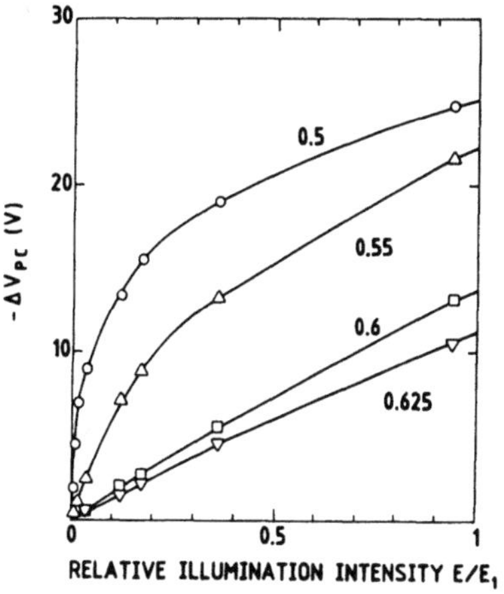

Fig. 9 : Influence of the methane concentration in the gas during the n layer deposition on the "photo response" of the PC film in the MISIM structure (V_{PC} drop vs illumination intensity). The illumination level of reference E_1 is about 7,000 lux (λ = 577 nm).

First, in dark conditions simulating an initially OFF sate, we applied transient voltage pulses, simulating writing pulses, on top of a continuous trapezoidal waveform - the sustain voltage. We measured an increase of V_{PC} limited to a few volts at the end of 100µs - long voltage pulses as high as 50 V. Therefore, in a PCEL device, more than 90 % of the writing pulse will be transferred to the EL structure itself in 100 µs.

Second, we studied the rise time of V_{PC} back to its dark value, upon removal of light excitation, using a hyper-red LED. The response time of V_{BI} is below our detection limit (≅ 1µs), whereas that of V_{SC}, τ_{SC}, depends strongly on the carbon content G in the n layer. Above G = 55 %, τ_{SC} is of the order of 1 ms. Below, it takes milliseconds or tens of ms to return to the dark value. In all cases, the time behaviour of V_{SC} after removal of the optical excitation is similar to that under a sudden voltage burst application after rest and in dark. This would mean that G main influence is on the electron trapping rate in the n layer : at higher values of G, carrier trapping is faster, hence space charge builds up in a shorter time either upon voltage application or upon light excitation removal. Furthermore, assuming a constant detrapping efficiency by photoexcitation, the stronger trapping rate at higher G values straightforwardly accounts for the lower net photo-response under steady-state conditions reported above. Tentative physical interpretations for the effect of carbon incorporation on photo-response are given in |12|.

Practically speaking, after our studies of the PC layer under dynamic conditions, we predict that, for a highly carbonated n layer (G > 55 %), the PCEL response time should be of the order of 1 ms. Moreover, knowing the steep L-V threshold and the short response time of an EL structure, switching ON a PCEL device should be possible by applying a few 100µs-short low voltage pulses on top of the sustain voltage. Indeed, we could demonstrate the switching ON of a PCEL cell with two 30 V-high and 100µs-short writing pulses, one for each polarity (fig.10).

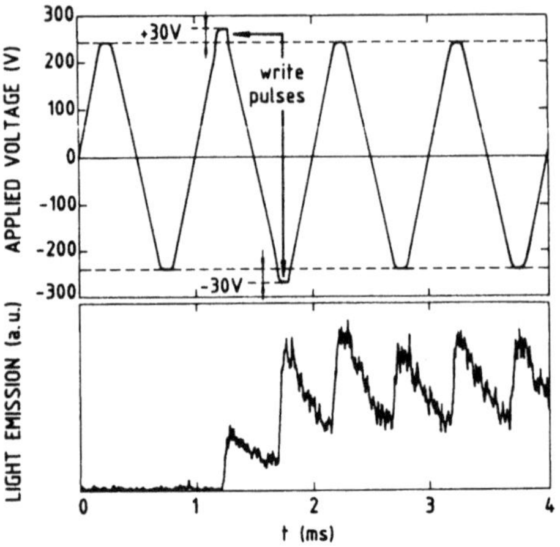

Fig. 10 : Switching ON of a PCEL pixel (example) : two writing pulses of 30 V amplitude and 100µs duration are superimposed onto the sustain voltage (trapezoidal waveform), one for each polarity.

VII - Conclusion

We presented the structure and fabrication of a thin film PCEL device with memory. We demonstrated the high visual performance and the high resolution capability of an improved PCEL structure. The memory capability of the device depends primarily on the electrical behaviour of the a-$Si_{1-x}C_x$:H photoconducting layer. The carbon incorporation has a strong impact on various important characteristics of the PC film such as the injection process from the n^+ layers into the n film, the voltage across the PC layer in the dark, the photo-response and the response time to an electrical or optical excitation. A high carbon content in the n layer (G ≅0.6) is found beneficial in many respects. We measured that the characteristics of an optimized PC film are compatible with a high speed and low voltage switching of PCEL displays. A detailed study of the electrical behaviour of PCEL cells in dynamic conditions is underway. Additional work on the fabrication conditions of the PC and the EL films should enable to reach the performance of our quasi-ideal model for a memory pixel. Our prediction of a high quality and potentially very low cost EL display could then be validated. Ultimately, the implementation of the thin film PCEL concept to multi-colour displays is very promising.

Acknowledgment

We wish to thank Prof. I. SOLOMON for his invaluable help in the optimization of the PC film fabrication and in the interpretation of measurements, P. GABELOTAUD and H. TRAN QUOC for the device fabrication, and D. COUNALI for her assistance in measurements.

References

1. O. Sahni, P.M. Alt, DB. Dove, W.E. Howard, and D.J. McClure, IEEE Trans. Electron Devices ED-28, 708 (1981).
2. P. Thioulouse and I. Solomon, IEEE Trans. Electron Devices ED-33, 1149 (1986).
3. P. Thioulouse, US patent application 905345 (1986).
4. P. Thioulouse, C. Gonzalez and I. Solomon, Appl. Phys. Lett. 50, 1203 (1987).
5. P. Thioulouse and I. Solomon, Proceedings of the 87' Eurodisplay Conf. (The Institute of Physics and the Society for Information Displays, London, 1987) p 25.
6. P. Thioulouse, French Patent Application 8611808 (1986).
7. P. Thioulouse and J.P. Budin, French Patent Application 8617985 (1986).
8. P. Thioulouse, Thesis, Ecole Nationale Supérieure des Télécommunications, 1987.

9. P. Thioulouse and I. Solomon, French Patent Application 8614715 (1986).
10. I. Solomon, R. Benferhat, and H. Tran-Quoc, Phys. Rev. B, Vol. 30, 3422 (1984).
11. M.P. Schmidt, J. Bullot, M. Gauthier, P. Cordier, I. Solomon, and H. Tran-Quoc, Phil. Mag. B51, 581 (1985).
12. J. Bullot, M. Gauthier, M. Schmidt, Y. Catherine, and A. Zamouche, Phil. Mag. B49, 489 (1984).

Tunable Color Electroluminescence Display Operated by Pulse Code Modulation

Y. Hamakawa, H. Fujikawa, M. Nakamura, T. Deguchi, and R. Fukao

Faculty of Engineering Science, Osaka University, Toyonaka, Osaka 560, Japan

ABSTRACT

As a new driving system of the tunable color electroluminescence (TCEL) devices, a series of technical trials on the pulse code modulation has been investigated. It has been shown from the results that the control of number of pulse ratio driving system with a constant pulse height and a constant pulse width is the best system for the continuous color mixing and variations with constant luminance. A series of technical data on the dual coloring TCEL devices are presented and discussed. This TCEL devices open a wide varieties of applications in the field of information display systems.

1 INTRODUCTION

In recent years, extensive studies have been made toward the development of thin film electroluminescence(TFEL) device which has a number of attractive advantages such as low power dissipation, capability of large area flat-type display with multi-color and complete solid state modular plate possibility. With the recent progress in computer technology, there exists a tremendous potential demand for multi-color imaging display for information processing. To respond these strong potential needs, a great deal of R&D efforts have been made to improve the EL device performance. As a result , a reduction of the threshold voltage down to IC drive level[1], and multi-coloring[2] technologies are considerably in progress. On the basis of this technological progress, a new type of device having a function of TCEL device which could be able to change emission color spectra by controlling applied voltage has been developed firstly by Osaka University group in 1982[3]. Since this invention, a remarkable advance has been seen in a few years not only cell fabrication process but also their operating circuit system.

In this paper, a series of technical data on fabricating thin film TCEL cells have been presented. Basic characteristics and EL performances of the dual color TCEL cells are also demonstrated together with the pulse code modulation operating system. The luminous color conditioned to match human-eye visibility by controlling number of pulse ratios, and pulse height ratios are also investigated and discussed.

2 FABRICATION of TUNABLE COLOR EL DEVICE

TCEL device using optical transparency has been developed by color mixing of dual monochrome EL cell. In principle, there are two kind of cell constructions for color mixing methods for the dual color TCEL. One is vertically-stacked structure, another is horizontally aligning structure. We use the

Springer Proceedings in Physics, Vol. 38 **Electroluminescence**
Editors: S. Shionoya and H. Kobayashi

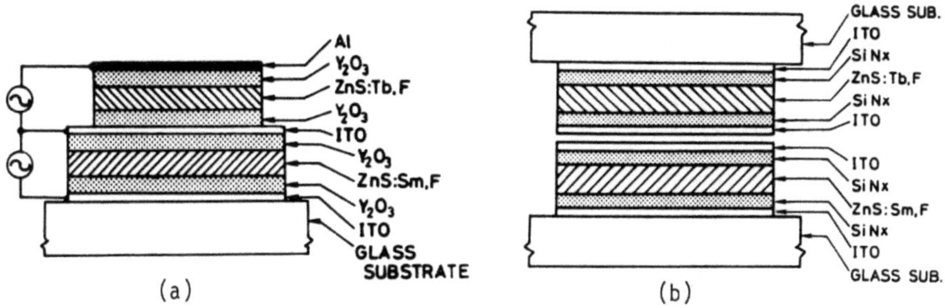

Fig.1. Schematic illustration of the cell construction of TCEL devices (a) Vertically stacked three-terminal device (b) Vertically stacked four-terminal device

vertically-stacked construction, because (a) all films forming ac TFEL cells are transparent, (b) a horizontal separation between two stacked cells can be made fine, since these films are thin, (c) various color can be realized with one address. It is seen from the Fig.1(a) that two kinds of primary color TFEL cells are stacked vertically on glass substrate using the middle ITO as the common electrode. In this case, ZnS:Tb,F and ZnS:Sm,F are used as emission layers, and two sources of driving voltage are applied across the top and the bottom electrodes to the middle ITO as the common electrode. Figure 1(b) shows an example of a four-terminal color mixing display.

3 DRIVING SYSTEM for CONTROLLING DUAL COLOR TCEL

There exist three kinds of color control methods to drive this color control system: (a) amplitude modulation, (b) pulse width modulation and (c) pulse number modulation.

Figure 2 shows the result of various emission spectra for several pairs using amplitude modulation. The voltage applied to the green cell(Vg) and to the red cell(Vr) are shown in the figure. Any colors between green and red would be

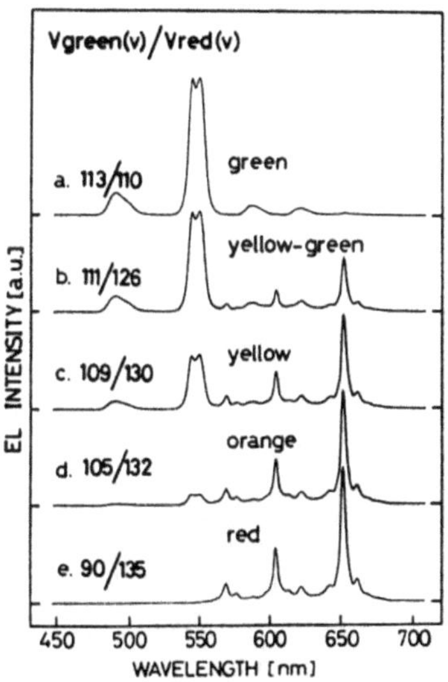

Fig.2. The emission spectra of the vertically stacked three-terminal TCEL device. Any color between green and red can be obtained by changing the modulation color voltage in the ranges from zero to 30 volts

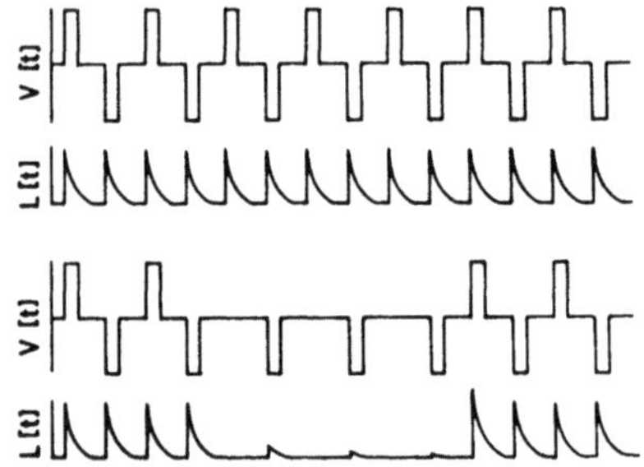

Δ
Fig.3.Wave form of applied pulse voltage V[t] and corresponding light emission wave form L[t] in the pulse number modulation

▷
Fig.4.The luminance versus number of positive pulses applied to the TCEL devices

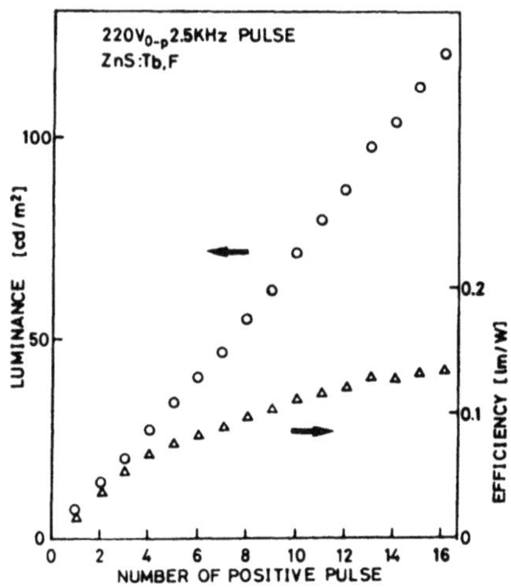

produced by controlling the color regulation voltage (0-30Vrms) superposed on a sustaining voltage of 100Vrms. Stable variable color emission of about 350cd/m^2 was observed on one address of this TCEL device. The emission color varies continuously from red to green.

In the case of three-terminal amplitude-modulation systems, the driving circuit system is rather simple, however, there are some technical difficulties in the color tuning, and also in the total luminance control. To solve these problems, the pulse number modulation system was developed[4]. Figure 3 shows an example of the applied voltage wave form and light emission wave form in the pulse number modulation system. As can be seen from the figure, the light emission can be skipped in the rest of the positive pulse duration. When the polarity of the succeeding pulse is the same, the electrons, which accumulate at the interface between the emission layer and the insulating layer, depress the inner field of the emission layer. So the light emission is negligibly weak. That is reason for the linear relationship between luminance and the number of positive pulses applied to the EL devices, as is shown in Fig.4. This pulse number modulation is suitable for operating three-terminal TCEL with the common ITO electrode and the dot matrix panel in which one electrode is used for line scanning.

4 CHARACTERISTICS of TCEL OPERATED by PULSE NUMBER MODULATION

Figure 5 shows the position of the TCEL emission operated by 8-bits pulse number modulation in the chromaticity diagram. The number of pulses applied to the red cell (PNred) and to the green cell (PNgreen) are shown in the fiqure. In this case, the total pulse number of PNred and PNgreen is 8. We arrange the luminance of pure green and pure red cell providing TCEL emission to 100cd/m^2 by tuning the pulse height. An example of the display is shown in Fig.6.

The nine tone of color between green and red are produced easily by controlling the pulse number ratio. When the ratio of PNred and PNgreen is 6/2, 3/5 and 1/7, we can obtain orange, yellow and yellow-green as neutral tints, respectively. Stable variable color emission of constant luminance (100cd/m^2) is always observed with one address of this TCEL cells. As can be seen in Fig. 5, the center color (Red 4 pulses/Green 4 pulses) looks yellow-orange and not pure yellow. This is due to the spectral sensitivity of the human eye. This kind of correction to match the human visibility factor can be made by controlling the pulse height ratio of the green to red pulse against the constant lunimance.

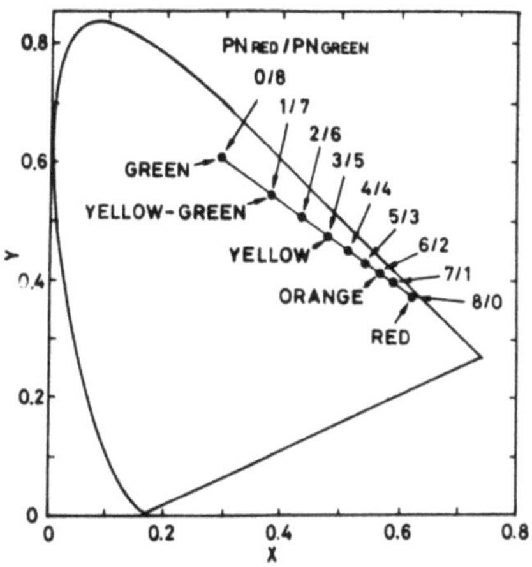

Fig.5.The position of emission color TCEL cell operated by pulse number modulation in chromaticity diagram

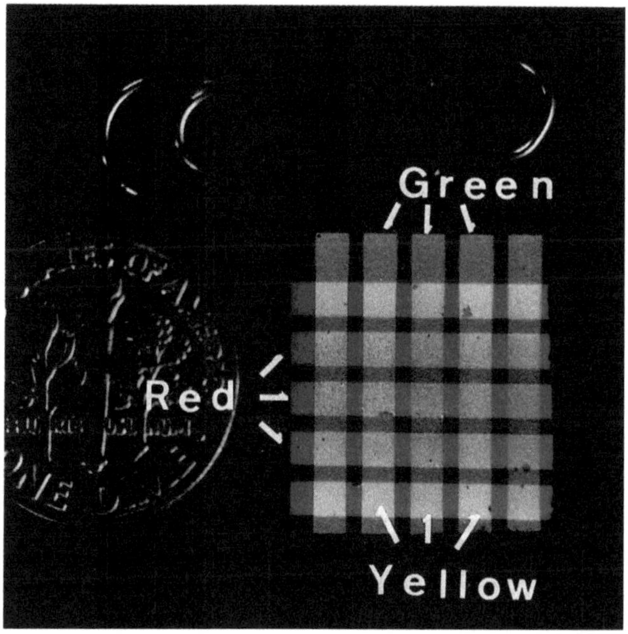

Fig.6. An example of the cross made in the vertically stacked four-terminal TCEL cell

5 CONCLUSION

A tunable color EL device having a constant luminance has been developed by an 8-bits pulse number modulation technique. The pulse number modulation method matches human visibility factor in tuning the color tone of TCEL. This TCEL devices open a wide varieties of application fields as active type information display systems in opto-electronics.

6 LITERATURE

1. K.Okamoto, Y.Nasu, M.Okuyama and Y.Hamakawa: Jpn. J. Appl. Phys. 20 Suppl. 20-1, 215 (1981)
2. T.Suyama, N.Sawara, K.Okamoto and Y.Hamakawa: Jpn. J. Appl. Phys. 21 Suppl. 21-1, 383 (1982)
3. Y.Oishi, T.Kato and Y.Hamakawa: Extended Abstract of the 15th Conf. on Solid State Device and Materials, 353 (Tokyo,1983)
4. R.Fukao, T.Yamamoto, and Y.Hamakawa: IEICE Technical Report, EID86-37, 5 (1986) (in Japanese)

Optical Behaviour of Electroluminescent Devices

R.H. Mauch[1], *K.A. Neyts*[2], *and H.W. Schock*[1]

[1]Universität Stuttgart, Institut für Physikalische Elektronik, Pfaffenwaldring 47, D-7000 Stuttgart 80, Fed. Rep. of Germany

[2]Rijksuniversiteit Gent, Laboratorium voor Electronica en Meettechniek, Sint-Pietersnieuwstraat 41, B-9000 Gent, Belgium

Abstract - In an optical multilayer thin film system, interference effects play an important role for the outcoupling of light due to the different refractive indices and thicknesses of the involved layers. A theoretical model has been investigated with the ability to calculate the outcoupled intensity of light for various wavelengths and viewing angles. $ZnS:TbF_3$ ac-TFEL devices have been prepared to verify the model and to check the applicability of the theory to electroluminescent devices. Experimental and calculated wavelength and viewing angle dependent intensity data show good agreement. The outcoupling efficiency η_{opt} of a TFEL device can be increased up to 25% compared with the value given by the wellknown formula of the simple theory /1/.

1. Introduction

Most papers on TFEL devices published in the past are dealing with fundamentals of physical and electrical properties or with materials research and technological problems. Only little attention was given to the optical behaviour of such devices. VLASENKO et al. /2/ have done first investigations on this subject with a ZnS:Mn film between two reflective electrodes. They show the effect of interference on the emission spectra of a three layer device. In this paper, interference effects for more complex multilayer systems were investigated, using the theory of LUKOSZ and KUNZ /3,4,5/. The total efficiency η_{tot} of an ac-TFEL device (usually a few lumen per watt) is given by the product of several efficiencies where each stands for one of the involved mechanisms, e.g. light production, light outcoupling and human eye sensitivity. The optical outcoupling efficiency η_{opt} is in the order of 10% and therefore an important limitation of η_{tot}.

TFEL devices on the base of rare earth doped ZnS show emission spectra with a couple of peaks with varying intensities /6/. Hence the use of $ZnS:TbF_3$ elements for the experimental verification of the theory gives the possibility to study the optical behaviour at various wavelengths in the visible range.

2. Theoretical Model

In the generalized optical thin film multilayer system in Fig. 1 light is isotrope emitted in layer e and partially reflected at every interface. The purpose of this work is to calculate the emission pattern in air. It will be assumed, that all layers are nonabsorbing and therefore the refractive indices n_e, n_a, n_0 are real.

2.1 Multiple Beam and Wide Angle Interferences

In contrary to the simplified model, it will be taken into account that light can be partially transmitted and reflected at the interfaces (Fig. 2a). The resulting

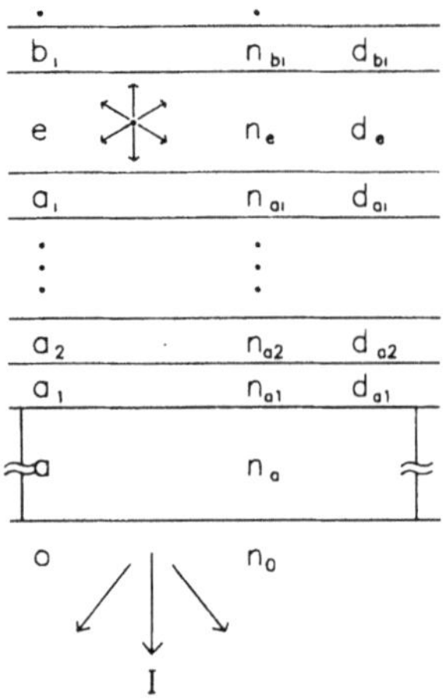

Fig. 1 Generalized optical thin film multilayer system

intensity I_1 is the coherent sum of the several light waves emitted in medium 0. This is called multiple beam interference. In the case of wide angle interference, the light waves S_1 and S_2 are emitted from one luminescent center (electric or magnetic dipole or multipole) in supplementary angles, leave the device under the same angle θ_0 and interfere as shown in Fig. 2b. The total intensity in medium 0 sums up to

$$I_{tot} = I_1 + I_2 + 2\cdot\sqrt{I_1\cdot I_2}\cos(\Delta\psi)$$

where $\Delta\psi$ is the phase difference depending on the depth x of the center. For further calculations it will be assumed, that the luminescent centers are homogeneously distributed over the depth x and $n_e\cdot d_e > \lambda$. Hence the interference term can be neglected. The total intensity is then $I_{tot} = I_1 + I_2 = I_1(1+R_{eb})$, where R_{eb} is the reflection coefficient of medium b. If the emitting layer e is infinitely thick, there are randomly oriented luminescent centers which would radiate with orthogonal (s) and parallel (p) polarisations equally represented: $I_\infty^s(\theta_e) = I_\infty^p(\theta_e) = I_\infty = \text{const.}$. The presence of interfaces and the noninfinite thickness of layer e disturb the geometrical isotropy so that the resulting interferences change the isotrope radiation significantly.

LUKOSZ and KUNZ /5/ developed a theory for the calculation of dipole and multipole radiation close to a flat interface by solving the electromagnetic equations. The theory can be nicely applied to the problem described here. A considerable simplification can be done by neglecting the wide angle interferences ($n_e\cdot d_e > \lambda$) and one finds for the wavelength dependent intensity in medium a:

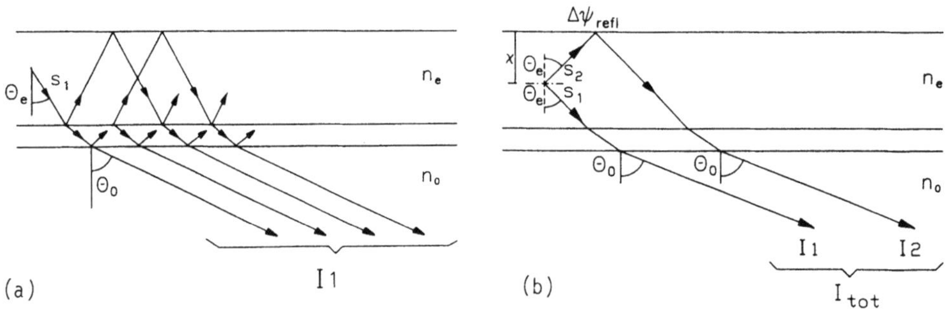

Fig 2 Interference mechanisms in a thin film multilayer system: a) multiple beam interference, b) wide angle interference

$$I_{\lambda a}^{(s,p)}(\theta_a,\lambda) = M^{(s,p)}(\theta_e)\cdot[1+R_{eb}^{(s,p)}]\cdot T_{ea}^{(s,p)}\cdot\frac{d\Omega_e}{d\Omega_a}\cdot I_{\lambda\infty}(\lambda) \qquad (1)$$

$$\text{with } M^{(s,p)}(\theta_e) = \left|1 - r_{ea}^{(s,p)}\cdot r_{eb}^{(s,p)}\cdot \exp\varphi_e\right|^{-2}$$

$$R_{eb}^{(s,p)} = \left|r_{eb}^{(s,p)}\right|^2, \quad T_{ea}^{(s,p)} = \frac{n_a\cdot\cos\theta_a}{n_e\cdot\cos\theta_e}\cdot\left|t_{ea}^{(s,p)}\right|^2$$

$$\frac{d\Omega_e}{d\Omega_a} = \frac{n_a^2\cdot\cos\theta_a}{n_e^2\cdot\cos\theta_e} \quad \text{and} \quad n_e\cdot\sin\theta_e = n_a\cdot\sin\theta_a$$

$$\varphi_e = \frac{(4\pi i)}{\lambda}\cdot n_e\cdot d_e\cdot\cos\theta_e$$

The reflection and transmission coefficients $r_{ea,eb}$, $t_{ea,eb}$ must be calculated by using the real or complex Fresnel coefficients of all intermediate interfaces. By neglecting all incoherent reflection at the glass-air interface, the radiation pattern in air can be calculated as

$$I_{\lambda}^{(s,p)}(\theta_0,\lambda) = I_{\lambda a}^{(s,p)}(\theta_a,\lambda)\frac{d\Omega_a}{d\Omega_0} .$$

If all reflection coefficients are zero and the transmission $T_{ea} = 1$, the formula simplifies to the result of the simplified model.

2.2 Optical Outcoupling Efficiency η_{opt}

The total emission in air (medium 0) $E_\lambda(\lambda)$ of an emitting device can be calculated by integrating the intensity over the solid angle 2π (half space). In an infinite layer e, the total emission $E_{\lambda\infty}(\lambda)$ is given after integrating the isotrope intensity over 4π to $8\pi\cdot I_{\lambda\infty}(\lambda)$. The optical outcoupling efficiency η_{opt} is now defined as the division of the two emissions $E_\lambda(\lambda)$ and $E_{\lambda\infty}(\lambda)$:

$$\eta_{opt}(\lambda) = \frac{E_\lambda(\lambda)}{E_{\lambda\infty}(\lambda)} = (4\pi)^{-1}\cdot\int_{2\pi}\frac{I_\lambda^s(\theta_0,\lambda) + I_\lambda^p(\theta_0,\lambda)}{2I_{\lambda\infty}(\lambda)}d\Omega_0 \qquad (2)$$

which can be calculated numerically, even if $I_{\lambda\infty}(\lambda)$ is unknown, by using formula (1). This relation is only valid during the radiating time of the luminescent center, but by assuming that the lifetime is not affected significantly by the presence of interfaces, integrating over the time will not change the result. These calculations do not include scattering effects which will increase η_{opt} due to the emission of internal reflected light /7/.

3. Experimental Verification

The ac-TFEL devices which were used for the experiments had been prepared on glass substrates in the layer sequence TC-Al_2O_3-ZnS:TbF_3-Al_2O_3-Al as shown in Fig. 3 /8/. The TC (transparent contact) has been realized either with normally used sputtered ITO or with thin thermally evaporated metal (Al, Ag, Au) with higher reflection coefficient at the interface to enlarge the influence of interference effects. The exact thicknesses of the layers which are necessary for the comparison of calculated and measured data have been determined with tally step and with an optical method by measuring the transmission curves $T(\lambda)$ sequentially after each deposition process. The optical thickness $n\cdot d$ (n ... refractive index of the layer) is then found by fitting a calculated transmission curve to the measured one, using $n\cdot d$ of the deposited film as fitting parameter. For the verification of the theory, the

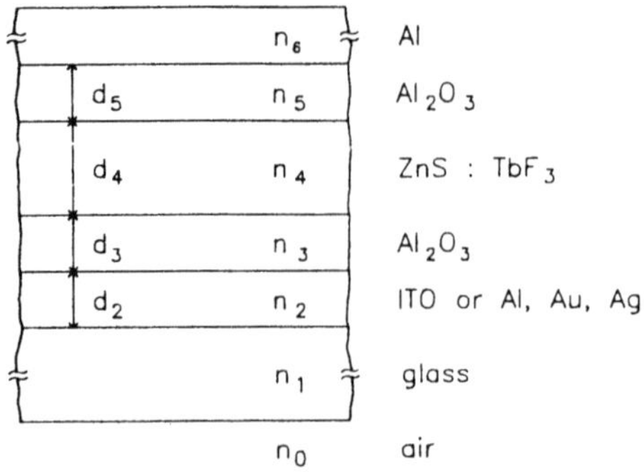

Fig. 3 Structure of an ac-TFEL device

outcoupling intensity $I_\lambda(\theta_0,\lambda)$ has been measured by lock-in technique and simultaneously turning the device between 0° and 90°. For the adjustment of the different wavelengths a double prism monochromator has been used. The comparison between the calculated and measured curves at 542 nm for both types of devices are shown in Fig. 4 (a ... thin Al as transparent contact, b ... ITO as TCO) where the dashed lines indicate the calculated curves. It can be deduced, that the presented theory describes the optical behaviour of TFEL devices very well.

As one application of the theory Fig. 4b shows angular dependent intensity behaviour $I(\theta_0,\lambda=542nm)$ of a device which is optimized for maximal outcoupling of the green 542nm emission line over all possible viewing angles. The calculated thicknesses for this device are d_2 = 149nm, d_3 = 252nm, d_4 = 452nm and d_5 = 239nm. Fig. 4 indicates a slight difference of the calculated (dashed line) and experimental (solid line) curve due to preparation tolerances.

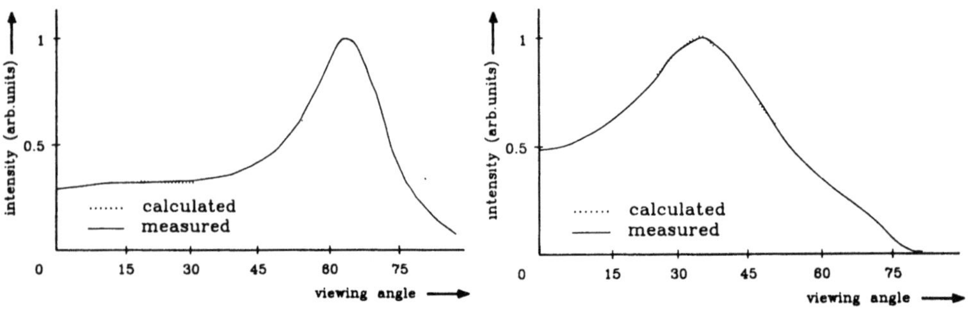

Fig. 4 Calculated and experimental data of the angular dependent 542nm outcoupling of $ZnS:TbF_3$ ac-TFEL devices with a) thin Al and b) ITO as TC

4. Conclusions

A theoretical model has been proposed for the calculation of interference effects in a thin optical multilayer system. The experimental verification was done with $ZnS:TbF_3$ TFEL devices and it can be shown, that the model describes the experiment excellently. The model is generally applicable for TFEL devices and can be used as a tool for optimization. Other criteria than the proposed ones, e.g. constant color for different viewing angles or $ZnS:TbF_3$ devices with different colors, can be calculated numerically.

5. References

1. R. Mach and G.O. Müller, phys.stat.sol.(a) 81 609(1984)
2. N.A. Vlasenko, S.A. Zynyo, Zh.A. Pukhlii, Opt.Spec.,Vol 28,1970 p.68
3. W. Lukosz and R.E. Kunz, J.Opt.Soc.Am.,Vol 67,12,1970 p.1607
4. W. Lukosz, Phys.Rev.B,Vol 22,6,1980 p.3030
5. W. Lukosz, J.Opt.Soc.Am.,Vol 71,6,1981 p.744
6. K. Okamoto and Y. Hamakawa, Appl.Phys.Lett. 35, 508(1979)
7. P.M. Alt, SID 25/2,1984 p.123
8. R.H. Mauch, R. Menner and H.W. Schock, J.Cryst.Gr. 86 (1988) p.885

Current Filaments in ZnS:Mn DC Thin Film Electroluminescent Devices

M.I.J. Beale, J. Kirton, and M. Slater

Royal Signals and Radar Establishment,
St. Andrews Road, Malvern, Worcestershire, UK

ABSTRACT

We describe an investigation into 'kinetic electroluminescence' (KEL) in ZnS:Mn direct current thin film electroluminescent (DCTFEL) devices. KEL consists of regions of anomalously high emission which move non-destructively around the device [1]. We demonstrate that, although KEL appears in widely varying forms, the current invariably flows in filaments, the radii of which are, at most, a few microns, and that the KEL, normally seen, consists of the envelope of many such filaments.

1 Introduction

DCTFEL devices have potential advantages over their AC counterparts arising from their relative simplicity of manufacturing and driving. However, the passage of a direct current through an insulator gives rise to a number of problems including localised destructive breakdown (LDB) and current-voltage characteristics which drift with time. To overcome such problems, we are currently engaged in a programme to elucidate conduction mechanisms in DC (and also AC) TFEL devices; an important part of this programme has been a study of the many manifestations of a phenomenon referred to as kinetic electroluminescence (KEL).

In KEL, regions of anomalously bright electroluminescence move rapidly around the active area of some DCTFEL devices. One manifestation of the phenomenon was previously briefly mentioned [1] and there is a substantial literature describing closely related effects in ACTFEL devices (eg [2]) and referred to as domain electroluminescence. Apart from its intrinsic interest and the potential for memory effect AC devices, we believe that KEL offers a vital clue to the problem of understanding the conduction mechanisms in these devices.

2 Experimental Conditions

Devices were fabricated according to the process previously reported [1]. They consist of a glass substrate, through which the EL is observed, a transparent conducting oxide (cadmium stannate), a sputtered film of ZnS doped with approximately 0.6 wt% Mn and an evaporated aluminium contact. The ZnS:Mn thickness was nominally 0.75 µm leading to operation in the region of 75 volts. Relatively small active regions, down to 10^{-4} cm^2, were studied. It must be stated that, despite care in device fabrication, the properties of nominally identical devices vary significantly, particularly with regard to KEL and LDB behaviour.

The devices were operated in vacuum at nominally room temperature unless otherwise stated, operation being possible within the range 100K to 600K. Devices were driven from a voltage supply through a 100kohm series resistance. We define forward bias as having the transparent conducting oxide negative, this

Springer Proceedings in Physics, Vol. 38 Electroluminescence
Editors: S. Shionoya and H. Kobayashi

polarity resulting in the lower LDB rate. Devices were imaged through a x 50 microscope objective lens by their own electroluminescence using a CCTV system based on a silicon intensified target (SIT) tube. The resolution limit was approximately 1 µm and the minimum light levels imaged could barely be detected by the dark adapted eye looking directly at the device.

3 Kinetic Electroluminescence and Current Filaments.

KEL regions move rapidly around the device at speeds in the regions of 0.1 to 1 mm per second and are non-destructive. KEL appears to be a common phenomenon in these ZnS:Mn DCTFEL devices even when manufactured with a wide range of parameters and in several different fabrication plants. Many devices, however, show KEL only when cooled, typically to 200K. We have observed KEL in both forward and reverse bias and with a variety of different top contact materials.

The first point we illustrate is that the appearance and behaviour of the KEL varies widely between different samples, there being no single 'typical' KEL. Figures 1 to 5, a series of micrographs, each a single 20 msec field, illustrate this diversity.

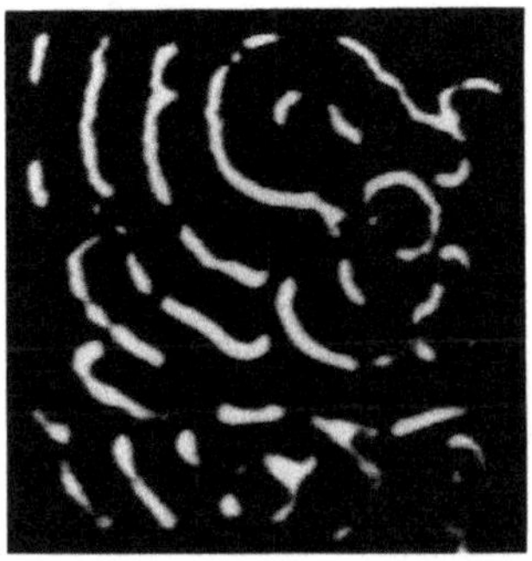

Fig. 1

'Wave patterns', with a velocity of 100 µm per second, on a 800 µm square sample operating in forward bias. The patterns are strongly reminiscent of those generated in a class of re-generative chemical reactions (see discussion). If they collide, the two wavefronts are annihilated, this is only visible on the moving images.

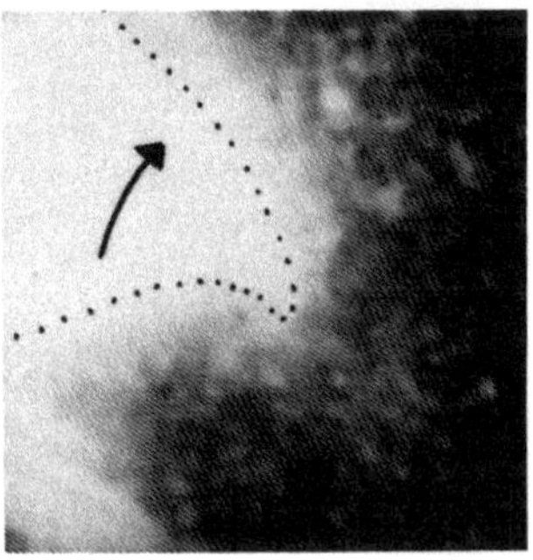

Fig. 2

KEL here manifests itself as a rotating sector, the dotted lines and arrow clarifying its shape and motion. It performs approximately three revolutions per second, sometimes in one sense and sometimes the other. The active area is a 150 µm square. The organised motion can be interpreted as follows: the KEL has a finite life-time after which it re-appears in the area unaffected by KEL for the greatest length of time.

Figures 1 and 2 demonstrate the apparent long range nature of KEL. Figure 3 shows the same sample as fig. 1 but at higher magnification and different recording conditions; filaments of diameter 5 µm or less are now revealed in the wake of the main KEL envelope.

We now turn to the effect of drive voltage on KEL. The background EL emission is very voltage dependent, there being a very sharp rise in current; so too is the KEL. Figure 4 illustrates the effect on the KEL of increasing the total current. Initially there is no KEL and then small point-like filaments with lifespans of a few tenths of a second appear (a); then areas of KEL develop which

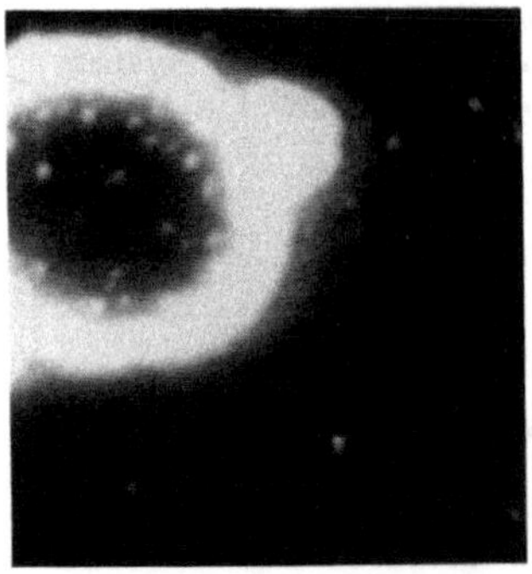

Fig. 3

A KEL wavefront that expands approximately as a circle; a secondary wavefront has developed near its periphery. The image is 150 μm across and the velocity of the wavefront is approximately 100 μm per second. In addition we see, left behind the wavefront, bright filaments which emit EL for periods ranging from less than 20msec up to several seconds before fading away.

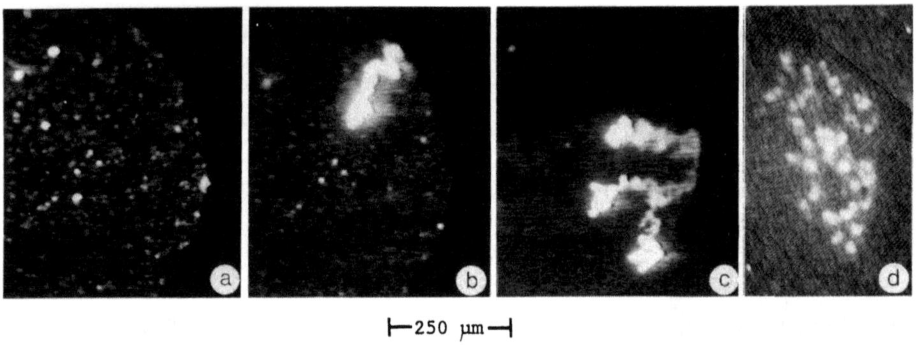

Figure 4. A 1 mm² active area passing a total current of a) 1μA, b) 5μA and both c) and d) 25μA. Image d) was recorded with a 2 msec exposure.

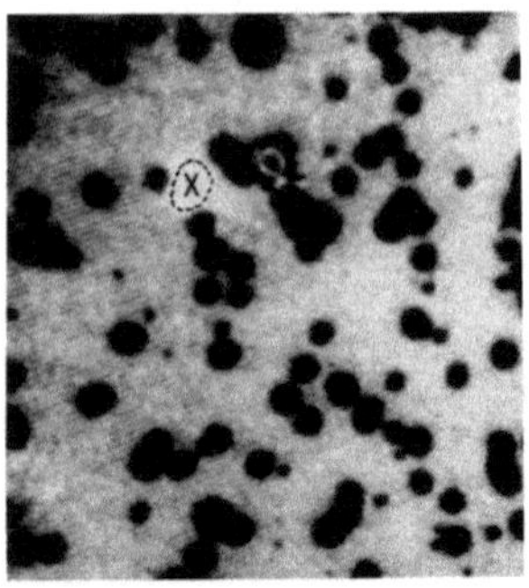

Fig. 5

An example of a further complication in KEL behaviour. There is highly non-uniform background EL against which a single bright KEL area appears (marked 'x'). The KEL always appears in the light regions and, after its passage, converts them into dark regions. These dark regions are not, however, permanently damaged and appear bright on reversal of the drive polarity. The sample is 300 μm across.

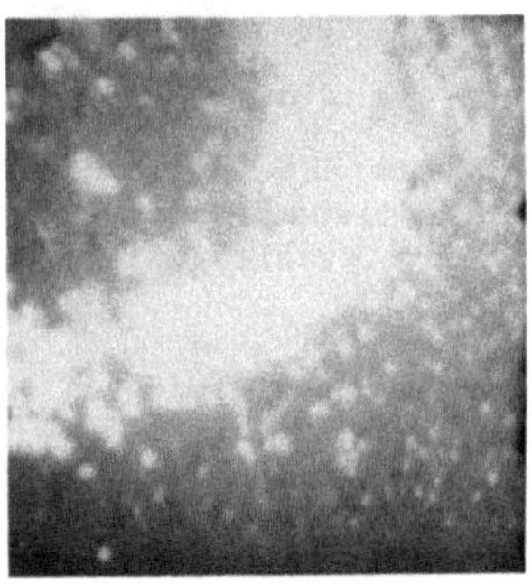

Fig. 6

The effect of applying a localised source of heat at 300C at the top left handside of the sample is to generate a cloud of filaments, the envelope of which moves out as an expanding wavefront with a velocity of approximately 100 μm per sec. The sample is 150 μm across.

move rapidly around the sample (b). Next there is KEL covering a larger area and splitting into separate areas (c) and finally the entire active area becomes covered by KEL. In these images we see no evidence of filamentary structure within the KEL envelope. However, (d) does reveal that even this type of KEL is composed of filaments, here a few μm in diameter. This image was taken with the relatively short exposure of 2 msec.; all other images had 20 msec. exposures. Using multiple exposures with a stroboscopic system (not reproduced here) shows that the filaments have a life-span of approximately 10 msec during which they move approximately 10 μm.

5 Discussion

We have shown there to be many varieties of KEL, all of which are composed of high current density filaments. These filaments have a wide range of life-times and, in general, one sees only their envelopes unless millisecond exposures are used. The filaments are non-destructive and are often mobile; they may travel tens of microns during their lifespans which range from milliseconds or less to seconds. Many of the KEL types are very photogenic when seen in motion.

The role of heat in the formation of filaments has been demonstrated (fig. 6); the similarity of the wave structures in fig. 1 to those seen in demonstrations of the Belousov-Zhabotinsky reaction and the computer simulation of it ('hodgepodge') [3] is remarkable. In this regenerative chemical reaction, the heat evolved changes the state of a catalyst which then allows the reaction to begin anew. In our case, we suggest that the apparently organised motion arises from local heating due to the high current densities in the KEL areas.

These observations of kinetic electroluminescence in DCEL devices and their obvious relationship to domain electroluminescence [2] suggest strongly that such behaviour is a general characteristic of the EL phenomenon in ZnS:Mn and not just highly specific to the boundary conditions in hysteretic AC devices. Furthermore, the observations are consistent with the hypothesis [1] that LDB was the result of a fundamental tendency for EL to become filamentary.

It is well known [4] that filamentary behaviour can arise from a tendency to current controlled negative differential resistance (CCNDR) and the proposition has been frequently put that such CCNDR arises in EL as a result of carrier injection by tunnelling being regeneratively increased by the build-up of trapped positive space charge near the cathode. Confirmation (or otherwise) of such a mechanism seems inherently difficult to obtain in AC structures and this may be the reason for its absence in the literature. However, it is probably less difficult to carry out the crucial experiments in DC devices and there are already strong indications from our own work that tunnelling does play a part in the injection process.

Although the lower mobility of holes can lead to CCNDR [1], this seems likely only at much higher current densities than are encountered in normal EL. This is why the build-up of trapped positive space charge has been invoked in CCNDR modelling so far; such mechanisms imply charge trapping at defects and this would be consistent with some of the other observed characteristics of KEL. A variety of as yet unpublished work on conduction behaviour in DCTFEL devices confirms the likely existence and importance of trapped positive charge.

At this stage, there are still many unanswered questions. Is CCNDR occurring in EL as a general phenomenon? If so, does it arise from an electronic or thermal mechanism, or from both? Why are some filaments destructive and others non-destructive? What is the role of trapped space charge and which traps are involved? We anticipate hopefully that studies on DC structures will allow us to answer these questions.

Our thanks are due to Prof. Brian Ridley, Mr Nick Harper, Mrs Jenny Blackmore and our partners in an Alvey-JOERS-supported consortium. We would like to thank Dr Norman Apsley for drawing our attention to reference 3.

REFERENCES

[1] J.M.Blackmore, A.F.Cattell, K.F.Dexter, J.Kirton and P.Lloyd, J. Appl. Phys.61,714 (1987). A.F.Cattell, J.C.Inkson and J.Kirton, J. Appl. Phys.61,722 (1987)

[2] A.Onton and V.Marrello 137-197 in Advances in Image Pickup and Display Vol 5 Academic Press 1982

[3] A.K.Dewdney in Scientific American 86 Aug. 1988.

[4] B.K.Ridley. Proc. Phys. Soc. 82, 954, 1963.

Cathode-Zinc Sulphide Barrier Heights and Electron Injection in Direct Current Thin Film Electroluminescent Devices

M.J. Davies and R.H. Williams

University College, P.O. Box 913, Cardiff, CF1 3TH, Wales, UK

1. Introduction

DC electroluminescent devices with the transparent conducting oxide (TCO) - ZnS:Mn - aluminium structure exhibit destructive breakdown. A detailed understanding of the electrical conduction mechanism is central to an improvement in device performance.

It is thought that the current is limited by the rate of electron tunnelling through an electric field thinned potential barrier at the cathode-ZnS interface[1]. It will be shown using ultraviolet photoelectron spectroscopy (UPS) and current-voltage measurements that this mechanism is not compatible with experiment.

2. Experimental

Prior to the UPS measurements, thin films of cadmium stannate (CDS), indium tin oxide (ITO) and zinc oxide (ZnO), three possible TCOs for device manufacture, were cleaned by argon ion bombardment in ultra high vacuum. A thin (several 10s of angstroms) layer of ZnS was then evaporated in a controlled manner onto the room temperature substrate from a tungsten filament.

The conduction (valence) band offset between two materials forming a heterojunction is of prime importance in determining the passage of an electron (hole) across the interface. The valence band offset ΔE_V at an abrupt heterojunction can be found using UPS by measuring the difference in valence band edge positions on either side of the junction[2]. With knowledge of the bandgaps of the two materials involved the conduction band offset, ΔE_C, can then be calculated.

All I-V measurements to be described were performed on devices employing CDS as the cathode and 3mm dots of aluminium as the anode. Characteristics were measured at room temperature in the dark, under roughing pump vacuum - ca 0.01 torr.

3. Results and Discussion

In fig 1 UPS spectra are presented of the valence band of a clean ZnO thin film and after evaporation of about 40 angstroms of ZnS. On evaporation, the overlayer valence band edge is seen to lie about 1.7 eV to higher kinetic energy than that from the substrate indicating that the ZnO - ZnS heterojunction has a valence band offset of about 1.7 eV. Addition of the difference in bandgaps results in a conduction band offset of 2.15 eV. The results of similar measurements on the other interfaces of interest are given in table 1.

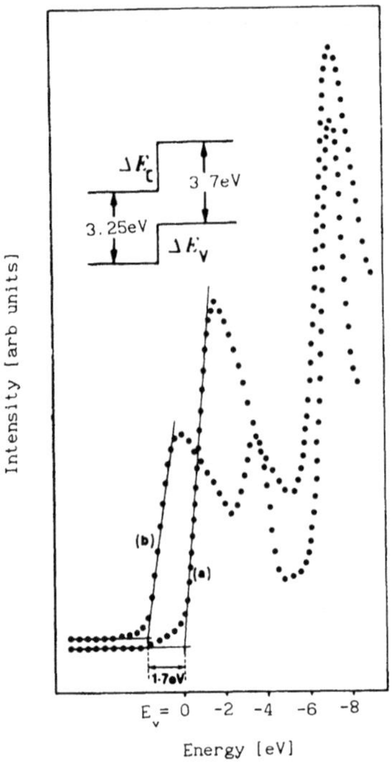

Fig 1. UPS spectra of (a) clean ZnO and (b) a 40 angstroms thick overlayer of ZnS

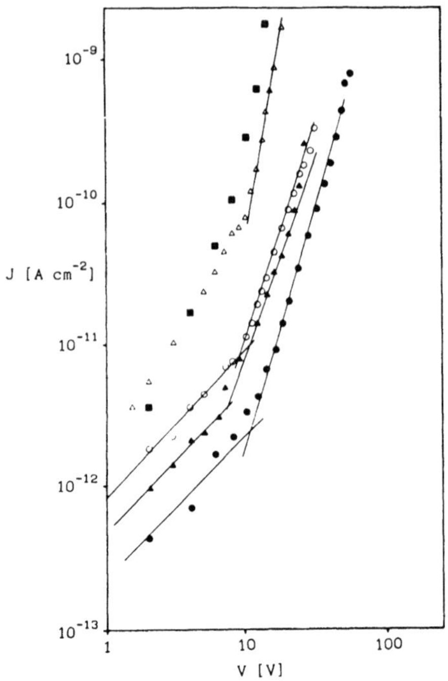

Fig 2. J-V characteristics at low voltage for several devices with different thicknesses. (△) 0.34µm; (■) 0.35 µm; (▲) 0.60 µm; (○) 0.68 µm, and (●) 1.15 µm

Table 1. Band offsets for TCO - ZnS heterojunctions

Material	Bandgap (eV)	ΔE_V (eV)	ΔE_C (eV)
ZnO	3.25	1.70	2.15
CDS	2.06[a]	0.74	2.38
ITO	3.70[b,c]	2.19	2.19

a A.J. Nozik, Phys. Rev. B6, 453 (1972)

b E. Shanthi, V. Dutta, A. Banerjee and K.L. Chopra, J. Appl. Phys. 51, 6243 (1980) have measured a value of 3.95 eV for the bandgap of SnO_2, while:

c J. Fan and J.B. Goodenough, J. Appl. Phys. 48, 3524 (1977) report a 3.5 eV bandgap in In_2O_3. ITO has here been given an intermediate value

The expression describing quantum mechanical tunnelling through a field thinned barrier[3] is incompatible with the existence of a 2 eV barrier between cathode and ZnS. Generally the device conducts current densities of about 1 mA cm^{-2} at a field of about 1.10^6 V cm^{-1}. With a 2 eV barrier the field would have to be about 5 times higher. Additionally it has been observed that devices can be run in either polarity with very similar I-V characteristics. The very sensitive dependence of tunnelling current on barrier height[3], with results to be presented below, precludes tunnelling as the rate limiting mechanism.

In fig 2, current density - voltage (J-V) characteristics are plotted on a log-log scale for several devices of different thicknesses. For thicker devices a power law dependence, $J \propto V^n$ is seen with an n value of 1 at the lowest voltages and a transition to n about 3 as the voltage is increased. In table 2, n is listed for the various devices of fig 2.

This data is characteristic of space charge limited conduction (SCLC) in a semi-insulator with an exponential distribution of trapping states $N_t(E)$, caused by structural discorder or impurities, extending below the conduction band[4]:

$$N_t(E) = N_o \exp\{(E - E_c)/kT_t\} = N_n \exp\{(E - F_o)/ kT_t\} \qquad (1)$$

$$N_n = N_o \exp\{(F_o - E_c)/ kT_t\}$$

where N_o is the density of trapping states at the conduction band edge E_c, F_o is the thermodynamic Fermi level, N_n is the density of trapping states at the Fermi energy and T_t is a temperature parameter characterising the extent of the trap distribution.

This distribution leads to a J-V characteristic of the form[4]:

$$J = e\,\mu\, N_c\{\epsilon/eN_okT_t\}^l\{V^{(l+1)}/d^{(2l+1)}\} \qquad T_t > T \qquad (2)$$

where ϵ is the dielectric constant of the insulator, N_c is the effective density of states in the conduction band and $l = T_t/T$, T is the lattice temperature.

The data of table 2 indicates that $l \approx 2$ for the thicker devices and is larger for the thinner ones. Since experiments were carried out at room temperature, where $kT \approx 0.025$ eV, the data indicates that $kT_t \approx 0.05$ eV.

The voltage of intersection between the Ohms law and the SCLC characteristic yields a value of N_n, the trap density at the Fermi energy[4]. This is also listed in table 2 for each device.

The $0.35.10^{-4}$ cm thick device did not give a straight line on the log-log plot, fig 2. This is probably due to the parameter T_t being too large in this case. Although the derivation of (2) assumes $T_t > T$, if the deviation of the

Table 2. The value of n (see text) and trap density as a function of d

d / x 10^{-4}(cm)	n	$N_n/10^{17}$($cm^{-3}eV^{-1}$)
0.34	5.64	3.48
0.35	-	7.96
0.60	2.95	2.66
0.68	2.97	1.78
1.15	3.18	0.62

electron quasi Fermi level away from F_0 for the current range of interest is less than T_t, the trap distribution is effectively uniform. In that case [4]:

$$J = 2 e \mu n_0 \{V/d\} \exp \{2\epsilon V/N_n kT e d^2\} \quad (3)$$

where n_0 is the thermally generated free carrier concentration.

A plot of ln (J/V) against V yielded a straight line for this device and N_n was calculated from the gradient. This is also listed in the table.

The very small magnitude of the current in the ohmic region indicates that the Fermi level is towards midgap. Thus all devices have considerable trap densities extending well into the gap. Clearly, thinner devices have more traps and they are more evenly distributed as is expected for a ZnS growth process (rf-sputtering) where the considerable power dissipation causes the substrate to heat during growth.

The thickness dependence of the current density, which provides confirmation of SCLC is plotted in fig 3 for data at 10 V.A power law dependence of $J \propto d^{-m}$ has been fitted to the data and m found to be 3.8. Ideally, if the mechanism of (2) is correct, m should be about 5. This discrepancy is probably explained by the varying trap densities. Figure 3 also shows a correction to the data calculated using the trap density in the thickest ($1.15.10^{-4}$cm) sample. This yields m = 5.4, much closer to the expected value.

SCLC necessitates an ohmic injecting contact. It is interesting to recall earlier work on SCLC in ZnS.ALFREY and COOKE[5] obtained an indication of SCLC when indium contacts were annealed to a crystal which they interpreted as being due to formation of an ohmic contact by indiffusion of indium creating a doped surface layer. RUPPEL[6], working with single crystal material, obtained an impressive characteristic consisting of an ohmic region followed by a current steeply rising with applied voltage, indicative of the filling of a discrete trapping level. LAWTHER[7], has observed SCLC in ZnS which had been manganese

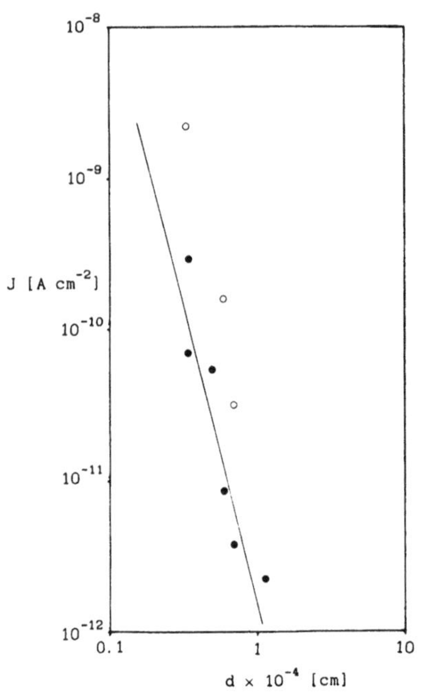

Figure 3. (●) The variation J with d at 10V; (○) corrected for different trap densities

doped by ion implantation. He argues in terms of surface doping by interstitial zinc displaced by manganese.

It has been shown [8] that there is considerable cadmium diffusion into ZnS on annealing the CDS-ZnS heterojunction, a device processing step. It seems likely that the other TCO - ZnS interfaces are not abrupt either and indiffusion of interstitial atoms (Zn or Cd) or dopants (In) may well act to form an ohmic contact in this device.

4. Conclusions

Results discussed in this paper indicate conduction in dc electroluminescent devices to be bulk, rather than contact limited and to be controlled by a density of traps of the order $10^{17}cm^{-3}eV^{-1}$ extending a considerable distance into the bandgap.

Acknowledgements

We wish to thank the staff of RSRE, Malvern, England; Pilkington Bros, Ormskirk, England; and Durham University for providing the samples used in this work.

References

1. A F Cattell, J C Inkson and J Kirton, J Appl Phys 61, 722 (1987).
2. I T McGovern, D Norman and R H Williams: In Handbook on Synchrotron Radiation, ed by G V Marr, Vol 7 (North-Holland, Amsterdam 1987) p 528.
3. R Fowler and L Nordheim, Proc Roy Soc A119 173 (1928)
4. M A Lampert and P Mark: In Current Injection in Solids, (Academic Press, London, 1970).
5. G F Alfrey and I Cooke, Proc Phys Soc 70, 1096 (1957).
6. W Ruppel, Helv Phys Acta 31, 331 (1958).
7. C Lawther, PhD Thesis, University of Kent, Canterbury, England, 1976.
8. M J Davies and R H Williams, Proc 1986 International Electroluminescence Workshop, Oregon.

Low Voltage Driven Electroluminescent Devices with Manganese-Doped Zinc Sulfide Thin Film Emitting Layer Grown on Insulating Ceramics by Metal Organic Chemical Vapor Deposition

T. Minami, T. Miyata, K. Kitamura, H. Nanto, and S. Takata

Electron Device System Laboratory, Kanazawa Institute of Technology,
7-1 Oogigaoka, Nonoichimachi, P.O.Kanazawa-South 921, Japan

1. Introduction

A doubly insulated ac thin film electroluminescent (EL) device is now ready in practical use in flat-panel lamps and displays because it has good luminance and life. However, EL devices of this type still have the most serious problem which is their high driving voltage. In order to overcome this problem, new thin film EL devices using a laminated dielectric ceramic [1] or a thick dielectric ceramic [2] as the insulating layer have been recently proposed. We have reported that a maximum luminance of 6300 [nt] and luminous efficiency of 11 [lm/W] can be observed in the manganese-doped zinc sulfide (ZnS:Mn) thin film EL device (ICEL device) using an insulating ceramic sheet of barium titanate ($BaTiO_3$) as the insulating layer [3]. In this paper, the relationship between EL characteristics and dielectric properties of the $BaTiO_3$ ceramic sheet used as the insulating layer of the ICEL device is investigated.

2. Experimental Procedure

Figure 1 shows the cross-sectional structure of the ICEL device. The $BaTiO_3$ ceramic sheet is used not only as the substrate but also as the insulating layer. The $BaTiO_3$ ceramic sheet was formed by the cold press method [3] or the doctor blade method [2]. In order to obtain ceramic sheets with a thickness of 0.15 - 0.5 [mm], ceramics sintered at a temperature of 1000-1300 [℃] were ground and then polished. The color of the ceramics fabricated by the cold press method regardless of the sintering temperature was white, while that of the ceramics fabricated by the doctor blade method was dark brown. In this paper, $BaTiO_3$ ceramic sheets fabricated by the cold press method were used as the insulating layer as far as it is not indicated. ZnS thin films with a thickness of about 400 [nm] were grown on the ceramic sheets with diethylzinc (DEZ)-carbon disulfide

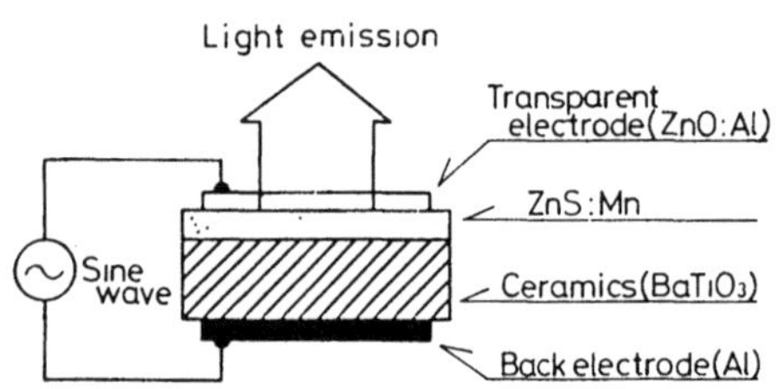

Fig. 1 Cross sectional structure of the ICEL device.

Springer Proceedings in Physics, Vol. 38 Electroluminescence
Editors: S. Shionoya and H. Kobayashi

(CS_2) gas system using metal organic chemical vapor deposition (MOCVD) method. Details of the MOCVD system have been described elsewhere [3,4]. The ceramic sheet substrates set on a susceptor in a quartz reactor were heated at a temperature of 375 [℃]. Manganese (Mn) as a luminescent center was thermally diffused into the ZnS thin films at 700 [℃] in an atmosphere of argon. An aluminum-doped zinc oxide (ZnO:Al) transparent conducting film as a top electrode was deposited by the rf magnetron sputtering [5]. An aluminum backed electrode was evaporated in a vacuum. The measurement of EL characteristics in the ICEL devices was carried out under sinusoidal wave (frequency of 5 [kHz]) voltage driving. The luminance was measured using a conventional luminance meter (Tokyo Kogaku Kikai BM-3). The dielectric properties of ceramic sheets were measured at a frequency of 1 [kHz] using an alternating current bridge.

3. Results and Discussion

We reported that EL characteriastics of the ICEL device are strongly dependent on preparation conditions of a ZnS:Mn emitting layer [3]. On the other hand, it was also found that the EL characteristics of the device with the emitting layer prepared under the same condition are strongly dependent on preparation conditions of $BaTiO_3$ ceramic sheet as the insulating layer. Figure 2 shows the dependence of luminance (L) and luminous efficiency (η) on the applied voltage (V) of typical ICEL devices using the insulating layer of $BaTiO_3$ ceramic sheets fabricated at a sintering temperature of 1050 and 1250 [℃]. It was found that the dielectric constant (ε_s) and dielectric loss ($\tan\delta$) of $BaTiO_3$ ceramic sheets are increased and decreased, respectively, with an increase in the sintering temperature.

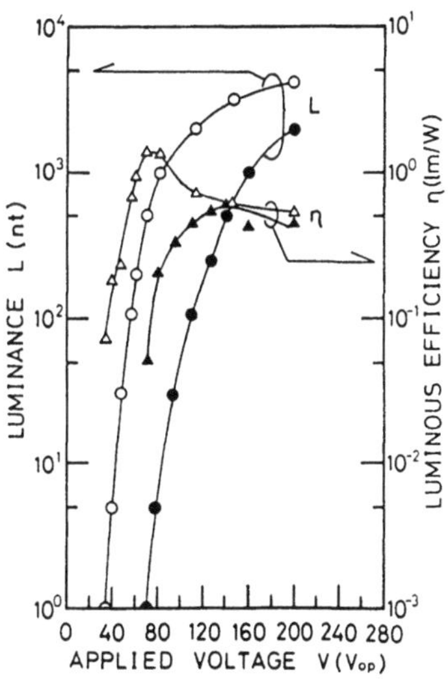

Fig. 2 L-V (○,●) and η-V (△,▲) characteristics of the device with insulating layer of $BaTiO_3$ ceramic sheets fabricated at a sintering temperature of 1050 (solid) and 1250 [℃] (open).

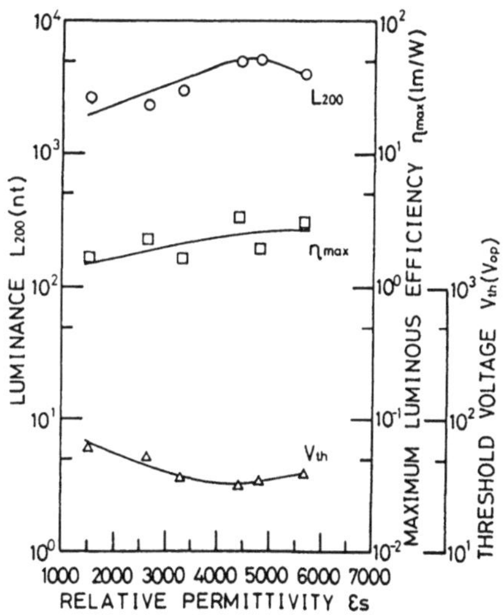

Fig. 3 Dependence of L_{200}, η_{max} and Vth on ε_s of $BaTiO_3$ ceramic sheets.

Therefore, the dependence of L and η on the sintering temperature may be related to the change of these dielectric properties of the ceramic sheets. Figure 3 shows the luminance (L_{200}) at an applied voltage of 200 [V], maximum luminous efficiency (η_{max}) and applied voltage (Vth) at the onset of a light emission of 1 [nt] obtained in the ICEL devices as a function of the ε_s of the $BaTiO_3$ ceramic sheets used as the insulating layer. It should be noted that the Vth exhibits the lowest value at ε_s =4500-5000. This result suggests that the Vth is not controlled by only the ε_s, because the Vth must be monotonically decreased with increasing the ε_s if it is controlled by only the ε_s. The L_{200}, η_{max} and Vth as a function of the $\tan\delta$ of the ceramic sheets are also shown in Fig. 4. It can be seen that the minimum Vth is obtained at $\tan\delta$=3-4 [%], and that the maximum L_{200} and η_{max} are also obtained at $\tan\delta$=2-4 [%]. These results as shown in Figs. 3 and 4 suggest that the low Vth obtained in the ICEL devices is related to the dielectric properties of $BaTiO_3$ ceramic sheets. In order to distinguish the contribution for EL characteristics between ε_s and $\tan\delta$, we fabricated ICEL devices using $BaTiO_3$ ceramic sheets prepared by the doctor blade method. The $\tan\delta$ of the ceramic sheets without change in the ε_s ($\varepsilon_s \approx 1800$) was changed from 0.8 to 7 [%] by the heat-treatment in hydrogen gas. Figure 5 shows the L_{max}, η_{max} and Vth obtained in the ICEL devices as a function of the $\tan\delta$ of the ceramic sheets. The thermal diffusion of Mn impurities into ZnS layer was carried out at 600 [℃]. It was confirmed that dielectric properties of the ceramic sheets are not changed at this diffusion temperature. Note that the minimum Vth is obtained at $\tan\delta \approx 3$ [%] and both the maximum L_{200} and η_{max} are also obtained at $\tan\delta \approx 3$ [%]. These results indicate that the Vth in ICEL devices is mainly influenced by dielectric properties of $BaTiO_3$ ceramic sheets, especially the $\tan\delta$. However, we think that in the more detailed discussion, effects of the color and surface roughness of the ceramics and interdiffusion at interface between the ceramics and ZnS:Mn films on the EL characteristics should be also considered. On the other

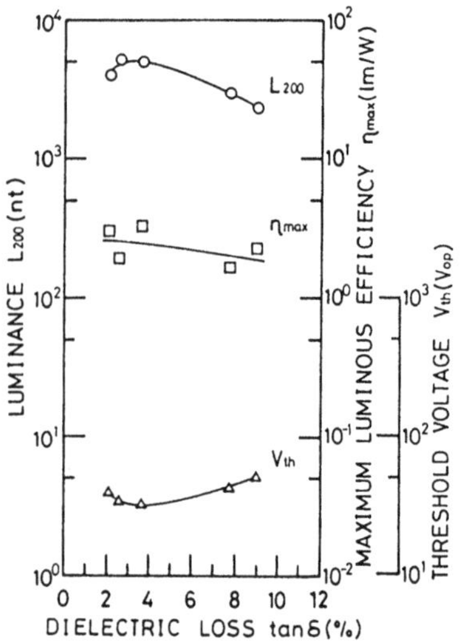

Fig. 4 Dependence of L_{200}, η_{max} and Vth on $\tan\delta$ of $BaTiO_3$ ceramic sheets.

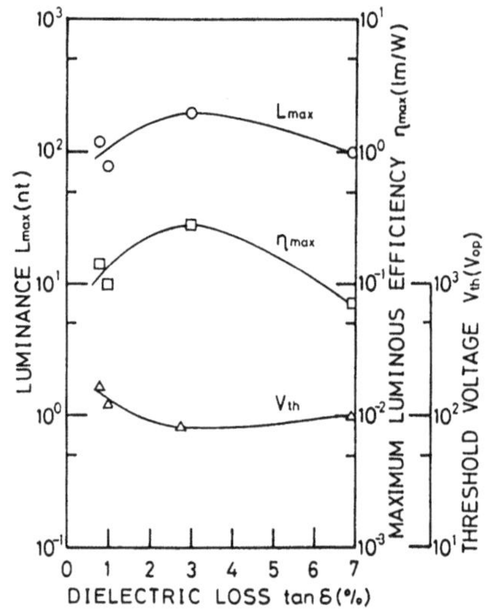

Fig. 5 Dependence of L_{max}, η_{max} and Vth on $\tan\delta$ of $BaTiO_3$ ceramic sheets fabricated by the doctor blade method.

hand, it was found that the Vth is gradually increased with a thickness of $BaTiO_3$ ceramic sheets, whereas the L_{200} and η_{max} are independent of the thickness from 0.15 to 0.5 [mm]. However, L-V characteristics were softened with increasing the thickness. The Vth of 31 [V] and 61 [V] was attained in ICEL devices using $BaTiO_3$ ceramic sheets with the thickness of 0.15 and 0.49 [mm], respectively.

4. Conclusions

The dependence of EL characteristics on preparation conditions of a $BaTiO_3$ ceramic sheet in ICEL devices was investigated. The following results were obtained: (1) The EL characteristics are strongly dependent on dielectric properties of $BaTiO_3$ ceramic sheet as the insulating layer. (2) The Vth is mainly influenced by the dielectric loss of the ceramic sheet as well as by the dielectric constant. (3) The good EL characteristics are obtained in devices using $BaTiO_3$ ceramics with a $\tan\delta$ of 2-4 [%]. (4) The Vth and L-V characteristics are increased and softened, respectively, with increasing the thickness of $BaTiO_3$ ceramic sheets. Thus we can conclude that in ICEL devices without a breakdown of the insulator, the insulating layer is best made from materials with a suitable dielectric loss of 2-4 [%].

Acknowledgement

The authors wish to thank M. Ikeuchi, H. Uesugi and Y. Honda for their technical assistance during the experiments. The authors are also grateful for preparation of ceramics fabricated by the doctor blade method of Dr. K. Tanaka of Murata MFG. Co., Ltd. and for preparation of $BaTiO_3$ fine powder of Central Glass Co., Ltd.. This work was supported in part by Grant-in-Aid for Scientific Research No. 62550019 from the Ministry of Education, Science and Culture of Japan.

References

1. Y. Sano, K. Nunomura, N. Koyama, K. Utsumi and H. Sakuma: Proc. 1985 Int. Display Research Conf., San Diego, California, 1985, Society for Information Display, California, 1985, p.173.
2. T. Minami, S. Orito, H. Nanto and S. Takata: Proc. Society for Information Display, vol.29/1 (1988) 83.
3. T. Minami, T. Miyata, K. Kitamura, H. Nanto and S. Takata: Jpn. J. Appl. Phys., 27 (1988) L876.
4. S. Takata, T. Minami, T. Miyata and H. Nanto: J. Crystal Growth 86 (1988) 257.
5. T. Minami, H. Nanto and S. Takata: Jpn. Appl. Phys. 23 (1984) L280.

Sound Emitting Thin Film Electroluminescent Devices Using Piezoelectric Ceramics as Insulating Layer

T. Minami, T. Miyata, K. Kitamura, H. Nanto, and S. Takata

Electron Device System Laboratory, Kanazawa Institute of Technology,
7-1 Oogigaoka, Nonoichimachi, P.O.Kanazawa-South 921, Japan

1. Introduction

A novel electroluminescent (EL) device which emits a sound as well as a light is proposed. This sound emitting EL device has the same structure as the insulating ceramic EL (ICEL) device reported in recent years [1,2]. The EL device consists of a thin film light emitting layer and a piezoelectric ceramic plate. The ceramic plate which is substrate of the device acts not only as a sound emitter but also as an insulating layer. In this paper, we demonstrate the sound emitting EL device fabricated using a manganese-doped zinc sulfide (ZnS:Mn) thin film as the light emitting layer and a barium titanate ($BaTiO_3$) ceramic plate as the insulating layer. Details of the response of electroluminescence and sound emission in this device to an applied alternating current (ac) voltage in the audio frequency are also described.

2. Device Preparation

A cross sectional structure of the sound emitting EL device which has metal-insulator-semiconductor (MIS) structure is shown in Fig. 1. A $BaTiO_3$ fine powder was molded into a disk with a pressure of 4 [ton/cm^2] using a conventional cold press method. The disk was then sintered at a temperature of 1300 [℃] in air. In order to obtain a ceramic plate with a thickness of 0.2-0.5 [mm], the sintered ceramic was ground and then polished. The ZnS thin film was grown on the ceramic plate at 375 [℃] with diethylzinc (DEZ)-carbon disulfide (CS_2) gases using metal organic chemical vapor deposition (MOCVD) method. The gas pressure during the growth was 3.5×10^4 [Pa]. The atomic ratio of sulfur to zinc (S/Zn) was 20. Details of the MOCVD system have been described elsewhere [2,3]. After the growth of ZnS thin film, Mn impurity as a luminescent center was thermally diffused into

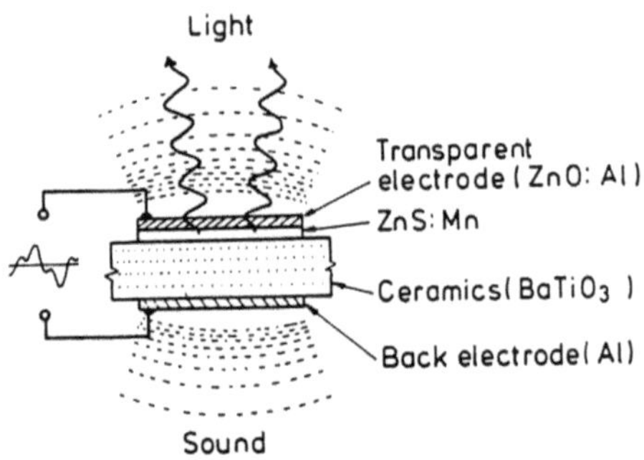

Fig. 1 Cross sectional structure of the sound emitting EL device.

Springer Proceedings in Physics, Vol. 38 Electroluminescence
Editors: S. Shionoya and H. Kobayashi

the ZnS thin film at 700 [℃] in an atmosphere of argon. An aluminum-doped zinc oxide (ZnO:Al) transparent conducting film as a top electrode was deposited on the ZnS:Mn thin film by the rf magnetron sputtering [4] and an Al electrode was vacuum-evaporated on the back surface of the $BaTiO_3$ ceramic plate. A poling treatment was carried out for some of the EL devices in order to obtain the $BaTiO_3$ ceramic with a high piezoelectricity. In the poling treatment, the temperature of the device was lowered from 200 [℃] to room temperature in air under an applied direct current voltage of 200 [V]. On measurements of acoustic characteristics, the devices were supported at each one point on both electrodes. All sound emitting EL devices were driven by sinusoidal wave voltage in the frequency range of 0.1-10 [kHz]. The luminance was measured using a conventional luminance meter. The sound pressure was measured using a sound level meter with a flat sensitivity over the audio frequency. The measurements were carried out in the anechoic room keeping the noise level below 40 [dB]. The distance between the sound level meter and the device was 5 [cm].

3. Results and Discussion

Figure 2 shows the frequency dependence of the sound pressure (Sp) for a typical EL device driven at various applied voltages (V). The thickness of ZnS:Mn emitting layer and ceramic plate was 370 [nm] and 0.26 [mm], respectively. Since the Sp increases with increasing frequency in the range of 0.2-7 [kHz], it can be seen that the device is useful as a sounder at high frequency in the audio range. The wavy spectra of Sp may be ascribed to spurious structures resulting from the resonance between the holder and the device. The frequency dependence of luminance (L) for the same device as shown in Fig. 2 is shown in Fig. 3 for various applied voltages. Note that the L is directly proportional to the frequency below 5 [kHz]. It can be, therefore, seen from Figs. 2 and 3 that both

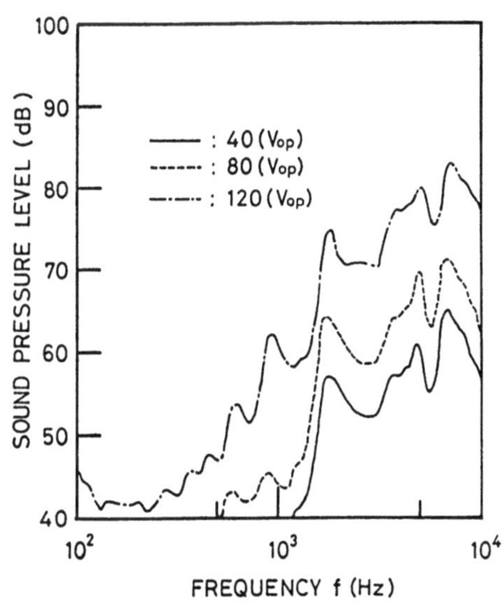

Fig. 2 Frequency dependence of Sp for devices driven at various voltages.

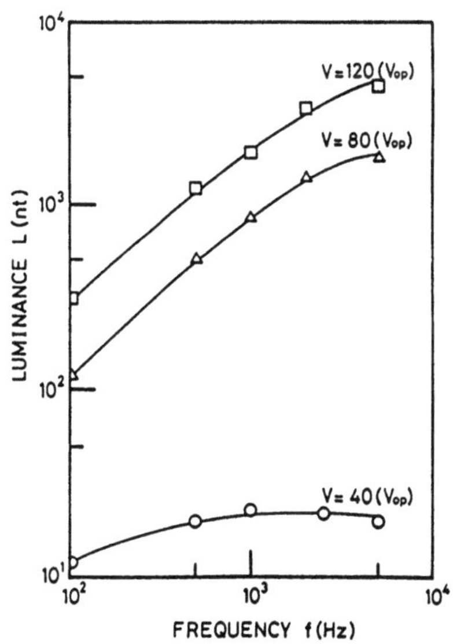

Fig. 3 Frequency dependence of L for devices driven at various voltages.

L and Sp increase with increasing frequency in the range of 0.2-5 [kHz]. Figures 4 and 5 show the Sp-V and L-V characteristics of an EL device driven at voltages with various frequencies, respectively. The thickness of ZnS:Mn emitting layer and $BaTiO_3$ ceramic plate was 580 [nm] and 0.27 [mm], respectively. It can be seen that the sound is emitted at a lower side of V as the frequency increases because the Sp increases with increasing the V and frequency. However, the L-V characteristics were strongly dependent on the frequency, whereas the applied voltage (threshold voltage) at the onset of light emission was nearly independent of the frequency. Note that the sound compared with the electroluminescence is emitted at a low V, and the sound emission does not exhibit the threshold voltage. The high luminance of above 8500 [nt] and sound pressure of above 85 [dB] were obtained for the EL device driven at V=200 [V] with a frequency of 5 [kHz]. It can be, therefore, concluded that the sound emitting EL devices always emit a loud sound when the devices emit an intense light. On the other hand, it was found that the Sp-V and L-V characteristics as well as the frequency dependence of Sp and L in the devices are strongly dependent on preparation conditions of the $BaTiO_3$ ceramic plate. It was also found that the acoustic characteristics of the devices are improved using $BaTiO_3$ ceramics with the poling treatment. The Sp of the devices after the poling treatment was increased by a factor of about 1.3, whereas the EL characteristics of the devices were not changed. Figure 6 shows waveforms of the V and the corresponding I and L of a typical sound emitting EL device driven by sinusoidal wave voltage. It should be noted that the light emission of the device is observed only when the ZnO:Al transparent electrode is positively biased, whereas the light emission at both polarities of applied voltage has been generally observed in thin film EL devices having MIS structure [5,6]. The response of L on the V, as shown in Fig. 6 was usually observed regardless of thicknesses of ZnS:Mn emitting layer and the $BaTiO_3$ ceramic plate and the driving

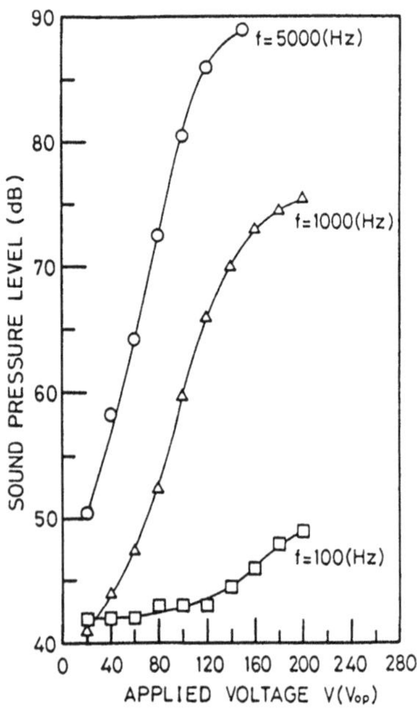

Fig. 4 Sp-V characteristics of devices for various driving frequencies.

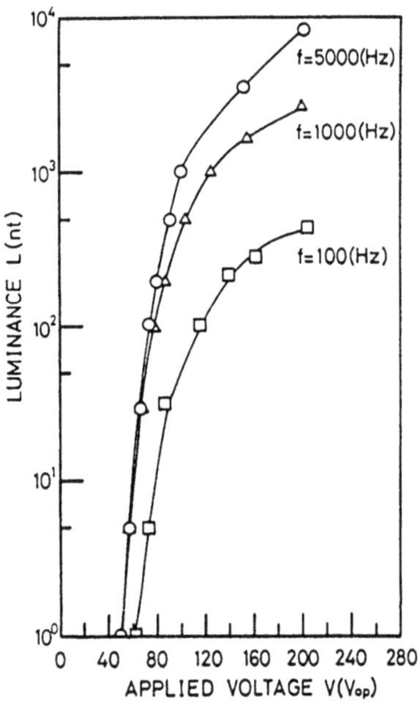

Fig. 5 L-V characteristics of devices for various driving frequencies.

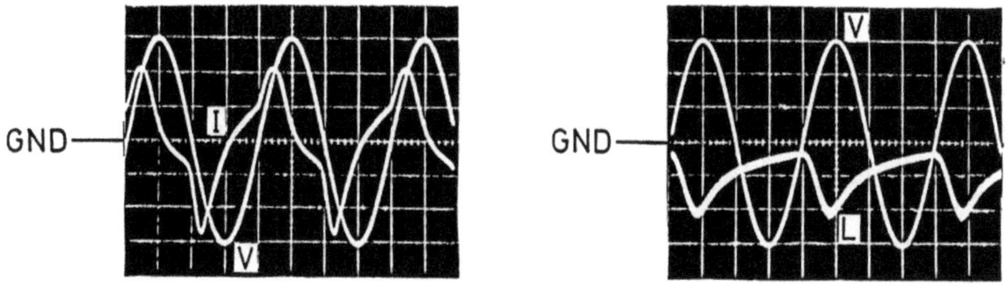

Fig. 6 Waveforms of V, I and L under driving at a frequency of 500 [Hz].

voltage and frequency. It is, however, thought that the waveform of L, as shown in Fig. 6 is a typical response as expected for ac EL devices having the MIS structure. This might suggest that any potential barriers have not been formed on the interface between ZnS:Mn emitting layer and ZnO:Al transparent electrode.

4. Conclusions

The novel sound emitting EL device which consists of a thin film light emitting layer and a piezoelectric ceramic plate has been proposed. The ceramic plate which is substrate of the device acts not only as a sound emitter but also as an insulating layer. The EL and acoustic characteristics have been described for the device fabricated using a ZnS:Mn thin film as the light emitting layer and a $BaTiO_3$ ceramic plate as the insulating layer. The sound pressure increased with increasing the frequency in the range of 0.2-7 [kHz], whereas the luminance was directly proportional to the frequency below 5 [kHz]. The devices always emitted a loud sound when the devices emitted an intense light. The high luminance of above 8500 [nt] and sound pressure of above 85 [dB] were obtained for the device driven at a voltage of 200 [V] with a frequency of 5 [kHz]. Thus, a newly proposed EL device which emits the light as well as the sound is useful as an EL lamp with loudspeaker.

Acknowledgements

The authors wish to thank Y. Asada, O. Hasegawa and T. Muramoto for their technical assistance in the experiments. The authors are also grateful for preparation of $BaTiO_3$ fine powder of Central Glass Co., Ltd.. This work was partially supported by a Grant-in-Aid for Scientific Research No.62550019 from the Ministry of Education, Science and Culture of Japan.

References

1. T. Minami, T. Miyata, K. Kitamura, H. Nanto and S. Takata: Jpn. J. Appl. Phys., 27 (1988) L876.
2. S. Takata, T. Minami, T. Miyata and H. Nanto: Jpn. J. Appl. Phys., 27 (1988) L247.
3. S. Takata, T. Minami, T. Miyata and H. Nanto: J. Crystal Growth 86 (1988) 257.
4. T. Minami, H. Nanto and S. Takata: Jpn. J. Appl. Phys., 23 (1984) L280.
5. K. Okamoto and Y. Hamakawa: Appl. Phys. Lett., 35 (1979) 508.
6. H. Kozawaguchi, J. Ohwaki, B. Tsujiyama and K. Murase: SID International Symposium Digest of Technical Papers, 13 (1982) 126.

Characteristics of ZnO:Al Transparent Conductive Films

H. Kawamoto, R. Konishi, H. Harada, and H. Sasakura

Department of Electronics, Tottori University, Koyama, Tottori 680, Japan

Conduction mechanism of sputtered ZnO:Al thin film was studied for developing heat resistant transparent electrodes of EL devices. Strong evidence of the formation of the Al donors in ZnO:Al films was obtained by XPS measurements. X-ray diffraction measurements showed that the decrease of the mobility with increasing Al concentration above 1.3% was partly due to the low crystallinity of the films.

1. Introduction

It is well known that performance of a thin film electroluminescent(EL) display is improved with increasing heat treatment temperature during the formation of an emission layer[1]. As transparent electrodes are prepared before the formation of the emisson layer, they are also exposed to high temperature during heat treatment. Therefore, heat-resistant material is desirable for the transparent electrodes. From this view point, ZnO:Al is worthwhile to study as well as conventional indium tin oxide(ITO).

Several studies have been done on the conductivity or the crystal structure of ZnO:Al thin films[2], [3]. Although both of Al atoms and oxygen vacancies are considered to be donors in the ZnO:Al semiconductor films, it is not yet clear which are predominant. As the Al donors are presumably stable compared with the oxygen vacancies, it may be important to establish the conditions for forming the Al donors in preference to the oxygen vacancies. Using various techniques of surface physics, we try to clarify the conditions.

2. Experimental

The ZnO:Al thin films were sputtered onto a glass substrate by facing targets type sputtering[4] at O_2 pressure 1×10^{-1} Pa and thickness of the films were 250 nm. Concentration of Al was adjusted by changing the ratio of areas of Zn and Al in the target. Structure and Al-bonding states of the films were measured by X-ray photoelectron spectroscopy(XPS) and X-ray diffraction, respectively.

3. Results and Discussions

The prepared ZnO:Al films showed no conductivity immediately after the sputtering. By increasing the temperature of the films in vacuum, their conductivity slowly increased at first(region A), and it abruptly increased at 400°C as shown in Fig. 1(region B). The increase of the conductivity of region A below 400°C was observed similarly both in the ZnO and ZnO:Al films. On the other hand, the conductivity of the ZnO:Al films rapidly increased compared with that of the ZnO films at 400°C . This suggests that the increase below 400°C was due to the formation of oxygen vacancies, whereas the increase at 400°C was mainly due to Al donors.

Springer Proceedings in Physics, Vol. 38 **Electroluminescence**
Editors: S. Shionoya and H. Kobayashi

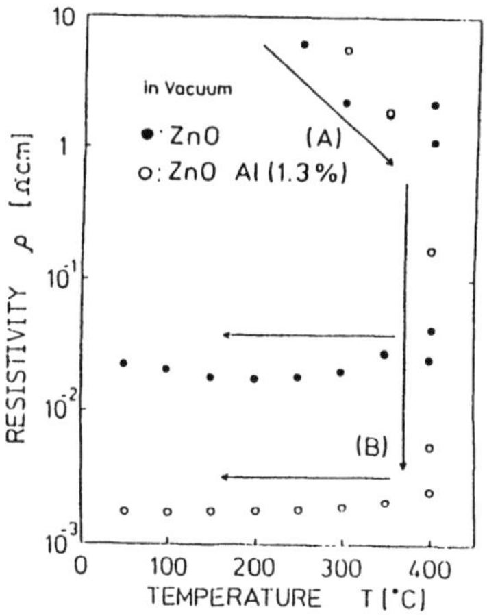

◁ Fig. 1 Change of the conductivity of the ZnO:Al films with heat treatment temperature

▽ Fig. 2 XPS spectra of Al 2p in the ZnO:Al film (Al concentraton 5%)

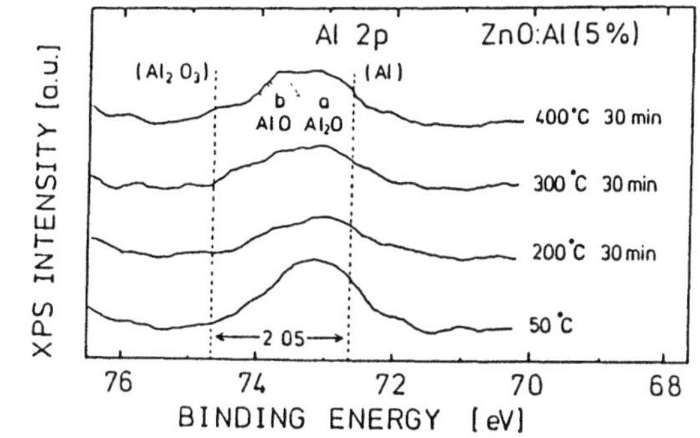

In order to confirm the above suggestion, XPS measurements were made for the ZnO: Al films of Al concentration 5%. Figure 2 shows XPS spectra of Al 2p. Immediately after the deposition by sputtering, the spectra has a peak at 73.20 eV (peak a), which is between peaks of Al(72.65 eV) and Al_2O_3(74.70 eV)[5]. After the heat treatment at 400°C, 30 min., a new peak appears at 73.75 eV(peak b) in the spectrum. From the consideration of peak shifts, peak a and peak b are attributed to Al_2O and AlO, respectively. This consideration seems to be important. When Al atoms occupy the substitutional points of Zn atoms in ZnO crystals, the formed Al compounds are regarded to be AlO, because valency of Al atoms is two in this case. Therefore, present results give the direct evidence that Al atoms substitute Zn atoms in ZnO:Al films after the heat treatment and they should act as Al donors.

Figure 3 shows the dependence of the resistivity of the ZnO:Al films upon the heat treatment time at 400°C. The film of Al concentration 1.3% showed minimum resistivity 1.2×10^{-3} Ω·cm. With increasing Al concentration, saturation of resistivity seems to occur at the shorter heat treatment time. This also suggests that conduction due to Al donors is more predominant than intrinsic conduction due to oxygen vacancies. Figure 4 shows dependences of the resistivity, Hall mobility and carrier concentration upon the Al concentration. Both the carrier concentration and Hall mobility increase at first with an increasing Al concentration up to 1.3%. The reason for the increasing mobility is not explained yet. Increasing the concentration further, we notice the Hall mobility decreases while carrier concentration seems to saturate. Therefore, the resistivity decreases at first and reaches the minimum, and it increases after that.

X-ray diffraction patterns of the films had the (0002) peak of ZnO crystals as shown in Fig. 5, when the Al concentration was from 1.3% to 5%. The target was Fe and a small (0002) peak due to Kβ appeared in lower angle. Immediately after the deposition, the peak was asymmetric and it had a tail in the higher side of the diffraction angle. After the heat treatment, the tail disappeared and the peak shifted to the higher side of the diffraction angle. This results can be explained as follows. Immediately after the deposition, the ZnO:Al films are composed of strained crystallites. Therefore, the X-ray diffraction pattern gives the asymmetric peak, which can not be attributed to the mixture of Al_2O_3 and ZnO [3], because XPS measurements do not show the presence of Al_2O_3. After the heat treatment, strain of the crystallite is removed and the X-ray diffraction pattern gives the shifted symmetric peak. For the ZnO:Al films of Al concentration 5%, X-ray diffraction pattern showed the broad peak. This indicates that the crystallinity of the film was extremely low. Therefore, the decrease of the mobility above Al concentration 1.3% is partly attributed to the impurity scattering and partly attributed to the low crystallinity of the ZnO:Al film.

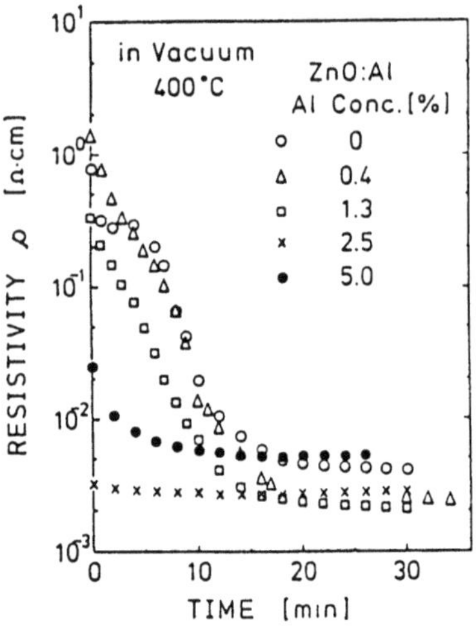

Fig. 3 The dependence of the resistivity of the ZnO:Al films upon the heat treatment time at 400 °C

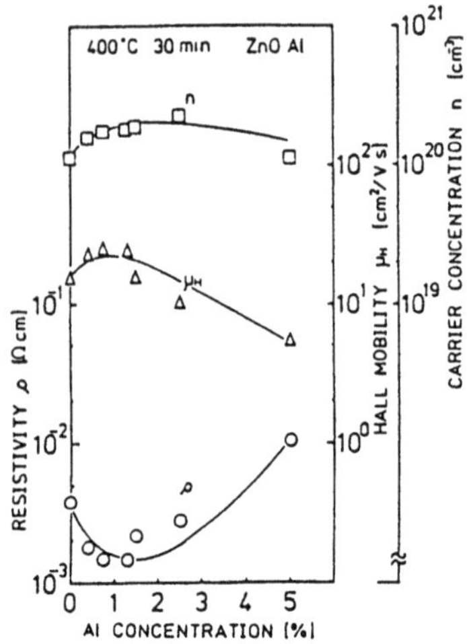

Fig. 4 The dependences of the resistivity, Hall mobility and carrier concentration upon the Al concentration

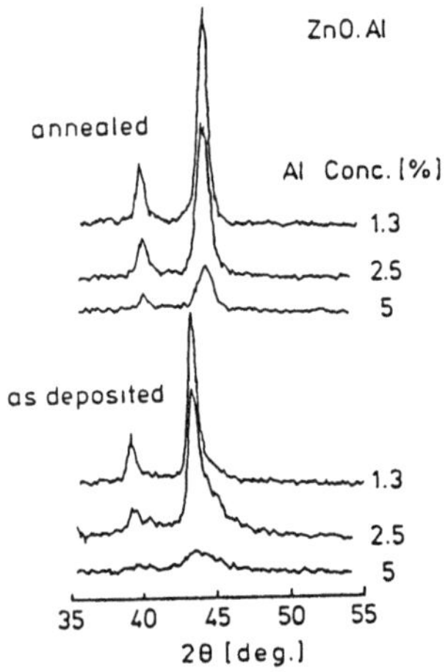

Fig. 5 X-ray diffraction patterns of the ZnO:Al films
(Al concentration 1.3%, 2.5% and 5%)

4. Conclusion

(1) After the heat treatment of sputtered ZnO:Al films, their conductivity increased mainly because of the increase of the carrier concentration. The increase of the carrier concentration below 400°C was mainly attributed to the formation of oxygen vacancies, and the increase at 400°C was due to the formation of Al donors for the most part.

(2) The strong evidence of formation of the Al donors was obtained by XPS measurements. The optimum condition for the formation of the Al donors in the present study was (a) Al concentration 1.3% and (b) heat treatment 400°C, 30 min.

(3) Changing the Al concnetration from zero to 1.3%, both of the carrier concentration and the mobility increased. The reason for the increase of the mobility is not explained yet.

(4) Increasing the Al concnetration above 1.3%, the carrier concentration saturated and the mobility decreased. The decrease of the mobility was partly due to the impurity scattering and partly due to the low crystallinity of the films. X-ray diffraction measurements supported this conclusion.

References

1. H.Sasakura, H.Kobayashi, S.Tanaka, J.Mita, T.Tanaka and H.Nakayama: J.Appl.Phys. 52 (1981) 6901.
2. T.Minami, H.Nanto and S.Takata: Jpn.J.Appl.Phys. 23 (1984) 280.
3. Y.Igasaki, N.Inoue, S.Tsutsui and G.Shimaoka: Shinku 31 (1988) 511. (in Japanese)
4. M.Naoe, S.Yamanaka and Y.Hoshi: IEEE Trans.Magn. 16 (1980) 646.
5. C.D.Wagner, W.M.Riggs, L.E.Davis, J.F.Moulder: Handbook of X-ray Photoelectron Spectroscopy, G.E.Muilenberg, (Perkin-Elmer, Minnesota, 1979).

Recent Developments and Trends in Thin-Film Electroluminescent Display Drivers

S. Sutton and R. Shear

Texas Instruments, P.O. Box 655303, MS 3672, Dallas, TX75265, USA

1. Introduction

The development of the thin-film electroluminescent display from a laboratory technology to a viable commercial product has in large part been made possible by advances in integrated circuit technology. Prior to the early 1980's, driving a display required thousands of discrete transistors in a configuration that was bulky, unreliable and costly. Integrated circuits pioneered by Sharp, Supertex, and Texas Instruments enabled system designers to overcome these shortcomings and led to the first commercial TFEL products. Those early products have since matured, responding to market demands for increased reliability, lower power, and reduced cost. The driver ICs have responded to these demands as well with symmetric row drivers, low power column drivers, and innovative new packaging.

2. Asymmetric vs. Symmetric Row Drive

Asymmetric row drive has been the industry standard from 1983 to 1986. Its use requires a positive refresh pulse with a magnitude just below the panel's threshold. This pulse is used solely to reverse the charge on the capacitance associated with all the rows so that light emission can occur during the next frame. Power is wasted since none of the energy expended during the refresh pulse emits light. Also, since the on pixels are charged to a lower potential during the refresh pulse than during the write portion of the cycle, a net charge imbalance occurs. This charge imbalance can cause the "burned-in pixel effect" which results in the retention of a faint image of any pattern that remains on the panel for a long period of time. Asymmetric drive requires 225V open-drain row drivers and is still in use in many systems.

Light is emitted during both positive and negative row pulses when symmetric row drive is used. During the negative row pulses, the selected columns add to the magnitude of the row driver voltage as in asymmetric drive. However, instead of one refresh pulse in which all the rows are taken positive, in symmetric drive each row is addressed separately. In this case the selected columns subtract from the row driver

Springer Proceedings in Physics, Vol. 38 Electroluminescence
Editors: S. Shionoya and H. Kobayashi

voltage. Therefore, inverted data is run through the columns during the positive row pulses. In this manner light is emitted during both the charging and discharging of the pixels. Also, no net charge imbalance results. These advantages make symmetric drive the most popular drive scheme today.

Symmetric drive requires a 225V tri-state row driver or complementary pair of n-channel and p-channel open-drain drivers. A two chip symmetric drive solution, while solving the problems associated with asymmetric drive and offering the simplest IC fabrication, has the drawbacks of high cost and increased consumption of circuit board area. High cost results from the redundant logic required on both the P-channel and N-channel devices and the need for two packages. The single chip 225V symmetric row driver eliminates these drawbacks and is a significant development in driver technology. This device addresses both reliability and power reduction market demands at a significantly lower cost than two chip solutions. Its 225V tri-state outputs have doubled what previously had been the limit for high voltage capability for this type of integrated circuit.

One way to fabricate single chip symmetric row drivers is to use N-channel devices only for the upper and lower stages. N-channel devices are inherently more efficient than P-channel devices. Therefore, these devices occupy less die area than their P-channel counterparts. This is significant as the output transistors are necessarily the largest components on the die in order to support the current and voltage requirements. However, this approach requires the most complex interface circuitry between the low and high voltage sections of the chip and moderate process complexity. Another approach is to use a N-channel lower stage device and a P-channel upper stage device. This approach allows simple interface circuitry but requires a more complex process.

Single chip symmetric row drivers have been fabricated with self, junction, and dielectric isolation resulting in a tradeoff between cost and latch-up immunity. These approaches are illustrated in Fig. 1. The self isolated technique requires the fewest number of process steps; however, it is very difficult to prevent parasitic bipolar transistor action. The process is compatible with standard MOS processes. A junction isolated approach requires a moderately complex process with a very thick epitaxial layer and double-diffused isolation but does not have the susceptibility to latch-up. It is most compatible with standard linear processes. Oxide isolation offers excellent immunity from parasitics and is the best approach to minimize component area. However, the processing is complex and can be extremely costly.

Row drivers that are built with processes other than oxide isolation have inherent output diode parasitics. A parasitic PNP that is associated with the upper stage diode causes on chip power dissipation and creates undesired substrate currents that can lead to debiasing. The parasitic transistor

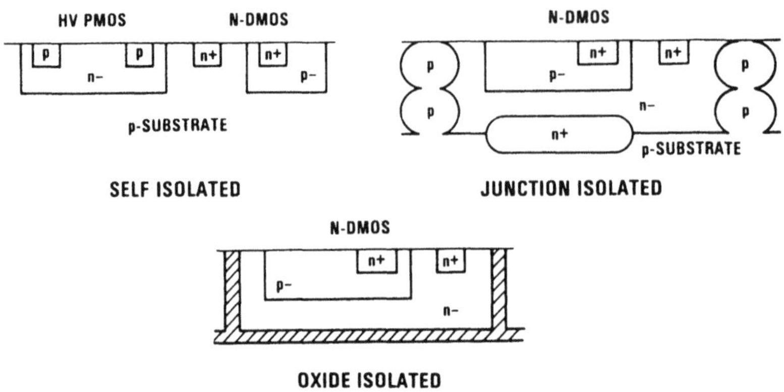

Figure 1. Row Driver Technology Configurations

gain must be minimized through process and layout techniques. A parasitic NPN that is associated with the lower stage diode can draw current from adjacent epi tanks and thus turn on devices that should remain off. The parasitic NPN is in a VBCES mode thus reducing the effective diode breakdown voltage. One way to eliminate this problem is to use both a N-type buried layer and a P-type buried layer and an epi tank tied to a low voltage potential that surrounds the diode tank. Thus, the transistor action is confined to the collector at low voltage potential which causes no harm. These parasitic devices along with the parasitic NPN solution are shown in Figure 2.

A row driver issue that affects reliability is output ruggedness. One must insure that the spreading resistance of the backgate pinch resistor is low enough to avoid turning on the parasitic NPN base-emitter junction between the backgate and source. This NPN may be turned on by an IR drop that results from avalanche injection current flowing in the channel area. If the backgate does not become debiased from the source, the parasitic NPN VBCBO limits the device breakdown. However, if the backgate does debias, then the transistor may go into a VBCES mode which results in a lower breakdown and can cause second breakdown and thermal runaway currents that lead to device destruction. This event can occur during a fault condition when an output is suddenly turned on with full voltage across it. Increased backgate contact and minimization of the parasitic resistor is key to increasing the SOA.

3. New Column Drivers Offer Power Savings and Speed Gains

Column driver advances have been in the area of reduced power consumption and increased data rates. Power consumption is proportional to frequency and the square of the logic voltage. Early metal gate CMOS designs offered a shift register speed of 8 Mhz at 12V and 2 Mhz at 5V. The system data rates are double the shift register data rates since column data is

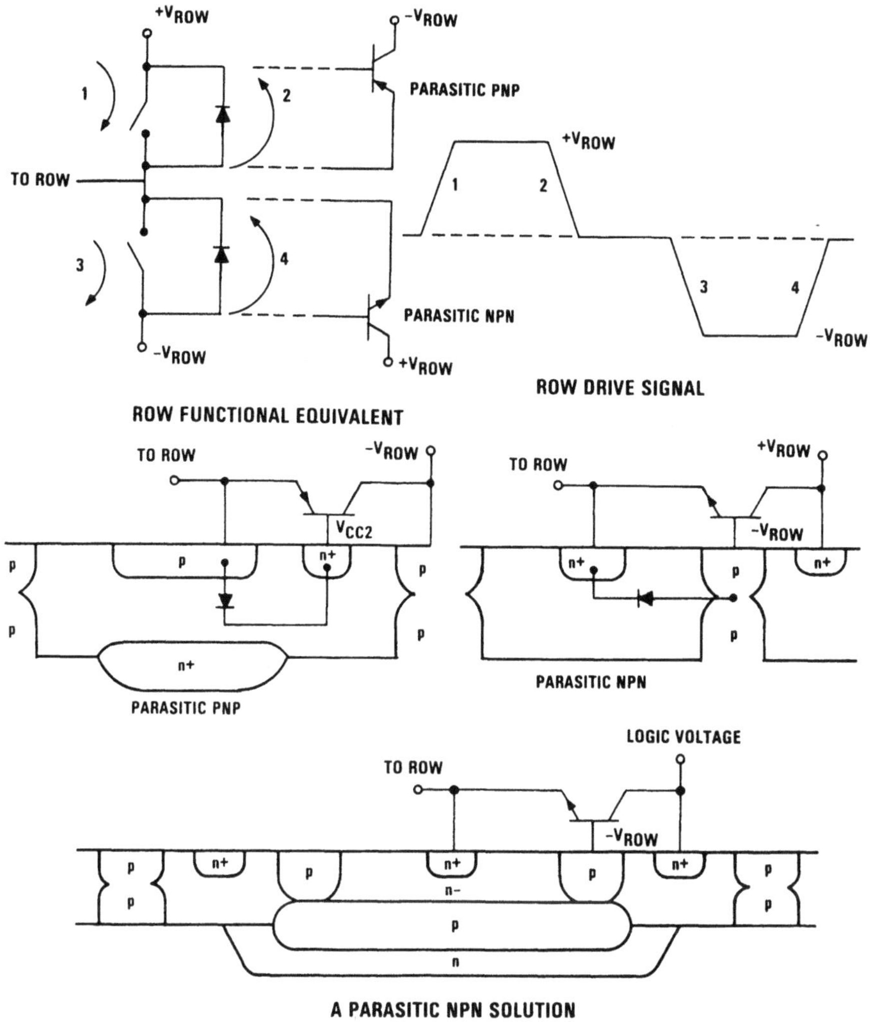

Figure 2. Row Driver Output Parasitics

typically loaded into drivers on both sides of the panel simultaneously. Second generation designs in silicon gate CMOS allow 20 Mhz. shift register speeds at 12V and 10 Mhz. speeds at 5V. This design is typically used in 5V systems to reduce power consumption. Another option that increases data rates is dual shift registers within a column driver. This dual data path approach requires the column data to be split into odd and even data on both sides of the panel. This approach is generally used in large panels where speed or power is critical. The reduced frequency cuts the power in half. 5V shift register logic and parallel data paths reduce system power dissipation by up to 2 watts. Shift register power vs. the clock frequency and the logic voltage for a 640 column system is shown in Figure 3. Zero bias level shift

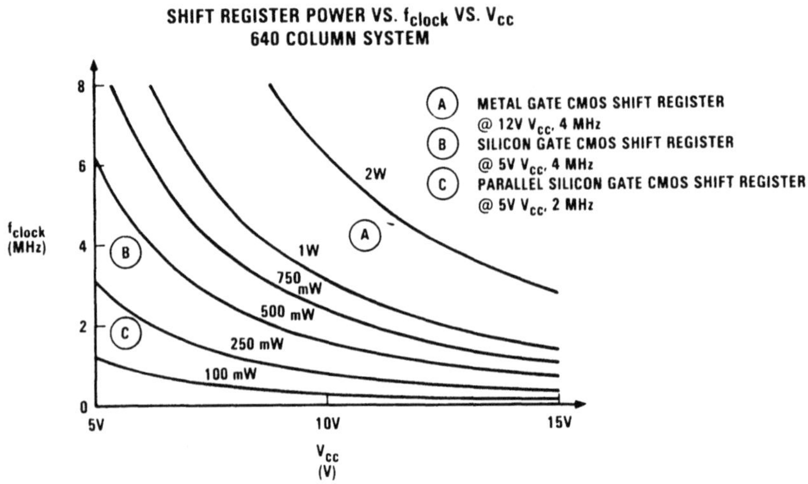

Figure 3. Shift Register Power vs. Logic Voltage for a 640 Column System

schemes also contribute to reduced system power with the drawback of high susceptibility to ground current induced latch-up. Passive level shift schemes have been proposed as a compromise.

Output clamp diode performance has been optimized to enable energy recovery drive schemes capable of multi-watt system power reduction. System power is reduced by pumping charge back into the power supply capacitor during the column discharge. In the past the parasitic PNPs to substrate and parasitic NPNs to epi tanks tied to adjacent outputs have reduced the efficiency of such energy recovery schemes. The efficiency is defined by the ratio of the current that charges the power supply capacitor divided by the total current as is illustrated in Figure 4. However, new column driver designs are eliminating such parasitics by using oxide isolation or reducing the parasitics with buried layers and sinker diffusions. Future column driver development will address output amplitude modulation for gray scale applications.

4. Package Developments

The first package that was used for row and column drivers is the 44 pin PLCC. This package is in use today in many systems and is still the favored package for some small panels. PCB layout is facilitated by offering devices pairs with clockwise and counterclockwise pinouts of the outputs or a control pin that controls right/left shift. New package development has centered around ease of PCB layout and assembly. 3-sided 64 pin packages are the choice for new driver designs offering single sided PCB layout, even thinner system assemblies, and clearly visible solder joints that can be inspected during assembly. Efforts to package drivers using tab technologies

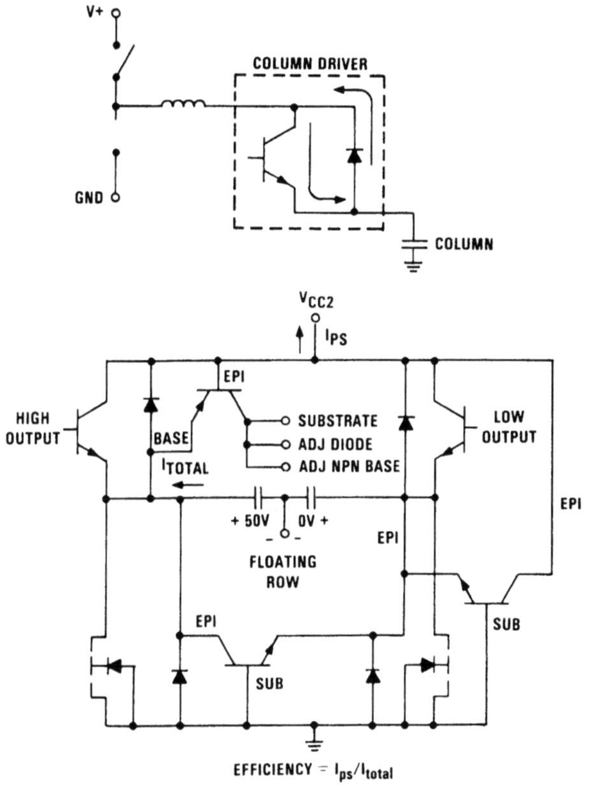

Figure 4. Column Energy Recovery

or "chip on glass" configurations have to date proven disappointing with cost and reliability cited as the major shortcomings.

5. Summary

These developments in driver and packaging technology are only the beginnings of a continuing trend to address key market demands which are reliability, low power, and reduced cost. Drivers in the future will reflect these trends with emphasis placed on output ruggedness, energy conservation techniques, and reduced functionality. It is only through developments in these areas that TFEL systems can continue to compete effectively in the flat panel marketplace.

Bidirectional Push-Pull Symmetric Driving Method of Thin Film Electroluminescent Display

K. Shoji, T. Ohba, H. Kishishita, and H. Uede

LCD Division, Sharp Corporation, Koriyama, Nara, Japan

1. Introduction

The TF-EL display has a wide range of applications in the high technology field such as computer, controller and electronic measurement. By the way, field refresh driving method [1] and p-n symmetric driving method [2],[3], already in practical use, are designed for driving EL displays of relatively small capacity. However, the demand of larger display capacity and lower cost is increasing. It has urged to develop a new driving method which realizes the following points.

1. Shortened driving time for increased number of scanning lines.
2. Reduced power consumption for increased number of scanning lines and increased capacity.
3. Assured long term reliability and luminance uniformity.

Focusing on these requiremensts, we have developed a new driving method called "Bidirectional push-pull symmetric driving method"(abbreviated as p-p symmetric driving method). We have applied high voltage push-pull driver IC to data side and scanning side driver circuits. We report here on the main developed points.

2. Construction of Driver Circuit

Fig.1 illustrates the construction of an X-Y matrix TF-EL panel and of high voltage driver circuits used in the p-p symmetric driving method. The output stage diode of the data side driver IC takes out the electric charge accumulated in the EL panel to be returned to an external capacitor. This diode is not just parasitic or just for protection. It also serves as a recovery element. Therefore, the driver IC has a bidirectional function. "3" is connected to the high voltage power line of the scanning side IC and "4" is connected to the IC GND line. Therefore the scanning side IC is used in a floating potential, and control signals are input through the isolator. "5" is connected to the high voltage power line. The recovery driver is recovering electric charge accumulated in the EL panel. Control signals of the data side IC are input on a logical voltage level.

Springer Proceedings in Physics, Vol. 38 Electroluminescence
Editors: S. Shionoya and H. Kobayashi

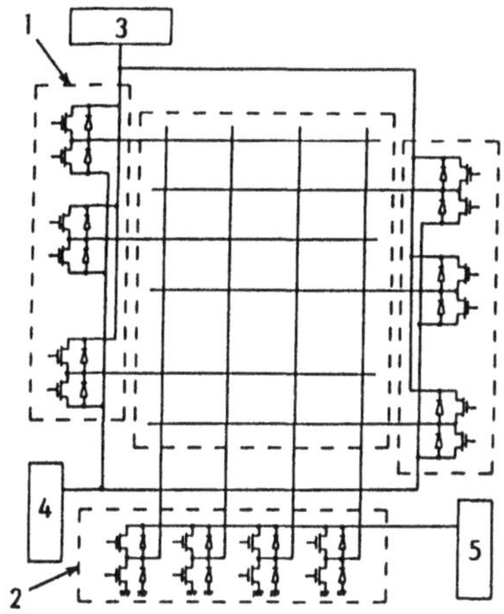

Fig.1 Drive circuit diagram of p-p symmetric drive

1 High voltage push-pull driver IC at the scanning side
2 High voltage push-pull driver IC at the data side
3 Positive writing driver
4 Negative writing driver
5 Modulation and recovery driver

3. High Voltage Driver IC

A high voltage driver IC is the most important device to determine the performance and cost of EL display. With the recent progress in high voltage IC technology, a high voltage push-pull driver IC for EL display has been developed on the basis of dielectric isolation technology and Bi-CMOS technology.[4],[5],[6] This driver IC has been already put to practical use. Table 1 shows the basic specifications of the column and row driver IC required in our new driving method. Of all the items in the specifications, breakdown voltage, output current and ON resistance of the row driver for the scanning side IC may be the three major parameters directly affecting reliability and display quality of the EL display.

The emission threshold voltage Vth of EL panel is 200 V (max).The modulation voltage V_M which determines the emission of each pixel is set in the saturation region of the luminance-to-voltage characteristic.($V_M \fallingdotseq$

Table 1 Basic specification of driver IC

	Column	Row
Breakdown voltage	70 v min	250 v min
Output current	15 mA min	40 mA min
ON-Resistance	———	300 Ω max
Data transfer speed	12 MHz max	100 KHz max
Output number	32 bit or 40 bit	34 bit or 40 bit
Logic power supply	5 V	5 V

50 v) Therefore, breakdown voltage is 250 V (min). Output current setting is 40 mA or higher, based on the load capacity for each line of the EL panel and on the charging period. While an EL panel is emitting light, it is assumed the luminous layer is short-circuited, so that the apparent capacity increases 2 to 2.5 times. Therefore, luminance decreases in proportion to the number of lit pixels in a line, thus resulting in poor display quality. To avoid this inconvenience, ON resistance in the IC operating region after start of emission should be as small as possible. In our new driving method, however, ON resistance is set at 300 Ω or less for feasibility and reasonable cost of IC.

4. Driving Method

Beside the push-pull IC, the following two points are important in driving a large capacity EL display. The first is to shorten the driving time for increased number of scanning lines. In our new driving method, the driving time is shortened in two methods. For one method, push-pull IC is used as data side driver IC, making it possible to apply modulation voltage in one stage, compared with two stages in the conventional driving methods (precharge/discharge and boost). For the other method, the modulation pulse applied time is allowed to overlap the writing pulse applied time. The second is to reduce power consumption for increased number of scanning lines and for increased capacity. If power consumption was allowed to increase with display capacity, driver ICs and their peripheral circuits would generate a larger amount of heat, resulting in less reliability and a lager burden on the power supply. Our p-p symmetric driving method has three methods to reduce power consumption.

First is the recovery method. In this method, the electric charge for modulation voltage accumulated in the EL panel is taken out and returned through the recovery diode of the data side driver IC to an external capacitor for recycled use. This method can save 25% of power consumed in a simple charging and discharging method. Second is the floating method. In this method, the data side driver IC uses push-pull type, and non-selected scanning electrodes are floated. This method can reduce the capacity load to 1/4 or less in which two capacitors are connected equivalently in series to form a charging circuit between data side electrodes. Third is the step method. In this method, modulation and writing voltage are applied in two steps. This method can make power consumption 25% lower than in the simple charging and discharging method. By using these three methods, power consumed for modulation voltage can be reduced to 9/64 in total and that consumed for writing voltage to about 4/5.

The p-p symmetric driving method is described in detail. The TF-EL panel is AC drive type, so the method relies on symmetric drive to ensure long term reliability and luminance uniformity. In order to minimize flickers caused by luminous intensity variation in applying positive and negative pulses to the EL panel, writing pulse is inverted for every line.

In the following, the first and fourth stage are modulation drive, the second and fifth stage are writing drive, the third and sixth stage are discharge and recovery drive. P-drive refers to the drive process of applying positive writing pulse, and N-drive refers to the drive process of applying negative writing pulse. In the P-N field, P-drive from first through third stage and P-drive from fourth through sixth stage are repeated alternately for all the odd and even number scanning electrodes. Here, the V_W voltage is less than the emission threshold voltage Vth.

In the first stage, the sink transistor of the data side IC for the lit pixel is turned ON to set data side electrode potential at 0 V. At the same time, the source transistor of the data side IC for the non-lit pixel is turned ON and modulation driver is allowed to operate, so as to increase data side electrode potential to 1/2 V_M and then to V_M . At this time, the source transistor of a selected scanning side IC is turned ON and the positive writing driver is allowed to operate, thus setting the potential of the selected scanning side electrode at $(V_W + V_M)/2$. Meanwhile, the sink transistor of a non-selected scanning side IC is turned ON and the negative writing driver is turned OFF to set the potential of the non-selected scanning side electrode at floating level.

In the second stage, the data side IC maintains the same state as in the first stage. The potential of the selected scanning electrode is pulled up from $(V_W + V_M)/2$ to $V_W + V_M$ by the positive writing driver. Thus, the pixel applied $V_W + V_M$ voltage, is allowed to emit light, whereas the pixel applied V_W voltage does not emit light.

In the third stage, the positive writing driver is turned OFF and sink transistor of the selected scanning side IC is turned ON to discharge writing voltage of $V_W + V_M$. At the same time, modulation driver is stopped to reduce the voltage to 1/2 V_M , thus allowing a part of the accumulated electric charge in the EL panel to return to the external capacitor in the recovery driver. The sink transistors of the data side IC are all turned ON to set the potentials of the data side and scanning side electrodes at 0 V. This completes P-drive.

In the fourth stage, the source transistor of the data side IC for the lit pixel is turned ON and modulation driver is allowed to operate, so as to increase the data side electrode potential to 1/2 V_M and then to V_M . At the same time, the sink transistor of the data side IC for the non-lit pixel is turned ON to set the data side electrode potential at 0 V. Then the sink transistor of the selected scanning side IC is turned ON and negative writing driver is allowed to operate, thus setting the potential of the selected scanning side electrode at $-V_W/2$. The source transistor of the non-selected scanning side IC is turned ON and the positive writing driver is turned OFF, to set the potential of the non-selected scanning electrode at a floating level.

In the fifth stage, the data side IC maintains the same state as in the fourth stage. The potential of the selected scanning electrode is pulled down from $-V_W/2$ to $-V_W$ by the negative writing driver. Thus, the pixel applied $V_W + V_M$ voltage, is allowed to emit light, whereas the pixel applied V_W voltage, does not emit light.

In the sixth stage, the negative writing driver is turned OFF and the source transistor of the selected scanning side IC is turned ON to discharge writing voltage of $-V_W$. At the same time, modulation driver is stopped to reduce the modulation voltage to 1/2 V_M , thus allowing a part of the accumulated electric charge in the EL panel to return to the external capacitor in the recovery driver. The sink transistors of the data side IC are all turned ON to set the potentials of the data side and scanning side electrodes at 0 V and complete N-drive.

The P-drive and N-drive are repeated alternately for the first through the last scanning lines to complete the driving of the P-N field.

In the N-P field, the odd number scanning electrodes are subjected to the N-drive and the even number scanning electrodes to P-drive.

Fig.2 shows a voltage waveform applied to the data side electrode, the scanning side electrode and the pixel. The solid line represents lit state, whereas the broken line represents non-lit state. As indicated, positive and negative writing voltage are applied to each pixel in the two fields: P-N field and N-P field.

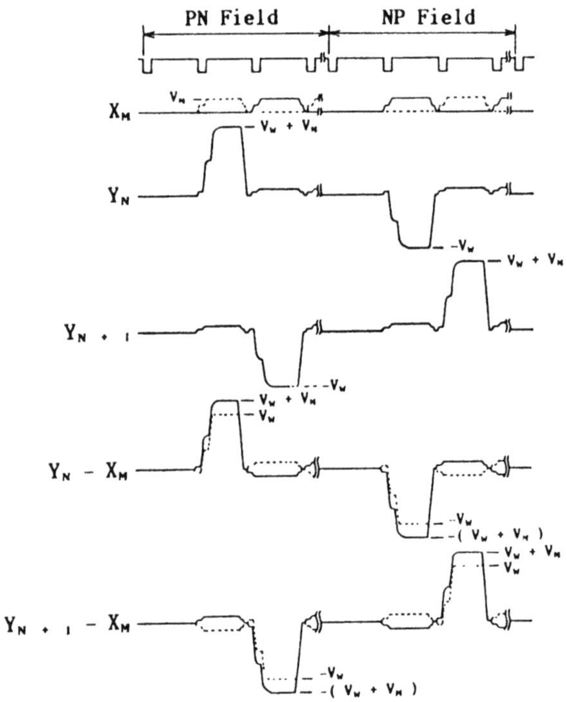

Fig. 2 Drive waveform

5. EL Display Unit Using p-p Symmetric Driving Method

The driving method we have described can be applied to EL panels of 640 x 400 or 640 x 480 dots. Table 2 shows the general specifications of a 640 x 400 dot EL unit.

Table 2 EL display unit specification

Display characteristics	
Display area (H/W)	120/192 mm
Pixel pitch	0.3 mm
Pixel size	0.22 mm square
Luminance	35 ft-L typ
Operational specification	
Supply voltage	5 v , 24 v
Supply current at 5 v	200 mA max
Supply current at 24 v	600 mA max
Power consumption	15 w max
Operating temperature	0 ~ 55 ℃

Fig.3 shows the relation between power consumption and lit pixel percentage in a 640 x 400 dot EL panel and shows the power consumption in the conventional p-n symmetric driving method. As appreciated from this figure, the maximum power consumption in the p-p symmetric driving method is smaller than that in conventional driving method. The advantage of this system is less power consumption at the actual operation state.

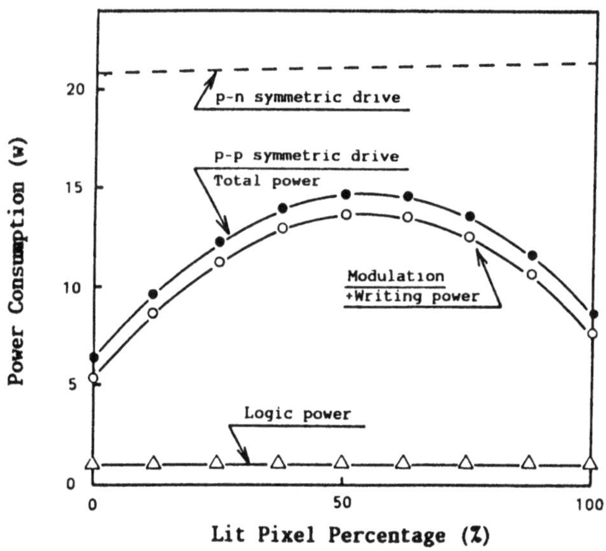

Fig.3 EL display unit power consumption

6. Conclusion

The emergence of high voltage push-pull driver IC has enabled us to develop the bidirectional push-pull symmetric driving method. Taking over an ideal AC drive by the conventional p-n symmetric driving method, our new driving method permits to be driven a large capacity EL display with less power consumption, high luminance and long term reliability.

We believe the p-p symmetric driving method will be widely used in driving EL panels of various sizes including color display panels and contribute to the popularization of EL displays.

7. Acknowledgement

We would like to express our sincere thanks to Mr.I.Washizuka, Mr.O.Kawaguchi and Mr.T.Mikami for their support of this work, and to EL engineering and producing staff of Sharp Corporation for their cooperation.

References

1. Y.Kanatani et al. : NIKKEI Electronics, vol.209, (1978) 118.
2. T. Ohba, Y.Fujioka et al, : Television Society Technical Report, March 26 (1985) 7.
3. S.Harada, H.Uede et al. : Japan Display 86 (1986) 238.
4. K.Fujii et al. : ISSCC,Digest of Technical Paper (1981) 46.
5. Technical data sheet "ECN2002/2012"
6. Technical data sheet "TD62C936AF"

Part VII

Powder Electroluminescent Panels

Analysis of the Lifetime of Powder Electroluminescent Phosphors

R.H. Marion, H.A. Harris, and W.A. Tower

Loctite Luminescent Systems, Inc., Etna Road, Lebanon, NH03766, USA

1. INTRODUCTION

It is well known that the luminance of an AC powder electroluminescent (EL) lamp decreases slowly with increasing operating time. Recent advances in phosphor materials (phosphors now exist with useful lifetimes exceeding 10,000 hours) and the countless new applications for powder EL lamps require a new methodology for evaluating the lifetime of EL phosphors.

FLEMING [1] recently reviewed various methods for accelerating life testing and proposed a methodology based on computerized curve fitting to an empirical equation. Unfortunately, he concluded that a "minimum testing time of around 5000 hours was needed" for long-life phosphors. This paper reports on our continuing studies to develop a highly accelerated method of ranking the lifetime of phosphors with varying composition, color, etc. A method which is valid for a limited color range is considered acceptable.

Although we will use the frequently quoted term "halflife," a word of caution is necessary. In the powder EL literature, "halflife" is defined as the time required for the brightness to decrease to one-half of its initial value. The "halflife" of an EL phosphor is very dependent on its starting brightness and it can increase substantially as the "starting brightness" decreases, as described by FLEMING [1]. It is more appropriate to know the useful life, which is application dependent, or the brightness after a given operating time.

2. EXPERIMENTAL

All testing was performed using standard plastic EL lamps constructed in the manner described in References [1,2]. Ideally, one would like to eliminate all extraneous sources of degradation. This is difficult because the degradation of phosphor particles is influenced by lamp construction, initial brightness, moisture, resin degradation, etc. To minimize the influence of these variables, lamp construction was held constant (layer thicknesses, resins, phosphor loading, lamp capacitance, etc.) and exterior series resistance was excluded. Any potential influence of moisture was minimized by the use of standard control lamps and post-test lamp analysis for phosphor blackening, color shifts, etc.

The phosphor types tested included commercially available green, blue-green and yellow-orange, dyed white, blended white and our proprietary long-life yellow-green phosphor, designated YGLL.

3. RESULTS

3.1 Acceleration with Temperature

Several instances of accelerating life testing through elevated temperatures are described in the Literature [3,4]. Degradation due to temperature for green, blue-

Springer Proceedings in Physics, Vol. 38 Electroluminescence
Editors: S. Shionoya and H. Kobayashi

green, blue, yellow-orange, and yellow-green long-life phosphors has been studied, with typical results for YGLL given in Fig. 1. The graph depicts degradation after 300 hours at 115V/400Hz, relative to ambient operation at the same voltage and frequency. The primary problem with acceleration at temperatures greater than 70°C is a slow deterioration of the lamp resulting in color shifts toward yellow. Although not a large shift, the filtering effect is enough to mask true lamp brightness and confuse degradation analysis.

Our results indicate that temperature is a viable acceleration method if temperature is kept below 70°C. However, as can be seen in Fig. 1, the acceleration factor is less than 2X.

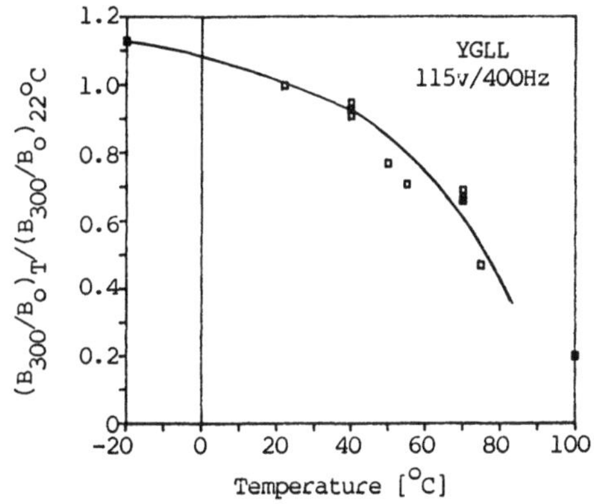

Fig. 1 Degradation of Brightness (B) with Temperature (T). B_0 = Initial Brightness, B_{300}= Brightness after 300 hrs. The ordinate has been normalized to the value at 22°C.

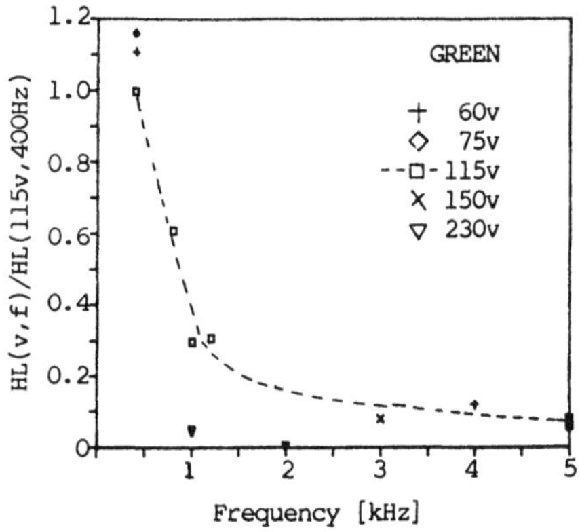

Fig. 2 Halflife (HL) at the indicated Voltage and Frequency -- normalized to halflife at 115V/400Hz.

3.2 Acceleration with Frequency

Results are given in Fig. 2 which depict the dependence of ageing on frequency, showing that the degree of acceleration is not linear. These results are similar to the two regions reported by THORNTON [3]. There are three problems that must be considered when testing at high frequencies:

1. As shown in the Literature [5], the volume of material degraded at the tips of the Cu_2S decorated lines decreases as the frequency increases. Therefore, all degraded brightness measurements must be made at the same frequency used in the degradation test.

2. With increasing test frequency, many phosphors exhibit a change in spectral energy distribution which affects the relationship between the total radiance and the photometric brightness. For example, a blue-green phosphor may have a lower photometric brightness at 5kHz than at 2kHz, even though the total radiance increased. Since brightness is not a true measure of emitted energy, its use in obtaining an acceleration factor between low and high frequency testing may be misleading. When performing a degradation test at a single frequency, any change in shape or peak location of the spectral energy distribution during the test may cause a change in the relationship between brightness and total radiance. This will also cause errors in assessing degradation. This problem may be avoided by using radiometric measurements.

3. Testing at higher frequency results in increased lamp temperature. As shown in Section 3.1, this causes further lamp degradation by an additional mechan-

ism of temperature. Figures 3 and 4 show that the external lamp temperature decreases with time and the large decrease within the first ten hours correlates with a large degradation in brightness. (Note that the actual phosphor particle could be much hotter.) The results in Fig 4 also indicate that the temperature at higher frequencies continues to decrease slowly until it reaches ambient. This changing degradation due to temperature variation causes a varying acceleration factor which complicates data interpretation.

Our results on the use of frequency to accelerate life testing of plastic lamps can be summarized as follows. Frequencies up to 1 kHz may be utilized because they only produce a minor temperature increase and a small color shift. However, this does not provide enough test acceleration.

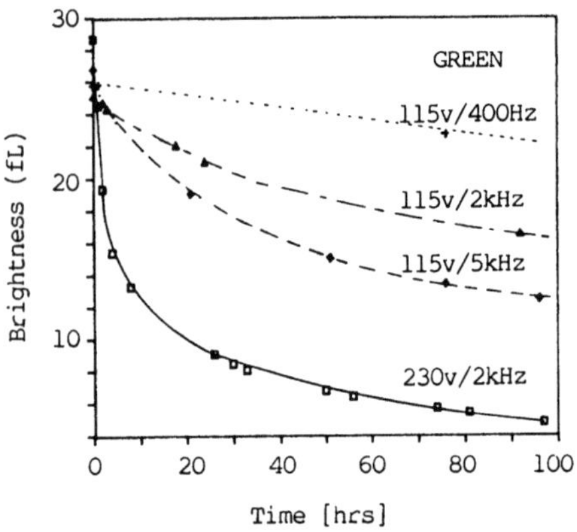

Fig 3. Brightness degradation for varying Voltage and Frequency.

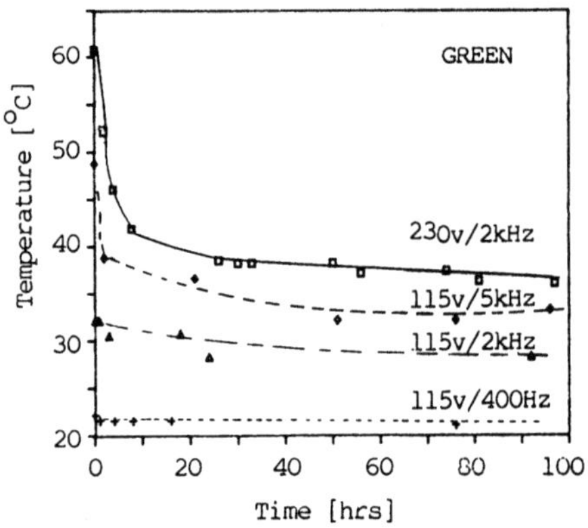

Fig 4. External Lamp Temperature vs. Time for same conditions as in Fig. 3.

3.3 Acceleration with Voltage

Figures 3 and 4 also show data for higher voltage. It can be seen that higher voltage causes the same temperature-related problems described in the preceding section. Although a small increase in voltage would not increase the lamp temperature significantly, it would also not provide much test acceleration. For this reason, this approach was not considered useful.

3.4 Acceleration with Combination of Temperature and Frequency

The results given above suggest that a combination of temperature below 70°C and frequency below 1 kHz should be evaluated. A method devised by British Aerospace has been used for several years at Lucas Aerospace to test materials supplied to them. This test consists of operating the lamp at 115V/800Hz while being subjected to a 72 hour profile ranging from 22°C to 65°C. The final test brightness is divided by the initial brightness to determine a Figure of Merit. Figure 5 gives a plot of this Figure of Merit (Degraded %B) versus halflife for 217 batches of standard and long-lived phosphors, indicating a reasonable ranking. There is considerable uncertainty for halflives greater than 5000 hours, as can be seen by difficulties differentiating between 10,000 and 14,000 hours, but the method is considered useful due to the relatively short (72 hrs) test period. Further plans are to investigate additional temperature-time cycles at temperatures less than 70°C and frequencies of 1 kHz.

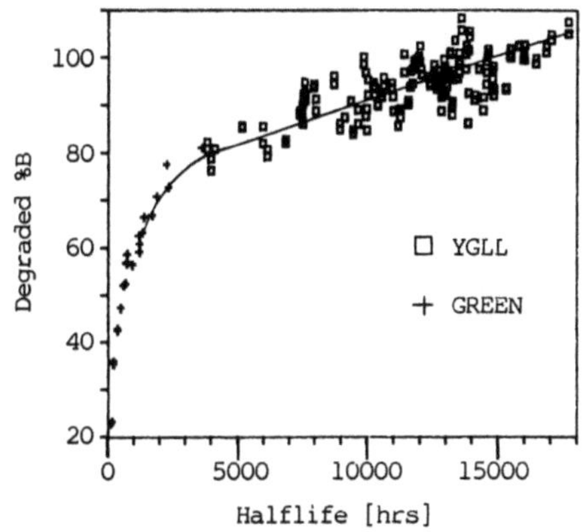

Fig. 5 Correlation between Figure of Merit for 72-hour/800Hz Test and the halflife. (See Text.)

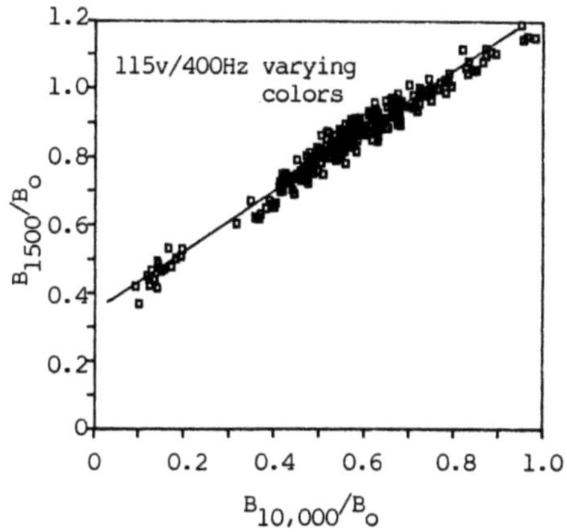

Fig. 6 Comparison of the Brightness after 1500 hrs (B_{1500}) with the Brightness after 10,000 hrs ($B_{10,000}$). Both values are normalized to the Initial Brightness (B_0).

3.5 Ranking by Analysis of 115V/400Hz Data

An extensive analysis of life data from over 500 tests was performed to see if we could avoid the need for long-term data cited previously [1]. Studies utilizing various types of extrapolation and fitting techniques did not yield positive results. The parameters of the fit to the modified ROBERTS equation [6,1] were either too sensitive to the initial brightness, did not properly account for the "anomalous" region, or required data for too long a time. However, this analysis suggested that rankings might be obtained by comparing the degree of brightness degradation at a fixed time. Life data from 270 tests consisting of both standard and long-life phosphors was examined for relative brightness degradation at various times. Typical results are shown in Figs. 6 and 7 indicating that a reasonable assessment of long-time behavior can be obtained in less than 1500 hours, at 115V/

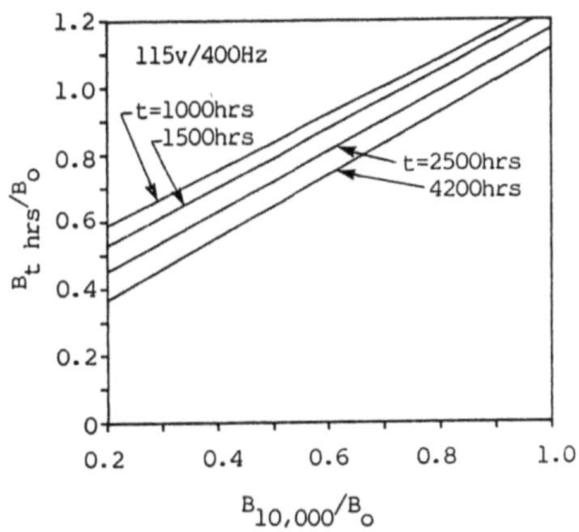

Fig 7 Summary of the linear regression lines to Brightness data after t hrs ($B_{t\ hrs}$) such as is shown in Fig. 6.

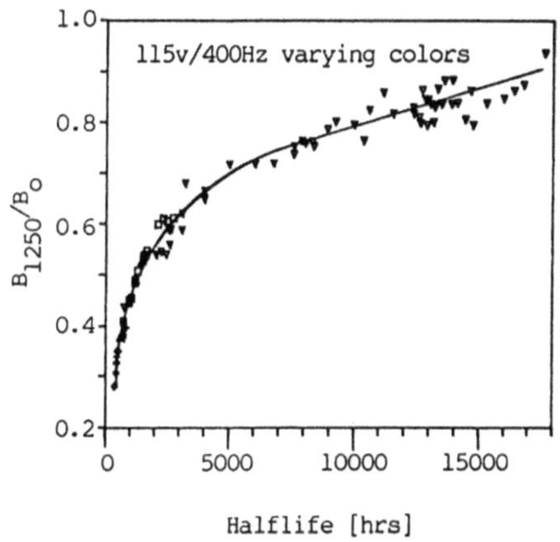

Fig. 8 Correlation between the Brightness after 1250 hrs (B_{1250}) with the halflife. The Brightness is normalized to the Initial Brightness (B_0).

400Hz. Figure 8 demonstrates that this type of parameter (Percent Initial Brightness at 1250 hours) correlates with the halflife of the phosphor.

4.CONCLUSIONS

The data presented in this paper demonstrate that temperature, frequency and voltage can only be varied within a limited range to accelerate ageing of phosphor in plastic lamps without changing the degradation mechanism. This results in a limited amount of acceleration. The best methods developed to date consist of a combination of temperature and frequency for 72 hours, or a ranking of the % Initial Brightness at 1000-1500 hours. The latter procedure does not provide sufficient acceleration.

REFERENCES

1. G.R. Fleming: To be published in J. Electrochemical Soc. (1988)
2. L.E. Tannas, Jr.: Flat Panel Displays and CRT's, Chapter 8, ed. by L.E. Tannas, Jr., (Van Nostrand, Reinhold, New York 1985) p.237
3. W.A. Thornton: J. Electrochemical Soc., 107, 895 (1960)
4. K. Hirabayashi and Y. Itoh: J. Electrochemical Soc., 129, 362 (1982)
5. A.G. Fischer: Luminescence of Inorganic Solids, Chapter 10, ed. by P. Goldberg (Academic Press, New York 1966) p. 541
6. S. Roberts: J. Applied Physics, 28, 262 (1957)

On the Mechanism of "Forming" and Degradation in DCEL Panels

S.S. Chadha and A. Vecht

Department of Materials Science and Physics, Thames Polytechnic,
Wellington Street, London SE18 6PF, UK

Introduction

Forming and degradation phenomena in direct current powder electroluminescent (DCEL) zinc sulphide manganese copper (ZnS:Mn,Cu) phosphor devices have been studied for over two decades and various models have been proposed /1/. The major drawback of powder DCEL devices has been their stability or maintenance. It was noticed very early on that the device ambient significantly affected the degradation rate. The use of dry ambients was found to lead to better maintenance, /2/. Despite these early results showing the effect of the ambient and the fact that panels are usually treated in an inert gas prior to lifetesting /3/, very little further research was carried out along these lines until recently /4/. However, stable DC devices can now be produced by the use of constant power circuitry /5/. These systems are now being produced for VDU application.

Recently we have carried out a systematic study of the effects of the binder /6/ and atmosphere on the DCEL ZnS:Mn,Cu powder systems. These cast some doubt on the established theories of forming/failure mechanisms operating in DCEL powder and possibly in thin film devices.

Construction of the DCEL Powder Displays

DCEL panel construction was described recently /7/, along with the procedure for the initial forming process and optoelectronic characterisation techniques. A structure of such a device is shown in Fig. 1.

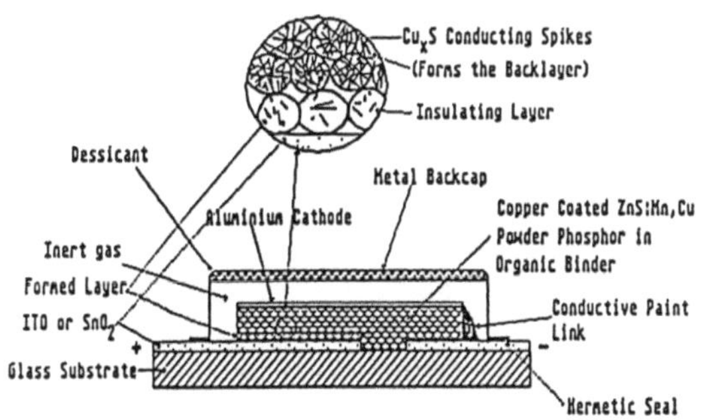

Fig. 1. Structure of a DCEL powder test panel

A. The Nature of the Forming Process

We will now describe the forming process, summarise the models and give conditions which either increase or decrease the rate of this phenomenon.

Under the influence of an undirectional field, the region nearest to the anode of the conductive powder layer is converted to an insulating region. This is generally described as 'forming'. In this resistive region, electrons are accelerated and then impact excite the Mn^{2+} luminescent centres. There are many explanations which attempt to explain 'forming'. Among these are the following:

a) The migration of Cu into the ZnS:Mn,Cu powder particles /8/.
b) The migration of Cu over the ZnS:Mn,Cu powder particles.
c) $Cu_{2-x}S$ phase changes due to Joule heating during the forming process leading to more resistive phase(s) /9/.
d) Oxidation of the $Cu_{2-x}S$ to $Cu_{2-x}O$, $CuSO_4$ etc. /10/.
e) Oxidation of the powder phosphor particles /4/.

In initial forming, it can be shown that:

(i) Forming only occurs next to the positive electrode.
(ii) The conductivity of the copper coated phosphor decreases with time and temperature.
(iii) The forming current is dependent on the forming rate (see Fig. 2).
(iv) The forming power is reduced if a dielectric thin film is deposited between the transparent anode and the phosphor layer /11/.
(v) The forming occurs in all ambients at room temperature even after a very prolonged outgassing or in binderless panels.

In secondary forming, usually from 30-70V,

forming is extremely rapid,

(i) when devices are unencapsulated, i.e. without any backcap.
(ii) when the ambient temperature is high.
(iii) with certain binders, e.g. poly vinyl butyral.

forming is significantly slowed down,

(i) when a panel is encapsulated with dessicant in the backcap.
(ii) when H_2S or S vapour, Ar, CH_4, H_2 are present in the backcap ambient.
(iii) when forming is carried out in vacuum.
(iv) at reduced temperature.
(v) with certain binders, e.g. poly alpha methyl styrene, poly methyl methacrylate and silicone oils.

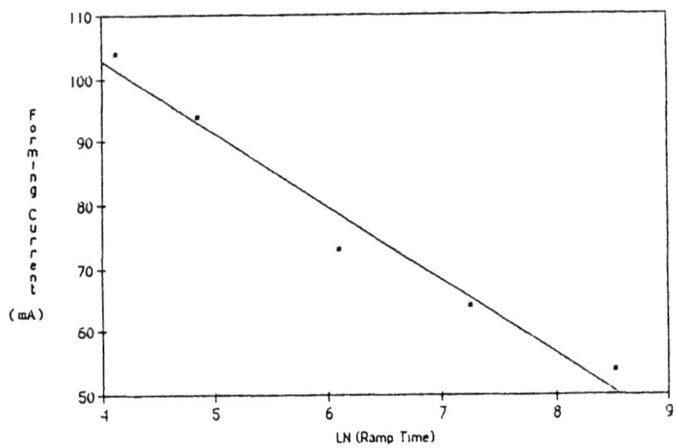

Fig. 2. Graph showing the dependence of forming current on ramp rate

B. The Effect of Gas Ambient within the Panel

Panels were formed in wet and dry air, argon, 10% CH_4 in Argon, town gas (impure CH_4), hydrogen, hydrogen sulphide, sulphur vapour and sulphur dioxide.

Table 1. The effect of various gases and vacuum on panel characteristics

Gas Ambient	Initial Resis (kΩ)	Forming Power (Wcm-2)	Forming Time (Mins)	Initial Lumin (ft-L)	Lumin at 5000 hr (ft-L)
Air	4 3	8	31 90	90	13
Vacuum*	2 8	10	77 (30 days)	75	6
CH_4/Ar	3 4	8	79	100	12
CH_4(Town Gas)	4 9	7	66	45	3
H_2	1 4	4	104	86	10
H_2S	10 2	4	103	12	14
SO_2	1 7	8	500	1	0
Ar	6 7	14	80	56	5
S	9 8	5	110	12	9

*Vacuum of only 10^{-2} mBar Figures in bracket refer to panel at 10^{-7} mBar with P_2O_5 dessicant.

Table 2. Forming time as a function of ambient conditions

Material/Condition	Forming Time 30 - 70 Volts
Belco binder in wet air	1-2 mins
Belco binder/dry air	30 - 40 mins
Belco binder/vacuum(10^{-7}mBar)	30 days

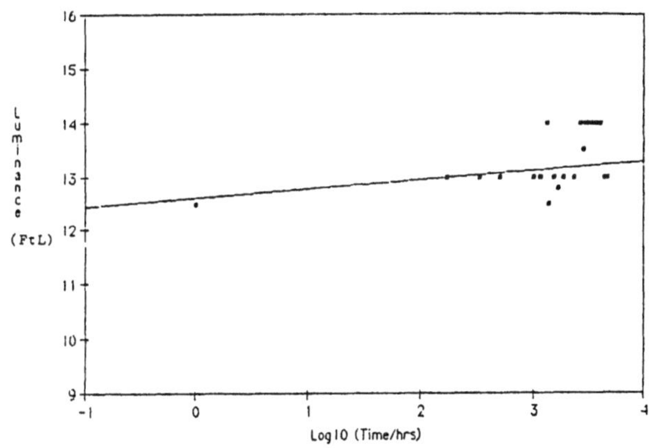

Fig. 3. Maintenance plot of a device treated with H_25 Gas

The effect of these gases is summarized in Table I. It should be noted that H_2S and sulphur vapour gave similar results, whilst the panels showed reduced efficiency, the stability improved significantly (see Fig. 3). The most pronounced and irreversible change was caused by the presence of moisture. Table II shows the effect of moisture or oxygen using Belco binder, (Belco is an ICI trade mark).

The effect of moisture and/or oxygen on the panel was tested by the following experiment:

Panels were placed in a vacuum chamber with freshly activated P_2O_5 and the system pumped down to 10^{-7} mBars for 100 hours prior to the initial forming. A normal initial forming trace was obtained up to 30V. The secondary forming at constant power showed a rate of 0.084 V/hr after 50 hours. Using a capillary valve, dried air was bled into the vacuum chamber and the pressure was increased to 10^{-2} mBar. Initially, the forming rate increased, but after about 10 minutes, the forming rate was again decreased to 0.08V/hr (Fig. 4). Reducing the pressure in the chamber again resulted in the panel formed voltage decreasing to almost that at which it had been prior to exposure to dry air. The panel was then exposed to standard wet air at various pressures. The forming rate increased as function of pressure (Fig. 5) and with time, and did not revert back to a slower rate.

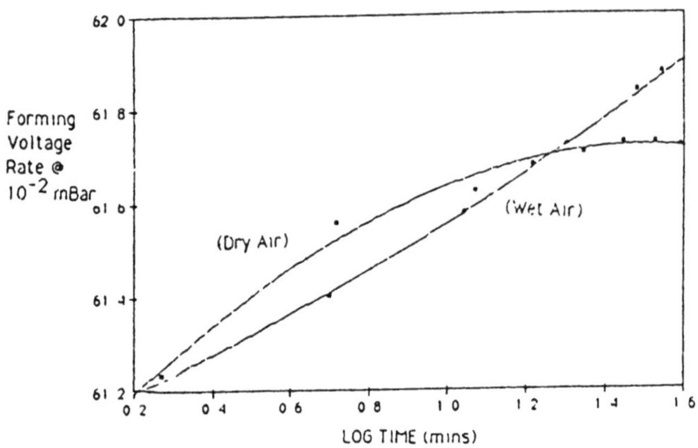

Fig. 4. Forming rates exhibited by DCEL powder device during exposure to dry and wet air

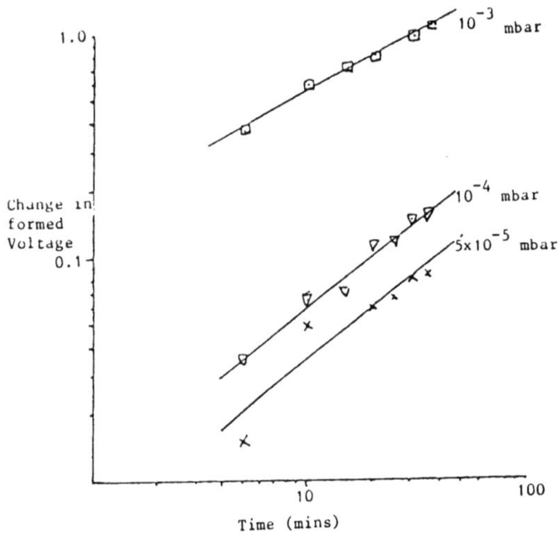

Fig. 5. Forming rate exhibited by a DCEL powder device during exposure to wet air at various pressures

Discussion

The results obtained are not consistent with many of the theories outlined above. The complicated picture that emerges points to two or three effects which may explain forming.

The principal cause of initial forming appears to be migration of Cu ions aided by traces of adsorbed water on the particle surface. The simple oxidation theories can be dismissed as we have observed no unambiguous or irreversible effects due to O_2 alone. We may also conclude from previous work /9/ and results reported here that copper migration with or without phase change in the Cu_xS predominates in the secondary forming process. In powder phosphor systems, surface copper is essential as it promotes the formation of a high field region and makes electron injection possible.

The forming process may be explained in part by local thermal gradients /12/ but the effects of the polarity of the cell are also important. It is well known that if the polarity of a DCEL panel is reversed, forming takes place below the rear aluminium electrode /13/. We observed that this still takes place even if a dielectric layer is deposited between the transparent electrode and the phosphor layer. (This layer reduces the forming power and localises heating effects to regions near to the transparent front electrode /11/). This indicates that forming is polarity-sensitive and that the original view regarding copper migration away from the positive electrode is still valid. Our results using a range of binders are being reported elsewhere /6/.

The role of S or H_2S in increasing the stability as well as the conductivity may be explained by either:

a) S fills sulphur vacancies in ZnS, /11/,
b) S enables more stable and conductive phases to be formed, or,
c) prevents the adsorption of H_2O. Zn excess is said to promote H_2O adsorption on the ZnS surface /14/.

We hope further research will point to which of these is the most dominant process.

Acknowledgements

We wish to acknowledge our colleagues in the Thames DCEL Research Team, especially Mr M Salem and Mr A J Staple. In addition, we wish to acknowledge the financial support from the Department of Trade and Industry through the Science and Engineering Research Council under the UK Alvey Initiative.

References

1) A Vecht and N J Werring, J Phys: Appl Phys, 3 (No. 2) 105 (1970).
2) P H J Beatty, MSc Thesis, Thames Polytechnic (1970).
3) C J Alder, A F Cattell, K Dexter, J Kirton and M S Skolnick, SID Biennial Display Research Conf Digest, Cherry Hill Meeting, 1 (1980).
4) P W Alexander, C Sherhod and M J Stowell, to be published.
5) D Channing and L Chiang, SID Intern Symp Digest 1987, paper 16.3 295 (1987).
6) S S Chadha, M Salem, A J Staple and A Vecht, to be published.
7) S S Chadha, C V Haynes, A Vecht, SID Intern Symp Digest, 1988, 35-38 (1988).
8) D May, Electronic Displays Conference Records, September (1981).
9) A Vecht, N J Werring, R Ellis and P J F Smith, Brit J Appl Phys (J Phys D) Ser 2 2 953 (1969).
10) P H J Beatty, PhD Thesis, Brunel University (1974).
11) A Vecht, S S Chadha, R Hayes and R Ellis, SID Intern Symp Digest (1982).
12) E Hirahara, J Phys Soc (Jap) 6 422 (1951).
13) A Vecht, J Lumin, 7 213 (1973).
14) J A Hedwall and S Nord, Z Electrochem. 49 467 (1949).

Application of Sol-Gel Technique to the Preparation of AC Powder Electroluminescent Device

R. Igarashi[1], *M. Jimbo*[1], *Y. Nosaka*[1], *H. Miyama*[1], *and M. Yokoyama*[2]

[1]Department of Chemistry, Technological University of Nagaoka, Nagaoka 940-21, Japan

[2]Nippon Seiki Co. Ltd., Nagaoka 940, Japan

1. Introduction

This paper describes the fabrication and performance of a new AC powder electroluminescent (EL) device, which has been successfully developed using a sol-gel technique. Organic binders such as cyanoetylcellulose have been used for the conventional powder EL devices. The softening point of organic binders is normally less than 100°C, and their thermal resistivity is low. When an organic binder is operated or kept above its softening temperature, the binder is fluidized. Then, the contents of the device such as phosphor and ferroelectric powders migrate making a heterogeneous layer. This migration of the contents causes an irregular luminous plate and short circuit of the device. Since a dry process in the preparation of the conventional powder EL devices must be carried out at low temperature because of low thermal resistance of binder, the removal of solvent and moisture from the binder may be incomplete. Then, conventional powder EL devices usually have a short life.

We attempt to use an inorganic binder for a powder EL device. The binder consists of silica gel synthesized by means of the sol-gel technique. The device was fabricated by spreading ZnS:Cu phosphor mixed with silica gel binder on an ITO glass substrate. One of the advantages for the inorganic binder is that the device does not appreciable amount of impurities because of high purity of starting materials. Another advantage is that solvent and water are left in the device and moisture because of the dry process at high temperature. In this way, the EL device is expected to have long life.

2. Experimental

Figure 1 shows the procedure of preparation for powder EL device by using sol-gel technique. Tetraethoxysilane, $Si(OC_2H_5)_4$, was used as a starting material for silica gel formation [1,2]. Twenty ml of $Si(OC_2H_5)_4$, 20 ml of ethyl alcohol, and 5 ml of 0.001-M CH_3COOH were put into a 300 ml flask. The hydrolysis for gelation was carried out at 40°C under reflux for 24 hours. In order to prevent excess gelation, 20 ml of propylene carbonate was introduced into the solution, and then ethyl alcohol and water were removed by distillation under reduced pressure. ZnS:Cu phosphor was mixed with the solution. An ITO glass substrate was coated with the mixture and then another ITO glass was placed on the layer of the mixture. Finally, the solvent was removed by heating under vacuum. Thus a test device consisting of ZnS:Cu phosphor and silica gel was prepared. A crack appears on the device if ethyl alcohol is not replaced with propylene carbonate before the coating process [3].

Springer Proceedings in Physics, Vol. 38 Electroluminescence
Editors: S. Shionoya and H. Kobayashi

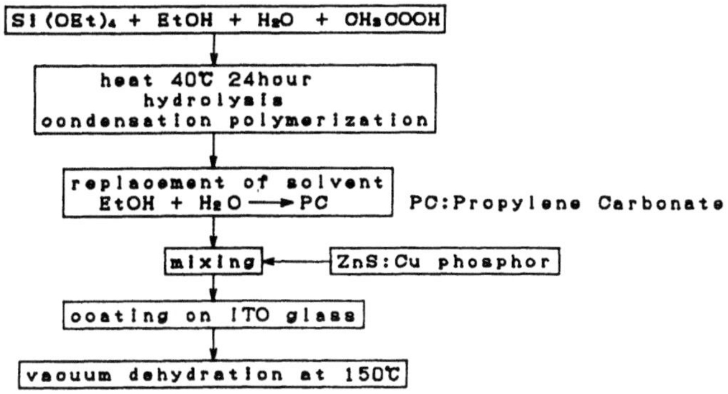

Fig. 1 Flow sheet of preparation for powder EL device

Luminescence was passed through a grating monochromator (Ritsu, MC-25N) and detected by a photomultiplier (Hamamatsu Photonics, R928). In the EL device fabricated using a sol-gel technique, luminance was measured as a function of applied voltage and operating frequency at various temperature.

3. Results and Discussion

The relationship between luminance and applied voltage is shown in Fig. 2, and represented by the following equation.

$$B = B_0 \exp(-CV^{-1/2}) \qquad (1)$$

where B is brightness, V is applied voltage, B_0 and C are constants determined by the materials of phosphor, structure of device, and exciting condition [4]. The experimental results shown in Fig. 2 agree with the (1), because logarithm of B directly proportional to $V^{-1/2}$. This means that the EL device fabricated by sol-gel technique has similar B-V characteristics of conventional EL devices. But, threshold voltage for the EL device made by using inorganic binder is higher than that for organic binder. It may be due to lower dielectric constant of silica gel. This problem may be solved by adding ferroelectric powders such as barium titanate as done for conventional powder EL [5]. Furthermore, the thermal resistivity of inorganic binder is not impaired by adding inorganic ferroelectrics in the EL device. Use of ferroelectrics is not presented in this report and will be reported in near future.

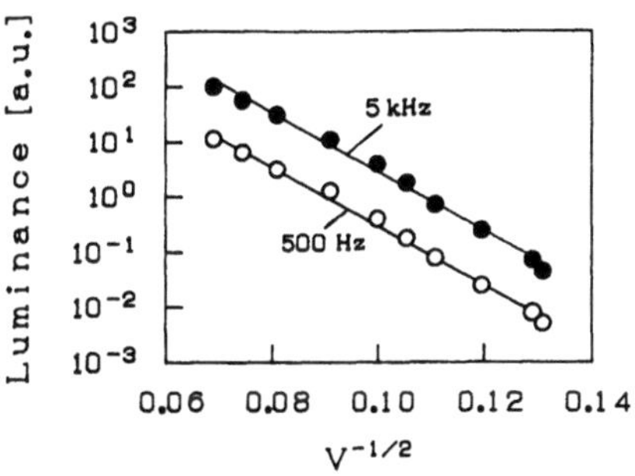

Fig. 2 Luminance vs. applied voltage

The magnitude of luminance depended on operating frequency. In the conventional EL device, luminance is proportional to operating frequency at lower frequency, but it is gradually saturated at high frequency. Dependence of luminance on the operating frequency for the sol-gel EL test device are shown in Fig. 3. In the figure, (a) shows that at 13 ℃ the luminance intensity is directly proportional to the frequency over wide range up to at 10 kHz. Further, the dependence of luminance on operating frequency was measured at 130 ℃ and shown as (b) in Fig. 3. Here, the luminance is proportional to the frequency up to 10 kHz. No saturation at high frequency as observed for conventional device was observed even at high temperature.

EL spectra were measured at various temperatures for operating frequency of 1 kHz and applied voltage of 180 V. The result is shown in Fig. 4. Noise is caused by passing the light through an optical fiber. The luminance decreased and the emission maximum shifted to longer wavelength with increasing temperature. In general, luminance depends on temperature, particularly at high operating frequency. Figure 5 shows that the intensity decreases with increasing temperature at operating frequency of 5 kHz. When applied voltage increased, the slope of temperature dependence increased. The luminescence was not observed at 160 ℃ in the any voltage. However, stable luminance was observed even at 130 ℃, and degradation of inorganic binder was not observed. The magnitude of luminance was recovered by cooling it to room temperature again. Therefore, the EL device produced by using the inorganic binder is relatively stable for operation and storage at the high temperature.

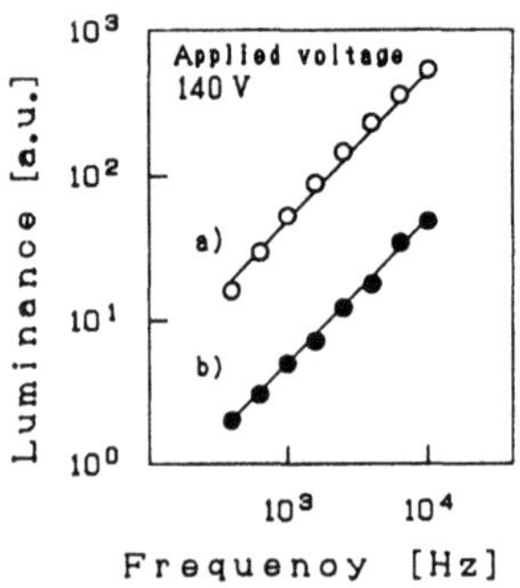

Fig. 3 Dependence of luminance on operating frequency at 13°C (a) and 130°C (b)

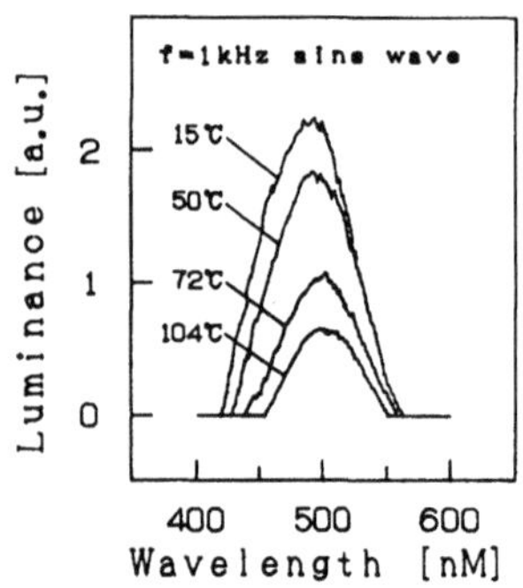

Fig. 4 EL spectra measured at various temperatures

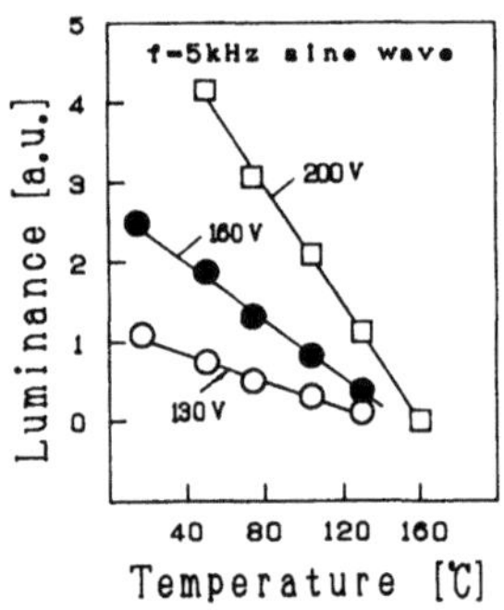

Fig. 5 Luminance vs. temperature at various voltages

4. Conclusion

The AC powder EL device consisting of a silica gel binder is expected to have long life, because the binder is hardly contaminated by impurities. Furthermore, its thermal resistivity is higher than that made with organic binder. Because of these characteristics, application of the device to automobile environment seems promising.

5. Acknowledgment

We are grateful to Dr. S. Furuuchi of Nippon Seiki Co. Ltd., for valuable discussion, to Mr. Okawa for his technical assistance, and to Mr. S. Ohta and

Mr. K. Kashiro of Nippon Seiki Co. Ltd., for their support. This work was supported by Nippon Seiki Co. Ltd.

6. Literature references

1. Y. Yamamoto, K. Kamiya, and s.sakka: Yogyo Kyokai Shi, 90(1982)328
2. H. Scheder: Physics of Thin Films, 5(1969)87
3. J. Zarzycki, M. Prassas, and J.Phalippou: J. Mater.Sci., 17(1982)3371
4. J. I. Pankove: Electroluminescence (Springer, Berlin, Heidelberg 1977)
5. P. W. Ranby and D. W. Smith: IEEE Proc., 127(1980)196

Multicolor ac-Electroluminescent Display Panel Using Red Electroluminescent Phosphors

Ge Baogui

Electroluminescent Laboratory, Department of Physics,
Shanghai Teachers University, 10 Guiling Road, Shanghai, P.R.China

1. Introduction

We have been studying the development of red electroluminescence (EL) powder phosphors so that we can complete a large display panel of multicolor which can be operated with ac drive. One of the difficult points is that the luminance level of red EL phosphor powder is low, about 10 times lower than that of green. None of our previous studies has been successful. Recently, however, we have succeeded in developing a red EL powder phosphor with luminance level not being bright but satisfactory enough for real panel use. Using this red EL powder phosphor, We also have developed a large display panel. Here in this paper, we report on our experimental results.

2. Red EL Powder Phosphors

2.1 Red EL phosphor

As starting EL phosphors, we use ZnSe, ZnS and CdS being activated by Cu and Eu. By changing the mixing ratio of these phosphors, we have found a desirable mixing ratio. The results are shown in Table 1. As can be seen in Table 1, the average relative luminance falls with increase in the emission peak. Consequently, we have chosen a phosphor of emission peak 610 nm and luminance 100 (ZnSe:ZnS:CdS =7:2:1) by balancing emission peak and luminance level.

Table 1

ZnSe:ZnS:CdS	5:4:1	6.5:2.5:1	7:2:1	7.5:2:0.5	8:1:1
Emission peak	580nm	600nm	620nm	630nm	650nm
Average relative luminance	280	120	100	80	50

2.2 EL phosphor particle size

To develop a uniform display panel, first, we must make the EL phosphors the same size. For this purpose, we used a sieve with mesh size of 400. Forty percent of the phosphor passes through the sieve, But sixty percent is left. This rest is ground by a special grinding method . It is again shaken and is sieved through the mesh size of 400. We can save much phosphor powder loss with this method. The properties of electroluminescence are not much changed after

Springer Proceedings in Physics, Vol. 38 **Electroluminescence**
Editors: S. Shionoya and H. Kobayashi

grinding. Relative luminance levels before grinding are between 17 and 34, and of after grinding are between 18-45, as shown in Table 2. The emission peaks remain at 610 nm for all phosphors.

Table 2.

Phosphors Sample No.	Relative luminance		Peak wavelength of EL	
	before grinding	after grinding	before grinding	after grinding
R 903	37	45	620 nm	620 nm
R 1001	30	35	620 nm	620 nm
R 1011	50	46	620 nm	620 nm
R 001	17	18	620 nm	620 nm

3. Red-Green EL Panels

The luminance levels are quite different between green and red phosphors, as shown in Table 3. For green, we have the relative luminance of 66 - 79 and its average is 68. Contrary to this ,for red, we have 3.7 -5.0 and its average is 4.5. The differences of luminance levels are about 15 times. This means the following. It is impossible to make a color EL display panel by mixing these green and red phosphors on one glass plate. In order to conquer this point, we have employed a new fabrication method.

Table 3.

Phosphors	Sample number and relative luminance								Average relative luminance
Green phosphor	Sample No.	G936	G924	G902	G905	G907	G903	G935	
	Luminance	79	76.5	71	66	58	68	55.5	67.7
Red phosphor	Sample No.	R725	R918	R101	R102	R106	R111	R10	
	Luminance	3.65	4.8	4.6	4.1	3.65	4.95	5.9	4.52

3.1 Thin screen method

In order to obtain brighter red EL emission, we make the thickness of red EL layer thinner. We call this fabrication method; "Thin screen process". The electric-field E in the red EL layer is determined by $E=V/d$, where V is the applied voltage and d the distance of electrodes. Smaller d and higher V lead to higher electric-field, that is, lead to higher luminance. However, at the same time, EL panels tend to breakdown. By increasing a packing density of luminescent layer and the medium insulating layer, we could improve the breakdown voltage and the luminance levels. The results are shown in Table 4.

Even by employing this process, the difference in the luminance levels between green and red phosphors are 4 - 5 times. However, as a sensitivity of human eye is different to green and red, developed EL panel is satisfactory as a display panel.

Table 4.

	Green EL screen		Red EL screen			
Phosphors to used make EL screen	Sample	Luminance	Sample	Luminance	Sample	Luminance
	G936	63 checks	R721	6 checks	R721	6 checks
Fabrication	Normal process		Normal process		Thin screen process	
Relative luminance of EL screen	Screen	Luminance	Screen	Luminance	Screen	Luminance
	G-EL31 G-EL32 G-EL33	505 515 510	R-EL14 R-EL15 R-EL16	44 46 47	R-EL11 R-EL12 R-EL13	96 85 95
Average luminance		510		45.6		92

3.2 Mixing color method

For this method, EL phosphors are mixed at the ratio to obtain the different colors. Then, each mixed phosphor of different color is printed by silk-printing method. A disadvantage of this method is poor uniformity. Therefore, this method was not employed.

The EL panels were also made by spraying method. The EL panels consist of two phosphor layers. Each layer emits different EL emission color. The results obtained are summarized in Table 5.

Table 5.

EL screen Sample No.	Color of EL layer		Vision color	Peak wave length	Relative luminance	Voltage
	Layer 1	Layer 2				
EL01	Green	Green	Green	510-520nm	100	300v
EL02	Yellow	Yellow	Yellow	578nm	60	300v
EL03	Green	Yellow	G-Y	540-550nm	70-80	300v
EL04	Yellow	Green	Y-G	565nm	60-70	300v

We also made EL panels with green and red EL layers. The green EL layer showed EL with a peak at 510-520 nm, and the red EL layer showed EL with a peak at 620 nm. The EL panel with the green and red layers showed yellowish-green EL with a peak at 550-560nm.

4. Summary

We have succeeded in fabricating the multi-color EL display panels. In addition, we succeeded in increasing in EL luminance levels, simplifying the fabrication process of EL panels, saving on the amount of phosphor and so on. Characteristics of the EL panels are shown below.

EL screen size 3.2 × 2.45 м2

Effective display size 2.8 × 2.00 м2

Display colors and Luminance

Color	Peak wavelength		Luminance
Green	510	nm	80 cd/м2
Yellow-Green	550-560	nm	60 cd/м2
Yellow	578	nm	40 cd/м2
Orange-red	620	nm	20 cd/м2

Display function; Digital display, Fixed letter, Analogous pattern, Double colors alarm display

References

1. Ge Baogui: Luminescence and Display, 3, 95 (1981)
2. Luoxi: Electroluminescence (Scientific Publishing house,1981) p.7
3 Ge Baogui: Theory and application of Electroluminescence (Mapping Publishing House, 1985) p.89
4. M.B.Fork: Practical Electroluminescence (Scientific Publishing House, 1984) p.137
5. Jing Qicheng et al.: Knowledge of Color Degree (Scientific Publishing House, 1979) p.272

Electroluminescence in Calcium Sulphide

B. Ray

Faculty of Applied Science, Coventry Poltytechnic,
Priory Street, Coventry, CV1 5FB, UK

1. Introduction

Interest in alkaline earth II-VI chalcogenides has been stimulated by the extended range of band gaps available compared to that of conventional II-VI compounds. LEHMANN [1], in reviewing the luminescence characteristics of calcium sulphide, drew out the versatility of this compound in its capacity to accept a wide range of dopants, including the rare earth elements, which could emit in different segments of the visible spectral range. Much of the early interest in calcium sulphide was centred on its cathodoluminescent characteristics and these have been exploited recently in the commercial context as $CaS_{1-x}Se_x$ photocathodes in cathode ray tubes. Fluorescence and thermoluminescence are other characteristics that have been explored in colour controlled emitters and dosimeters respectively; the latter application has been developed to some degree using CaS.

Much of the early work on electroluminescence in calcium sulphide was focussed on powders under a.c. excitation conditions although one of the first substantive reports on red EL in CaS: Eu,Cu by WACHTEL [2] was a forerunner to exploratory work on its exploitation under dc excitation conditions by VECHT [3]. Whilst ac powder EL represents the major part of the literature, much of the recent work has been on thin film a.c. EL device structures. This paper aims to draw out the major features on both powder layer and thin film EL and to suggest some future developments that are needed to consolidate work in this field.

2. Preparative Methods

2.1 Powder Samples

There are a number of methods reported for the preparation of CaS with or without dopants. These range from simply the reduction of hydrated calcium sulphate with charcoal at 850 to 950°C to carefully controlled cycles using the flow of high purity H_2S and H_2 over sulphates, nitrates, carbonates or hydrosulphides at between 1000 and 1200°C, followed by slow cooling in flowing hydrogen or other reducing atmosphere. It is extremely important to minimise the risk of oxidation at elevated temperatures since even at room temperature CaS can be extremely susceptible to water oxidation. Often too little attention is paid to the detail of the preparative method and its reporting.

2.2 Thin Films

Carefully prepared starting materials, often in doped form, are employed in most instances. Thin film device structures are commonly deposited in situ within an electron beam evaporation system and normally, for CaS deposition, sulphur coevaporation is applied to control stoichiometry. The substrate temperature is considered to be critical in improving crystallinity of the CaS layer with

Springer Proceedings in Physics, Vol. 38 **Electroluminescence**
Editors: S. Shionoya and H. Kobayashi

temperatures of 350 - 450°C, OGAWA [4], KOBAYASHI [5] and TANAKA [6], giving crystal orientation similar to powder grains and more efficient EL.

3. Experimental EL Structures

3.1 Powders

In general, the test structures have consisted of an active CaS layer of 150 to 300 μm thickness sandwiched between glass conducting and metallic electrodes, sometimes with a thin mica or polymeric insulating layer between the glass electrode and the sample. The calcium sulphide is dispersed normally in a dielectric medium such as araldite or castor oil. Exciting voltages up to 2.5kV are used in the frequency range 50Hz to 25kHz.

For dc electroluminescence studies, much narrower electrode separations (20 to 30 μm) have been used and accordingly demand fine particle size material. In order to operate at relatively low voltages (∿ 100V), the inclusion of conducting copper sulphide layers over the phosphor surface facilitates this possibility. However, an initial forming process is required starting at a voltage of 10-15 volts to cause migration of the copper and to get the conditions for efficient EL emission.

3.2 Thin Films

A variety of structures on indium-tin oxide substrates have been used ranging from metal - CaS - conducting glass to metal - insulator - CaS - insulator - conducting glass. The fact that, where the measuring device is formed in situ, it becomes effectively sealed and in consequence makes it less vulnerable to moisture attack subsequently, is an attractive feature of this approach, although the higher cost of such processing sets limits to commercial applicability.

4. Electroluminescence

4.1 Brightness-Voltage/Brightness - Frequency Characteristics

Much of the work on a.c. electroluminescence in powders was centred on determining brightness - voltage and brightness - frequency characteristics. Comparisons between different studies are not simple because of the variations in the fabrication of the active layers and of the test cells.

However, the different regions of the brightness - voltage characteristics in all the reported measurements can be fitted either to the relationship

$$B = B_o \exp(-b / V^{\frac{1}{2}})$$

or

$$B = A \ (V-V_o)^n$$

Mixed order kinetics appear to prevail in the overall operation of most of the powder samples investigated. Values of the index n vary greatly from 2 up to 6 depending upon excitation frequency, nature of doping and also the nature of the phosphors investigated; LAWANGAR [7] investigating CaS : Bi, Pd system found n = 2 where only Bi was added but n = 4 for the doubly doped samples. A universal description cannot be provided for the B - V characteristics because of the 'ad hoc' nature of the investigations. The dc powder electroluminescence investigated by VECHT [3] indicates that the exponential relationship is followed over many orders in CaS : Ce and is distinct from the mixed behaviour observed in ZnS : Mn,Cu, in which saturation effects at high voltages and deviations from an exponential relationship at low voltages occur.

Logarithm brightness - frequency measurements tend in the main to follow a linear relationship with some saturation occurring at higher frequencies and higher voltages. PANDEY [8], investigating CaS : Ce, found near linear logarithm brightness - frequency characteristics up to 10 kHz and 770 V, as did RENNIE [9] in studies of CaS : Cu, Cl.

With a.c. thin film EL studies, whilst clearly the brightness - voltage and brightness- frequency characteristics have been reported, a major focus has been on the influence of the preparative conditions on the brightness achievable. The influence of crystallinity has figured prominently in such studies. TANAKA [6] has reported some particularly interesting results on brightness - frequency measurements in CaS : Eu. Using bidirectional square wave pulsed excitation, Tanaka observed three emission peaks, the first occurred on the rise of the exciting voltage pulse followed by a second peak after 5 μs and a third peak on the fall of the voltage pulse. The first and third peaks because of the low electric field conditions $\sim$ 10 kV mm^{-1} must be linked to electron transfer from the host lattice to the luminescent centre, whilst the field conditions $\sim$ 100 kV mm^{-1} for the second peak permit impact ionisation of the host lattice and luminescent centres to occur. This behaviour is not observed in rare earth doped ZnS, in which the single, impact ionisation derived, peak occurs. The reasons proposed for this difference in behaviour between CaS and ZnS is that the former has an indirect band gap; doping with Eu creates an excited state very close to the conduction band minimum at the X point permitting lower energy carrier transfer processes to occur.

4.2 Spectral Characteristics

Copper has frequently been used as a codopant in electroluminescent powder phosphors and particularly in the case of dc powder electroluminescence; here a copper sulphide layer is seen to be important to achieve high luminescence levels. It is interesting that the characteristic emission of copper in the blue-green is suppressed almost totally in doubly doped CaS; the copper is considered to act as a sensitiser for the rare earth centre transferring energy directly to it. RASTOGI [11] reported one of the earliest results on this effect in CaS : Nd, Cu, in which there was no trace of the Cu blue green emission but a

Table 1 Spectral Emission Peaks and Transition Types in Doped CaS

Dopant	Peak Wavelength/ nm	Transition	Ref.
Eu^{2+}	650	^7F — $^8S_{7/2}$	(6)
Sm^{3+}	560 600 650	$^4G_{5/2}$ — $^6H_{5/2}$ $^4G_{5/2}$ — $^6H_{7/2}$ $^4G_{5/2}$ — $^6H_{9/2}$	(14)
Ce^{3+}	505 570	^2D — $^2F_{5/2}$ ^2D — $^2F_{7/2}$	(6)
Mn^{2+}	580	^4G — ^6S	(10)
Tb^{3+}	400 550	$^5D_3$ — $^7F_6$ $^5D_4$ — $^7F_5$	(10)

characteristic Nd peak at 545 nm and a further peak at 600 nm linked to Cu and Nd interaction; BHUSHAN [12], however, did not observe this suppression of the blue-green copper emission in CaS : Nd, Cu and the difference between these two results might be related to the fact that the latter used Na_2Co_3 flux at 1200°C in the preparation process. BHUSHAN [13] did observe suppression of blue-green emission in CaS : Cu, Er, although there was no evidence of red peak at 600 nm.

In the more recent work, particularly in the context of thin film a.c. EL, detailed characterisation of the emission spectra in a number of differently doped CaS samples has been made. Table 1 provides some typical results for CaS.

Clearly, there is considerable substructure within the individual emission peaks, some of which are quite broad in character being dependent to an extent on the concentrations and interactions occurring.

The most striking feature of the thin film CaS devices has been the luminances and efficiencies obtained, eg CaS : Eu 170 cd m^{-2} at 1 kHz and 0.05 lm W^{-1} (6) and CaS : Ce, Cl 650 cd m^{-2} and 0.11 lm W^{-1} (5). VECHT [3] in his dc EL investigations of CaS : Ce has produced panels capable of exhibiting 850 cd m^{-2} in dc pulsed voltage address modes; panels operated under continuous dc excitation of 100V with appropriate series resistance have given sustained luminances of 35 cd m^{-2} for in excess of 1000 hours.

5. Future Developments

The work undertaken on thin film structures is highly promising and offers scope for extension on three fronts:

(i) the use of alternative, and possibly additional, insulating layers has been observed already to give extended life, albeit at the expense of higher switch on voltage (15), and merits further investigation;

(ii) recent fluorescence studies on $Ca_{1-x}Cd_xS$ have confirmed massive increases in emission efficiency for x between 0.01 and 0.04 (16), and suggest that enhanced EL in R.E. doped mixtures of the same might be achievable;

(iii) the greater versatility and higher brightness of $CaS_{1-x}Se_x$: Ce, Cl layers reported by OSETO [17] opens up some new challenges also.

More fundamental studies on pure and doped CaS prepared in single crystal form are needed to underpin the applied work being undertaken on electroluminescence in CaS. The crystal growth method developed by KANEKO [18] represents one starting point for these more basic investigations.

6. References

1. W. Lehmann: J. Luminescence 5, 87 (1972)
2. A. Wachtel: J. Electrochem. Soc. 107, 119 (1960)
3. A. Vecht, M. Waite, M. Higton and R. Ellis: J. Luminescence 24/25, 917 (1981)
4. M. Ogawa, T. Shimouma, S. Nakada and T. Yoshioka: Jap. J.Appl. Phys. 24, 168 (1985)
5. H. Kobayashi, S. Tanaka, V. Shanker, M. Shiiki and H.Deguchi: J. Crystal Growth 72, 559 (1985)
6. S. Tanaka: J. Luminescence 40/41, 20 (1988)
7. R. D. Lawangar, S. H. Pawar and A. V. Narlikar: Mat. Res.Bull. 12, 341 (1977)
8. R. Pandey and P. K. Ghosh: phys. stat. sol.(a) 93, K173 (1986)
9. J. Rennie and M. A. S. Sweet: Cryst. Res. Technol. 2, K119 (1987)
10. S. Tanaka, V. Shanker, M. Shiiki, H. Deguchi and H.Kobayashi: Proc. Soc. Inf. Display Conf. 26, 255 (1986)

11. A. M. Rastogi and S. L. Mor: Ind. J. Pure & Appl. Phys. 16, 779 (1978)
12. S. Bhushan and F. S. Chandra: Crystal Res. & Technol. 20, K15 (1985)
13. S. Bhushan and F. S. Chandra: J. Phys. D 17, 589 (1984)
14. S. M. Pillai and C. P. G. Vallabhan: phys. stat. sol.(b) 134, 383 (1986)
15. K. Tanaka, A. Mikami, T. Ogura, K. Taniguchi, M. Yoshida and S. Nakajima: Sharp Tech. J. 37, 17 (1987)
16. B. Ray, J. W. Brightwell, D. Allsop and A. G. J. Green: J.Crystal Growth 86, 644 (1988)
17. S. Oseto: Japanese Patent 62 79285, Ricoh Co Ltd (1987)
18. Y. Kaneko, K. Morimoto and T. Koda: J.Phys.Soc. Japan 51, 2247 (1982).

Part VIII

Light Emitting Diodes

Organic Electroluminescent Diodes*

C.W. Tang and S.A. VanSlyke

Corporate Research Laboratories, Eastman Kodak Company,
Rochester, NY 14650, USA

A novel electroluminescent (EL) device is constructed using organic materials as the emitting elements. The diode has a multilayer structure of organic thin films prepared by vapor deposition. In a basic two-layer configuration a hole transporting layer is in contact with a luminescent layer. Hole injection is provided by an indium tin oxide electrode and electron injection into the luminescent layer is by a low work-function alloy, Mg:Ag. Recombination of electron-hole pair occurs primarily at the interface between the organic layers and results in electroluminescence which is characteristic only of the luminescent material.

In another configuration consisting of three layers of organic thin films, the EL emission is localized in a narrow luminescent layer sandwiched between a hole transporting layer and an electron transporting layer. The luminescent layer consists of a doped guest-host system whereby the energy of the electron-hole recombination in the host material results in EL emission from the guest molecules. Through this guest-host system, the EL spectrum can be shifted easily from blue/green to red with the use of a single host material and various guest molecules. The EL efficiency of the guest-host system is also significantly increased due to the higher fluorescence quantum efficiency of the guest molecules.

The luminescent host materials used in the EL diode belong to the class of quinolinol metal chelates. The molecular structure is shown in Fig.1 (a). The tris quinolinol aluminum complex is a typical example. The luminescence of these metal chelates is rather broad, centering around the blue/green region. The hole transporting materials belong to the class of aromatic amines. The structure is shown in Fig.1 (b). The guest molecules in the guest-host system can be chosen from several classes of organic fluorescent materials, the criteria being that they act as acceptors of energy transfer (Forster type) from electron-hole recombination in the host.

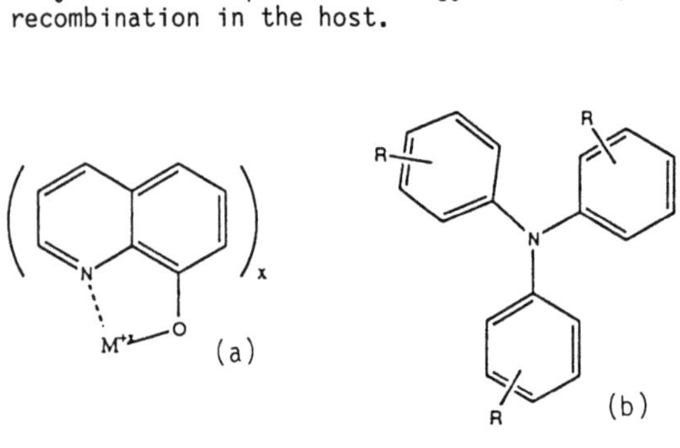

Fig. 1. Molecular structures of (a) luminescent host materials and (b) hole transporting materials

*Note: Details of this paper will be published in J. Appl. Phys.

Springer Proceedings in Physics, Vol. 38 **Electroluminescence**
Editors: S. Shionoya and H. Kobayashi

The organic EL cell behaves like a diode. The forward current appears to be limited by electron injection from the cathode whereas the reverse current is limited by leakage at high field. The EL emission is linearly proportional to the forward current and the external quantum efficiency is about 1% in the undoped system and 2% in the guest-host system. The voltage required for operation is quite low, typical below 10 volts. High brightness is achievable (>1000 cd/m^2) and the EL stability at a lower brightness level, (50–100 cd/m^2), can be maintained continuously for well over 1000 hours.

Reference

C.W. Tang and S.A. VanSlyke: Appl. Phys. Lett. 51, 913 (1987)

Electroluminescence in Vacuum-Deposited Organic Thin Films

C. Adachi, S. Tokito, M. Morikawa, T. Tsutsui, and S. Saito

Department of Materials Science and Technology,
Graduate School of Engineering Sciences, Kyushu University,
Kasuga-shi Fukuoka 816, Japan

Introduction

Electroluminescence in organic thin films, ideally suited for large-area display devices, has been investigated by many researchers. The fabrication of thin-film EL cells using organic materials that exhibited high-EL efficiency has not yet been successful. Recently, Tang and VanSlyke reported a two-layer organic EL device with superior EL characteristics./1/ Most recently, we proposed a new organic EL device with a three-layer structure; hole transport layer/emitting layer/electron transport layer./2,3/ The introduction of hole and electron transport layers made possible the enhancement of the efficiency of hole and electron injection. Moreover, it made the selection of emitter materials free from the intractable technical problems of the fabrication of thin film EL devices; a polycrystalline emitting layer was sandwiched between two pin-hole free homogeneous, amorphous carrier transport layers in the resulting three-layer cells.

Our provisional survey of organic emitter materials for EL devices resulted in an excellent emitter material, a perinone derivative, which gave bright yellow emission in our three-layer EL cell structure. In this report, we present the characteristics of the three layer EL cells with perinone derivatives for an emitter, and discuss the effects of the hole and electron injection layers. The mechanism of carrier recombination and emission will also be mentioned.

Experimental

The structure and the fabrication procedure of the three-layer EL cells were the same as those in the previous paper./2/ The cell structure was Au/hole transport layer/emitting layer/electron transport layer/Mg. For the hole and electron transport materials, N,N'-diphenyl-N,N'-(3-methyl phenyl)-1,1'-biphenyl-4,4'-diamine (TPD) and a perylene tetracarboxylic derivative (PV), respectively, were used. The emitter material was a perinone derivative (Pe). (Fig.1)/4/ The vacuum-deposited thin layer of Pe was polycrystalline and exhibited intense fluorescence at around 580 nm. All layers were deposited on a glass substrate by successive vacuum evaporation. All measurements of EL characteristics were carried out in a vacuum of 10^{-3} torr.

Springer Proceedings in Physics, Vol. 38 Electroluminescence
Editors: S. Shionoya and H. Kobayashi

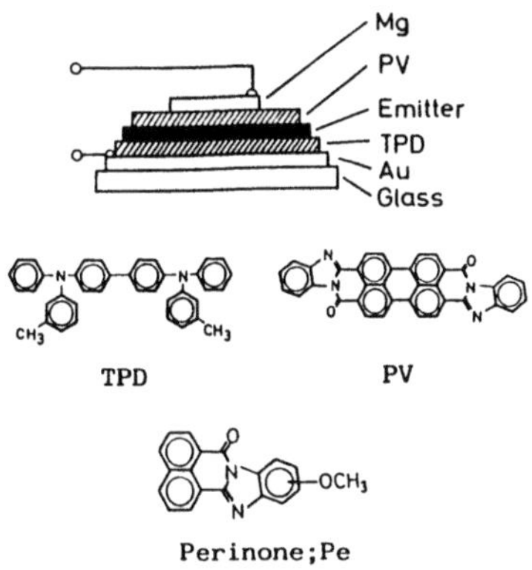

Fig.1: Structure of EL device and molecules used in this study.

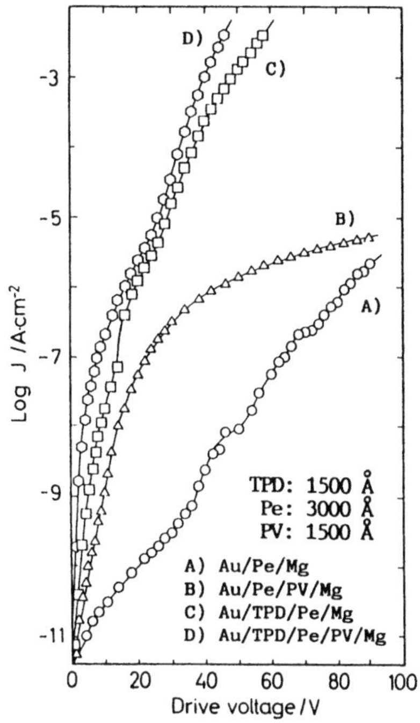

Fig.2: J-V characteristics for four different types of EL cells.

Results and Discussion

Figure 2 shows the relationships between current density and voltage for four different types of EL cells, when a bottom Au electrode was biased positive: A) Au/Pe/Mg, B) Au/Pe/PV/Mg, C) Au/TPD/Pe/Mg and D)Au/TPD/Pe/PV/Mg. One notices that the insertion of carrier transport layers is very effective for the increase of current densities. In particular, the insertion of the hole transport layer caused a large injection current into the emitting layer and resulted in bright EL emission. In the case of the EL cell without carrier transport layers (case A), a strong voltage dependence of the current was observed at voltages higher than 40V which corresponds to the onset of a carrier injection. However, no EL emission was observed at the drive voltage of 90V. The introduction of the electron transport layer brought about a small increase of a current density above 10V. Also no emission was observed. In contrast, in the EL cells having the hole transport layer (cases C and D), large injection current was observed and EL emission was observed at the voltage higher than 10V. The insertion of the hole transport layer was found to give rise to the drastic reduction of the potential barrier for hole injection into the emitting layer, in other words, the decrease of resistance for the hole injection. Moreover, this may bring about the increase of electron injection as the result of increase of the local bias-field crossing the electron injection electrode. The electron transport layer also contributed to the increase of current density. The reduction of the potential barrier for the electron injection by the insertion of the electron transport layer was proved. However, little difference in the efficiency of the EL emission was found between two cells, (C) and (D), suggesting that emission efficiency was not determined by the amount of injected electrons in our EL cells.

Figure 3 shows how the drive voltage at J(current density)=5mA/cm ^{2} depends on the thickness of the hole transport layer in Au/TPD/Pe/Mg and Au/TPD/Pe/PV/Mg cells. Apparently the insertion of the PV layer lowers the drive voltage. The reduction of the thickness of the TPD layer enables the lower-voltage driving without losing EL brightness. This result shows that the optimal thickness of the hole transport layer is a few hundred angstrom, if the fabrication of smooth pin-hole free layers are feasible. The drive voltage was also dependent on the thickness of the emitting layer. In contrast, it was almost independent of the thickness of the PV layer. Our EL cells can be driven at the voltage of below 20V, when the thickness of each layer is properly designed.

Cathode metals affected the EL emission characteristics and the drive voltages. Figure 4 shows the relationship between the the drive voltage and the current density in the EL cells with different metal cathodes; Mg, Al, Ag and Au. The relationship between the brightness and the current was independent of the cathode metals at the region of current above J=0.1mA/cm ^{2}. The drive voltage were largely dependent on the cathode metals when compared at the same current; the metals with higher workfunction necessitated a higher drive voltage. Because the cathode with high workfunction prevents the electron injection into the PV layer, the metal with low workfunction is arequisite for the low voltage driving of the EL cells. The emission intensity of about 100 cd/m ^{2} was achieved for all the cells at the current of 10mA/cm ^{2}.

Finally, we discuss the location of carrier recombination sites. Figure 5 shows how the EL emission intensity changes with the thickness of the emitting layer at the current of 5 mA/cm ^{2}. The emission intensity was independent of the thickness of

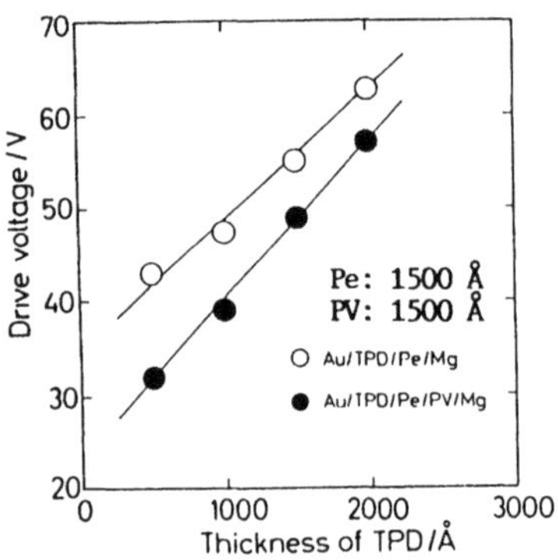

Fig.3: The drive voltage at J=5mA/cm^2 as a function of the thickness of the hole transport layer.

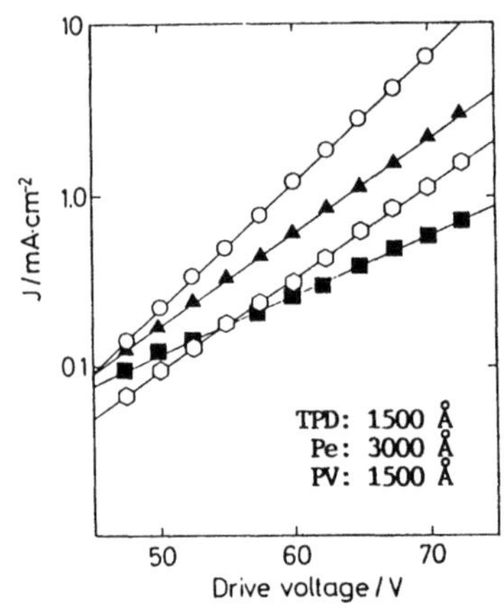

Fig.4: J-V characteristics in the EL cells with different cathodes; ○)Mg, ▲)Al, ⬡)Ag and ■)Au.

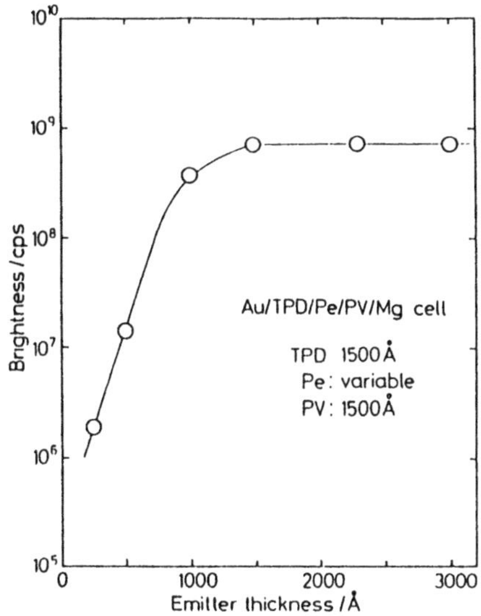

Fig.5: Emitter-thickness dependence of the EL emission intensity at J=5mA/cm^2.

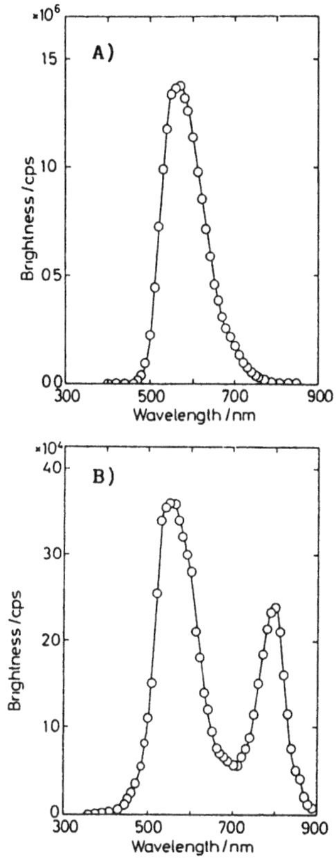

Fig.6: Emission spectra in two EL cells;
A) Emitter thickness; 1500 Å
B) Emitter thickness; 500 Å

the emitting layer at the emitter thickness larger than 1500 Å, but it decreased rapidly with the decrease of the thickness at the emitter thickness less than 1500 Å. This observation clearly indicates that the sites for carrier recombination and emission are located not in the narrow interface regions between the emitting layer and the transport layers but in the broad bulk region crossing the emitting layer. The EL spectra shown in Fig.6 support our interpretation: The EL spectrum coincided well with the fluorescence spectrum of a perinone evaporated film, when the emitter thickness was larger than 1500 Å. However, the EL emission of both perinone and PV was observed in the cells with the emitter thickness of 500 Å, indicating that recombination and emission sites penetrated into the PV layers due to the deficiency of recombination sites in the emitter layer.

References

1) C.W.Tang and VanSlyke: Appl. Phys.Lett. 51(1987)913
2) C.Adachi, S.Tokito, T.Tsutsui and S.Saito: Jpn.J.Appl.Phys. 27(1988)L269
3) C.Adachi, S.Tokito, T.Tsutsui and S.Saito: Jpn.J.Appl.Phys. 27(1988)L713
4) James Dassigny and Jean Robin: Fr. 1,111,620, Mar.2,1956.

Conductivity Control of ZnSe Grown by Metalorganic Vapor Phase Epitaxy and Its Application for Injection Electroluminescence

T. Yasuda, I. Mitsuishi, T. Koyama, and H. Kukimoto

Imaging Science and Engineering Laboratory, Tokyo Institute of Technology, 4259-Nagatsuda, Midori-ku, Yokohama 227, Japan

The control of n- and p-type conductivities in ZnSe has been achieved by metalorganic vapor phase epitaxy. Optical and electrical properties of these epitaxial layers and their application for blue injection electroluminescence are described.

1. Introduction

In achieving the full-color panel displays based on light-emitting diodes(LEDs), it is essential to develop blue LEDs to be utilized together with already available red and green LEDs. The materials for blue LEDs investigated in the past include GaN, SiC, ZnS and ZnSe. The fabrication of blue LEDs by using epitaxial layers of GaN and SiC (with a specific crystal structure of 6H, suited for blue emission) have already been demonstrated [1,2]. However, the technology used does not meet a requirement of mass production capability, because the large-area and low-cost substrates of sapphire and 6H-SiC are not available at present. For wide-bandgap II-VI compounds and related alloys, ZnS, ZnSe and ZnSSe, on the other hand, their epitaxial growth itself has long been a problem. However, the situation has changed dramatically since growth technologies of metalorganic vapor phase epitaxy (MOVPE) and molecular beam epitaxy (MBE) were used for growing these materials epitaxially on GaAs and GaP substrates.

In recent years, significant progress in the surface and crystalline quality of ZnSe and related alloys has been made by MOVPE based on the suitable choice of source materials and the lattice match to substrates. A major issue of current interest, therefore, is conductivity control, which is directed toward development of blue LEDs with p-n junction. In order to achieve conductivity control, the so-called self-compensation which is typical of wide-bandgap II-VI compounds should be taken into account. It is evident that low-temperature and non-equilibrium growth is required for reducing the self-compensation. MOVPE growth actually meets the requirement.

The purpose of this paper is to present an overview of the current state of MOVPE growth and conductivity control of ZnSe. A preliminary result of its application for injection electroluminescence is also described.

2. MOVPE Growth

During the last decade, ZnSe layers has been extensively grown on GaAs substrates by MOVPE using dialkyl zincs (dimethyl or diethyl zinc, DMZn or DEZn) and hydrogen selenide (H_2Se) [3-8]. A problem of the growth was a premature reaction between dialkyl zincs and H_2Se taking place even at room temperature, resulting in unsatisfactory surfaces with respect to morphology and uniformity. Recently, this problem has been solved by adopting appropriate source material combinations such as dialkyl zincs (DMZn or DEZn) and dialkyl selenides (DMSe or DEZn) or adducts

Springer Proceedings in Physics, Vol. 38 Electroluminescence
Editors: S. Shionoya and H. Kobayashi

of dialkyl zincs with dialkyl selenides (DMZn-DMSe, DMZn-DESe, DEZn-DMSe or DEZn-DESe) and H_2Se [9,10]. These source combinations are obviously better than the conventional combination of dialkyl zinc and H_2Se with respect to the reduction of premature reaction.

By the use of these sources, the mass-transport limited growth which is a feature of MOVPE has also been achieved [10]. When dialkyl zincs and alkyl selenides are used, a temperature-independent (and mass-transport limited) growth rate region is present at temperatures higher than 500°C. For the growth using adducts and H_2Se the temperature-independent growth rate region is located at lower temperatures, typically even at 300°C. Therefore, the latter source combination is better than the former from a point of the lower growth temperature which is more favorable for achieving conductivity control. In some cases, however, H_2Se in the latter source combination may react with dopants. An example is a case for the growth with p-type dopants of Li compounds, which will be described later. In such a case the former combination should be adopted.

Further improvements in the layer quality, at the surfaces and at the epilayer-substrate interfaces, has been accomplished by growing the ZnS_xSe_{1-x} alloys with x=0.06 and x=0.83 which are lattice matched to GaAs and GaP substrates, respectively [11,12].

3. n-type Control

Since the early pioneering work of Stutius [13], n-type doping in ZnSe by MOVPE has been extensively studied [14-16]. The n-type doping elements investigated include Al to be substituted on the Zn site of the ZnSe lattice, and Cl, Br and I to be substituted on the Se site. The typical results are summarized in Table 1. By using triethylaluminum ($(C_2H_5)_3Al$, TEAl), octylchloride ($1\text{-}C_8H_{17}Cl$), ethylbromide (C_2H_5Br, EBr), ethyliodide (C_2H_5I, EI) and normal-butyliodide ($n\text{-}C_4H_9I$, BuI), high carrier concentrations up to $10^{18}cm^{-3}$ or $10^{19}cm^{-3}$ have been obtained. It is noted that the controllability of iodine doping is quite satisfactory over a wide range of carrier concentration ranging form $10^{15}cm^{-3}$ to 10^{19} cm^{-3} [15,16].

Table 1. Summary of n-type impurity doping

Element		Dopant	Phase (at 1atm)	Carrier concentration $[cm^{-3}]$	
III	Al	$(C_2H_5)_3Al$	L	10^{16}-10^{18}	[13]
VII	Cl	$1\text{-}C_8H_{17}Cl$	L	10^{15}-10^{17}	[14]
	Br	C_2H_5Br	L	10^{18}-10^{19}	[this work]
	I	C_2H_5I	L	10^{15}-10^{19}	[15]
		$n\text{-}C_4H_9I$	L	10^{15}-10^{19}	[16]

The key point to avoid the self-compensation is to reduce the growth temperature. In view of this, we have grown n-type ZnSe using an adduct of DMZn-DESe and H_2Se as sources to reduce growth temperature as low as possible, typically down to 350°C, and TEAl, $1\text{-}C_8H_{17}Cl$, EBr, and EI as dopants. Our results have also confirmed the high doping efficiency of these dopants.

Above 400°C, it is difficult to achieve high conductivity. An example is shown in Fig. 1. The carrier concentration for Al-doped ZnSe layers is as high as 10^{18} cm^{-3} for growth temperatures of 300°C and 400°C, but tends to decrease dramatically above 400°C. One can assume that the observed decrease in carrier concentration is due to the self-compensation which becomes dominant at such a

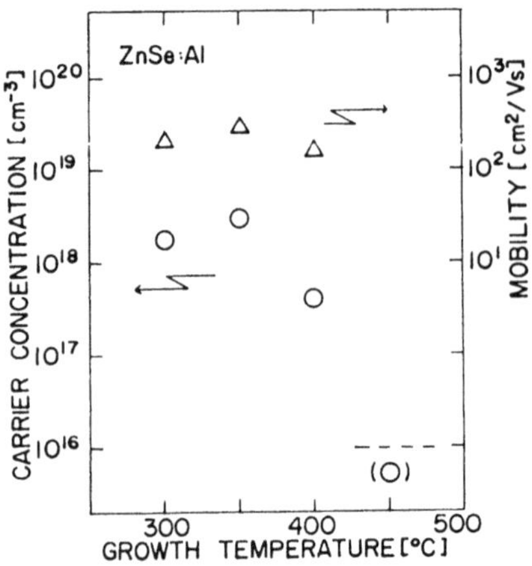

Fig. 1 Carrier concentrations (open circles) and mobilities (triangles) in Al-doped ZnSe layers grown at different growth temperatures

high growth temperature. This assumption was confirmed by the behavior of photoluminescence spectra of these samples. Namely, the so-called self-activated luminescence, which is due to donor(Al)-acceptor(a self-activated center in the form of a Zn vacancy-Al complex) pair transition, was dominant for the samples grown above 400°C. These electrical and optical properties of Al-doped ZnSe layers are quite similar to those of Al-doped ZnS layers grown by MOVPE [17]. Halogen-doped ZnSe layers also behave similarly, but it seems that a decrease in impurity incorporation efficiency with increasing growth temperature must be taken into account in addition to the self-compensation.

It should be emphasized again that the low temperature growth is of great importance in obtaining high conductivity n-type layers. It would be reasonable to assume that the situation is also true for p-type layers.

4. p-type Control

Extensive studies have been performed in the past on p-type conductivity control by doping the group V elements of N, P and As to be substituted on the Se site of the ZnSe lattice [18-20]. However, the results have shown until recently only high resistivity materials. In view of the fact that high quality epitaxial layers can be grown by MOVPE at high VI/II ratios, e.g., VI/II>10, we have tried to dope the group I_a elements to be substituted on the Zn sites [21]. The growth was carried out by using DMZn and DESe as source materials and lithium nitride (Li_3N) as the dopant. The dopant source of Li_3N heated at 400°C was carried by hydrogen gas onto the substrates kept at 450°C. As a results high conductivity p-type ZnSe layers with carrier concentrations ranging from low $10^{16} cm^{-3}$ to high $10^{17} cm^{-3}$ (a mobility of about 40 cm^2/Vs, and a resistivity of 0.2 Ωcm) were obtained. An activation energy of about 80 meV, estimated from the slope of the carrier concentration vs. 1/T straight line, was somewhat smaller than the reported Li acceptor depth of 114 meV. This suggests that hole binding energy decreases owing to the interaction among high concentration (compensated and uncompensated) acceptors. From low temperature photoluminescence spectra of these samples, it has been found that the so-called I_1 emission line, which is ascribed to the radiative recombination of excitons bound to neutral acceptors, becomes dominant with increasing doping level. At present it is difficult to identify the chemical species responsible for the acceptor, since the bound exciton lines due to Li and N acceptors are know to appear very closely to each other at around 2.79 eV. However, we believe it is quite probable that Li and N are simultaneously doped in our samples.

Current state of p-type impurity doping in ZnSe is summarized in Table 2. Cyclopentadienyl-lithium (C_5H_5Li) and cyclopentadienyl-sodium (C_5H_5Na) are in the

Table 2. Summary of p-type impurity doping (G, L and S indicate gas, liquid and solid, respectively.)

Element		Dopant	Phase (at 1atm)	Remarks	
I	Li	C_5H_5Li	S(100°C)	I_1 line	[22]
	Na	C_5H_5Na	S(100°C)	I_1 line	[this work]
V	N	NH_3	G	$\leq 10^{14}$ cm^{-3}	[23]
		$(C_2H_5)_2NH$	L	no indication	[this work]
		$(CH_3)_2N_2H_2$	L	I_1 line	[this work]
	P	PH_3	G	deep emission	[18]
	As	AsH_3	G	deep emission	[20, this work]
I+V	Li+N	Li_3N	S(400°C)	$\leq 10^{18}$ cm^{-3}	[21]
		$(CH_3)_2LiN$	S(100°C)	I_1 line	[this work]

solid phase at room temperature, but can be transported into the reactor when they are heated up to around 100°C. The layers grown by using these dopants have shown the I_1 line in their low temperature photoluminescence spectra, but no indication of p-type conduction. A p-type carrier concentration of about 10^{14}cm^{-3} was obtained by using ammonia (NH_3) as the dopant [23]. Other dopants of dienylamine ($(C_2H_5)_2NH$), dimethylhydrazine ($(CH_3)_2N_2H_2$), phosphine and arsine have shown at present either the I_1 emission line or no definite indication of effective acceptor doping. Dimethylaminolithium ($(CH_3)_2LiN$) which is also solid at room temperature, can be easily transported at 100°C, and is expected to become a useful dopant for increasing the controllability of p-type carrier concentration.

5. Injection Electroluminescence

On the basis of these results we have recently demonstrated blue emission from ZnSe p-n diodes fabricated on the GaAs substrate [21]. The device structure is inserted in Fig. 2. A p-type ZnSe layer 1 μm thick was grown on the Zn-doped p-type GaAs substrate, followed by an Al-doped n-type layer 3 μm thick. The In ohmic electrode were made on both sides of the diode chip. The diode showed good rectification, but current started to flow at somewhat higher voltage than expected for ideal p-n junction. This is probably due to the presence of an

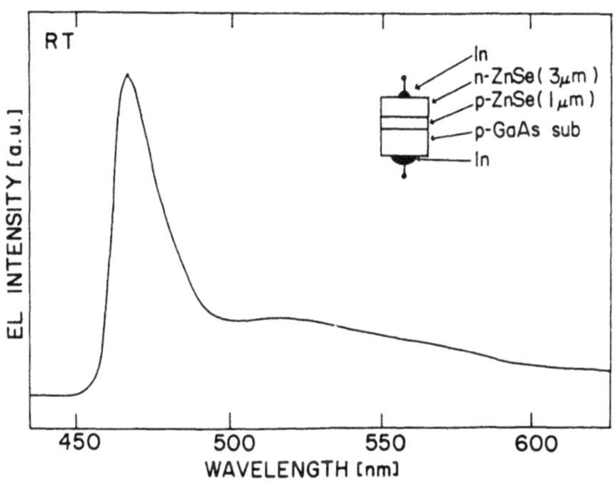

Fig. 2 Electroluminescence spectrum of a ZnSe p-n diode at room temperature

insulating layer at the junction, resulting from the so-called memory effect of impurities during growth. Typical electroluminescence spectrum is shown in Fig. 2. The emission band has a spectral peak at 467 nm and a long tail in the longer wavelength region.

Further improvements in efficiency, spectrum, and quality of junction interface is expected by optimizing growth conditions and device fabrication processes.

References

1. A. Shintani, S. Minagawa: J. Electrochem. Soc. 123, 1725 (1976)
2. A. Suzuki, M. Ikeda, N. Nagao, H. Matsunami, T. Tanaka: J. Appl. Phys. 47, 4546 (1976)
3. W. Stutius: Appl. Phys. Lett. 33, 656 (1978)
4. P. Blanconnier, M. Cerclet, P. Henoc, A. M. Jean-Lois: Thin Solid Films 55, 375 (1978)
5. P. J. Wright, B. Cockayne: J. Cryst. Growth 59 148 (1982)
6. F. A. Ponce, W. Stutius, J. G. Werthen: Thin Solid Films 104, 133 (1983)
7. S. Fujita, Y. Matsuda, A. Sasaki: Jpn. J. Appl. Phys. 23, L360 (1984)
8. A. Yoshikawa, K. Tanaka, S. Yamaga, K. Kasai: Jpn. J. Appl. Phys. 23, L773 (1984)
9. H. Mitsuhashi, I. Mitsuishi, M. Mizuta, H. Kukimoto: Jpn. J. Appl. Phys. 24, L578 (1985)
10. H. Mitsuhashi, I. Mitsuishi, H. Kukimoto: J. Cryst. Growth 77, 219 (1986)
11. H. Mitsuhashi, I. Mitsuishi, H. Kukimoto: Jpn. J. Appl. Phys. 24, L864 (1985)
12. I. Mitsuishi, H. Mitsuhashi, H. Kukimoto: (to be published)
13. W. Stutius: Appl. Phys. Lett. 38, 352 (1981)
14. A. Kamata, T. Uemoto, M. Okajima, K. Hirahara, M. Kawachi, T. Beppu: J. Cryst. Growth 86, 285 (1988)
15. N. Shibata, A. Ohki, S. Zembutsu: Jpn. J. Appl. Phys. 27, L251 (1988)
16. N. Shibata, A. Ohki, A. Katsui: J. Cryst. Growth 93, 703 (1988)
17. T. Yasuda, K. Hara, H. Kukimoto: J. Cryst. Growth 77, 485 (1986)
18. W. Stutius: Appl. Phys. Lett. 40, 246 (1982)
19. P. J. Dean, W. Stutius, G. F. Neumark, B. G. Fitzpatrick, R. N. Bhargava, Phys. Rev. 27, 2419 (1983)
20. M. Okajima, M. Kawachi, T. Sato, K. Hirahara, A. Kamata, T. Beppu: Extended Abstracts of the 18th(1986 International) Conference on Solid State Devices and Materials, Tokyo, 1986, p.647
21. T. Yasuda, I. Mitsuishi, H. Kukimoto: Appl. Phys. Lett. 52, 57 (1988)
22. A. Yoshikawa, S. Muto, S. Yamaga, H. Kasai: Jpn. J. Appl. Phys. 27, L260 (1988)
23. A. Ohki, N. Shibata, S. Zembutsu: Jpn. J. Appl. Phys. 27, L909 (1988)

Characteristics of an Efficient ZnS Blue Light-Emitting Diode with a High-Resistivity ZnS Layer Grown by Metalorganic-Chemical-Vapour-Deposition

K. Kurisu and T. Taguchi

Department of Electrical Engineering, Faculty of Engineering, Osaka University, Suita, Osaka 565, Japan

An efficient ZnS blue light-emitting diode which consists of MπS structure has been reproducibly fabricated. A high-resistivity ZnS epitaxial layer as a π layer can be grown by low-pressure metalorganic-chemical-vapour deposition. Diode characteristics have been evaluated by means of current-voltage and forward-biased electroluminescence spectra measurements. Room temperature external quantum efficiency as high as about 0.04% can be obtained. It is shown that the recombination process is originated from the free-to-bound deep acceptor transition at higher forward current densities.

1. INTRODUCTION

ZnS holds much promise for use in a variety of light-emitting device applications in blue portion of the visible spectra because of a zincblende semiconductor with a room temperature (RT) bandgap of about 3.7 eV. Very recently, we have successfully obtained a high-quality cubic-structured single crystal of about 1.5 cm^3 in size using the iodine-transport method without a particular seed crystal /1/. The crystal exhibits an intense blue photoluminescence in the vicinity of 2.65 eV at RT and the blue luminescence is stemmed from an electronic transition between an I donor and a complex acceptor center /2,3/.

Taguchi and Yokogawa /2/ have fabricated an efficient blue LED (external efficiency η of 0.08%) with MπS structure in which a layer consists of ZnS-ZnO mixed layer on n-ZnS:I crystal. In spite of a high efficiency, it has been suggested that it is difficult to control the thickness and quality of the layer.

Recently, Hirahara et al. /3/ have reported a new structure MπS diode with a ZnS epitaxial layer fabricated by metalorganic-chemical-vapour-deposition (MOCVD) on n-ZnS:Al crystal and the η was estimated to be about 0.008% at π layer thickness in the range of 500 to 3000 Å.

This paper is concerned with the fabrication and characterization of MπS blue diode with a high-resistivity ZnS layer grown on n-type ZnS:I crystal.

2. EXPERIMENTAL PROCEDURES

Fig. 1 shows (100) and (111) ZnS wafers cut from a bulk ZnS:I single crystal ingot which has etch pit densities as low as 10^4 cm^{-2}. After the molten Zn extraction at 900 °C for 240 hrs, the RT resistivity was much reduced and as a result rendered reproducibly the low-resistivity n-type conduction (about 3 ohm·cm).

The π layer thickness was calculated from the growth rate and the epitaxial growth was then confirmed by X-ray diffraction measurement.

Springer Proceedings in Physics, Vol. 38 **Electroluminescence**
Editors: S. Shionoya and H. Kobayashi

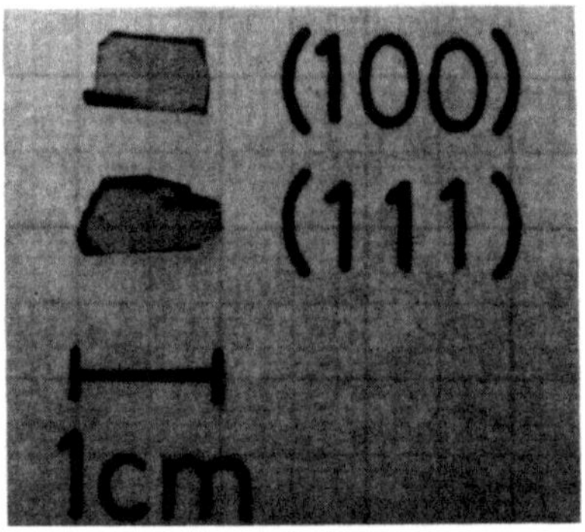

Fig. 1 ZnS wafers cut from a bulk crystal grown by the iodine transport method.

Table 1 Growth condition of ZnS epitaxial layer by MOCVD.

GROWTH CONDITION		
SUBSTRATE	ZnS:I (3 Ω·cm)	
SUBSTRATE TEMP.	350	°C
FLOW RATE OF DMZ	7.1×10^{-6}	mol/min
FLOW RATE OF H_2S	9.0×10^{-6}	mol/min
(VI)/(II) RATIO	12.7	
PRESSURE	3.0	TORR

3. RESULTS and DISCUSSION

Using a low-pressure MOCVD method /1/, an insulating layer of ZnS with an appropriate resistivity above 10^6 ohm.cm was deposited on the low-resistivity n-type ZnS:I crystal. Growth of the high-quality homoepitaxial layer as thin as 500 Å is extremely important to obtain a high quantum efficiency since it has been found that the η has a maximum value around a 300 Å layer thickness which is close to that of hole-diffusion length of ZnS /2,3/. Under these conditions, the hole tunnelling from the metal fermi level to the valence band of the ZnS is possible.

As shown in fig. 2, under an Xe-Cl excimer pumping capable of exciting the above band-gap, the n-ZnS:I crystal exhibits a sharp free-exciton emission at 3.669 eV and the well-known self-activated (SA) blue emission band at RT. After pulsed excitation, the SA band shifts toward longer wavelength. The time-resolved spectra show a characteristic of donor-acceptor pair recombination. The SA band locates at about 470 nm at RT.

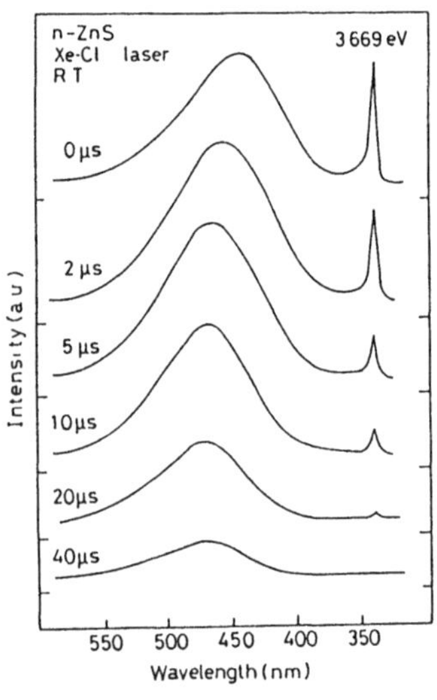

Fig. 2 Xe-Cl excimer laser excited time-resolved photoluminescence spectra of a n-ZnS:I crystal at RT.

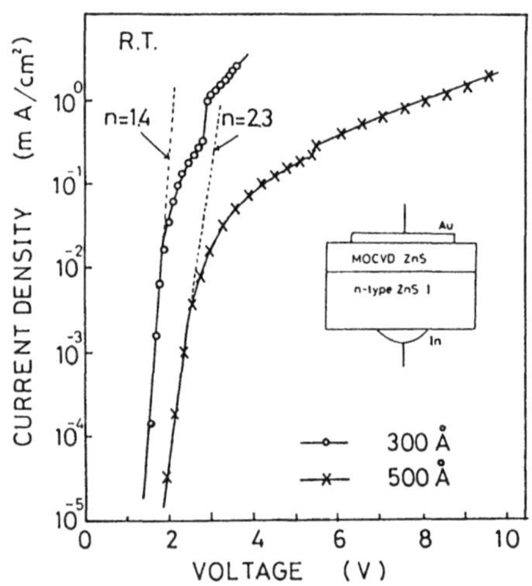

Fig. 3 I-V characteristics of MπS ZnS diodes with different π layer thicknesses (o:300 Å and x:500 Å).

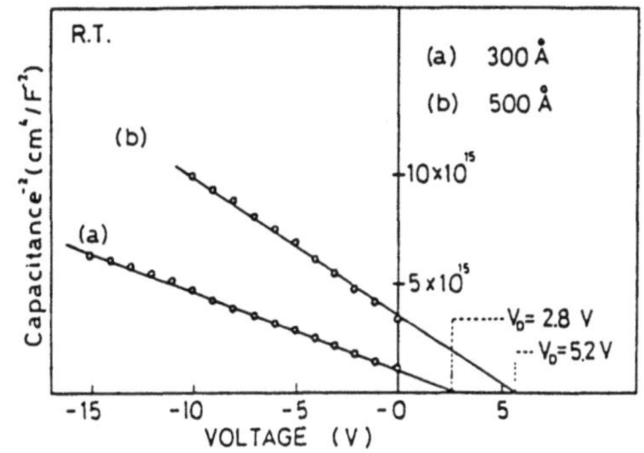

Fig. 4 C^{-2}-V characteristics at RT of two diodes with π-layer thickness of 300 and 500 Å.

Fig. 3 shows the voltage-current (I-V) characteristics of the blue diodes with different ZnS π layer thickness of 300 and 500 Å, respectively. The typical diode structure shown in the insert of this figure consists of Au-MOCVD ZnS π-layer-(110) n-ZnS:I crystal.

A good diode characteristic was obtained: till the forward bias reaches at 3 V, the forward current remarkably increases and is expressed by exp(eV/nkT). An ideal factor n at RT is estimated to be 1.4 and 2.3 for 300 and 500 Å thickness, respectively. The diffusion voltage (V_D) was obtained from the C-V characteristics as shown in Fig. 4 and is about 2.8 and 5.2 V for a 300 and 500 Å thickness, respectively. The energy density of interfaces is then calculated to be about 10^{11} $cm^{-2}.eV^{-1}$, which is a reasonable value. For each diode, there appears a "kink" portion in the I-V characteristics. This phenomena may relate to the tunnelling of electrons through the barrier /5/. The blue emission in the vicinity of 450 nm can be observed over a threshold voltage (~2 V), but never be observed when the reverse bias was applied.

Fig. 5 shows the relationship between the forward current and the emission spectra. The peak position of the blue-emission band around 450 nm shifts to-

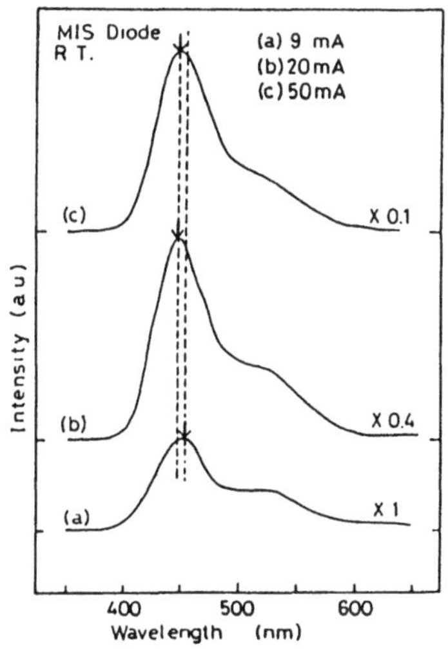

Fig. 5 Forward-biased electroluminescence spectra as a function of forward current at RT.
(a) 9, (b) 20 and (c) 50 mA.

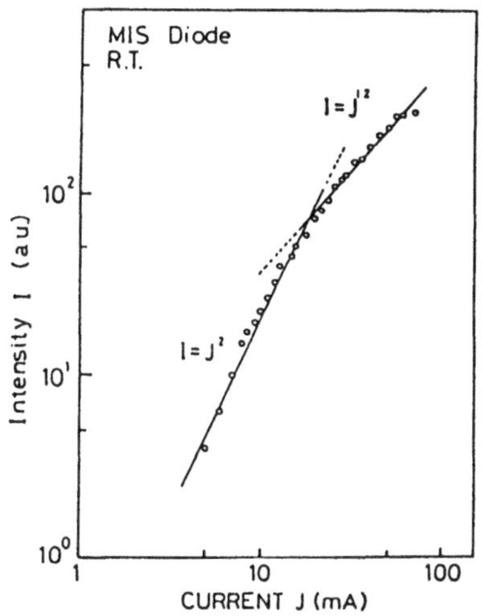

Fig. 6 Dependence of blue intensity (I) on forward current (J) at RT.

ward shorter wavelength with increasing forward current up to 20 mA. It is therefore considered that the emission band attributes to the donor-acceptor pair emission /1,2/. In the higher current regions above 20 mA, the peak position no longer moves, suggesting that the free-to-bound emission becomes dominant because of the ionization of the donor level by an electric field.

Fig. 6 indicates the dependence of blue-emission intensity on the forward current across the diode. This characteristic is approximately expressed by $I=J^n$, where I is the emission intensity, J is the current density and n is power. The bend appears around 20 mA, so that the recombination process seems to be changed from a bimolecular process (n=2) to a monomolecular process (n=1). This supports that the radiative recombination which takes place at higher currents is due to the free-to-bound transition encompassing the deep SA acceptor /2,3/. The present blue LED with a 300 Å thickness suitable for an MOCVD ZnS insulator layer, exhibits the high external quantum efficiency above 0.04% at RT.

4. CONCLUSION

We have reproducibly fabricated an efficient ZnS:I blue LED with an MπS structure by the low-pressure MOCVD method and this diode exhibits the blue emission with η above 0.04% at RT. It is shown that the energy density of interfaces in the MOCVD ZnS layer/n-ZnS:I diode is a great deal lower than that of an insulator NaI layer/n-ZnS diode /6/.

One of the authors (T.T) is grateful to Mitsubishi Kasei Ind., and to Mr. K. Matsumoto of Nippon Sanso Corp. for supplying MO gas and Mr. M. Hatsuta of Seitetsu Chemical Ind. for supplying H_2S gas. Part of this work was supported by a Grant-in-Aid for Scientific Research on Priority Area, New Functionality Material-Design, Preparation and Control, No. 62604583, from the Ministry of Education, Science and Culture of Japan.

1. T. Taguchi et al., J. Electrochemical Society, 172nd Society Meeting, Extended Abstract 87-2 (1987) p.1255.
2. T. Taguchi and T. Yokogawa, J. Phys. D 17 (1984) 1067.
3. T. Taguchi, JARECT. Vol. 19 Semiconductor Technologies (1986) ed. J.Nishizawa (Ohmsha LTD and North-Holland) 1986, p.309.
4. K. Hirahara, A. Kamata, M. Kawachi, T. Sato and T. Beppu, Extended Abstract of the 15th Conference on Solid State Devices and Materials, (1983) p.49.
5. N. V.Gorbenko, L. A.Kosyachenko, V. P. Makhii and M. K.Sheinkman, Sov. Phys. Semiconduct. 20 (1986) 393.
6. L. G.Walker and G. W.Pratt Jr., J. Appl. Phys. 47 (1976) 2129.

Electron Injection and Electroluminescence in Graded II–VI Compound Hetero-Junctions

W. Lehmann

Phosphor Consultant, 14 Surrey Run, Hendersonville, NC28739, USA

1. Introduction

We have investigated the possibility to inject electrons from n-type conducting CdS into insulating ZnS or ZnSe, and of holes from p-type ZnTe into insulating ZnSe [1]. This report is restricted to electron injection from CdS into ZnS.

Previous work on the luminescence of the II-VI compounds [2] had indicated that the energies of the conduction band edges are determined mainly by the cations (Zn, Cd) and those of the valence band mainly by the anions (S, Se, Te). This conclusion has been confirmed from measurements by Mead [3] on barrier energies between the compounds and several metals, combined with the known electronic workfunctions of the metals [4] and with the, also known, band gaps of the compounds. The result (Fig. 1) shows that the conduction band edges of ZnS and CdS differ by about 1.2 to 1.3 eV.

Since ZnS and CdS are completely soluble into each other, graded junctions between the two materials are possible in which the conduction band edge changes over a certain distance with a finite slope (Fig. 2). This slope corresponds to an electric field strength which can be counteracted by an applied voltage. When the latter cancels the band edge slope, electrons can freely move from the conducting CdS into the, originally insulating, ZnS.

There are two limitations. Firstly, the band edge slope in the unbiased junction must be shallower than the dielectric breakdown field strength of the material. The latter is about 120 V/μm for ZnS. With a band edge difference of about 1.2 eV between ZnS and CdS the minimum thickness of the graded part of the junction becomes about 0.01 μm. A junction narrower than that does not permit electron injection into the ZnS.

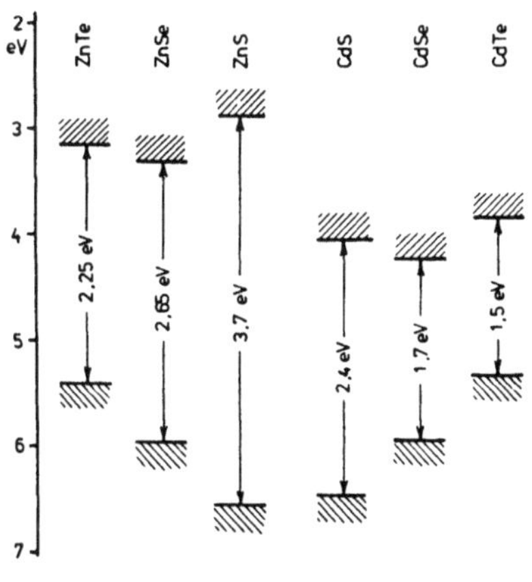

Fig. 1 Energies of the band edges of II-VI compounds

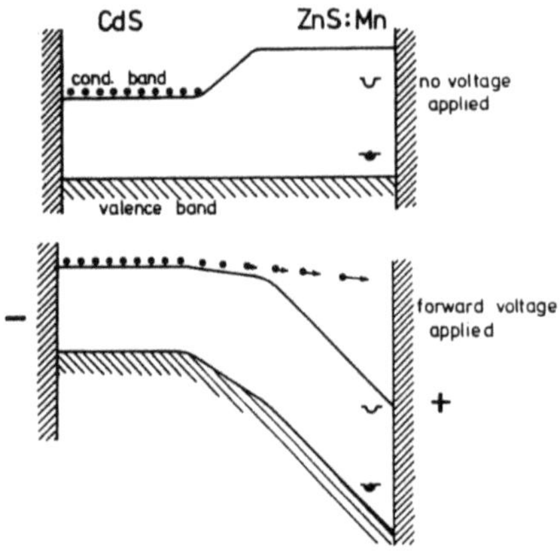

Fig. 2 Band diagram of a graded junction between conducting CdS and insulating ZnS

Secondly, ZnS unavoidably contains electron traps, and trapped electrons create space charges with blocking potentials which the applied voltage has to overcome. A rough estimate assuming a reasonable density of traps indicates about 10 μm to be the upper manageable thickness for the ZnS. That points to thin films. No such thickness limitation applies to the conducting CdS and, in fact, relatively thick (about 1 mm) single crystals of CdS were used in the beginning of this work. It was learned only later that comparable results can be obtained also from polycrystalline films of CdS.

2. Preparations

Films of ZnS and CdS were deposited from resistance heated crucibles by vacuum evaporation onto heated (400°C) SnO_x-coated conducting glass using the "hot chamber method" (Fig. 3). Since the sticking probability at elevated temperature is noticeably below unity, the evaporated particles (atoms, ions) of the materials may bounce several times from wall to wall before condensing into the final film. The films so prepared were very uniform in thickness and well adherent to the substrate.

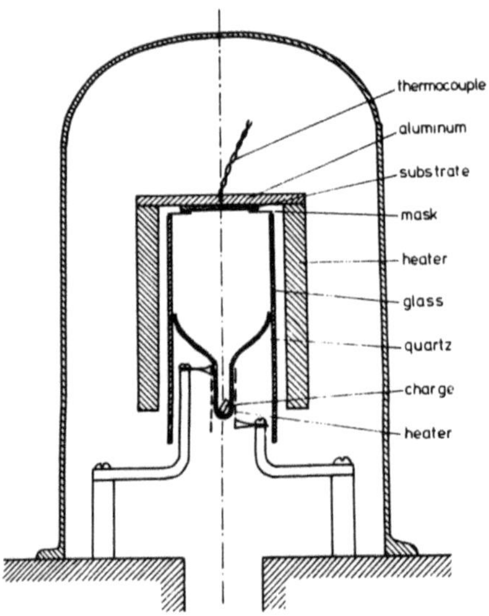

Fig. 3 Schema of the vacuum evaporation system used to prepare graded heterojunctions

The ZnS films were always highly insulating. Also CdS films, if pure, were insulating but well conducting (about 100 mhos/cm) CdS films were obtained by adding 0.1 mole-% of Ga or In to the charge. At the temperature used, some interdiffusion between ZnS and CdS takes place already during deposition. The result is a hetero-junction with a steady transition from one material to the other, and a corresponding steady variation of the band edge over the thickness of the interdiffused range.

The CdS was hexagonal (wurtzite) in structure, the ZnS cubic (zinc blende). Interatomic distances in the two compounds differ somewhat so that some lattice mismatch is unavoidable. Detrimental effects caused by this lattice mismatch were not observed.

Top contacts on the ZnS were metal spots of about 5 mm^2 area, the choice of the metal dose not seem to be critical. Mostly used were Al or Ni.

3. I(V)-characteristic

An example of a 60 Hz characteristic as displayed on an oscilloscope is shown in Fig. 4 demonstrating that current densities of more than one amp/cm^2 can be injected from n-type CdS via a graded junction into and through a thin film of originally insulating ZnS.

A more quantitative measurement reveals the characteristic to consist of two parts. At low applied voltage, the barrier between CdS and ZnS is reduced but not completely eliminated, and the current is limited by diffusion over the remainder of the barrier. An analysis shows that the quantity $\log_{10}(I/V)$ plotted over a linear scale of V should give a straight line with the slope 16.7 V^*/V where V^* is that part of the applied voltage, V, which drops across the graded part of the junction. Figure 5 shows an actual measurement (of the device of Fig. 4) with a slope close to 4.2 indicating that, in this case, about 1/4 of the applied voltage drops across the graded part of the junction, the rest mainly across the uncontaminated ZnS film.

The situation is different with an applied voltage high enough to let the barrier completely disappear. The major current limitation then seems to be the space charge of the electrons injected into the ZnS. A space charge-limited current in a solid is proportional to V^2 [5] so that a plot of the square root of the current over the voltage should yield a straight line. That has been observed, an example (again of the device of Fig. 4) is shown in Fig. 6.

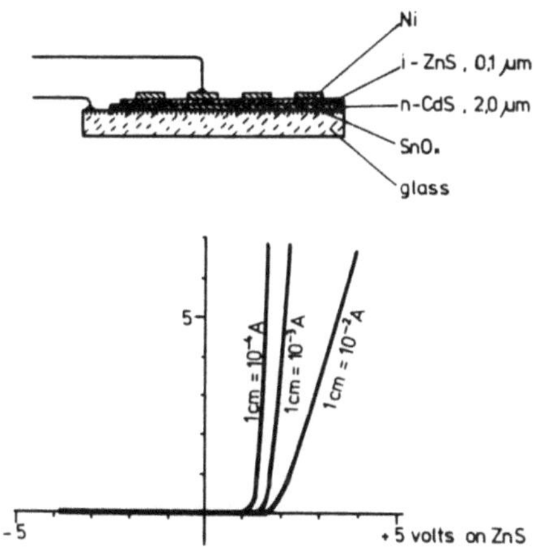

Fig. 4 Electron current injected from n-type CdS into and through a thin film of ZnS

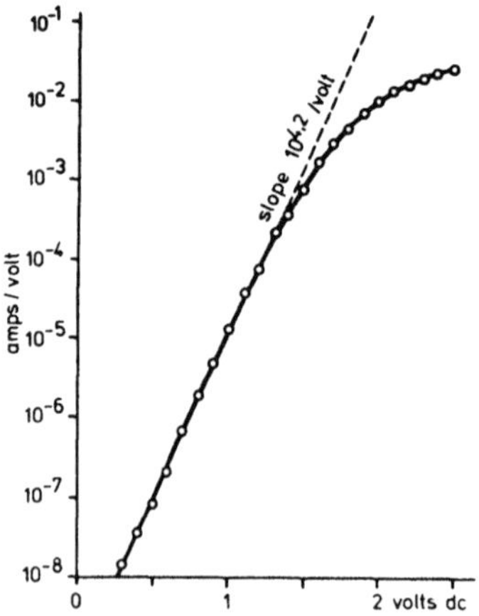

Fig. 5 Diffusion-limited current in an n-i hetero-junction between CdS and ZnS

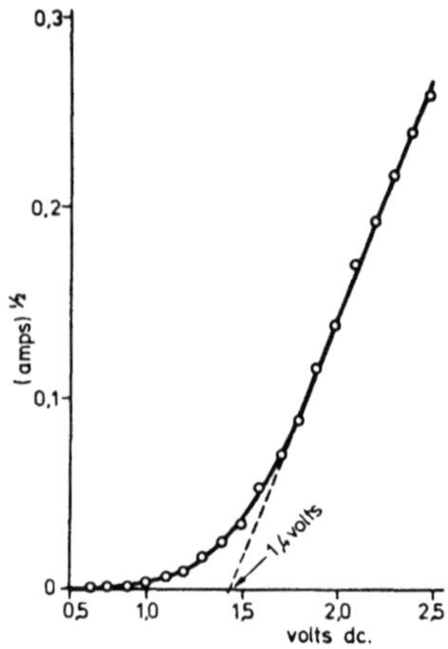

Fig. 6 Space charge-limited current in an n-i hetero-junction between CdS and ZnS

4. Electroluminescence

DC-electroluminescence appears in forward biased graded CdS-ZnS junctions provided that the ZnS contains about 1% MnS and that the slope of the junction corresponds to an electric field strength of about 10^5 V/cm or somewhat more. The excitation obviously is by collision of injected electrons with Mn^{2+}-ions.

Very thin ZnS:Mn films (about 0.1 µm) require only a few volts for visible light to appear but somewhat thicker films (about 0.3 to 0.5 µm) perform better. Typical brightnesses are 1 ftL at about 5-6 volts and 10 ftL at about 10-15 volts. The emission intensity increases rapidly with voltage and is nearly proportional to the current (Fig. 7). Quantum efficiencies in the range of about 10^{-5} to 10^{-3} have been measured but no attempt of optimization has been made yet.

The emission spectrum consists only of the well known yellow band of ZnS:Mn. The spatial emission distribution over the film area is very uniform (provided the film dose not contain visible defects). Luminescent "filaments" as reported by Rühle et al [6] are not observed. Neither DOES the device require the sort of "forming" reported by Vecht [7].

Reliable life test data are not yet available. One might perhaps expect the same stability over long operation times as is known for conventional TF-EL panels but that is still to be demonstrated.

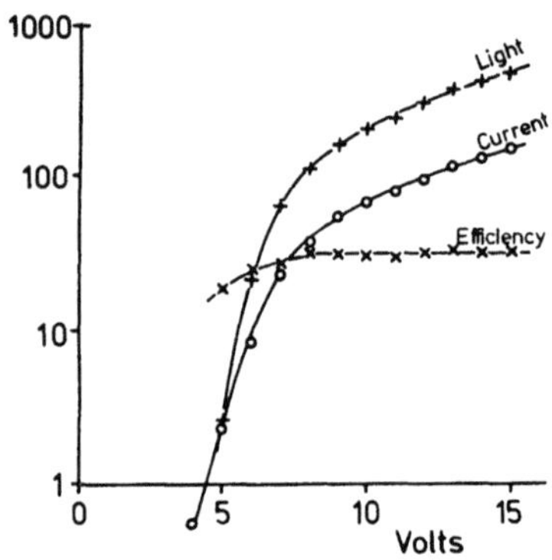

Fig. 7 Electroluminescence of an n-i hetero-junction between 2 µm CdS and 0.3 µm ZnS:Mn. The graded part of the junction in this case is estimated to be about 0.08 µm thick. Current in mA, light emission and quantum efficiency in arbitrary units.

References

1. Performed at Westinghouse Electric Corp., Pittsburgh, PA, USA., Part of the work was supported by Gov. Contract AFAL-TR-67-85.
2. W. Lehmann: J. Electrochem. Soc. 113, 449 and 788 (1966)
3. C.A. Mead: Solid State Electronics 9, 1023 (1966)
4. H.B. Michaelson: CRC Handb. Chem. Phys., page E-76 (CRC Press, 64.ed.)
5. H.B. Ivey: In Advances in Electronics and Electron Physics, Vol.6 (Acad. Press, New York 1954)
6. W. Rühle, V. Marello, A. Onton: J. Luminescence 18/19, 729 (1979)
7. A. Vecht: J. Vacuum Sc. Technol. 10, 789 (1973)

Free Exciton Emission in ZnS_xSe_{1-x} MIS Diodes with High Pulse Current Density

Dezhen Shen and Xiwu Fan

Changchun Institute of Physics, Academia Sinica, Changchun, P.R.China

The electroluminescence emission spectrum of forward-biased $ZnS_{0.22}Se_{0.78}$ MIS diodes excited by pulse current density from 30 to 300mA/mm^2 have been studied at 77K. We first found the inelastic collision of two free excitons in $ZnS_{0.22}Se_{0.78}$ MIS diodes with high pulse current density. On the basis of the distribution of kinetic energy of free excitons, we discussed the line shape of 2LO phonons. The results indicated that the effective temperature of free exciton is higher than the lattice temperature under high pulse current density, it was due to the inelastic collision of two free excitons.

1 INTRODUCTION

ZnS_xSe_{1-x} is a ternary compound , in which the band gap energy changes continuously with the composition x. The increase in composition x causes the high emission rate of free exciton[1]. Therefore, in order to obtain blue and violet spontaneous and stimulated emission, it is important to study free exciton emission of ZnS_xSe_{1-x}. In this paper, we first report free exciton emission in forward-biased ZnS_xSe_{1-x} MIS diodes with high pulse current density.

2 EXPERIMENTAL

ZnS_xSe_{1-x} (x=0.22) crystals used in this study were grown with sublimation method under a controlled partial pressure of selenium and sulphur corresponding to the minimum total pressure. The composition x was measured by X-ray diffraction. Dice with dimensions of 3x4x1mm^3 were cut from the crystalline boules and heated in molten zinc for about 100 hours at 900°C to reduce their resistivities. The MIS structure was performed by making an In ohmic contact on one large area surface and evaporating an insulating layer and gold electrode on opposite surface.

The $ZnS_{0.22}Se_{0.78}$ MIS diodes were excited by pulse current in forward bias with the pulse width of 10μs, repetition frequency of 3kHz and current density from 30 to 300mA/mm^2. EL spectra were measured using a grating monochromator of Model Spex 1404 with a RCA-C31034 cooled photomultiplier.

3 RESULTS AND DISCUSSION

Fig.1 shows the EL spectra at 77K of $ZnS_{0.22}Se_{0.78}$ MIS diodes excited by pulse current density from 30 to 300mA/mm^2 in forward bias. In Fig. 1, when the current density is 30mA/mm^2, the band at 4215Å(2.9415eV) labelled E_x and the band at 4313Å (2.875eV) labelled E_x-2LO are due to free exciton emission with zero and 2LO phonons, respectively. The energy separation between the two bands is closed to twice 33meV, which corresponds to LO phonon energy in $ZnS_{0.22}Se_{0.78}$[2]. The band labelled E_s[3] is associated with free exciton emission following scattering

Springer Proceedings in Physics, Vol. 38 Electroluminescence
Editors: S. Shionoya and H. Kobayashi

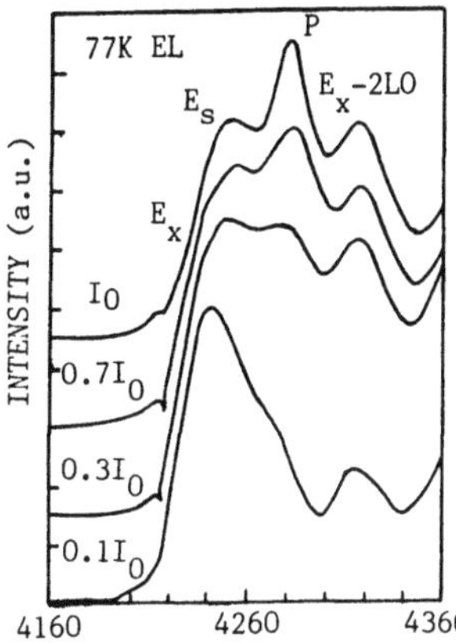

Fig.1 EL spectra in $ZnS_{0.22}Se_{0.78}$ MIS diodes with different pulse current densities (I_0=300mA/mm^2)

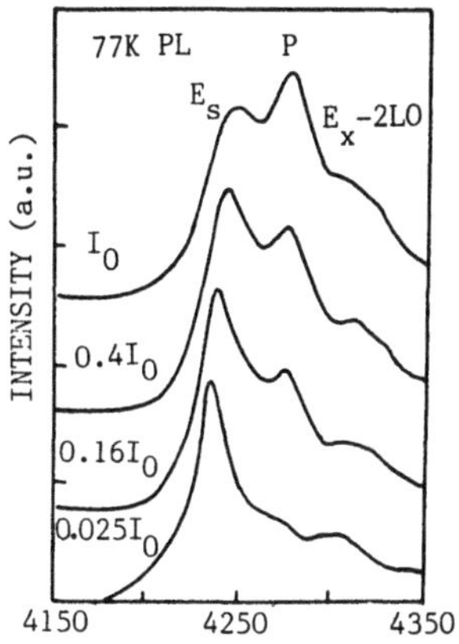

Fig.2 PL spectra in $ZnS_{0.22}Se_{0.78}$ crystal with different excitation densities (I_0=2MW/cm^2)

from electron in conduction band[3,4]. On increasing the pulse current density, we found that a new emission band p in Fig.1, which does not exist under low current density, appears in the low energy side of E_s band and increases more rapidly than other bands on increasing the pulse current density.

Fig. 2 shows the PL spectra at 77K in $ZnS_{0.22}Se_{0.78}$ crystal excited by a N_2 Laser of 3371Å with pulse width of 10ns, repetition frequency of 20Hz and excitation density from 0.025 to 2 MW/cm^2. In Fig.2, when the excitation density was low such as 0.025 MW/cm^2 there are two bands labelled E_s and E_x-2LO, respectively. When the excitation density increases a new p band appears and its energy position is the same as that in EL in $ZnS_{0.22}Se_{0.78}$ MIS diodes. Increasing the excitation density the p band increases rapidly. Comparing the new p band in Fig.1 and in Fig.2, the nature of the new p band is quite similar in EL and PL, in other words it does not associate with excitation method. They appear only at higher excitation density and increase rapidly with increasing the excitation density.

One of the most interesting aspects of the work described is concerned with the nature of new p band in $ZnS_{0.22}Se_{0.78}$ particularly in $ZnS_{0.22}Se_{0.78}$ MIS diodes. Saito et al [5] and our earlier work [3] attributed the p band, which was observed in PL and EL, to E_x-E_x interaction. Comparing the new p band in $ZnS_{0.22}Se_{0.78}$ MIS diodes obtained here with the p band in ZnSe obtained by Saito et al[5] and our work[3], it is reasonable to think that the new p band in $ZnS_{0.22}Se_{0.78}$ MIS diodes is the same as that of the p band in PL and EL in ZnSe. So it can also be ascribed to the inelastic collision of two free excitons.

In EL and PL in ZnSe, it is concluded that the p band is produced by E_x-E_x interaction under high excitation and the effective temperature of free exciton is higher than the lattice temperature[3,6]. On the basis of the results of ZnSe in EL and PL, we consider that in $ZnS_{0.22}Se_{0.78}$ MIS diodes when the new p band appears under high excitation density the effective temperature of free exciton should be higher than the lattice temperature. According to Gross's semiclassical fromula [7,8]

$$P_m \propto \Delta_m^{5/2-m} \exp\left(-\frac{\Delta_m}{KT_x}\right) \quad (1)$$

$$\Delta_m = h\nu + mh\nu_{LO} - h\nu_{ex} \quad (2)$$

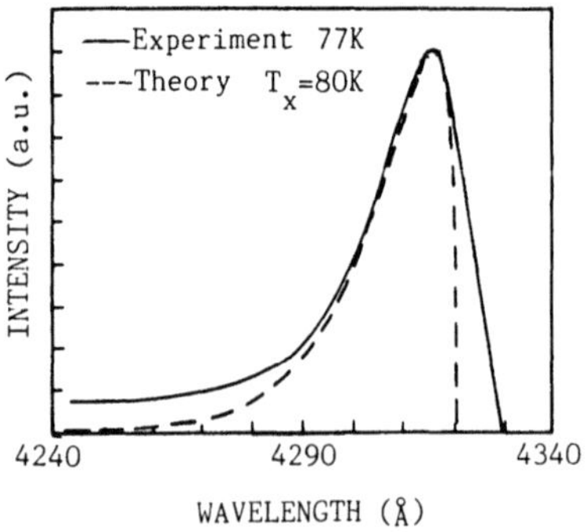

Fig.3 Comparison of the shape of the E_x-2LO (solid) measured at 77K with the theoretical curve (broken), calculated according to equation (1) with T=80K and m=2 for $ZnS_{0.22}Se_{0.78}$ MIS diode with pulse current density of $30mA/mm^2$

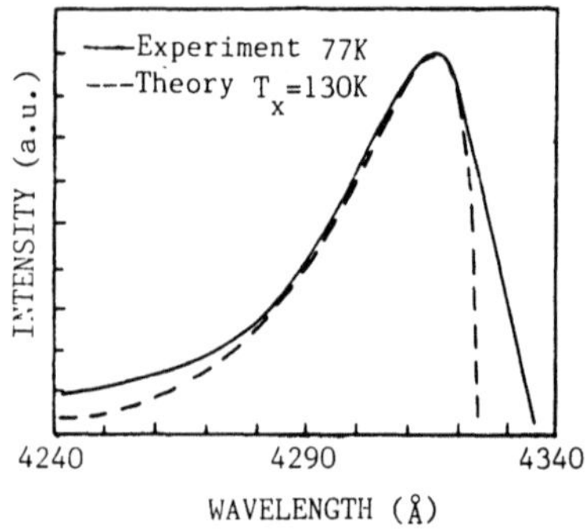

Fig.4 Comparison of the shape of the E_x-2LO (solid) measured at 77K with the theoretical curve (broken), calculated according to equation (1) with T=130K and m=2 for $ZnS_{0.22}Se_{0.78}$ MIS diode with pulse current density of $300mA/mm^2$

Where T_x is the effective temperature of free exciton, P_m is the recombination rate of mLO phonons assisted free exciton; $h\nu$ is the energy of photon recombination, Δ_m is the energy of free exciton kinetic, and $h\nu_{ex}$ and $h\nu_{LO}$ are the energy of recombination of free exciton and of LO phonon, respectively. Comparing the shape of the E_x-2LO band (solid curve) measured at 77K with the theoretical curve (broken curve) for $ZnS_{0.22}Se_{0.78}$ MIS diodes with pulse current density of 30 and $300mA/mm^2$ are showen in Fig.3,4, respectively. Reasonably good agreement between experiment and theory is obtained, when the effective temperature as an adjustable parameter is taken to be 80 and 130K, respectively. The mentioned results further identified that the new p band in EL in $ZnS_{0.22}Se_{0.78}$ MIS diodes under high excitation should be attributed to E_x-E_x interaction, and under low excitation density (such as $30mA/mm^2$) there is no p band, the effective temperature of free exciton is close to the lattice temperature. This result in EL in $ZnS_{0.22}Se_{0.78}$ MIS diodes as shown in Fig.1 is the same as that of ZnSe MIS diodes-[4].

In summary, the free exciton emission in $ZnS_{0.22}Se_{0.78}$ MIS diodes with high pulse current density is first studied. Increasing the pulse current density, a new p band appears in the low energy side of the free exciton emission band. Evidence is produced to demonstrate that the p band associated with E_x-E_x interaction. We believe that the pulse current density increases further, it is possible to obtain the stimulated emission in $ZnS_{0.22}Se_{0.78}$ MIS diodes.

ACKNOWLEDGMENTS

This work was supported by the National Natural Science Foundation of China. The authors wish to express their thanks to Mr. Tian Hua for his assistance.

REFERENCES

1. Ximin Huang et al.: Chinese J. Lumin. 6(3),186(1985)
2. Kenzo Ohmri et al.: J. Appl. Phys. 49(8), 4506(1978)
3. D.Z. Shen and X.W. Fan: Chinese J. Lumin. 9(1), 26(1988)
4. X.W. Fan and J. Woods: J. Phys. C14(13), 1863(1981)
5. H. Saito and S. Shionoya: J. Phys. Soc. Japan., 37, 2, 423(1974)
6. R. Baltramiejunas et al.: Solid Stat. Commun., 44, 6, 955(1982)
7. E.F. Gross et al.: Sov. Phys. Solid. Stat., 8, 1180(1966)
8. E.F. Gross et al.: J. Phys. Chem. Solids, 27, 1647(1966)

Electroluminescence of ZnSe:Mn MS Diodes in the High Electric Field

Xiwu Fan and Xurong Xu

Changchun Institute of Physics, Academia Sinica, Changchun, P.R.China

ZnSe:Mn MS diodes are made from two types of ZnSe crystals grown by iodine vapour transport method (A) or sublimation method (B), respectively. Relations between EL intensity J, current I, reverse bias V and EL efficiency η for A and B diodes are measured. It is found that for A diodes the curves can be distinguished by three different regions. The most important result obtained here is that both J and η increase with excitation simultaneously when electric field is sufficiently high.

1 INTRODUCTION

Yellow-orange light emission from a ZnSe:Mn MS Schottky diode in the reverse bias was described in many papers[1-6]. In general, the characteristic yellow--orange luminescence is produced by the internal $^4T_1-^6A_1$ transition within the manganese d^5 shell; Three possible mechanisms for exciting the manganese luminescence are impact excitation, impact ionization and e-h recombination[6]. The highest brightness of ZnSe:Mn MS diode reported was about 860Ft-L[5]. In this paper we are interested in the behavior of the light emission intensity and efficiency at different reverse biased excitation conditions, especially when the electric field is sufficiently high.

2 EXPERIMENTAL

Two types of ZnSe crystals grown in this laboratory were used in the present experiments. These crystals were grown by iodine vapour transport method (A) and sublimation (B), respectively. The starting material was also prepared here by direct vapour phase synthesis from the high-purity elements. Dice with dimension of $3x3x1mm^3$ were cut from the boule crystals, and annealed in molten zinc at 850°C for 100h to reduce their resistivities. After polishing and etching, one large area face of a die was provided with an In ohmic contact, and a circular Au electrode, 1mm in diameter was evaporated on the opposite surface. Thus a ZnSe MS diode was performed.

The current I and reverse bias V were measured using digital voltmeter Model PZ8 and light intensity J was measured using a photomultiplier Model EMI 9558B.

3 EXPERIMENTAL RESULTS

Fig.1 and 2 show the dependence of light intensity J and current I on the reverse bias V, in Fig.3 the variation of EL efficiency η with V is also shown. It is found that for A diodes the J and I change quite abruptly as the applied voltage increases continuously. The slope of curves can be distinguished by three different values, and the corresponding voltage region can be divided into three parts. In the intermediate bias region, I-V and J-V curves have the forms:

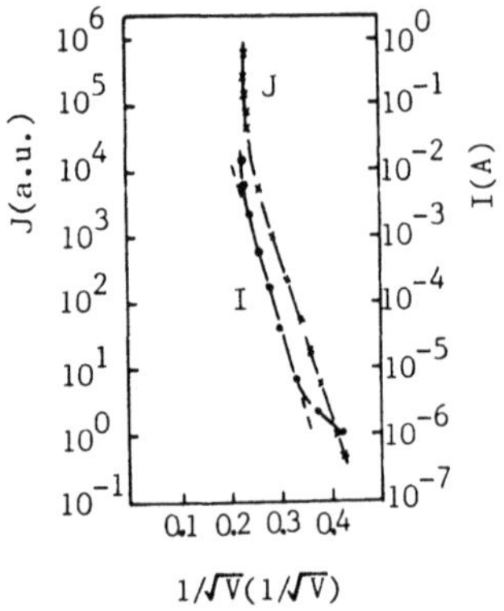

Fig.1 J–V and I–V characteristics of A diode

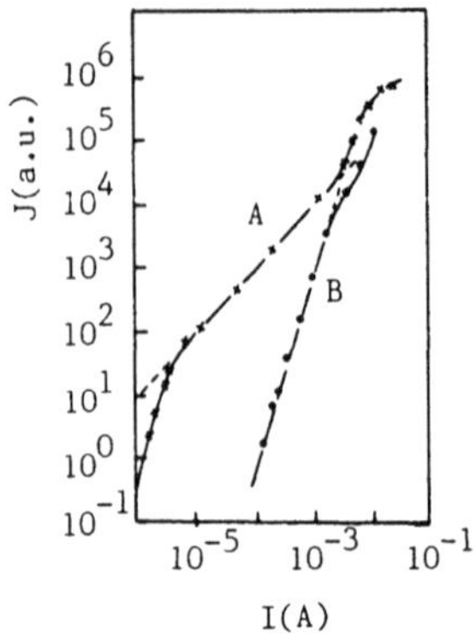

Fig.2 J–I characteristics of A and B diodes

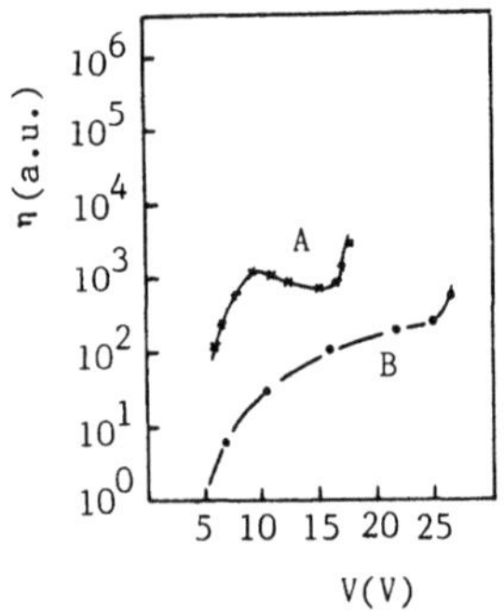

Fig.3 η–V characteristics of A and B diodes

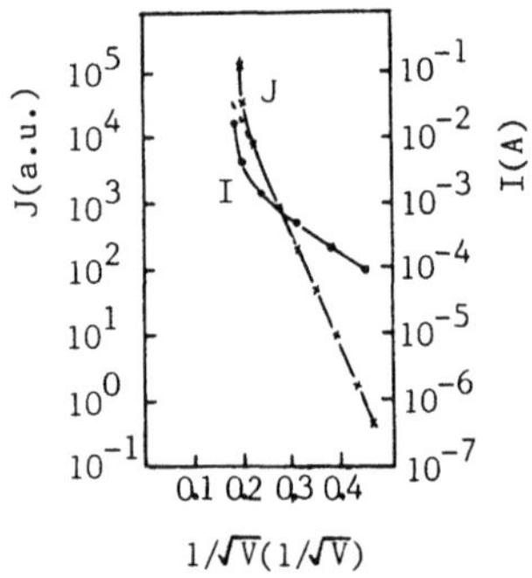

Fig.4 J–V and I–V characteristics of B diodes

$$I \propto \exp(-b/\sqrt{V}) \qquad (1)$$

$$J \propto \exp(-b'/\sqrt{V}) \qquad (2)$$

where b and b' are constants depending on ionized energy, and J–I curve has the form:

$$J \propto I^{n} \qquad (3)$$

where n=1; in the low bias region, equation (1) fails to be satisfied, while n becomes larger than 1 in equation (3), and η increases more rapidly with V. In the high bias region , I increases more rapidly with V than in the other two regions. The most important result obtained here seems that both J and η increase with excitation simultaneously. For instance the A diode has a light flux of 17 mlum (Schottky contact 1mm in diameter) and a power efficiency of 2×10^{-4}. But the results for B diodes reveal quite different picture, no turning point can be found in its luminescent characteristics in Fig.2,4.

4 DISCUSSION

The ground state of Mn^{2+} centers is taken to be localized at approximately 0.68eV above the valence band [7]. The impact ionization of Mn^{2+} centers will play a dominat part at lower value of bias, but when the bias becomes larger and larger, the band to band impact ionization will turn to play the leading role. As to the variation of the light intensity with excitation, it can be interpreted on the basis of impact ionization mechanism, and the I-V, J-V and J-I relations can be derived to have the forms of the equations (1), (2) and (3), respectively. This is actually the case in the intermediate bias region. For the variation of the efficiency with voltage, our early work[8] has suggested that if an acceleration collision mechanism is assumed, then the efficiency is limited by some factors. Such as the recombination probability of optical electrons (i.e. hot electrons) is less than that of thermal electrons[9]. This effect prevails also in the preavalanche process in our ZnSe:Mn diodes.

In the low bias region, EL behavior is probably related to leakage current in the diodes from a perfect single crystal. In B diodes, the donor impurities (e.g. chlorine) may accumulate in the crystal boundaries by diffusion. The leakage current should be sufficiently significant. The influence of the leakage current may be prolonged to the intermediate excitation region, for A diodes the influence of leakage current is restricted in the low bias region.

In high bias region, the sharp variation of current I with voltage shows the beginning of an avalanche process, which can be checked by the value of breakdown voltage calculated from donor concentration on the basis of C-V measurements. The emission spectra before or after an avalanche of an A diode have been meassured (e.g. 0.4, 6, and 8mA). It is shown that under these conditions the emission centres remain to be Mn^{2+}. The above experimental results indicate that the excitation and the recombination of Mn^{2+} centres are more efficient than those of other centers in the bias region. The details of this interesting behavior are to be reviewed in the futher works.

This work was supported by the National Natural Science Foundation of China.

REFERENCES

1. Y.S. Park and B.K. Shin: In Book: Electroluminescence J.I. Pankove(ED) P. 133-170 Berlin (1977)
2. J.W. Allen: J. Lumin. 7, 228 (1973)
3. J.I.B. Wilson and J.W. Allen: Solid Stat. Electron 19(6) 433(1976)
4. M.E. Ozsan and J. Woods: Solid Stat. Electron 18(6) 519(1975)
5. M.E. Ozsan and J. Woods: J. Phys. D. 9(18) 2613(1976)
6. N.T. Gordon: IEEE Trans. ED ED-28(4) 434 (1981)
7. H.G. Grimmeiss, C. Ovren and J.W. Allen: J. Appl. Phys. 47(3) 1103(1976)
8. X.W. Fan, X.R. Xu and S.H. Xu: Acta Physica Sinica 17(2) 99(1961)
9. X.R. Xu: AH CCCP 103, 585(1955)

Photo- and Electro-Luminescence of Rare Earth (Er, Yb)-Doped GaAs and InP Grown by Metalorganic Chemical Vapor Deposition

K. Takahei, P. Whitney, H. Nakagome, and K. Uwai

NTT Basic Research Laboratories,
3-9-11 Midori-cho, Musashino-shi, Tokyo 180, Japan

Rare earth (Er, Yb)-doped GaAs and InP epitaxial layers are grown by metalorganic chemical vapor deposition, and light emitting diodes are fabricated from these materials. Rare earth-related electroluminescence of Er-doped GaAs is observed up to room temperature, while that of Yb-doped InP is only observed up to about 140 K. The temperature dependence of efficiency and decay time constant of photoluminescence indicates that, for both materials, the decrease of luminescence intensity at elevated temperature is mainly due to de-excitation of excited rare earth ions in the host crystals, and not due to a decrease of energy transfer efficiency from host to rare earth ions.

1. INTRODUCTION

Rare earth-doped Ⅲ - V compound semiconductors may offer a new class of temperature-stable optical devices. They have the combined advantages of temperature-independent sharp rare earth luminescence, and the ability of semiconductors to activate emitting rare earth centers by minority carrier injection. Since most rare earth ions tend to have a valence of three, the Ⅲ - V compound semiconductors seem to be suitable hosts for rare earth atoms, as far as charge compensation is concerned. However, because of the larger ionicity of Ⅲa atoms compared to Ⅲb atoms, an arsenic compound of Er forms an NaCl structure, while an arsenic compound of Ga forms a zinc-blende structure. Therefore, solubility of rare earth atoms in Ⅲ - V compound semiconductor hosts is expected to be poor and it will be difficult to grow the materials by methods such as liquid phase epitaxy (LPE) [1]. This difficulty may be overcome by growth methods such as metalorganic chemical vapor deposition (MOCVD) [2, 3] and molecular beam epitaxy (MBE) [4] which are more flexible in terms of growth conditions.

This paper briefly reviews and discusses optical properties of rare earth-related photoluminescence (PL) in Yb-doped InP and Er-doped GaAs grown by MOCVD in relation to the application of the materials to light emitting devices. Then, the cause of thermal quenching of rare earth-related electroluminescence (EL) is discussed based on the experimental results for PL efficiency and PL decay time constant.

2. EXPERIMENTAL RESULTS

2.1 Photoluminescence

Yb-doped InP and Er-doped GaAs were grown by low pressure MOCVD using organometallic rare earth compounds [2, 3]. For Yb-doped InP, the low temperature

Springer Proceedings in Physics, Vol. 38 Electroluminescence
Editors: S. Shionoya and H. Kobayashi

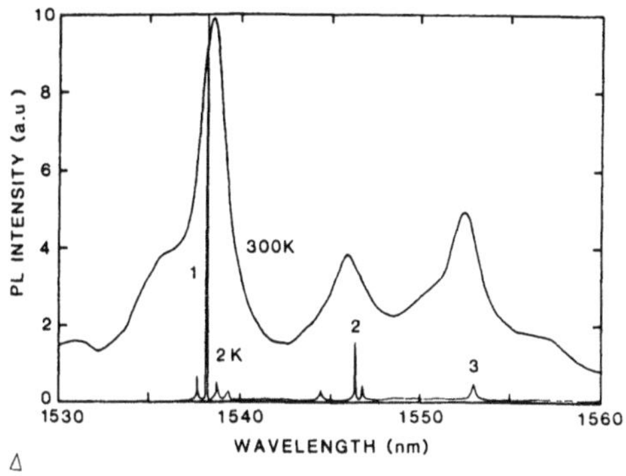

Fig. 1. Photoluminescence spectra of Er-doped GaAs at 2 K and at 300 K

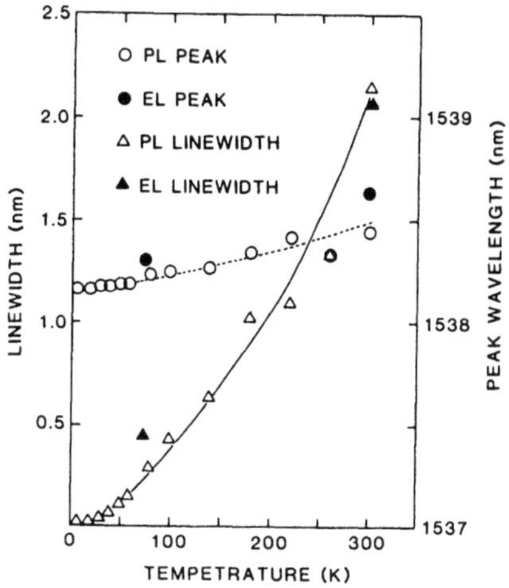

Fig. 2. Temperature dependence of linewidth and peak wavelength for peak 1 in Fig. 1

(2 K to 6 K) Yb-related PL spectra around 1000 nm reported so far are essentially independent of the synthesizing method such as LPE [1], MOCVD [2], or ion implantation [5].

On the other hand, Er-doped GaAs shows PL spectra around 1540 nm, that depend not only on the synthesizing method, but also on the synthesizing conditions. In MOCVD, epitaxial crystals grown at higher temperatures (650°C) show broad spectra consisting of many lines [6]. Recently, we have observed extremely sharp Er-related PL in epitaxial crystals grown at relatively low growth temperatures (below 550°C) and low arsenic pressure [7]. Figure 1 shows such PL spectra at 2 K and 300 K. The linewidth of the main peak 1 is less than 0.03 nm at 2 K. A similarly narrow spectrum, but with different peak wavelengths, has also been observed in a crystal grown by MBE at a relatively low growth temperature [4]. Figure 2 shows the temperature dependences of linewidth and peak wavelength for the main peak 1 in Fig. 1, details of which will be reported in a separate paper [8].

2.2 Electroluminescence

Light emitting diodes (LEDs) were fabricated from MOCVD-grown epitaxial crystals which show Yb-related PL and sharp Er-related PL as described in the previous section. Rare earth concentrations, measured by secondary ion mass spectroscopy, were about $10^{17} cm^{-3}$ and pn junctions were formed within the epitaxial crystals by Cd diffusion for Yb-doped InP and by Zn diffusion for Er-doped GaAs [9].

EL spectra consist of band-related broad luminescence and rare earth-related sharp luminescence. The rare earth-related EL and PL luminescence spectra are essentially the same at corresponding temperatures for each material . The linewidth and peak wavelength of the EL peak corresponding to the PL peak 1 in Fig. 1 are plotted in Fig. 2. EL of a Yb-doped InP LED was observed only up to about 140 K, while EL of a Er-doped GaAs LED was observed up to room temperature.

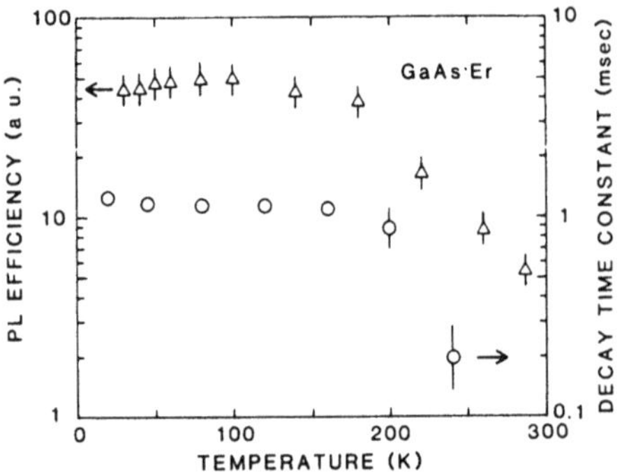

Fig. 3. Temperature dependence of PL efficiency and PL decay time constant for Er-doped GaAs

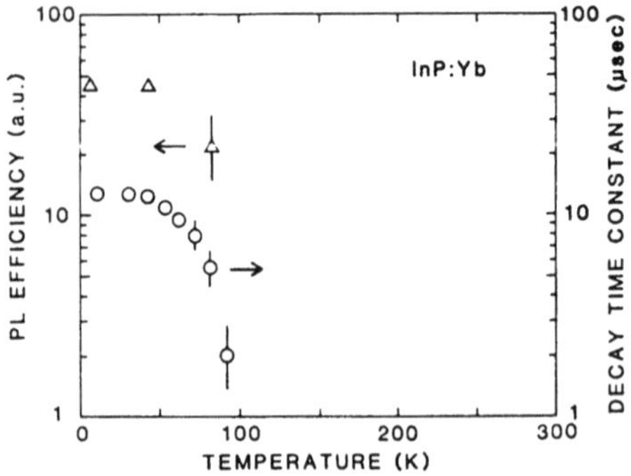

Fig. 4. Temperature dependence of PL efficiency and PL decay time constant for Yb-doped InP

2.3 Temperature Dependence of PL Efficiency and PL Decay Time Constant

Efficiency and decay time constant of PL were measured as a function of temperature to clarify the cause of the EL thermal quenching. Doping concentration of the rare earth ions was about 10^{17} to 10^{18} cm^{-3}. The PL efficiency was measured under excitation by a 633 nm He-Ne cw laser. The magnitude of the product of peak intensity and linewidth of a specific line was defined as the relative efficiency. For Er-doped GaAs, line 1 in Fig.1 was measured and the efficiency is plotted in Fig.3. Similarly, the PL line at 1.001 nm [2] was measured for Yb-doped InP and its relative efficiency is plotted in Fig.4.

PL time decay was measured using a 532 nm SHG Nd:YAG pulse laser. Peak 1 in Fig.1 was selected by a spectrometer with a band-pass of 4 nm centered at 1538.3 nm. This condition is suitable for this experiment, since the linewidth of the peak is less than 2 nm and the peak wavelength shift is less than 0.3 nm in the temperature range of present measurement. The decay of PL intensity measured as a function of time is exponential at low temperatures, becoming non-exponential at higher temperatures [10]. The decay time constant of the line 1 in Fig.1 is plotted as a function of temperature in Fig.3. A similar plot for the 1.001 nm line of Yb-doped InP is shown in Fig.4. Note that the decrease in the PL efficiency and in the decay time constant take place at nearly the same temperature for each material. A similar observation has been reported for Yb-implanted InP [11].

3. DISCUSSION

The PL and EL spectra of Er-related luminescence in Er-doped GaAs show that, in spite of the fact that the host is dominantly covalent crystal, the temperature dependences of linewidth and peak wavelength are of the same order as a rare earth ion doped in an insulating crystal such as Nd in YAG [12]. The rate of wavelength shift due to temperature is nearly three orders of magnitude smaller than that of band-to-band related luminescence of a typical semiconductor [8] and the linewidth of the main peak is less than 2 nm even at room temperature. ThusEr-doped GaAs is an attractive material for light emitting devices requiring temperature stability.

The Er-related luminescence intensity, however, decreases rapidly above 200°C. Such thermal quenching of luminescence intensity is a serious problem for application of the material to light emitting devices. The similarity in the temperature dependence of PL efficiency and PL decay time constant shown in Figs.3 and 4 indicates that the dominant thermal quenching process of the luminescence in both materials under study is the de-excitation of excited rare earth ions, and not a decrease in energy transfer from host crystal to rare earth ions. The mechanism of de-excitation has not been identified yet, but it may be the free carrier Auger effect [13] due to thermally activated carriers. If this is the case, the carrier concentration in the material would have to be controlled and excess carrier injection avoided. The difference in temperature at which effective thermal de-excitation of rare earth ions begins to take place for Yb-doped InP and Er-doped GaAs indicates that a suitable choice of rare earth ion and host is also essential for achieving high luminescence efficiency at high temperatures.

In the materials under study, the efficiency of rare earth ion-excitation by electron and hole pairs is rather small even at low temperatures where the de-excitation of excited rare earth ions is still negligible. Quantum efficiency of MOCVD-grown Yb-doped InP LED is on the order of 10^{-4} even at 77 K [9] and is even smaller in Er-doped GaAs LED at 77 K. Such low excitation efficiency is due to the presence of efficient non-radiative recombination channels which compete with the energy transfer from host to rare earth ions. It is very important to clarify whether or not such recombination channels are intrinsic to this class of materials.

ACKNOWLEDGMENT

The authors are grateful to Dr. Tatsuya Kimura and Dr. Hiroshi Kanbe for their helpful discussion.

REFERENCES

1. H. Nakagome, K. Takahei and Y.Homma: J. Cryst. Growth **85**, 345 (1987)
2. K. Uwai, H. Nakagome and K. Takahei: Appl. Phys. Lett. **50**, 977 (1987)
3. K. Uwai, N. Nakagome and K. Takahei: Appl. Phys. Lett. **51**, 1010 (1987)
4. H. Ennen, J. Wagner, H.D. Muller and R.S. Smith: J. Appl. Phys. **61**, 4877 (1987)
5. H. Ennen, G. Pomrenke and A. Axmann: J. Appl. Phys. **57**, 2182 (1985)
6. K. Uwai, H. Nakagome and K. Takahei: to be published in Pro. 4th Int. Conf. on Metalorganic Vapor Phase Epitaxy, Hakone, Japan (1988)
7. H. Nakagome, K.Uwai and K.Takahei: to be published in Appl. Phys. Lett.
8. K. Takahei, P. Whitney, H. Nakagome and K. Uwai: to be published in J. Appl. Phys.
9. P. Whitney, K. Uwai H. Nakagome and K. Takahei: presented at the 46th Device research Conf., Boulder, Corolado (1988)
10. K. Takahei, H. Nakagome and K. Uwai: to be submitted to J. Appl. Phys.
11. P.B. Klein: in Mat. Res. Soc. Symp. Proc. **104**, 437 (1988)
12. T. Kushida, Phys. Rev. **185**, 500 (1969)
13. J.M. Langer: J. Lumin. **40&41**, 589 (1988)

Ultraviolet Light-Emitting Diode of Cubic Boron Nitride PN Junction

K. Era, O. Mishima, Y. Wada, J. Tanaka, and S. Yamaoka

National Institute for Research in Inorganic Materials,
Tsukuba-shi, Ibaraki-ken 305, Japan

The present paper deals with light-emitting diode action of a cubic boron nitride(cBN) pn junction; after explaining our substantiation of the LED action[1], we show recent experimental results, then discuss light-emitting mechanisms and potentialities of the junction as light emitting devices.

A compound, cBN, has been used as a hard material for an abrasive. However, it has not been used as electronic material; correspondingly, little has been revealed experimentally concerning its electronic properties, though there have been quite a few band calculations, showing that the compound has an indirect gap. The mean of the calculated gaps is about 7 eV[2].

1. pn Junctions

Crystals of cBN(zincblende) were first synthesized under high pressure and temperature by Wentorf[3]. He also showed the existence of both p- and n-types[4]. Recently Mishima et al. developed a temperature difference method to grow large cBN single crysals[5]. They also made pn junctions and observed their diode action up to 650 °C[6]. The juncton was made under a pressure of about 55 kbar at about 1700 °C using a $LiCaBN_2$ solvent in a molybdenum cell. An n-type crystal was grown epitaxially on a p-type seed crystal. The acceptor dopant was Be forming a level of 0.23 eV, andthedonor Si forming a level of 0.24 eV. Resistivity of the p-type was estimated to be $1 \sim 10^2$ Ωcm, while for n-type it was $10 \sim 10^3$ Ωcm. These facts indicate that the concentration of the acceptor, i.e. of free holes, is one to two orders of magnitude higher than that of the donor and free electrons. Formation of the pn junction was confirmed by rectification characteristics and the existence of a space charge layer at the pn interface. The latter was shown by an EBIC image.

2. LED Action

A pn junction used in the present study is similar to that used in the previous work[6] mentioned above. Electrodes were formed with silver-paint. They were non-ohmic. The electrode was broken down when currents exceeded $2 \sim 3$ mA.

Springer Proceedings in Physics, Vol. 38 **Electroluminescence**
Editors: S. Shionoya and H. Kobayashi

Generation of light by the recombination of injected minority carriers with majority carriers is verified as follows: The light emission occurs only under the forward bias, and its intensity grows with increasing injection currents: This emission was confirmed by microscope observation to occur in a narrow range close to the space charge layer.

3. Luminescence Spectra

The emission spectra of the junction are shown in Fig.1. Curves, A and B, are injection luminescence spectra under the forward bias with junction currents as indicated. The observation was made, down to 185 nm, in such an arrangement that the emission generated at the junction region directly entered a spectrometer. Two spectral patterns, A and B, were obtained on the same junction for different observations with electrodes mended after every breaking of electrodes. These two patterns are considered to indicate the existence of two spatially different emission patterns in the junction. The two spatial patterns seem to have resulted from lack of crystal uniformity and lack of reproducibility in forming the electrode, and the asymmetrical shape of the junction.

The shortest wavelength observed was 215 nm in the case A. The peak occurs at 260 nm for A, and 310 nm for B. These bands are located in an energy range more than about 0.5 eV lower than the estimated energy, $(E_g - E_D)$ or $(E_g - E_A)$. The half width of the bands was about 1 eV. The emissions of A and B extend towards the longer wavelengths.

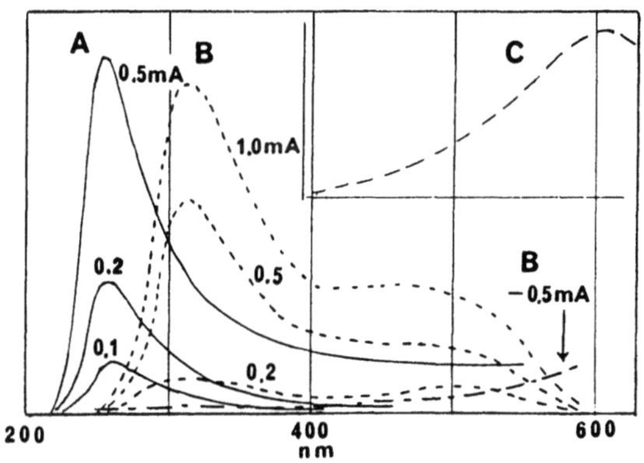

Figure 1. Injection(A and B) and cathode(C) luminescence spectra

The cathodoluminescence spectrum of the n-type portion is shown as curve C. Excitation was made by a 25 keV electron beam of a scanning electron microscope. The spectrum of the p-type portion was similar to that of the n-type, but the emission intensity was lower than that of the n-type because of self-absorption due to bluish black body color of the p-type portion. The spectrum resembles that of the reverse biased junction.

4. Discussion of Characteristics

The recombination emission must occur mainly in the n-side near the depletion layer because the much higher hole concentration brings about the hole injection rather than the electron injection. The Si donor level is the most probable initial level of the emission. The facts that the emission bands are located at the substantially lower energy position than the expected position, and also the wideness of the band widths, suggest participation of unknown imperfections in the luminescence transition. The emissions in the long wavelengths are also due to unknown imperfections. A low power efficiency of the junction measured to be around 10^{-6} indicates the presence of a considerable amount of nonradiative imperfections. As such undesirable imperfections, the following are candidates considering the preparation conditions: a nitrogen vacancy, oxygen, carbon, calcium, lithium, magnesium, and molybdenum impurities. If molybdenum is present in the crystal, it may act as a killer center. Molybdenum is an isoelectronic element of chromium.

The fact that the emission peaks do not shift with a bias change suggests that the photon-assisted tunneling and the usual donor-acceptor(d-a) pair recombination are not responsible for the emission. Considering the depth of the donor and acceptor, the degeneracy of carriers can not take place at room temperature and the radii of trapped carriers are too small for d-a overlapping for a wide range d-a separation to occur.

The fact that the cathodoluminescence spectrum is quite different from that of the injection luminescence suggests that in the injection emission region near the depletion layer, luminescence centers and recombination mechanisms are quite different from those in the other region. The acceptor may diffuse into the n-region and thus play the role of a terminal state of the emission. To investigate the origin of this difference is important for revealing the emission mechanisms of the present junction.

5. Potentialities of cBN pn Junctons as Light-Emitting Devices

For the present junction, no particular attention was paid to purity, perfection and junction structure. If the junction were prepared by paying appropriate attention to these points, a substantial improvement in efficiency would be achieved; further, injection emission might occur in the vacuum ultraviolet and spectral width would be narrow.

The material of cBN can be regarded as the widest band gap semiconductor which possesses both p- and n-types. The large band gap energy is advantageous for light-emitting devices, which is shown below. A factor ν^3 of cBN in the spontaneous emission probability is more than an order of magnitude larger than those of indirect gap LED materials such as SiC. This may compensate to some extent for the difficulty of cBN in obtaining better efficiency. Although the uniqueness of cBN is exhibited as ultraviolet electroluminescence devices, visible devices could be made of cBN by either appropriate doping or hybridization with phosphors. In this connection, development of thin films[7], particularly realization of dopable amorphous films would be important. The large gap bears also potentialities for electrically excitable lasers. Although cBN has indirect gap, optical gain range may be set far from the loss range of free carrier absorption. However, in order to attain the degeneracy of carriers and high conductivity, it is necessary to find a way to introduce much shallower donor and acceptor states. It might be possible to introduce into cBN a color center emitting in the visible or ultraviolet; then, electrically excitable short wavelength, even tunable, lasers could be realized. For these high power devices, the high thermal conductivity of cBN would be favorable.

6. Concluding Remarks

We have shown that cBN is a promising material for electroluminescence devices, in principle. To make such devices practicable, much further refinement of the present material preparation technology is necessary in the two directions: ultra-high pressure synthesis and thin film preparation; a method of forming ohmic electrodes must be established; moreover, the fundamental properties of the material need to be revealed. The development of such devices would open new areas in optoelectronics. The fact that both p- and n-types can be made of cBN which has such an extremely large band gap seems very peculiar. Understanding the essence of this nature of cBN would be the key to solving the problem of whether both p- and n-types can be made of a wide gap material or not.

7. Literature

1. O. Mishima, K. Era, J. Tanaka and S. Yamaoka: Appl. Phys. Lett. 53, 962 (1988)
2. As one of the latest reports, M. J. Huang and W. Y. Ching: J. Phys. Chem. Solids 46, 977 (1985)
3. R. H. Wentorf,Jr.: J. Chem. Phys. 26, 956 (1957).
4. R. H. Wentorf,Jr.: J. Chem. Phys. 36, 1990 (1962).
5. O. Mishima, S. Yamaoka and O. Fukunaga: J. Appl. Phys. 61, 2822 (1987).
6. O. Mishima, J. Tanaka, S. Yamaoka and O. Fukunaga: Science 238, 181 (1987).
7. For example, M. Satou, and F. Fujimoto: Jpn. J. Appl. Phys. 22, L171 (1983).

Electroluminescence Spectrum of Manganese-Doped $CuAlS_2$

K. Sato, K. Ishii, K. Tanaka, S. Matsuda, and S. Mizukawa

Faculty of Technology, Tokyo University of Agriculture and Technology,
2-24-16, Nakacho, Koganei, Tokyo 184, Japan

1. INTRODUCTION

The authors have been working with optical studies on a number of chalcopyrite-type crystals doped with transition elements and have clarified electronic energy levels introduced by such impurities[1]. In the previous studies[2], we observed a red photoluminescence(PL) band in a single crystal of $CuAlS_2$:Mn and assigned the luminescence to the ligand-field transition from the lowest excited state $^4T_1$ to the ground state $^6A_1$ in the $3d^5$ manifold of Mn^{2+} ion, just like the orange luminescence in ZnS:Mn. It was also found that the similar luminescence can be excited by the electron beam, as well as by the electric field.

Recently, we carried out studies of the time-resolved PL spectra on this material to clarify the mechanism of luminescence. We also measured the electroluminescence(EL) spectrum in a simple MIM diode using single crystals of $CuAlS_2$:Mn. This paper describes the results of these optical studies.

2. EXPERIMENTAL

2.1 Sample Preparation

Single crystals of $CuAlS_2$ doped with 5 mol% Mn were grown by the chemical transport technique. The obtained crystals were needle-like with a typical dimension of 5 mm x 1 mm x 0.5 mm. These crystals were analyzed by X-rays. The major surface of the crystal was determined as {112} plane of the chalcopyrite structure.

Samples for EL measurements were prepared as follows: Crystals were polished with lapping films and etched slightly by the HNO_3 solution. Thin aluminum film was evaporated on one side of the crystal as an electrode. The sample was pasted to a copper plate by a Ga-In soldering alloy.

2.2 Photoluminescence Spectrum Measurements

A CW-PL spectrum was measured at room temperature using the 488nm line of an Ar^+ ion laser as an excitation source. The emitted light was dispersed by a JASCO CT-25C monochromator and detected by a photomultiplier with the S-20 response. A lock-in detection system was used.

Springer Proceedings in Physics, Vol. 38 Electroluminescence
Editors: S. Shionoya and H. Kobayashi

The PL-decay characteristics and the time-resolved PL spectra were measured using a nitrogen laser(for 10ns-30µs) and a Xe flash lamp (for 10µs-10ms) as light sources. Boxcar integrator was employed to average the photo-signal. Measurements and data-processing were performed with the help of a microprocessor-controlled data acquisition system.

2.3 Electroluminescence Spectrum Measurements

An ac. voltage with a frequency 5 kHz from an audio-frequency oscillator was boosted to 150-300 Vrms by a transformer and was applied between the copper plate and the aluminum electrode at room temperature. The spectroscopic measurements were performed using the same system as described in the previous section.

3. EXPERIMENTAL RESULTS

3.1 Photoluminescence Spectrum

Figure 1 illustrates a PL spectrum observed in a single crystal of $CuAlS_2$:Mn at room temperature. The spectrum is broad with a peak at 1.96 eV and a small shoulder at 1.9 eV. This spectrum agrees completely with that measured with an excitation by a 356 nm line of a high pressure Hg-lamp excitation reported earlier[2].

Time-resolved spectra measured with a nitrogen laser are given in Fig. 2 for several delay times (from 20 to3000 ns) after excitation. The spectrum has been corrected for the spectral dependence of the measuring system. The spectral shape is essentially the same as the CW-PL spectrum. The peak energy showed a negligible shift throughout the measured time interval.

A semi-logarithmic plot of the decay curve of the PL peak at 1.96 eV measured with a nitrogen laser and a Xe flash lamp is illustrated in Fig. 3, from which the recombination lifetime was estimated: As seen in the figure the curve consists of three exponential decay curves with time constants of 54 µs, 111 µs and 152 µs.

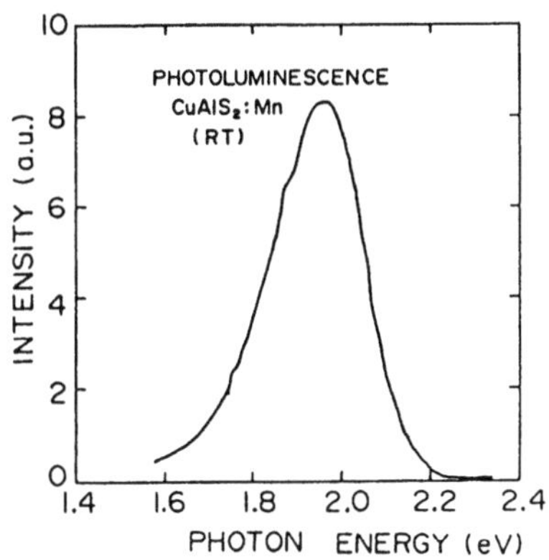

Fig.1 Photoluminescence spectrum of $CuAlS_2$:Mn measured at room temperature using an Ar^+ ion laser as an excitation source.

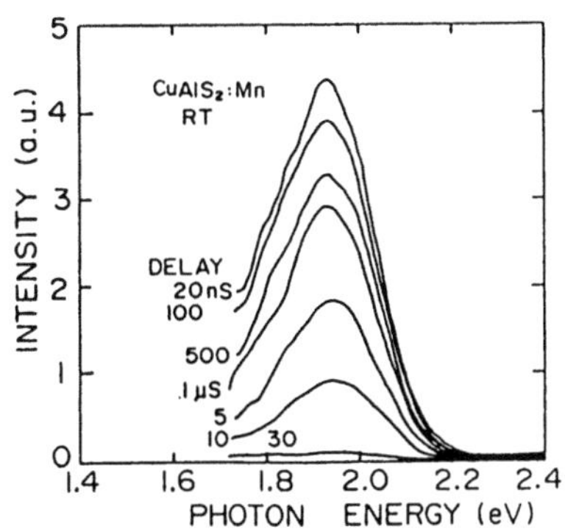

Fig. 2 Time-resolved PL spectra in a single crystal of $CuAlS_2$:Mn measured at room temperature with a nitrogen laser excitation.

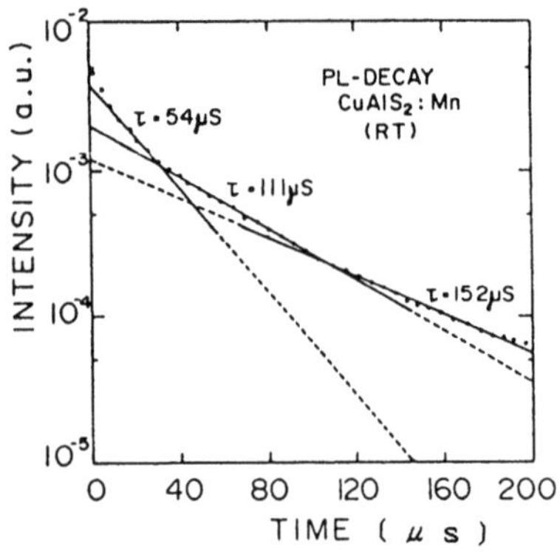

Fig. 3 A semi-log plot of the decay curve of 1.96 eV PL-peak measured with pulsed excitations by a nitrogen laser and a Xe flash lamp at room temperature.

3.2 Electroluminescence Spectrum

A red EL was observed in most of the samples investigated. The threshold voltage differed from sample to sample between 150 V to 250 V. Some of the specimens had poor stability and deteriorated in a few minutes. EL spectrum was measured on a stable sample with the applied voltage as small as possible. Crystals with strong coloration showed only a weak electroluminescence. Considerable increase of the brightness was observed, when crystals were annealed in vacuo at 200°C for 24 hrs prior to making them into the EL cell structure.

A typical example of the EL spectrum is shown in Fig. 4. Since the spectral feature of EL is very close to that of PL, it may be considered that the same kind of transition as in PL is involved in the observed EL.

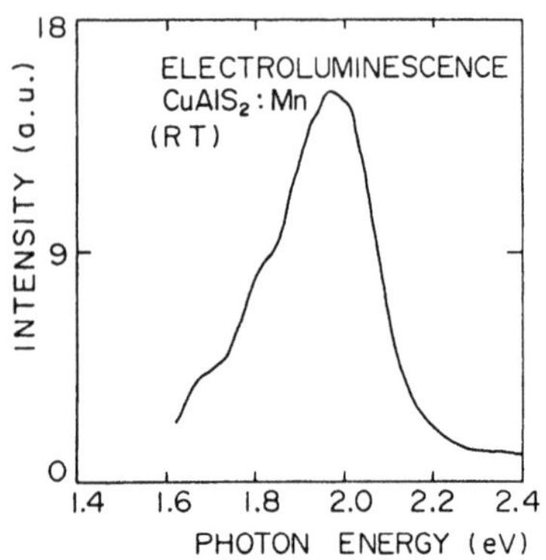

Fig. 4 Electroluminescence spectrum in a MIM diode of $CuAlS_2$:Mn.

4. DISCUSSION

In the case of ZnS:Mn, the lifetime of the orange PL associated with the single Mn^{2+} center has been determined to be about 2 ms. The PL decay lifetime in the material of the present study is much shorter than that of ZnS:Mn; i.e. less than 200 μs. According to the work of Busse and co-workers[3] on ZnS:Mn, recombination with the decay time of 90-200 μs can be ascribed to the Mn-Mn pair emission. Taking account that our sample contains Mn ions as much as 5 mol%, we assume that the red emission in our $CuAlS_2$:Mn sample is caused by the Mn-Mn pair emission.

The result that the spectral shape of EL is very close to that of PL implies that the same recombination process is involved in both phenomena. The impact excitation by the electric field will cause excitation to the higher excited states in the Mn-Mn pair, which becomes a recombination center. Measurements of decay curves in samples with different Mn concentration will be helpful to verify this assumption.

5. CONCLUSION

Electroluminescence spectrum of $CuAlS_2$:Mn MIM diode is observed for the first time. The spectral feature is very close to that of photoluminescence spectrum, indicating the involvement of the same center as in the photoluminescence. Time-resolved PL spectrum showed no change of line-shape, indicating that only one type of radiative recombination is involved. PL decay time was less than 200 μs and recombination in the Mn-Mn pair is suspected. The observed EL is not so strong in the present stage. Thin film EL device is now under investigation to improve the EL characteristics of this material.

ACKNOWLEDGMENT

The authors are much indebted to Dr. E. Nakazawa and Mr. S. Okamoto of NHK Science and Technical Laboratories for helpful technical advice and fruitful discussions. This work is partially supported by a Grant-in-Aid for Scientific Research on Priority Areas (No. 62604006) from the Ministry of Education, Science and Culture.

References

1. K. Sato, H. Tsunoda and T. Teranishi : Proc. 7th Int. Conf. Ternary & Multinary Compounds, Snowmass 1986 (Mater. Res. Soc., Pittsburg 1987) p113.
2. K. Sato, S. Okamoto, M. Morita, A. Morita, T. Kambara and H. Takenouchi: Prog. Cryst. Growth & Char. 10, 311 (1984).
3. W. Busse, H.-E. Gümlich, B. Meissner and D. Theis: J. Lumi. 12/13, 693-700 (1976).

Index of Contributors

MIX
Papier aus verantwortungsvollen Quellen
Paper from responsible sources
FSC® C105338

If you have any concerns about our products,
you can contact us on
ProductSafety@springernature.com

In case Publisher is established outside the EU,
the EU authorized representative is:
Springer Nature Customer Service Center GmbH
Europaplatz 3, 69115 Heidelberg, Germany

Printed by Libri Plureos GmbH
in Hamburg, Germany